# Acid Stimulation

# Acid Stimulation

Syed A. Ali, Leonard Kalfayan, and
Carl T. Montgomery
*Editors*

Society of Petroleum Engineers

Printed in the United States of America.

**Disclaimer**

This book was prepared by members of the Society of Petroleum Engineers and their well-qualified colleagues from material published in the recognized technical literature and from their own individual experience and expertise. While the material presented is believed to be based on sound technical knowledge, neither the Society of Petroleum Engineers nor any of the authors or editors herein provide a warranty either expressed or implied in its application. Correspondingly, the discussion of materials, methods, or techniques that may be covered by letters patents implies no freedom to use such materials, methods, or techniques without permission through appropriate licensing. Nothing described within this book should be construed to lessen the need to apply sound engineering judgment nor to carefully apply accepted engineering practices in the design, implementation, or application of the techniques described herein.

ISBN 978-1-61399-426-9

16 17 18 19 / 9 8 7 6 5 4 3 2 1

Society of Petroleum Engineers
222 Palisades Creek Drive
Richardson, TX 75080-2040 USA

http://www.spe.org/store
service@spe.org
1.972.952.9393

# Preface

The idea behind this monograph was generated by me at my first technical advisory group meeting as standing Society of Petroleum Engineers (SPE) Director for Drilling and Completions. It was felt that the original *Acidizing Fundamentals* monograph that was published in 1979 and edited by Bert Williams, John Gidley, and Bob Schechter was a marvelous document, but it needed to be updated to include developments that had occurred since its publication. This monograph was prepared to be a companion to *Acidizing Fundamentals* and is intended to serve, along with the earlier volume, as a basic reference for engineers who design, execute, and evaluate acid stimulation treatments and conduct research in the area of oilfield acidizing technologies.

The contents and structure of this monograph are well outlined in the Table of Contents and Chapter 1 and will not be repeated here. Putting together a document such as this was a challenging and time-consuming process that took approximately 5 years to achieve. This effort was coordinated, driven, and pounded into place by my two coeditors, Syed A. Ali with Schlumberger and Leonard J. Kalfayan with Hess Corporation. It is difficult to describe the issues and problems that occur during the preparation of a document such as this, so exceptional recognition must go out to these two individuals for holding on to the effort and bringing it to completion.

The technical content was reviewed by an SPE Monograph Review Committee, coordinated by David A. Curry with Baker Hughes, with members including Dean M. Bilden, Joel L. Boles, and Dr. Qi Qu, all with Baker Hughes. Jennifer Wegman, Editorial Services Manager for SPE, provided clear guidance to the editors on what was required in the preparation of the book. Special thanks also go out to all of the authors who contributed to the monograph. They are the ones who, at the end of the day, made all of this possible.

**Carl T. Montgomery**

# Table of Contents

Chapter 1

# Introduction

**George King,** Apache Corporation

The history of acidizing can be traced back more than 100 years when the first US patents (Frasch 1896; Van Dyke 1896) were issued to Herman Frasch and John W. Van Dyke of Standard Oil. Though well-known for their invention of removing sulfur from crude oil produced in Ohio, these two inventors proposed the acidizing methods that have impacted upstream oil and gas exploration and production for a century. Working together, the two inventors filed the two patents on the same date on 27 June 1895, and both were issued patents on 17 March 1896. Strictly speaking, Frasch proposed using hydrochloric acid (HCl), whereas Van Dyke proposed using sulfuric acid ($H_2SO_4$) to increase flow of oil wells in limestone formations. Nonetheless, Frasch and Van Dyke each assigned one-half of the inventor rights to the other. The books by Williams et al. (1979), Kalfayan (2008), and Kalfayan and Martin (2009) summarized these inventions, as well as the later developments in field treatments using corrosion inhibitors with HCl. This treatment solution was supplied by Dow Chemical Company, which later created Dowell Incorporated. Application of these early acid jobs was rudimentary at best, but HCl provided improvements in flow in enough cases to create interest in nonexplosive well stimulation.

As pointed out in the first SPE monograph on acidizing fundamentals (Williams 1979), stimulation of oil, gas, and injection wells with acid is almost as old as the industry itself. These early efforts sought to increase production and recognized some effects of formation variance and even some forms of completion damage as a barrier to flow. Although initially successful, acidizing soon decreased in popularity after the first few trials, probably because of corrosion and handling difficulty, and only in 1928 was the acidizing process again investigated and a few years later applied in a wide scope, this time to remove scale in pipe.

From the earliest application of acid, the desire was to use the acid to improve the flow of oil and gas to the wellbore. In 1936, Plummer and Newcome described the principal factors of acidizing at the time as (1) type of acid, (2) amount of acid, and (3) method of acid introduction into the well. They document the spread of acid to nearly every field in the US, Canada, and Mexico, where the producing horizons are in limestone and dolomitic rocks and to several formations where reservoir sandstones are held together by calcite or limey cements. Acid was widely popularized in the early 1930s after a workable corrosion inhibitor was found. By this time, HCl had come to be the acid of general choice, although several other inorganic and organic acids had been tried. The benefit of hydrofluoric acid (HF) was recognized, and work was progressed by Wilson (1935) in improvements in handling and understanding some elements of byproduct-precipitation control, but the uptake of HCl/HF acid was slow, taking more than 20 years before it was widely accepted. As more laboratory and well experiments followed, the use of $H_2SO_4$ and nitric acids was discontinued because of damaging byproducts or dangerous reactions.

Although the concept of unequal flow in comparison of wells appears to be at least known in the 1890s, the cause of reduced flow in both initial completions and during production did not appear to be well-understood. Laboratory and field assessments of matrix and fracture acidizing, described by Krueger (1973) from the mid-1930s to the early 1970s, gradually identified factors important to acidizing performance both as a stimulation and as a tool for removing formation damage. Smith and Hendrickson (1965) described HCl/HF stimulation in the field, Harris et al. (1966) showed the benefits of higher concentrations of HCl in carbonates, Smith et al. (1969) documented the effects of iron precipitation, while Gidley (1971) demonstrated the benefits of mutual solvents and Williams and Nierode (1972) started pulling all the pieces into order with a design approach. The concepts of acid spending and forming higher-conductivity channels was communicated from laboratory work, and the thought of "acid wormholes" in limestone from acid reaction took hold (Kalfayan and Martin 2009). This idea eventually led to advances in understanding both formation heterogeneity and the source, effect, and treatment of some forms of formation damage.

Early work on formation damage focused on the most obvious damage mechanisms such as paraffin, scales, and drilling mud damage. These more obvious damage mechanisms were considered to be a product of naturally occurring production or workover mechanisms. Understanding of rock/fluid chemistry did not evolve until much later. The wider field of formation damage is one of the most important areas in the modern practice of acidizing and is covered in detail in Chapter 2 of this monograph.

In the 1979 SPE monograph *Acidizing Fundamentals*, Williams et al. (1979) cite the birth of the modern era of acidizing in 1932, when Pure Oil Company, using laboratory and operational information from Dow Chemical Company, acidized a series of wells with 500 gal of acid and included the first early inhibitor. The treatments produced solid gains in production but also triggered the first appearance of emulsion as an acid byproduct. Within 3 years, a number of companies were offering the acid service to operators. During these first years of commercial service, problems were noted on a number of wells in which acid interfered with formation strength (disaggregation), and created flow upsets, and precipitates. Production improvements from acidizing were promising enough to offset most problems, but laboratory work was needed to resolve the upset problems and improve the application success.

Acidizing chemistry describes the reaction characteristics of acids and helps determine factors for acid type and concentration selection. During the 1950s through the 1970s, concepts of stoichiometry, thermodynamic equilibrium, and reaction rate were introduced into the acidizing design, providing an early technical basis to replace what was primarily a trial-and-error approach. These concepts were from laboratory measurements on pure test samples and often did not easily align with the highly heterogeneous world in the subsurface. Williams et al. (1979) laid out the framework of acid type, advantages and disadvantages, and how the acids interacted with the various components in reservoir rock, adding working explanations of methods of retarding acid reactions. This work greatly assisted the industry for years in design of acid treatments.

The importance of formation-composition and depositional-environment factors was considered by the early 1970s, as were the principles of acid contact and mechanics of reaction in a heterogeneous system with orders-of-magnitude permeability difference. Laboratory concepts of acid dissolving potential were introduced as a design variable, although discussion was still mostly laboratory-result dominated. Reaction-equilibrium considerations made their way into the design, as reaction and precipitation controls were recognized in acidizing design. Byproduct control became a major issue with the recognition of damage produced by iron, sulfur, and silicate precipitation.

Reaction kinetics for each acid system were researched, fluid and byproduct diffusion and convection (flow) mechanisms were described, and a number of spending models based on laboratory measurements were proposed. Temperature, pressure, concentration, and activity were advanced as controls, and weak and strong acid reaction differences were considered. Most of the work in these areas was on linear core (often with fractures), parallel reactive plates, circular tubes, or rotating disks. The focus was mainly on carbonate fracture acidizing, although some work, chiefly with HF mixtures, advanced sandstone acidizing. Application principles, primarily wetting, surface modification, oil coating, and other contact limiting properties, were discussed in only general terms at this point because of both significant rock variability and difficulty in achieving repeatable laboratory results.

Division of acidizing into near range (wellbore treating), critical flow area (near-well matrix), and deep formation (fracture acidizing) evolved over several decades from field applications. As with any field-derived approach, the recipe for acidizing in an area was usually unique and required trial and error to find the right formula with changes in temperature, pressure, produced fluid, and rock type.

Acidizing first and foremost is a chemical reaction rather than an applied-force process such as hydraulic fracturing. Acidizing has a varied set of controls that depend on presence or absence of materials or conditions. Time is a factor, and knowledge of analog situations is often the first step in a design. The reaction rates of acid in reactive formations may be first-order (mass-transfer controlled) in a near-pure carbonate matrix in which the area/volume ratio may be approximately 20,000:1. Acid normally spends rapidly where reactive materials are present, although field results suggested that some oil coatings could effectively slow acid reactions. Acid spending and leakoff and the exponentially increasing pore area in divergent radial flow are the major controls in acid penetration into the reservoir. Acidizing chemistry must consider all parts of the acid reaction from reaction kinetics to the stability of acid byproducts. Acid reaction modeling and predictive reaction behaviors are introduced in Chapter 3.

Carbonate acidizing, as detailed in Chapter 4, may be a fracturing-style stimulation or has potential to improve matrix permeability and remove damage when injected through the matrix in the highly restricted near-wellbore region. Fracture acidizing takes advantage of the channel-etching behavior of HCl in carbonates to generate short but highly conductive flow passages. The benefits of stimulation in this first ½ to 1 m of radius in an unfractured, high-permeability well cannot be overstated. Unlike the sandstones in which acidizing tends to be functional only in removing acid-soluble damage, acidizing carbonates usually produces open, narrow channels as acid flows and rapidly reacts to enlarge high-permeability channels into distinct long (1 cm to 3 m) tunnels or "wormholes."

From the first realization that HF acid may be useful in stimulating sandstone reservoirs, there has been interest in improving conductivity in the near-wellbore region of sandstones (Chapter 5). By simple modeling using the Darcy's law beds-in-series equation, it is quickly demonstrated that the most economic target for matrix acidizing is removal of near-wellbore formation damage in high-production-potential unfractured wells and even in some hydraulically fractured and propped sandstones where silt and other materials have accumulated or precipitated in the last few feet of the fracture near the wellbore. Matrix acidizing of sandstones employs a variety of acids, chemicals, and application techniques, but the identity and other aspects of the damage mechanism are often unknown. Matrix acidizing has attained a poor reputation, often because the treatment has been applied indiscriminately when damage is suspected but not confirmed or because of the notion that an acid job is a form of diagnostic in itself. However, there is a significant wealth of chemical knowledge that can explain many, if not most, design and application mistakes and definitely can improve acid stimulation in sandstones.

Acid placement in the well, Chapter 6, discusses application operations that are responsible for successfully moving acidizing from the laboratory to the field and methods of successfully diverting acid from the most-open flow paths and into the damaged or low-permeability sections. Delivery of acid to the zone to be treated has always been challenged by acid reactivity with debris, coatings, reaction products, and other contaminants on the walls of the pipe used to deliver the acid to the formation. The amount of debris from this pipe/acid reaction may be a 100 lb. or more of reactive and nonreactive contaminants per 1,000 ft of pipe. Failure to remove this material before the acid job has caused many failures of otherwise properly designed acid jobs because the acid loosens and carries the debris into the formation, creating shallow and deep formation damage. Avoidance of this self-inflicted damage is the first critical operation in acidizing. Acid-injection methods are often required to place a controlled amount of acid into a specific injection face or perforated interval, and an array of tools and techniques is used to achieve the delivery.

Acidizing has a long and undignified history of improper design and placement and of false hope that an acid job will cure the problem without first knowing the identity of the problem. Treatment Evaluation and Real-Time Diagnostics, Chapter 7, explores the methods of describing, designing, and monitoring an acid job with a view toward incorporating learning into both the design and application areas. Effects of acid on various damage scenarios are detailed and data-gathering opportunities are described, with examples and case histories.

Acid fracturing, Chapter 8, is a competitor for some forms of hydraulic fracturing and shares some of the fracturing principles with proppant fracturing. Because acid reactions form the channels through which production flow can be increased, improved understanding of the reaction is essential for improving the process. Recent advances in acid-fracture-etching experiments have sharply improved understanding of acid in open fractures. Acid transport, diffusion, reactivity, and byproduct removal and stabilization have also improved.

Acid additives, Chapter 9, have been developed for a great many reasons and may be beneficial or harmful, depending on conditions. The importance of knowledge about the acid target and the interactions of acids and additives has frequently been a weak point in acidizing. The need for additives in an acid treatment must be justified by a need to control or a reason to promote an action within the specific environment into which the acid is introduced. Multiple additives used for "insurance" may not be a good idea from a performance standpoint as well as an economic one. The intent of an acid additive is to modify the acid behavior, the placement, the reactive surface, the byproducts, or the flowback to increase the chances of a good final result. Regardless of the purpose, chemical and/or physical additives are reactive with many fluid and rock components and should never be combined in a system without careful consideration of the outcome. Test protocols for these additive-package tests vary with the applier but should always involve representative formation components and live and spent acids.

Acids may also enter into reactions with formation fluids, usually to the detriment of production. Iron and asphaltene components or water and oil may catalyze the formation of sludges. Iron content increases in most acids to a point at which solubilized metals begin to precipitate as the acid spends and pH rises. If no oil is present, iron oxides can precipitate as a gelatinous mass, but iron in spent acid can catalyze sludge formation in oil with even a small amount of asphaltenes. Development of acid additives that control emulsion formation, iron precipitation, sulfur precipitation, and scale development reflects the recognition of byproduct problems.

Acid additives that are more biodegradable and less toxic are a welcome addition to the industry. A significant amount of work has driven the additive market to a point at which greener products are widely available.

Corrosion control, Chapter 10, is a constantly evolving technology that must manage several forms of corrosion in wells from low to high temperatures and with alloys that range from low-carbon steel to the exotic CRA and duplex alloys with high percentages of chrome and nickel. There is no uniform additive package for all applications. Some alloy pipes can easily be destroyed by even brief exposure to mineral acids if the right corrosion-inhibitor package is not provided and properly applied. Many of the components of inhibitors, for example, are not soluble in acid and are usually much lower density, requiring constant circulation in the acid tank with suitable equipment that ensures all sections of the tank are mixed with the dispersed inhibitor. If not properly mixed, some inhibitors will separate quickly from static fluid leaving raw (noninhibited) acid to be pumped in a well followed by pure inhibitor at the end of the job. Numerous cases of this problem have been seen, with both severe pipe corrosion and highly damaged formations as the outcome. Additionally, inhibitors slow but do not stop the attack of acid on the steel and alloy pipes in the well. Every additive has a microcosm of enabling and competing reactions. For corrosion inhibitors, an additive common to every acid job, the performance is influenced by time (mixed shelf-life), acid concentration, effect of temperature and pressure, reactivity on certain surfaces, interference of other chemicals such as solvents, and even physical forces.

Acidizing-treatment economics, Chapter 11, may range from successful payout in days to complete loss of the well. There has been much discussion in papers and forums over the value of acidizing. In the simplest terms, economics will depend on how successfully the acid and additive selection, design, application, and evaluation are executed. Properly designed and applied acid jobs have a high success rate. Many acid failures are a failure to get the needed information and failure to provide the right level of monitoring. That is not the fault of the acid. The approach to economics in Chapter 11 is a consistent use of each piece of the acidizing treatment and how the pieces all fit together on an economic scale.

Chapter 12 presents a needed discussion of the challenges and the potential of acidizing. Each segment of acidizing, from safer chemical additives to improved effectiveness, needs to be reviewed and challenged. The use of acid and/or specific additives where it has not demonstrated success must be questioned. The importance of education and improvements in acidizing cannot be overstated. Field-based quality-control checks at each step of design and application are important learning tools, as well as checks on prudent and safe application.

## References

Frasch, H. 1896. Increasing the Flow of Oil Wells. US Patent No. 559,669.

Gidley, J. L. 1971. Stimulation of Sandstone Formations with the Acid-Mutual Solvent Method. *J Pet Technol* **23** (05): 551–558. SPE-3007-PA. http://dx.doi.org/10.2118/3007-PA.

Harris, O. E., Hendrickson, A. R., and Coulter, A. W. 1966. High-Concentration Hydrochloric Acid Aids Stimulation Results in Carbonate Formations. *J Pet Technol* **18** (10): 1,291–1,296. SPE-1654-PA. http://dx.doi.org/10.2118/1654-PA.

Kalfayan, L. 2008. *Production Enhancement with Acid Stimulation*, second edition. Tulsa: PennWell Corporation.

Kalfayan, L. I. and Martin, A. N. 2009. The Art and Practice of Acid Placement and Diversion: History, Present State and Future. Presented at the SPE Annual Technical Conference and Exhibition, New Orleans, 4–7 October. SPE-124141-MS. http://dx.doi.org/10.2118/124141-MS.

Krueger, R. F. 1973. Advances in Well Completion and Stimulation During JPT's First Quarter Century. *J Pet Technol* **25** (12): 1,447–1,462. SPE-4702-PA. http://dx.doi.org/10.2118/4702-PA.

Plummer, F. B. and Newcome, R. B. Jr. 1936. Laboratory Investigations on Acid Treatment of Oil Sands. In Transactions of the AIME, Vol. 118, Issue 1. SPE-936100-G, 100–115. Richardson, Texas: Society of Petroleum Engineers.

Smith, C. F. and Hendrickson, A. R. 1965. Hydrofluoric Acid Stimulation of Sandstone Reservoirs. *J Pet Technol* **17** (02): 215–222. SPE-980-PA. http://dx.doi.org/10.2118/980-PA.

Smith, C. F., Crowe, C. W., and Nolan, T. J., III. 1969. Secondary Deposition of Iron Compounds Following Acidizing Treatments. *J Pet Technol* **21** (09): 1,121–1,129. SPE-2358-PA. http://dx.doi.org/10.2118/2358-PA.

Van Dyke, J. W. 1896. Increasing the Flow of Oil Wells. US Patent No. 559,651.

Williams, B. B. and Nierode, D. E. 1972. Design of Acid Fracturing Treatments. *J Pet Technol* **24** (07): 849–859. SPE-3720-PA. http://dx.doi.org/10.2118/3720-PA.

Williams, B. B., Gidley, J. L., and Schechter, R. S. 1979. Acidizing Fundamentals, second edition, Monograph Vol. 6. Dallas: Henry L. Doherty Series, SPE of AIME.

Wilson, J. R. 1935. Well Treatment. US Patent No. 1,990,969.

Chapter 2

# Formation Damage*

**Mukul M. Sharma,** University of Texas at Austin

## 2.1 Introduction

Any unintended impedance to the flow of fluids into or out of a wellbore is referred to as formation damage. This broad definition of formation damage includes flow restrictions caused by a reduction in permeability in the near-wellbore region, changes in relative permeability to the hydrocarbon phase, and unintended flow restrictions in the completion itself. Flow restrictions in the tubing or those imposed by the well partially penetrating a reservoir or other aspects of the completion geometry are not included in this definition because, although they may impede flow, they have either been put in place by design to serve a specific purpose or do not show up in typical measures of formation damage such as skin.

The primary purpose of acidizing is to remove near-wellbore formation damage. It is important, therefore, to understand the different mechanisms of formation damage so that we may be able to better tailor acid formulations to remove the specific type of damage that is suspected to be reducing the well's productivity.

Some of the types of damage mentioned above can effectively be removed by the use of acids, while others cannot. For example, a reduction in well productivity associated with near-wellbore permeability reduction because of fines and clays can be removed by acidizing, while productivity impairment caused by a reduction in hydrocarbon relative permeability may not be removable with acid treatments and may require alternative stimulation treatments such as solvents.

It is important to state that prevention is always better than cure, and preventing formation damage is clearly better than trying to remove it by use of acidizing or any other technique. This can be accomplished by a careful review of drilling, completion, and production operations. This allows us to make operational changes, minimize the extent of formation damage induced in and around the wellbore, and have a substantial impact on hydrocarbon production. In most instances, however, despite our best efforts, formation damage will occur and some sort of stimulation treatment will be necessary to remove it.

Being aware of the formation-damage implications of various drilling, completion, and production operations can help in better diagnosing the problem, which leads to better remediation recommendations.

In this chapter, we discuss methods to measure and quantify the extent of formation damage and provide criteria that can be used to identify various types of formation damage. Our goal is to better define the mechanisms involved so that an operator can recommend and design the correct remedial action and/or make changes to drilling, completion, and production operations to minimize damage in the future.

---

*Adapted from Sharma, M. M. 2007. Formation Damage. In *Petroleum Engineering Handbook, Vol. IV—Production Operations Engineering*, ed. J. D. Clegg, Chap. 6, 241–274. Richardson, Texas: Society of Petroleum Engineers.

## 2.2 Quantifying Formation Damage

A commonly used measure of well productivity is the productivity index, $J$, (bbl/psi)

$$J = \frac{q_o}{P_R - P_{WF}} \quad (2.1)$$

The most commonly used measure of formation damage in a well is the skin factor, $S$. The skin factor is a dimensionless pressure drop caused by a flow restriction in the near-wellbore region. It is defined as follows (in field units):

$$S = \left(\frac{kh}{141.2\, q\mu B}\right)\Delta P_{\text{skin}} \quad (2.2)$$

**Fig. 2.1** shows how flow restrictions in the near-wellbore region increase the pressure gradient, resulting in an additional pressure drop caused by formation damage ($\Delta P_{\text{skin}}$). Standing (1970) introduced the important concept of well flow efficiency $F$, which he defined as

$$F = \frac{\bar{P}_R - P_{WF} - \Delta P_{\text{skin}}}{\bar{P}_R - P_{WF}} = \frac{\text{ideal drawdown}}{\text{actual drawdown}} \quad (2.3)$$

Clearly, a flow efficiency of 1 indicates an undamaged well with $\Delta P_{\text{skin}} = 0$, a flow efficiency greater than unity indicates a stimulated well (perhaps because of a hydraulic fracture), and a flow efficiency < 1 indicates a damaged well. Note that to determine the flow efficiency, we must know the average reservoir pressure, $P_R$, and the skin factor, $S$. Methods to measure these quantities are discussed in Section 2.3.

The impact of the skin on well productivity can be estimated by the use of inflow performance relationships (IPRs) for the well such as those proposed by Vogel (1968), Fetkovich (1973), and Standing (1970). These IPRs can be summarized as follows:

$$\frac{q}{q_{\max}} = FYx\left(\frac{x+1}{x} - FY\right), \quad (2.4)$$

where

$$Y = 1 - \frac{P_{WF}}{\bar{P}_R} \quad (2.5)$$

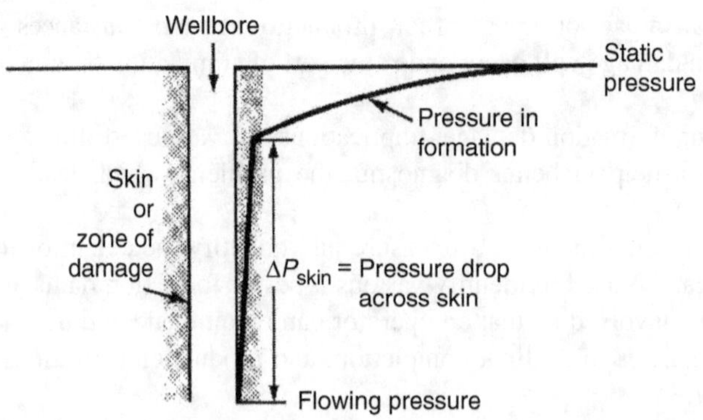


**Fig. 2.1—Pressure profile in the near-wellbore region for a well with formation damage (fixed oil rate) (Hurst 1953).**

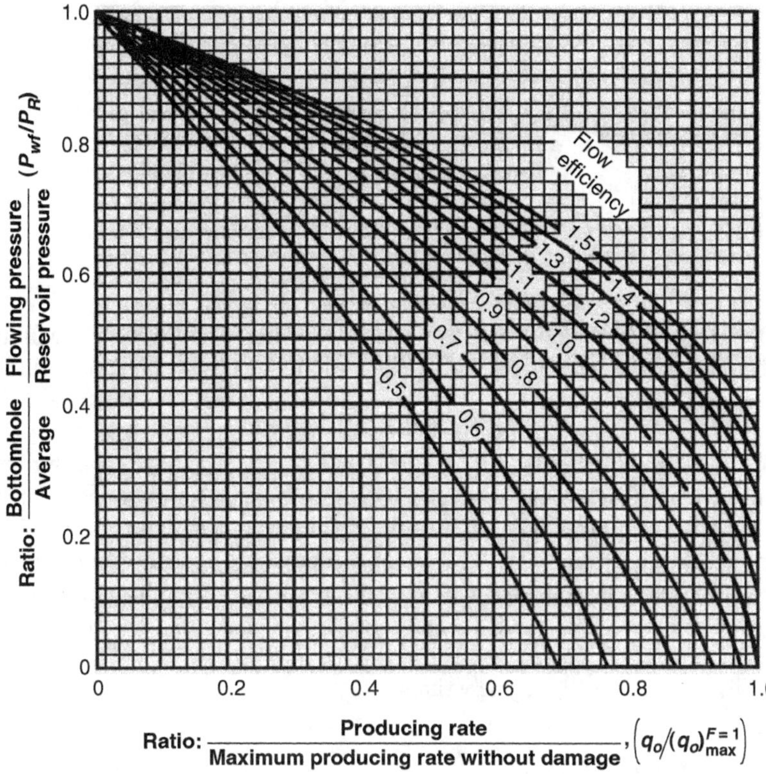


**Fig. 2.2—IPRs for different flow efficiencies (F) (Brown and Beggs 1977). Courtesy of PennWell Publishing Inc.**

When $x = 0$, a linear IPR model is recovered; when $x = 0.8$, we obtain Vogel's IPR; and when $x = 1$, Fetkovich's IPR model is obtained. An example of a plot for the dimensionless hydrocarbon production as a function of the dimensionless bottomhole pressure (IPR) is shown in **Fig. 2.2** for different flow efficiencies. It is evident that as the flow efficiency decreases, smaller and smaller hydrocarbon rates are obtained for the same drawdown $(\bar{P}_R - P_{WF})$.

The choice of the IPR used depends on the fluid properties and the reservoir drive mechanism. Standing's IPR is most appropriate for solution-gas-drive reservoirs, whereas a linear IPR is more appropriate for waterdrive reservoirs producing at pressures above the bubblepoint and for hydrocarbons without substantial dissolved gas. A more detailed discussion of this is provided in Brown and Beggs (1977).

## 2.3 Determination of Flow Efficiency and Skin

It is evident that to quantify formation damage and to study its impact on hydrocarbon production, one must have reasonable estimates of the flow efficiency or the skin factor. Several methods have been proposed to evaluate these quantities for oil and gas wells. The most common methods are

1. Multirate tests
2. Isochronal gas-well tests
3. Transient well tests (pressure-buildup analysis)

**2.3.1 Multirate Tests.** Multirate tests can be conducted on both oil and gas wells. In these tests, several stabilized flow rates, $q_i$, are achieved at corresponding stabilized flowing bottomhole pressures $(P_{WF})$. The simplest analysis considers two different stabilized rates and pressures. The IPR can be written as

$$\frac{q_1}{q_2} = \frac{FY_1(x+1-FY_1x)}{FY_2(x+1-FY_2x)} \qquad (2.6)$$

Simplifying and solving for the flow efficiency, $F$, we obtain

$$F = \left(\frac{x+1}{x}\right)\left(\frac{q_1Y_2 - q_2Y_1}{q_1Y_2^2 - q_2Y_1^2}\right), \qquad (2.7)$$

where $(x \neq 0)$.

Eq. 2.7 clearly shows that it is possible to obtain flow efficiency rather simply with two stabilized bottomhole pressures and two stabilized flow rates. A similar analysis can be performed to obtain an expression for a linear IPR ($x = 0$).

**2.3.2 Multirate Tests in Gas Wells—Inertial Effects.** For many gas wells and for some oil wells, flow rates are high enough that turbulent or inertial pressure drops near the wellbore can be significant. In such cases, the additional pressure drop measured by the skin can be confused with the pressure drop because of non-Darcy or inertial flow. It is very important to separate the pressure drop caused by turbulent flow from that caused by physical skin because it has a significant impact on the stimulation recommendations made on the well. To analyze high-rate gas or oil wells, the following analysis is needed (Jones et al. 1976).

Darcy's law for high-rate gas wells can be written as

$$\frac{q_{sc} = 703\times10^{-6}kh\left[m(\bar{P}_R) - m(P_{WF})\right]}{T\left(\ln\frac{r_e}{r_w} - 0.75 + S + Dq_{sc}\right)} \qquad (2.8)$$

Here,

$$m(P) = \int_{P_b}^{P}\frac{2P}{\mu_g Z}dP \qquad (2.9)$$

This equation can be rearranged to obtain

$$m(\bar{P}_R) - m(P_{WF}) = Aq_{sc} + Bq_{sc}^2 \qquad (2.10)$$

Here, $Aq_{sc}$ represents a laminar pressure drop and $Bq_{sc}^2$ represents an inertial or non-Darcy pressure drop (sometimes referred to as a turbulent pressure drop). Note that $A$ contains the physical skin $S$ and $B$ is directly proportional to the non-Darcy coefficient $D$. By plotting multirate-test data as a plot of $\frac{m(\bar{P}_R) - m(P_{WF})}{q_{sc}}$ vs. $q_{sc}$, we obtain $A$ and $B$ as an intercept and a slope, respectively. It is then possible to compare the magnitude of the pressure drop caused by $S$ with that caused by inertial effects, $Dq_{sc}$.

If $S > Dq_{sc}$, a stimulation treatment would be recommended. However, if $Dq_{sc} > S$, the well may need to be reperforated or fractured to increase the inflow area and reduce inertial effects.

**2.3.3 Isochronal Test in Gas Wells.** In gas wells that take a long time to achieve stabilized rates, wells are shut-in and produced for a fixed time interval ($\Delta t$) at several different rates. These isochronal tests are then interpreted by the following deliverability relation,

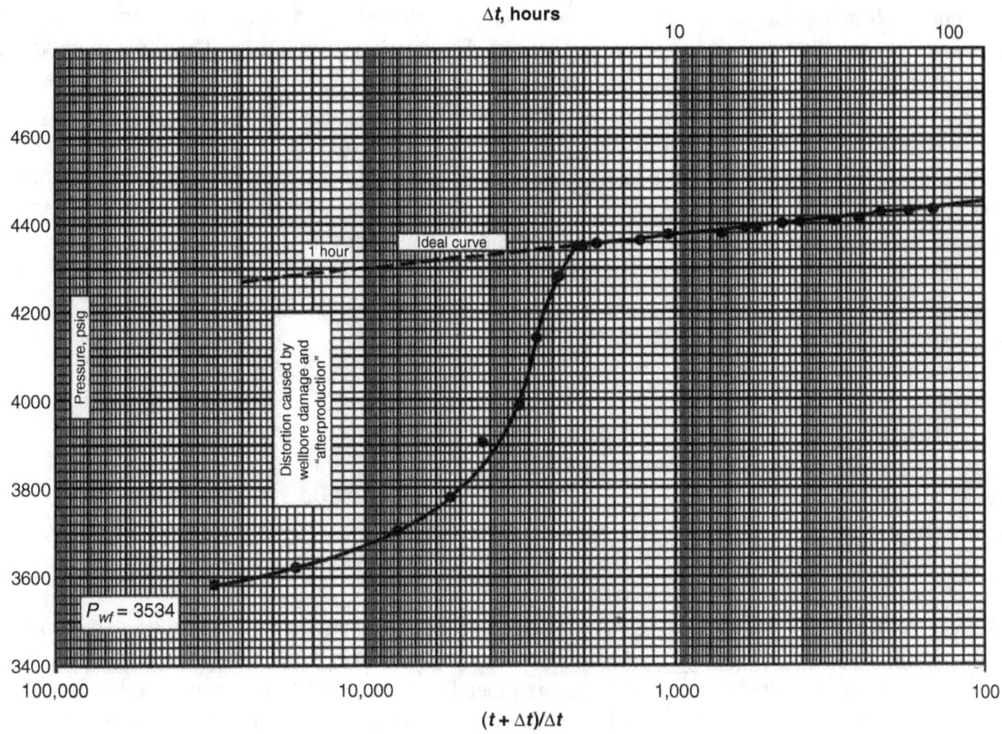


**Fig. 2.3—Horner plot from a pressure-buildup test (Matthews and Russell 1967).**

$$q_{sc} = c\left(\overline{P}_R^2 - P_{WF}^2\right)^n \qquad (2.11)$$

where the exponent $n$ lies between 0.5 and 1. An exponent closer to 0.5 indicates that non-Darcy effects are important, while an exponent close to unity indicates that they are not (Brown and Beggs 1977).

It should be noted that the deliverability equation is a variation of the equation derived in the previous section.

**2.3.4 Pressure-Buildup Analysis.** The most common method for determining skin is a pressure-buildup test (Matthews and Russell 1967; Horner 1951). In this test, a well that has been producing for a time, $t_p$, is shut in for time $\Delta t$. The pressure buildup is recorded as a function of time. By constructing a Horner plot (Horner 1951) such as one shown in **Fig. 2.3** we are able to compute the skin and the product of the permeability and formation thickness, $kh$, of the reservoir (in field units).

$$S = 1.151\left(\frac{P_{WF} - P_{ws,\,1\,hr}}{m} - \log\frac{k}{\phi\mu c r_w^2} + 3.23\right) \qquad (2.12)$$

$$kh = \frac{162.9 q\mu B}{m} \qquad (2.13)$$

Here, $m$ is the slope of the straight-line portion of the Horner plot and $P_{ws,1\,hr}$ is the extrapolated shut-in pressure at a shut-in time of 1 hour.

It is also possible to obtain the average reservoir pressure with the Matthews, Brons, and Hazelbrook (MBH) method from the pressure-buildup data (Matthews et al. 1954). Knowing both the average reservoir pressure and the skin, we can then calculate the flow efficiency of the well. This provides a direct and quantitative measure of the extent of formation damage in a well.

Methods following the same principle have been developed for deviated and horizontal wells. Equations for analysis are more complex and are not discussed in this chapter. The same methods can also be used to analyze data from gas wells and from wells on artificial lift.

The short discussion presented in the preceding shows how near-wellbore formation damage can be quantified by measurements made on oil and gas wells. Such measurements are essential to determine the extent and magnitude of the formation damage and its impact on hydrocarbon production. However, these measures do not provide us with any clues on the reasons for the formation damage. In subsequent sections in this chapter, reasons and mechanisms for formation damage and strategies to minimize the impact of drilling and completion operations on well productivity are discussed.

## 2.4 Formation Damage vs. Pseudodamage

Formation-damage mechanisms can be broken down into two broad classes:

1. Near-wellbore permeability reduction
2. Near-wellbore relative permeability changes

These changes can occur under a variety of different circumstances. The following sections deal with different ways in which the permeability and relative permeability in the near-wellbore region are altered by drilling, completion, and production operations.

Before we discuss formation-damage mechanisms, it is important to clearly distinguish formation damage from well-completion and reservoir effects that are a consequence of how the wellbore penetrates the reservoir and where the perforations are placed (sometimes referred to as pseudoskin effects) (Dake 1978; Jones and Watts 1971; Odeh 1968; Weeks 1974) and permeability loss as a result of depletion (Marek 1979). Reservoir engineering models for limited-entry flow in partially penetrating wells are presented in several reservoir engineering texts such as Dake (1978).

The second major cause of pseudoskin is high-velocity flows near the wellbore, which induce turbulence or inertial effects. As discussed in the preceding section, this can lead to an additional turbulent pressure drop that needs to be clearly distinguished from the pressure drop induced by a reduction in permeability. Finally, flow restrictions in the wellbore itself such as chokes, scale buildup, wax, or asphaltene deposits can often result in tubing pressure drops that are substantially larger than anticipated. This reduction in well productivity is not commonly referred to as formation damage. Other types of production impairment caused within the tubing are

- Collapsed tubing or flow restrictions caused by mechanical restrictions such as corrosion products
- Poor cement jobs, resulting in commingling of produced fluids from different zones
- Insufficient tubing diameter or improper design of artificial-lift systems

This partial list provides some examples of flow restrictions that are caused primarily in the tubing and that should not typically be categorized as formation damage. They do not show up in measures of formation damage such as skin, which are primarily a measure of flow restrictions in the near-wellbore region.

In this chapter, flow restrictions in the completion itself such as the compacted zone around perforation tunnels and plugged gravel packs are included in our discussion of formation damage because they are typically measured as a well skin (Section 2.3).

## 2.5 Drilling-Induced Formation Damage

Drilling fluids serve to balance formation pressures while drilling to ensure wellbore stability. They also carry cuttings to the surface and cool the bit. The drilling engineer traditionally designs drilling fluids with two primary goals in mind:

1. Ensure safe, stable boreholes, which is accomplished by operating within an acceptable mud-weight window.
2. Achieve high rates of penetration so that rig time and well cost can be minimized.

Note that these primary considerations do not include well-productivity concerns. Over the past decade, a growing recognition of the importance of drilling-induced formation damage has led operators to mesh the objectives of the drilling engineer with those of the production and reservoir engineer. This can only be achieved if the design of the drilling program is a coordinated effort between drilling and production engineers. The use of drill-in fluids (fluids used to drill through the pay zone) that minimize formation damage has become more widespread.

Drilling as well as well-productivity concerns are addressed when designing drill-in fluids. To meet well productivity objectives (i.e., to minimize formation damage), the drill-in fluid must meet the following additional objectives:

1. Minimize the extent of solids invasion into the formation by bridging across the pores and forming a thin, low-permeability filter cake.
2. Minimize the extent of filtrate and polymer invasion into the formation through the formation of an external filter cake.
3. Ensure ease of removal of the external filter cake during flowback to maximize the inflow area during production and to avoid plugging gravel packs.

To achieve the above goals, various strategies have been adopted. In this section, we address these strategies in terms of the basic mud formulations being used. Traditional water-based muds, oil-based muds, and some special formulations of drill-in fluids for fractured formations and unconsolidated sands are discussed. This is followed by a discussion of formation damage caused by drilling in deviated and horizontal wells and the use of drill-in fluids for such applications.

**2.5.1 Formation Damage Caused by Water-Based Muds.** The vast majority of drilling fluids used consists of bentonite mixed with polymers to enhance the rheology (or, more specifically, the cuttings-carrying capacity of the fluid), starches to control fluid loss, dissolved salts such as potassium or sodium chloride, and perhaps a pH buffer to maintain the pH of the mud to a desired level. A great deal of work has been performed in the last 3 decades on evaluating the formation-damage potential of water-based drilling fluids (Abrams 1977; Nowak and Krueger 1951; Glenn and Slusser 1957; Suri and Sharma 2001; Ladva et al. 2000). The following factors have been observed to have an impact on the depth of invasion of solids and filtrate and, therefore, on the extent and depth of formation damage or permeability impairment:

- The state of dispersion of solids in the mud
- The size and concentration of solids and polymers in the mud
- The pore-throat size or permeability of the formation
- The pH and salinity of the filtrate
- The water sensitivity of the formation

In most instances, the invasion of solids into the formation is limited to 2 or 3 in. from the wellbore wall, which implies that the productivity of perforated wells with relatively shallow depth of damage will not be significantly affected. **Fig. 2.4** shows the productivity index (PI) of a well for different depths of damage assuming an 8-in.-long perforation. It is evident that as long as the depth of damage is smaller than the perforation length, the well PI is not significantly impacted. Wells that are completed as open hole without stimulation are particularly susceptible to this kind of damage.

In some instances, deep penetration of drill solids can occur. **Fig. 2.5** shows the depth of invasion of formation damage when a 300-md Berea sandstone core is subjected to dynamic circulation of different water-based drilling fluids across its face (Jiao and Sharma 1992). It is evident that in overtreated muds (containing too much thinner or dispersant), dispersed bentonite particles can penetrate through > 8 in. of rock and cause severe and irreversible damage. The other extreme, flocculated muds (too little thinner or too much salt), will limit solids invasion but will result in thick, high-permeability filter cakes. This can result in problems such as stuck pipe and large filtrate loss. The use of salts and

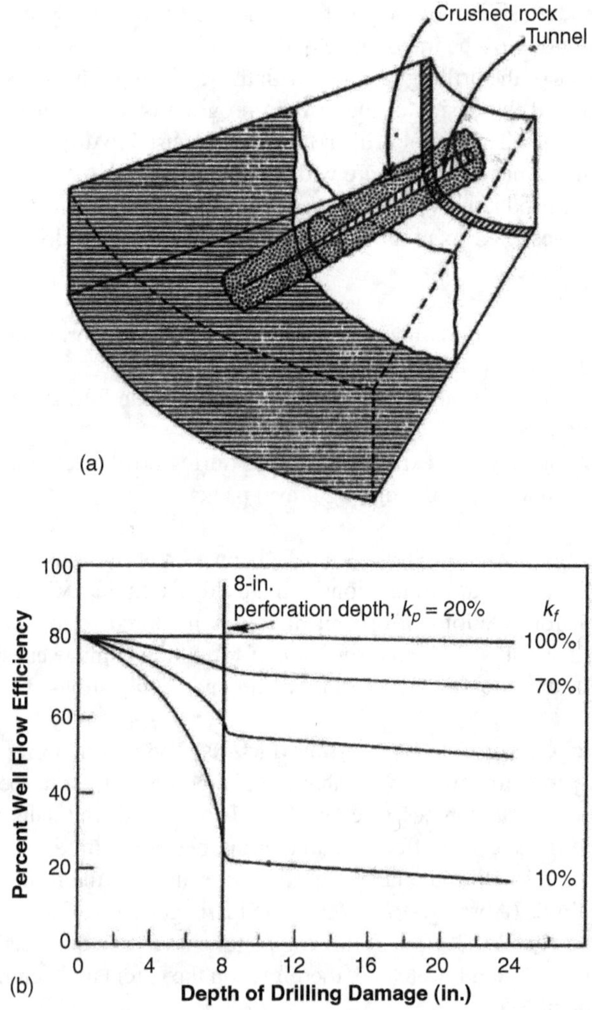


**Fig. 2.4—(a) Schematic of a gun perforation showing zone of crushed rock around tunnel. (b) Effect of damage by mud while drilling on well productivity when perforated with a nondamaging fluid is permeability of crushed zone around perforated tunnel as a percent of initial permeability (Klotz et al. 1974).**

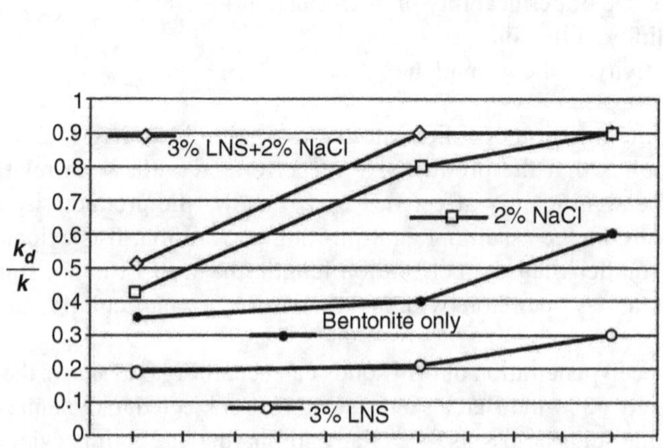


**Fig. 2.5—Depth of permeability damage caused by mud invasion for muds with different degrees of bentonite dispersion. All muds contain 4% bentonite by weight (Jiao and Sharma 1992).**

thinners is, therefore, a critical part in the design of drilling fluids for a given application. Appropriately conditioned muds must be used to eliminate the possibility of solids invasion and to minimize filtrate invasion. As discussed later, using sized bridging solids is a powerful tool for reducing solids and polymer invasion.

Although solids invasion clearly is detrimental to well productivity, filtrate invasion can also lead to substantial formation damage and to greater depths in some instances. It has been shown, for example (Jiao and Sharma 1992; Roy and Sharma 2001), that the use of freshwater muds can result in filtrates that can be damaging to water-sensitive sandstones. In such instances, the simple process of increasing the salinity of the filtrate can prevent fines migration induced by filtrate leakoff. The loss of aqueous filtrates also results in a reduction in the relative permeability to the hydrocarbon phases (Roy and Sharma 2001; Jiao and Sharma 1994). Such relative permeability effects are referred to as waterblocks and are discussed in Section 2.15.

Similarly, the use of polymers is widespread but can, in some instances, lead to formation damage. It has been shown that the use of improper mixing in dissolving polymers into brines can result in the formation of "fish eyes" or unhydrated aggregates of polymer that can be several micrometers in diameter. These particulate gels are effective as plugging agents and can lead to irreversible damage if not broken up and completely hydrated in the mud. Proper conditioning and dispersal of polymers is of critical importance in the field (Hodge et al. 1997; Browne and Smith 1994; Ryan et al. 1995; Francis 1997).

There is a limited database of information on the formation damage caused by starches and other polymers such as xanthan or carboxymethylcellulose (CMC). These data indicate that the flow of such polymers can induce a substantial reduction in permeability as a result of constriction of pore throats, particularly in low-permeability formations.

**2.5.2 Formation Damage Caused by Oil-Based Muds.** Oil-based muds consist of water droplets dispersed in a continuous oil phase. The water droplets are stabilized by emulsifiers and organophilic clays. Standard American Petroleum Institute (API) fluid-loss tests show that the fluid-leakoff rate in oil-based muds is lower than that for water-based muds. However, as Jiao and Sharma (1993a) showed, when tests are conducted on oil-saturated cores (not filter paper), leakoff rates for oil-based muds can be comparable with those for water-based muds. One of the important conclusions of this study is that API fluid-leakoff tests should not be used to determine filtration rates in oil-based muds. Instead, dynamic filtration tests conducted on oil-saturated cores are much more representative. The relative permeability to oil in oil-saturated zones is high, leading to large leakoff rates in the productive pay zone.

The invasion of solids and oil droplets into the formation is determined largely by the effectiveness of the external filter cake formed by organophilic bentonite and water droplets. The structure of the filter cake formed is substantially different from that of water-based muds. Here, water droplets bridge across the pore throats to form the external filter cake. Because the droplets are deformable, they can form very impermeable filter cakes, leading to good leakoff control. However, if the overbalance pressure exceeds the capillary pressure needed to squeeze the water droplets into the pores, a significant loss in productivity can result (Jiao and Sharma 1993b). To prevent this from happening, large overbalance pressures should be avoided.

Experimental studies have shown that the accumulation of drill solids in the mud results in the introduction of fines that can be much more damaging than clean mud. Drill-solids control, therefore, is an important issue in oil-based muds. In general, however, oil-based muds prove to be excellent (though expensive) candidates for drilling gauge hole and providing high-productivity wells (Jiao and Sharma 1993a; McKinney and Azar 1988).

It is important to recognize and identify damage caused by oil-based muds because the recommended treatment procedures for stimulating wells damaged by oil-based muds can be quite different from those for wells damaged with water-based muds. Acidizing wells with conventional acid formulations may not be successful and, in fact, may result in additional damage as a result of the presence of emulsifiers in the filtrate. Solvent preflushes may need to be designed on the basis of compatibility tests between the mud, the crude oil, and the acid formulation.

**2.5.3 The Concept of Minimum Underbalance Pressure.** It is clear from the preceding discussion that the formation of an external mudcake is important in protecting the formation from solids and filtrate invasion. Are there conditions under which an external mudcake will not form across the face of the formation? Yes, there are at least two situations in which an external filter cake does not form across of the face of the formation:

1. Lost circulation
2. Drilling overbalanced below the minimum overbalance pressure

When drilling through very-high-permeability rocks or fractured formations, solids present in the drilling fluid may not be able to bridge across the face of the pores or fractures, resulting in leakoff of whole mud into the formation (Ladva et al. 2000). This leakoff can result in very severe, irreversible damage to the fracture or matrix. In general, bridging solids are added to the drilling fluid to bridge across the pores or fractures. Sizing of these solids is discussed in more detail in Suri and Sharma (2001).

The second case where filter cakes do not form is less intuitively obvious. To form a mudcake, solids in the mud are pushed against the formation by a hydrodynamic force that is proportional to the leakoff velocity. In addition, because of mud circulation, particles are constantly being sheared away from the face of the external cake. This balance between the hydrodynamic shearing action resulting from the mud circulation and the fluid leakoff into the formation results in an equilibrium cake thickness as discussed in Jiao and Sharma (1996) and Singh and Sharma (1997). Because the leakoff is proportional to the overbalance pressure, smaller overbalance pressures will lead to smaller leakoff rates and thinner external filter cakes, resulting in a minimum overbalance pressure below which no external filter cake is formed at all. Alternatively stated, there is a minimum permeability for a fixed overbalance pressure below which no external filter cake will form. This suggests that we must always drill either underbalanced or above the minimum overbalance pressure to ensure that an external cake is formed and available to protect the formation when drilling through the productive pay zone. Additional details for calculating the minimum overbalance pressure are provided in Jiao and Sharma (1993b) and Singh and Sharma (1997).

**2.5.4 Mud-Induced Damage in Fractured Reservoirs.** When drilling through fractured formations, large quantities of whole mud can be lost to the fracture network, resulting in fracture plugging. Because fractures contribute almost all the productivity of such wells, it is important to keep these fractures open as much as possible. In such cases, underbalanced drilling is recommended and frequently used. Underbalanced drilling allows fluids from the fracture to flow into the wellbore, keeping the fractures relatively undamaged. If however, because of safety and regulatory constraints, underbalanced drilling is not possible, bridging additives need to be added to the mud system to ensure that large-enough particles are available to bridge across the fracture face. The bridging additives most commonly used to ensure the formation of a bridge across the face of the fracture are calcium carbonate and fibrous additives such as cellulosic fibers and acid-soluble fibers (Jiao and Sharma 1996; Singh and Sharma 1997). Sizing of these granular or fibrous additives has been discussed in more detail in Jiao and Sharma (1996) and Sharma (1997).

**2.5.5 Formation Damage in Horizontal Wells.** Horizontal wells are more susceptible to formation damage than vertical wells for the following reasons:

1. The pay zone in a horizontal wellbore comes into contact with a drilling fluid for a much longer period than that in a vertical pay zone (days compared with hours).
2. Most horizontal wells are openhole completions, which means that even shallow damage becomes significant—damage that in a cased-and-perforated completion would be bypassed by the perforations.
3. Because the fluid velocity and pressure gradient during flowback are usually small, cleanup of internal and external cakes is not as effective as in vertical wellbores. Thus, only a fraction of the wellbore contributes to flow when the well is put back on production.

4. Removing mud-induced formation damage by acidizing horizontal wells is often very difficult and expensive because of the large volumes of acid required and the difficulty in placing the acid in the appropriate wellbore locations.

Studies conducted on a simulated horizontal wellbore indicated that the heel is more damaged than the toe and that the upper part of the well is less damaged than the bottom of the wellbore where the drillpipe rests (Bailey et al. 1998; Thomas and Sharma 1998). The damage zone around the horizontal wellbore can, therefore, be modeled as an eccentric cone around the wellbore with a significantly larger depth of penetration at the heel and a shallower depth of penetration at the toe (Frick and Economides 1993).

Because the drilling fluid is in contact with the producing pay zone for an extended period of time, drill-in fluids have been devised to minimize the potential formation damage. Sized calcium carbonate and sized salt fluids are the drill-in fluids most often used in such applications. Oil-based muds have also been evaluated for this purpose. A more detailed discussion of their formation damage potential is provided in several sources (Ali et al. 1999; Browne and Smith 1994; Browne et al. 1995; Zain and Sharma 1999; Frick and Economides 1993; Jiao and Sharma 1996; Bruton 1995; Burnett 1998; Zain 1999; Zain et al. 2000).

## 2.6 Formation Damage Caused by Completion and Workover Fluids

When completion or workover operations are conducted on a well (e.g., perforating, gravel packing), the fluid present in the wellbore must minimize the impact on the near-wellbore permeability. Several decades ago, engineers realized that the use of drilling fluids during completions was inappropriate because that caused severe damage to the productive pay zone. A wide variety of fluids are now available as completion or workover fluids. A list of these fluids is provided in **Table 2.1.** Our discussion here focuses on formation-damage issues related to these different types of completion and workover fluids.

Water-based fluids usually consist primarily of clear brines. The only problem with clear brines is that they are not ever really clear (Morgenthaler 1986; Eaton and Smithey 1971; Azari and Leimkuhler 1990). They always contain some solids, including corrosion products, bacteria, and debris from the wellbore and surface tanks. The density of the brine is maintained large enough so that the bottomhole pressure exceeds the reservoir pressure by a safe margin (typically 300 to 600 psi). Substantial amounts of solids can be pushed into the formation, resulting in a loss of permeability in the near-wellbore region. **Fig. 2.6** shows the loss in permeability observed when brines with differing quantities of solids are injected into a core. Rapid reductions in permeability are observed even with relatively clean fluids. Surface filtration facilities are often used to clarify and filter completion brines, which can

Water-Based Fluids
- Clear brines
- Oil-soluble resins and waxes
- Polymer fluids
- Water-soluble solids (sodium chloride)
- Acid-soluble particles
- Biodegradable polymers

Emulsions
- Usually oil-in-water (12.5 lbm/gal)

Oil-Based Fluids
- Lease crude
- Invert-emulsion muds
- Asphalt pitches

Foams

Table 2.1—Completion and workover fluids.

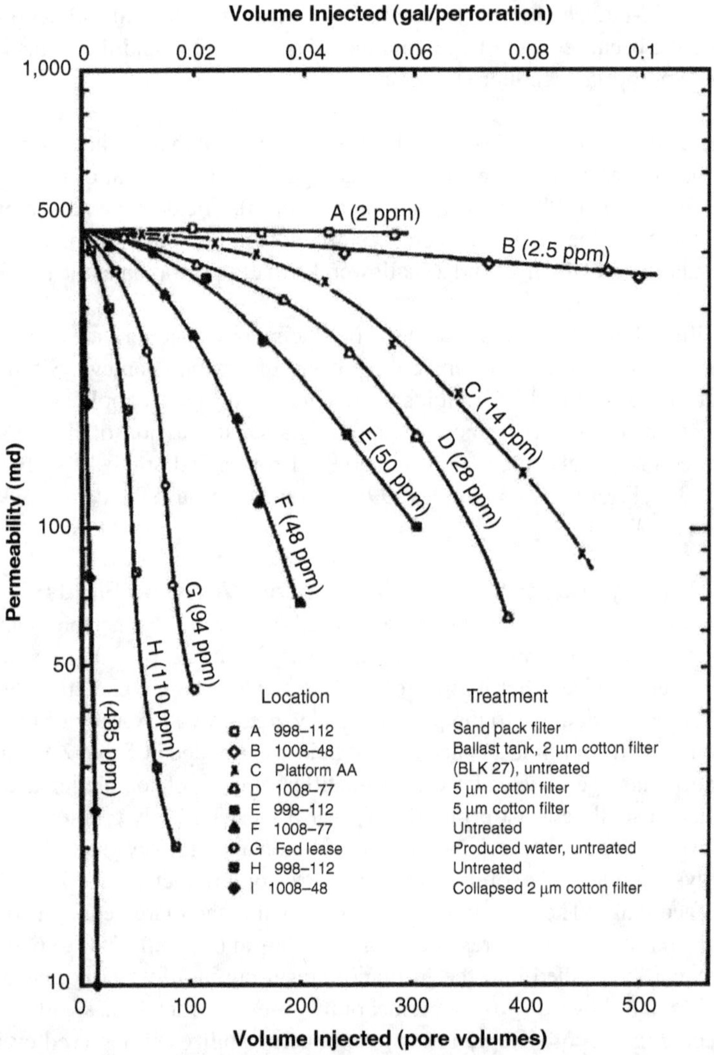


**Fig. 2.6—Apparent permeability reduction in Cypress sandstone cores injected with treated and untreated bay water from offshore Louisiana (Tuttle and Barkman 1974).**

help to reduce the permeability impairment substantially. Most of the high-density brines used can be quite expensive. Large volumes of fluid loss can add substantially to the cost of a completion operation. An important fact to keep in mind with completion and workover fluids is that, unlike drilling fluids, they do not contain drill solids. This means that there is no effective bridging material that is available to reduce fluid leakoff.

When fluid-leakoff rates are very high, fluid-leakoff-control additives may be used to minimize leakoff and formation damage. The use of acid-soluble granular additives such as calcium carbonate is the most common strategy (Tuttle and Barkman 1974; Mahajan and Barron 1980; Sloan et al. 1975). If this method proves to be ineffective, viscosifying polymers are used to reduce the amount of fluid loss. Hydroxyethylcellulose (HEC) is commonly used because it is soluble in hydrochloric acid. HEC is a poor viscosifier at higher (>250°F) temperatures, and unbroken and unhydrated HEC in the form of fish eyes can be damaging.

Polymer fluids suffer from similar drawbacks. Severe formation damage can occur if large amounts of polymer are lost to the formation. This problem is particularly acute if the polymer is not completely hydrolyzed in the brine.

If the density requirements of the completion fluid are relatively modest, emulsions can be used as completion fluids. In these instances, the droplets that form the dispersed phase act as a filtration-control agent. Both water and oil-external emulsions have been used when reservoir pressures are low.

Oil-based fluids such as crude oil and invert-emulsion muds can be used as completion fluids. It is important to ensure that the crude oil does not contain asphaltenes or paraffins that might precipitate under changes in pressure and temperature as the fluid is circulated into the well. Several sources (Tuttle and Barkman 1974; Sloan et al. 1975; Mondshine 1977; Priest and Morgan 1957; Priest and Allen 1958; Darley 1972; and Azari and Leimkuhler 1990; Chauvin 1976; Al-Riyamy and Sharma 2002; Patton and Phelan 1985) provide a more detailed discussion of some of the issues summarized in this section. In addition, crude oil is flammable and messy to handle.

## 2.7 Damage During Perforating and Cementing

When cement is bullheaded into the annulus to displace the mud, the differential pressure between the cement and the formation fluid can lead to a significant loss of cement filtrate into the formation. If, however, large volumes of cement filtrate invade the rock, the possibility of formation damage exists.

The major constituents in the aqueous phase in contact with hydrating cement are calcium silicates, calcium aluminates, calcium sulfates, calcium carbonates/bicarbonates, and alkali sulfates. Depending on the specific composition of the cement and its pH, the filtrate may be supersaturated with calcium carbonate and calcium sulfate. As the cement filtrate invades the formation and reacts with the formation minerals, its pH is reduced from > 12 to a pH buffered by the formation minerals. This rapid change in pH can result in the formation of inorganic precipitates such calcium carbonate and calcium sulfate.

Evidence of formation damage induced by cement filtrates has been clearly demonstrated in experimental studies presented by Yang and Sharma (1991), Cunningham and Smith (1968), and Jones et al. (1991). Cunningham and Smith (1968) investigated the influence of cement filtrates on formation permeability and concluded that there was little evidence of fines migration or clay swelling induced by the cement filtrate. They observed severe permeability reductions of 60 to 90% in cores invaded by cement filtrate. Yang and Sharma (1991) investigated the impact of cement additives such as lignin derivatives, cellulose derivatives, organic acids, and synthetic polymers on the extent of permeability reduction in cores exposed to cement filtrate. In that study, cement filtrate was injected immediately after filtration into a sandstone core. Reductions in permeability of 40 to 80% were observed up to 6 in. into the core. Most of the damage observed was attributed to the precipitation of insoluble material such as calcium carbonate and calcium sulfate in the core. The quantity of precipitate and the rate of precipitation relative to fluid convection were important factors that controlled the extent and depth of permeability damage. Cement filtrates that showed fast rates of precipitation tended to damage the upstream end of the core, whereas filtrates with slow precipitation rates tended to plug the downstream end of the core or not plug the core at all. The composition of the cement played an important role in determining both the quantity and the rates of precipitation. For example, the addition of lignin derivatives or polymer reduced the quantity of precipitate and resulted in less damage to the rock. The addition of cellulose derivatives, on the other hand, increased the rate and quantity of precipitation by an order of magnitude and resulted in more damage (Yang and Sharma 1991).

If the depth of invasion of the cement filtrate can be restricted to ≈ 4 in., cement-filtrate-induced damage should not be a major concern because the perforation tunnels will bypass the damage. However, in some situations in which large volumes of cement filtrate may be lost, this form of damage should be seriously considered. In such cases, the use of fluid-loss-control additives and polymers in the cement slurry needs to be evaluated carefully so that the cement is properly designed to minimize both the leakoff rate and the amount of insoluble precipitates formed in the formation.

The process of perforating is critical to the productivity of a well because the perforation is the only channel of communication between the wellbore and the formation. During underbalanced perforating, the surge flow of fluid into the wellbore should clean the perforation tunnel of all disaggregated rock and liner debris. Any remaining debris in the tunnel could plug gravel packs during production. Even clean perforation tunnels show a narrow region of reduced permeability around them. The nature of this crushed or compacted zone around perforation tunnels created during perforating has been widely studied (Klotz et al. 1974; Sharma 2000; Bartusiak et al. 1997; Crawford 1989; King et al. 1986; Tariq 1990; Behrmann 1996; McLeod 1983; Suman 1972; Petitjean et al. 1996; Halleck 1997; Hsia and Behrmann 1991; Halleck et al. 1995; Kooijman et al. 1996; Pearson and Zazovsky 1997; Zhang et al. 1998; Behrmann et al. 1991; Behrmann et al. 1992; Bird and Blok 1996; Pucknell and Behrmann 1991; Brooks et al. 1998; Behrmann et al. 1997; Halleck et al. 1999; Venkitaraman et al. 1997). It is now well-recognized that it consists of shattered grains and fines generated by the perforation charge and perhaps fines that flow in from the formation during underbalanced surge flow. The reduction in permeability in the compacted region is typically on the order of 20 to 50% but can be larger in some cases (Sharma 2000). Using an optimal underbalance pressure results in better perforation performance (King et al. 1986; Colle 1988; Krueger 1967). The reasons for this are not completely understood. It is likely that too low an underbalance results in insufficient perforation cleaning and that too large an underbalance results in the generation and migration of additional fines. This explanation is consistent with the observation that the optimum underbalance pressure in higher for lower-permeability formations.

## 2.8 Formation Damage Caused by Fines Migration

Fines migration is a recognized source of formation damage in some production wells, particularly in sandstones (Gray and Rex 1996; Muecke 1979). Direct evidence of fines-induced formation damage in production wells is often difficult to come by. Although most other forms of formation damage have obvious indicators of the problem, the field symptoms of fines migration are much more subtle. Indirect evidence such as declining productivity over a period of several weeks or months is the most common symptom. This reduction in productivity can usually be reversed by mud-acid treatments. A large number of wells around the world follow these patterns of reduction of productivity followed by significant improvements when subjected to a mud-acid treatment. This behavior most often suggests a buildup of fines in the near-wellbore region over a period of time. Field studies and laboratory experiments have indicated that the fines causing the permeability reduction include clays, feldspars, micas, and plagioclase. Because the mobile fines are made up of a wide variety of minerals, the clay content of the reservoir may not always be a good indicator of the water sensitivity of the formation.

Core flow tests conducted in the laboratory clearly show if low-salinity (<2%) brines are injected into water-sensitive rocks; large reductions in permeability (up to a factor 500) are obtained **(Fig. 2.7)** (Khilar an Fogler 1983; Valdya and Fogler 1992; Sarkar and Sharma 1990; Gruesbeck and Collins 1982). It is now well-established that this dramatic reduction in permeability is almost entirely a result of fines migration. Evidence is shown clearly in Fig. 2.7. Reversal of flow results in a temporary increase in the permeability as the fines plug pores in the reverse flow direction.

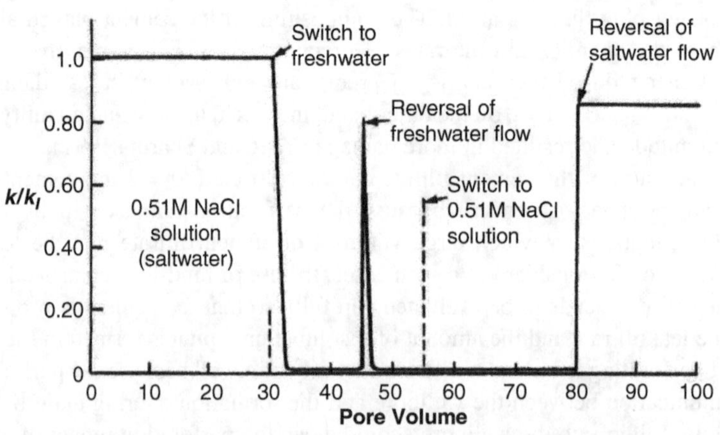


**Fig. 2.7—Permeability reduction. Temporary and permanent permeability gain illustrating fines migration in Berea sandstone (Khilar and Fogler 1983).**

Fine-grained minerals are present in most sandstones and some carbonates. They are not held in place by the confining pressure and are free to move with the fluid phase that wets them (usually water). They remain attached to pore surfaces by electrostatic and van der Waals forces. At high salt (>2%) concentrations, the van der Waals forces are sufficiently large to keep the fines attached to the pore surfaces. As the salinity is decreased, the repulsive electrostatic forces increase because the negative charge on the surfaces of the pores and fines are no longer shielded by the ions. When the repulsive electrostatic forces exceed the attractive van der Waals forces, the fines are released from pore surfaces (Sharma et al. 1985). There is a critical salt concentration below which fines are released (Khilar and Fogler 1983; Sharma et al. 1985). The typical magnitude of the critical salt concentration is in the range of 5,000 to 15,000 ppm (1.5%) sodium chloride. For divalent ions, this concentration is significantly lower. If a water-sensitive sandstone is exposed to brine with a salinity below the critical salt concentration, fines are released, and significant reductions in permeability are observed (Fig. 2.7).

Fines migration can also be induced by mechanical entrainment of fines, which can occur when the fluid velocity is increased above a critical velocity (Gruesbeck and Collins 1982; Sharma et al. 1985; Das et al. 1995; Sharma et al. 1992; Sharma and Yortsos 1987; Freitas and Sharma 1999). Gruesbeck and Collins (1982) among others, have measured the critical velocity for sandstones. Typical reported values of critical velocities are in the range 0.02 m/s. This translates into modest well flow rates for most oil and gas wells.

It is experimentally observed that critical flow velocities for fines migration are lower when the brine phase is mobile. Critical velocities are an order of magnitude higher when the brine is at a residual saturation. This implies that fines migration will be more important with the onset of water production in a well. This is indeed the case. It is often observed that well productivity declines much more rapidly after the onset of water production. In such instances, more-frequent acid treatments are needed to maintain production after water breakthrough.

The extent of permeability reduction observed is also a function of the wettability of the rock. More-oil-wet rocks tend to show less water sensitivity, maybe because the fines that are partially coated with oil are not as readily accessible to the brine. Significantly smaller reductions in permeability are observed when the rock is made less water-wet (Sarkar and Sharma 1990; Freitas and Sharma 1999).

The preceding observations imply that fines migration can be induced by any operation that introduces low-salinity (<2%) or high-pH (>9%) fluids into a water-sensitive formation. Fines migration can also be induced by high flow rates in the near-wellbore region, particularly in wells producing water. Examples of such operations include

- Loss of freshwater-mud filtrate or completion fluid to the formation
- Steam injection in a huff 'n' puff operation for recovering heavy oil
- Water injection from a freshwater source
- High well production rates (flow velocities above the critical velocity)
- Water breakthrough in production wells

## 2.9 Formation Damage Caused by Swelling Clays

Swelling clays, although relatively abundant in shales, do not occur as commonly in producing intervals. Thus, problems with swelling clays are not nearly as common as those associated with fines migration. The most common swelling clays found in reservoir rock are smectites and mixed-layer illites. Earlier, it was thought that much of the water and rate sensitivity observed in sandstones was caused by swelling clays. However, it is now well-accepted that the water-sensitive and rate-sensitive behavior in sandstones is more commonly the result of fines migration and only rarely of swelling clays (Jones 1964; Mungan 1965). Swelling clays reduce formation permeability by peeling off the pore surfaces and plugging pore throats, not by reducing porosity alone. Should this happen to any extent, large reductions in permeability are observed.

The presence of swelling clays is generally associated with drilling problems (i.e., hole quality and stuck pipe). This can result in poor cement jobs and sensitivity to completion fluids. Poor hole quality in the producing interval can result in significant migration of fluids behind pipe, resulting in loss of fluids control in the wellbore. These problems are encountered if either the producing formation or the intervening shales contain substantial quantities of swelling clays. When swelling clays are present in the producing interval, formation-damage problems can occur because of rate sensitivity or water sensitivity. Care must be exercised to ensure that production rates and drawdowns in such wells are maintained so that the critical velocity is not exceeded in the near-wellbore region.

Clay minerals, such as smectites and mixed-layer illites, can expand in volume up to 20 times their original volume through adsorption of layers of water between their unit cells. Such 2:1 clay minerals are particularly prone to swelling because there is no hydrogen bonding between the octahedral layers of the unit cells.

Swelling is known to occur in three steps. In the first step, referred to as crystalline swelling, layers of water enter the interlayer space in the clay mineral, resulting in an increase in the C spacing of the clay mineral in steps. The size of these steps is observed to be approximately equal to the diameter of the water molecule. Extremely large swelling pressures can be generated through such an expansion of the clay lattice. The next stage in swelling is referred to as hydration swelling. This is thought to occur through the hydration and dehydration of ions entering the interlayer region. Several theories have been proposed to explain the observed repulsive hydration force observed in the presence of different cations (Israelachvili 1992). Finally, when the interlayer spacing is ≈ 50 Å, free swelling occurs. This is driven primarily by the balance between electrostatic and van der Waals forces between the layers of clay. In this stage of swelling, the clay layers are sufficiently far apart that very little mechanical integrity exists in the clay. Such clay minerals are liable to be dispersed in the flowing fluid and plug pore throats.

To prevent fines migration and clay swelling, various chemical treatments have been designed. These include

- Polymers containing quaternary ammonium salts (Borchardt et al. 1984)
- Hydrolyzable metal ions such as zirconium oxychloride (Peters and Stout 1977)
- Hydroxy-aluminum (Coppel et al. 1973)
- Polymerizable ultrathin films (Sharma and Sharma 1994)

Each of these methods relies on coating the fines (which are usually negatively charged) with large polyvalent cations that can attach irreversibly to the mineral surfaces. When the electrostatic charges on the fines are neutralized, the likelihood of fines migration is reduced significantly. Fines-stabilizing chemicals have been used in treatments such as acidizing, gravel packing, and fracturing (Ezeukwu et al. 1998). The effectiveness of such treatments is discussed extensively in Borchardt (1989).

## 2.10 Formation Damage in Injection Wells

Water is commonly injected into formations for three primary reasons: pressure maintenance, water disposal, or waterflooding. In such projects, the cost of piping the water and pumping it is determined primarily by reservoir depth and the source of the water. However, water-treatment costs can vary substantially depending on the water quality required. In most cases, the well injectivity is a crucial factor in determining the cost of water injection. Maintaining high injectivities over long periods of time is extremely important for all water-injection projects.

Historically, a great deal of expense and effort has been expended in treating water to ensure very-high-quality water is being injected so that the injectivity of the well can be maintained over a long period of time.

There are two main properties of injection water that determine the formation damage or the injectivity of water-injection wells: (1) the total dissolved solids in the injection water and (2) the total suspended solids (solids and oil droplets) in the injection water (Barkman and Davidson 1972; Eylander 1988; Sharma et al. 2000; van Oort et al. 1993; Wennberg and Sharma 1997; Pang and Sharma 1997).

The salinity and ion content in the injection water control two types of formation damage in an injection well: (1) freshwater sensitivity of the formation and (2) precipitation of inorganic scale.

In water-sensitive formations, if fresh water is being injected from a nearby lake or river, caution must be exercised to ensure that fines migration is not a major factor. This can be achieved by ensuring that the salinity is above the critical salt concentration for the rock. Injection wells are usually less susceptible to fines-migration problems than production wells because the fines being generated are pushed away from the wellbore, leading to less severe impairment in the near-wellbore region and, therefore, relatively small losses in injectivity. In some instances in which the reservoir contains large proportions of clays and fines, severe injectivity losses may be experienced when injecting below the critical salt concentration.

The precipitation of inorganic scale is a major concern when injecting brines with a high concentration of divalent ions (Allaga et al. 1992; Read and Ringen 1982; Mahmoud 2014). The hardness of the injection water is a good indicator of its scaling tendency. Should the water analysis indicate large concentrations of calcium, magnesium, iron, or barium, a water-treatment facility that softens the water may be required. This is also an issue when injecting seawater into formations that contain brines with high salinity.

Large persistent drops in injectivity are expected when inorganic scales are formed in injection wells. Most field experience, however, indicates that the injection fluid quickly displaces the native brines away from the near-wellbore region with very little mixing. Inorganic-scale precipitation resulting from incompatibility between the injection and reservoir brine is, therefore, not usually an issue for most injection wells. Geochemical interactions between injected fluids and the reservoir minerals can sometimes result in the formation of insoluble precipitates. Scale precipitation can also be induced by changes in pH, temperature, and state of oxidation of the brine. The formation of insoluble iron precipitates as a result of corrosion is a common source of damage in injection wells. These precipitates, mixed with other organic material, can result in severe and irreversible reductions in well injectivity. Careful analysis of both the formation brines and injected fluids and a check of the reservoir mineralogy are necessary. Checking for compatibly and ensuring that inorganic-scale precipitation does not occur at reservoir temperature and pressure conditions are important when any water-injection program is planned.

The presence of solids and oil droplets in the injection fluid can result in severe and rapid declines of injectivity (Barkman and Davidson 1972; Eylander 1988; Sharma et al. 2000; van Oort et al. 1993; Wennberg and Sharma 1997; Pang and Sharma 1997). If the injection pressure is below the fracture gradient and if fracturing is undesirable from a reservoir engineering or environmental point of view, small concentrations of solids can result in rapid reductions of well injectivity. As an example, 5 ppm of solids being injected into a well at 10,000 B/D computes to 45 kg/d of solids being injected. This large volume of solids can result in severe and rapid plugging of the injection well in a relatively short duration. Field experience in many parts of the world suggests that matrix injection of clean brines containing 3 to 5 ppm of suspended solids results in injection-well half-lives (time it takes for injectivity to decline to half its value) of 3 to 6 months. **Fig. 2.8** shows the injectivity of a well in the offshore Gulf of Mexico. Seawater was being injected into this well at the rates indicated (Sharma et al. 2000). As the figure shows, despite the relativity good quality of the water, a rapid reduction in injectivity was observed in this and other wells in this field. This reduction led to costly stimulation and workover operations being required in these subsea wells.

In other field experiences, water has been injected into injection wells with minimal impact on injectivity. A good example of this type of injection-well behavior is the injection of produced water in Prudhoe Bay field in Alaska, where 2,000 ppm oil plus solids in the injection water have been routinely injected with relatively little impact on well injectivity. The apparent lack of formation damage is a consequence of thermally induced injection-well fractures that propagate hundreds of meters into the formation (Perkins and Gonzalez 1985; Martins et al. 1995; van den Hoek 1996; Paige and Murray 1994; van Oort et al. 1993; Suarez-Rivera et al. 2002; Detienne 1998). A great deal of work has been conducted to study the impact of water quality on the growth of fractures in water-injection wells and the impact of injection-well fractures on reservoir sweep and oil recovery (Gadde and Sharma 2001; Saripalli et al. 1999). This discussion is outside the scope of this chapter.

When fracturing injection wells is undesirable or unacceptable, the quality of the injection water plays an important role in determining well injectivity or formation damage in injection wells. Various water-clarification devices such as sedimentation tanks, sand filters, cartridge filters, flotation devices, or hydrocyclones are available. These facilities significantly prolong the life of water-injection wells and significantly reduce the formation damage. An economic analysis is thus necessary to ensure that the benefits are greater than the costs.

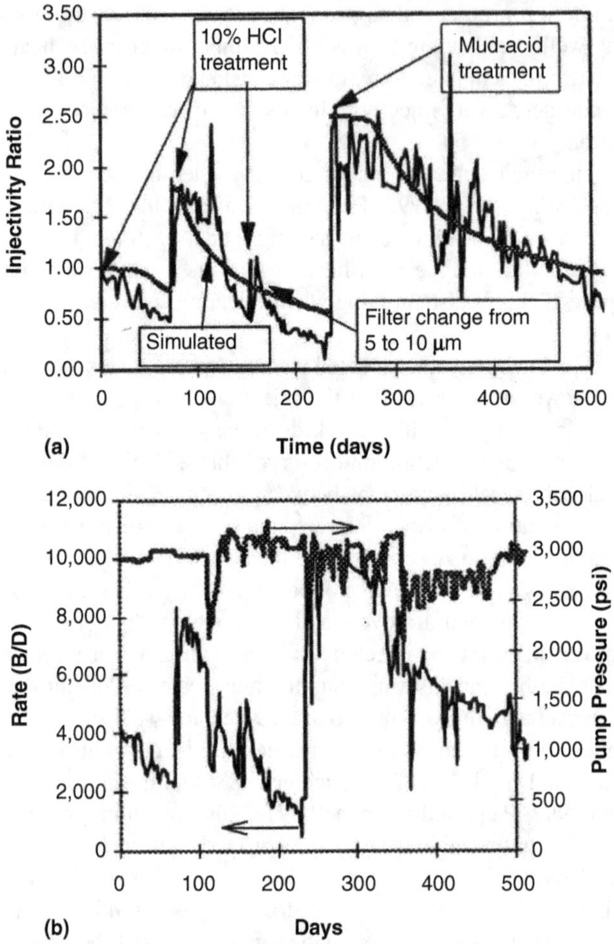


**Fig. 2.8—Behavior of Well A10: (a) injectivity decline; (b) pressure and rate data (Sharma et al. 2000).**

## 2.11 Formation Damage Caused by Paraffins and Asphaltenes

Perhaps the most common formation-damage problem reported in the mature oil producing regions of the world are organic deposits forming both in and around the wellbore. These organic deposits fall into two broad categories: paraffins and asphaltenes.

Crude oils contain three main groups of compounds: saturated hydrocarbons or paraffins, aromatic hydrocarbons, and resins and asphaltenes. **Table 2.2** shows the gross composition of crude oils, tars, and bitumens obtained from various sources. It is evident that crude oils contain substantial proportions of saturated and aromatic hydrocarbons with relatively small percentages of resins and asphaltenes. More-degraded crudes, including tars and bitumens, contain substantially larger proportions of resins and asphaltenes.

**2.11.1 Paraffin Deposition.** Paraffins are high-molecular-weight alkanes ($C^{20+}$) that can build up as deposits (e.g., in the wellbore, in feed lines). These organic deposits can act as chokes within the wellbore, resulting in a gradual decrease in production with time as the deposits increase in thickness. This can result in producing problems unless some remedial action is taken on a systematic and periodic basis. Deposits vary in consistency from soft accumulations to hard brittle deposits. Usually the deposits are firmer and harder as the molecular weight of the paraffin deposits increases. Sometimes paraffins and asphaltenes occur together in organic deposits.

The primary cause of wax or paraffin deposition is simply a loss in solubility in the crude oil (McClaflin and Whitfill 1984; Thomas 1988). The loss of solubility is usually a result of changes in

| Average Value of: (number of samples) | Normal Producible Crude Oil (517) | All Crude Oil Including Tars (636) | Disseminated Bitumen (1057) |
|---|---|---|---|
| Saturated hydrocarbon | 57.2 | 53.3 | 29.2 |
| Aromatic hydrocarbon | 28.6 | 28.2 | 19.7 |
| Resins + asphaltenes | 14.2 | 18.5 | 51.1 |
| Aromatic sulfur (% of aromatic fraction)* | 2.07 | | 1.85 |

*Number of samples for aromatic sulfur: 230 and 88, respectively.

Table 2.2—Gross composition of crude oils (wt% of the fraction boiling above 210°C).

temperature, pressure, or composition of the crude oil because of loss of dissolved gases. Paraffins that have the highest melting point and molecular weight are usually the first to separate from solution, with lower-molecular-weight paraffins separating as the temperature decreases further. For example, a $C^{60}$ alkane with a melting point of approximately 215°F will deposit at a much higher temperature than a $C^{20}$ alkane with a melting point of 98°F.

The ability of the crude oil to hold the paraffin in solution is generally quantified by use of two indicators:

- Pour point
- Cloud point

The procedure for measuring the pour point and the cloud point may be found in American Society for Testing and Materials (ASTM) manuals (ASTM D2500-11 for cloud points; ASTM D97-12 for pour points). The cloud point is defined as the temperature at which paraffins begin to come out of solution and a clear solution of hydrocarbons turns cloudy. Obviously, it is difficult to measure the cloud point for dark crude oil, because cloudiness is not visible. In such cases, the presence of paraffin crystals may have to be detected by using a polarizing light microscope. The pour point is defined as the temperature at which the crude oil no longer flows from its container. As the temperature is lowered, wax crystals form an interlocking network that supports the hydrocarbon liquid within it. This network of paraffin crystals is quite shear sensitive and loose when first formed but can harden and become extremely rigid as fluid is lost from it. Pour points are relatively easy to measure in the field and provide a good indication of conditions under which large quantities of paraffin will fall out of solution in crude oils.

The most common cause of loss of solubility of the paraffin in the crude oil is a decrease in temperature, which may occur for a variety of reasons (Newberry and Barker 1985): cooling produced by the crude oil and associated gas expanding through the perforations, gas expansion while lifting fluids to the surface, radiation of heat from the tubing to the surrounding formation being induced by intrusion of water into or around the wellbore, and loss of lighter constituents in the crude oil because of vaporization. Several other possible reasons for a decrease in temperature can be envisioned. In offshore installations, for example, paraffin problems are usually associated with the rapid change in temperature as the crude oil from the wellbore enters subsea pipelines that are immersed in seawater at 4°C. Large volumes of paraffins can be deposited on the surfaces of the pipelines, which require periodic pigging.

Pressure itself has little or no influence on the solubility of paraffin in crude oil. However, it does have a significant impact on the composition of the crude oil. Reductions in pressure usually lead to loss of volatiles from the crude oil and can induce the precipitation of paraffins. This is the primary reason that paraffin problems are more common in the more-mature regions of the world. As the reservoir pressure is depleted and the lighter components of the crude oil are produced in preference to the heavier fractions, the likelihood of paraffin precipitation is significantly increased.

For paraffin deposition to be a significant problem, the paraffin must deposit on the pore walls or on the tubing surface. If the paraffin remains entrained in the crude oil, it usually offers few production problems. Several factors influence the ability of paraffin to deposit on the pipe walls:

- The presence of water wetting the surfaces of the pipe tends to inhibit paraffin deposition. In addition, water has a higher specific heat than oil, which increases flowing temperatures.
- The quality of the pipe plays an important role. Rusty pipes with large surface area and numerous sites for paraffin-crystal formation offer an ideal location for paraffin deposition. Paraffin adheres to rough surfaces better than smooth surfaces.
- The temperature profile in the near-wellbore region or within the pipe plays an important role in determining whether the paraffin will deposit on the walls or continue to be entrained with the fluid.

The injection of fluids such as stimulation fluids or injection water into the wellbore can often induce paraffin-deposition problems. This is particularly true if the surface temperature is significantly colder than the reservoir temperature. Field cases documenting paraffin precipitation during fracture stimulation are provided in McClaflin and Whitfill (1984).

**2.11.2 Removal of Paraffin Deposits.** Paraffin accumulations are removed by methods that can be broadly placed in three categories:

- Mechanical removal of paraffin deposits
- The use of solvents to remove paraffin deposits
- The use of heat to melt and remove the wax

Mechanical methods such as scrapers, knives, and other tools are most commonly used to remove paraffin deposits in the wellbore. They can be very effective and are relatively inexpensive.

The most common solvent used for paraffin removal from tubulars and the near-wellbore region is crude oil. Hot oiling is the least-expensive method, commonly employed on stripper wells to remove paraffin deposits. Lease crude taken from stock-tank bottoms is heated to temperatures of 300°F or more. This heated oil is then injected or gravity-fed into the tubing or the annulus (more common). The high temperature induces solubilization of the paraffin deposits in the injected crude, which is then produced back to the surface. Hot oiling has been used successfully to remove paraffin deposition but can also result in formation damage. The use of hot salt water to melt the paraffin may be a safer approach.

Solvents, both organic and inorganic, have been used in the past. These include crude oil, kerosene, diesel, and surfactant formulations that can solubilize the paraffin. Organic solvents that consist of a blend of aromatics are usually used to remove mixtures of paraffin and asphaltenes deposits. However, the cost of such treatments can be significantly higher than that of hot-oil or -water treatments.

Steam has been used in a number of cases in which severe paraffin problems have resulted in plugged tubulars. The lack of solubility of paraffin in hot water necessitates the use of surfactants with steam or hot water so that the melted paraffin can be removed.

**2.11.3 Methods for Preventing Paraffin Deposition.** There are several mechanical adjustments that can be made in the production string that can minimize the likelihood of paraffin deposition. In general, these steps are designed to minimize the cooling of the crude oil as it is produced to the surface. This can be accomplished by designing pumping wells or tubing sizes and gas lift systems that maximize the flow of oil to the surface and minimize the heat lost to the surrounding formations. Use of more-expensive methods such as plastic coatings on tubulars and the use of electrical heaters is severely limited by economics.

Paraffin inhibitors are a class of compounds that consist of crystal modifiers that prevent the deposition of paraffin onto the surfaces of pipes. These surface-active materials retard paraffin deposition by inhibiting the adhesion of paraffin to sites on the tube walls. Surfactants used in these applications include wetting agents, dispersants, and crystal modifiers (Thomas and Sharma 1988; Newberry and Barker 1985; Houchin and Hudson 1986). Each of these chemicals needs to be tested for a specific crude oil to evaluate its effectiveness.

**2.11.4 Asphaltene Precipitation.** High-molecular-weight constituents of crude oil containing nitrogen, sulfur, and oxygen (N, S, and O) compounds are referred to as asphaltenes. This broad class of compounds is clearly not hydrocarbons because these compounds contain a large portion of heteroatoms in their structure. Lower-molecular-weight NSO compounds are referred to as resins. The separation of crude oil into resins and asphaltenes and other constituents is primarily based on solubility. Asphaltenes and resins are generally defined as the pentane-insoluble fraction of the crude oil (Yen 1974).

The average molecular structure of an example asphaltene fraction from a crude oil from Venezuela is shown in **Fig. 2.9.** It consists primarily of condensed aromatic rings associated with aliphatic tails. The polynuclear aromatic rings associate with each other through their $\pi$ electron systems to form clusters of stacked rings as shown in the figure. In crude oils, these asphaltene structures are dispersed and maintained in suspension by the action of resins. If sufficient quantities of resin molecules are present in the crude oil, the asphaltenes remain dispersed and in solution. However, the addition of large quantities of alkanes or the removal of the resin fraction can result in a loss of solubility because the asphaltene molecules associate with each other forming large aggregates or micelles, and precipitate out. These micelles or aggregates are visible under optical microscopes as dark, solid aggregates. Precipitation of asphaltenes occurs through the formation of such aggregates. The solubility of asphaltenes is, therefore, a function of temperature, pressure, and the composition of the crude oil. Any action that impacts the compositional balance of the crude oil can affect the ability of the oil to maintain the asphaltenes in solution.

A very common example of the change in composition of a crude oil is what occurs during pressure depletion in a reservoir. As shown in **Fig. 2.10,** the solubility of asphaltene is a minimum at the bubblepoint pressure (Hirschberg et al. 1984). This has important consequences for predicting where asphaltene precipitation will occur in a reservoir. As the reservoir is depleted and the bubblepoint pressure is achieved lower in the tubing or even in the formation itself, the possibility of asphaltene deposition occurs at these locations. Indeed, in studies published in the literature, the location of asphaltene deposition is observed to move from the top of the tubing to the bottom and into the reservoir over a period of time as the reservoir pressure is depleted and the location where the bubblepoint pressure is reached moves farther out toward the reservoir.

Asphaltene deposition can also be induced by changes in composition of the crude oil through injection of fluids such as carbon dioxide ($CO_2$) or lean gas. Several studies have documented the possibility

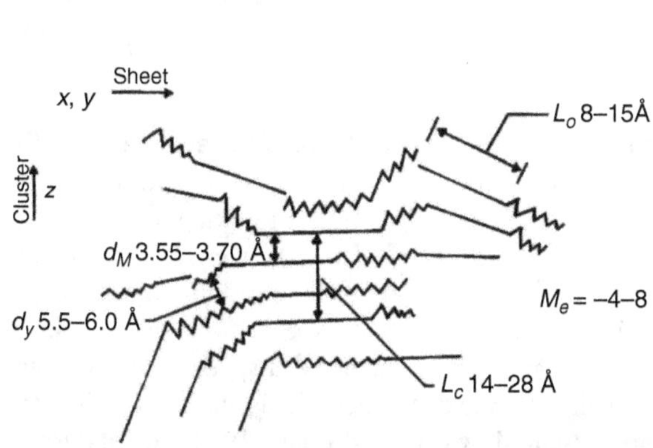


**Fig. 2.9—Cross-sectional view of an asphaltene model based on X-ray diffraction. Zigzag line represents configuration of saturated carbon chain or loose net of naphthenic rings; straight line represents the edge of flat sheets of condensed aromatic rings (Yen 1974).**

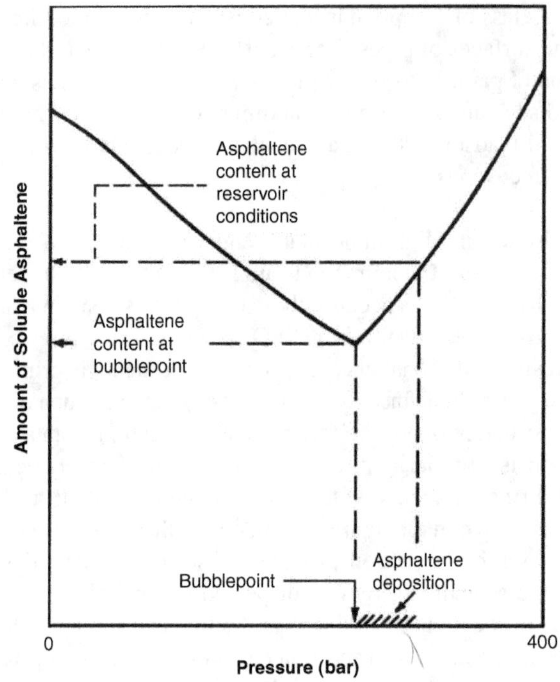


**Fig. 2.10—Pressure dependence of asphaltene solubility for a North Sea crude oil showing the possibility of asphaltene deposition in the well tubing (Hirschberg et al. 1984).**

of asphaltene precipitation during lean-gas and $CO_2$ injection (Monger and Fu 1987; Monger and Trujillo 1991; Leontaritis et al. 1992). Large changes in temperature can also induce asphaltene deposition (Leontaritis 1989; Kawanaka et al. 1991). In such cases, deposits of paraffin and asphaltene are commonly observed together. The asphaltene particles frequently act as nucleation sites for paraffin crystals.

**2.11.5 Removal of Asphaltene Deposits.** The removal of asphaltene deposits also requires the use of solvents or mechanical devices. However, the solvents used for asphaltene removal are quite different from those used for paraffins. Because asphaltenes are soluble in aromatic solvents, mixtures of aromatic solvents such as xylene have been used to remove asphaltene deposits (Schantz and Stephenson 1991). It should be noted that solvents such as diesel and kerosene that are primarily straight-chain alkanes should not be used because they may induce asphaltene precipitation. Alternatively, terpenes or terpene blends can be used to remove asphaltene deposits. Compared with aromatic solvents, terpenes have good solvency and are biodegradable, less toxic, and less flammable (Curtis 2003).

## 2.12 Formation Damage Resulting From Emulsion and Sludge Formation

The presence of emulsions at the surface does not imply the formation of emulsions in the near-wellbore region. Often, surface emulsions are a result of mixing and shearing that occur in chokes and valves in the flow stream after the fluids have entered the well. It is uncommon to have emulsions and sludges form in the near-wellbore region without the introduction of external chemicals (Schechter 1991). The mixing of two immiscible fluids at high shear rate in the formation can sometimes result in the formation of a homogeneous mixture of one phase dispersed into another. Such emulsions usually have a higher viscosity than either of the constituent fluids and can result in significant decreases in the ability of the hydrocarbon phase to flow (Soo and Radke 1984; Soo and Radke 1986).

Crude-oil/brine emulsions are stabilized by the presence of surfactants and colloidal particles such as clays, paraffins, and asphaltenes (Tambe and Sharma 1994). In general, organophilic particles such as paraffins and asphaltenes will favor the formation of oil-external emulsions and sludges. Water-wet solids such as clays will favor the formation of water-external emulsions. It is important to minimize the loss of surface-active materials into the near-wellbore region to ensure that emulsions do not form. For example, large volumes of surfactants are used as corrosion inhibitors and dispersants in acid treatments. A significant cause of failure of acid treatments is the formation of sludges and emulsions during an acid treatment as a result of the presence of these surfactants. The compatibility of crude oil with the acid package needs to be evaluated before it is pumped into the well. It has also been observed that the presence of iron enhances the formation of these sludges. It is, therefore, recommended that iron be removed from the tubing by circulating a slug of acid to the surface to ensure that the iron-rich acid is not squeezed into the formation during an acid treatment.

In general, it is difficult to remove emulsions and sludges once they are formed. Thus, it is imperative to prevent the formation of such emulsions. Use of mutual solvents such as alcohols and surfactants (demulsifiers) is the most common way of removing these deposits from the near-wellbore region. However, because of the unfavorable mobility ratio of the injected fluid, placing the treatment fluids in the plugged zones can be difficult. Again, laboratory tests with the crudes should be conducted to ensure compatibility.

## 2.13 Formation Damage Resulting From Condensate Banking

As shown in **Fig. 2.11,** gas/condensate reservoirs are defined as reservoirs that contain hydrocarbon mixtures that on pressure depletion cross the dewpoint line. In such instances, as the bottomhole pressure is reduced during production, the dewpoint pressure of the gas is reached in the near-wellbore region. This results in the formation of liquid hydrocarbons near the wellbore and in the reservoir. As the liquid-hydrocarbon saturation in the near-wellbore region increases, the gas relative permeability is decreased, resulting in significant declines in well productivity (Narayanaswamy et al. 1999a; Afidick et al. 1994). An example of this is shown in **Fig. 2.12.** Here, a substantial reduction in the well productivity is obtained as the average reservoir pressure declines below the dewpoint for a well in the Arun gas field. This mechanism of formation damage is related primarily to changes in fluid saturation in the near-wellbore region, resulting in decreases in gas relative permeability.

The buildup of the condensate bank and its consequences on well productivity have been well-studied in the literature (Barnum et al. 1995; Boom et al. 1995; Boom et al. 1996; Asar and Handy 1988; Hartman and Cullick 1994; Henderson et al. 1998; Wang and Mohanty 1999; Cole and Hartman 1998). Early predictions of productivity loss because of condensate dropout indicated that a loss in PI by a factor of 5 to 8 would be expected because of liquid buildup. However, the decline in PI observed in many of the fields is much smaller (a factor of 2 to 4). Further investigation of this problem indicated that the high gas-flow rates in the near-wellbore region can result in stripping out of the

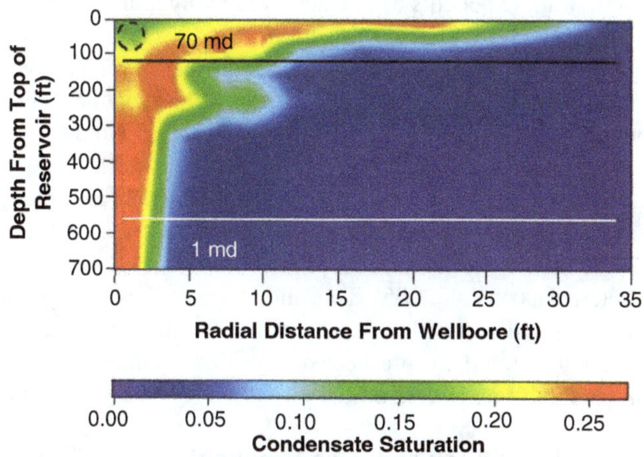


**Fig. 2.11—Illustration of condensate dropout in near-wellbore region. Buildup of liquid hydrocarbons in this region can cause large reductions in gas relative permeability (Narayanaswamy et al. 1999a).**

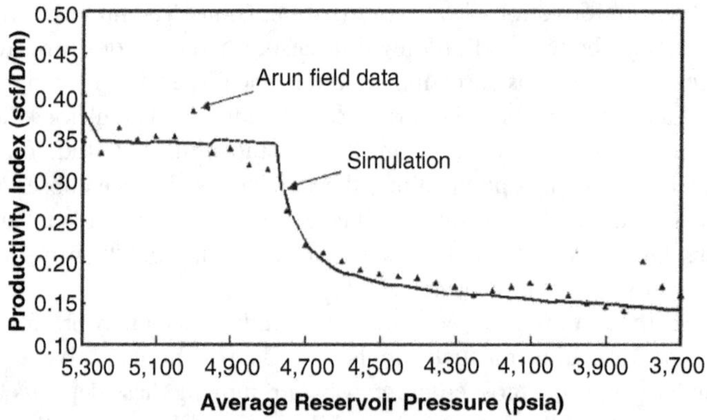


**Fig. 2.12—Reduction in well productivity caused by condensate buildup, Arun field, Indonesia (Narayanaswamy et al. 1999a).**

liquid-hydrocarbon phase in regions around the wellbore. This stripping-out effect has been quantified through capillary-number-dependent models for relative permeability of the gas phase (Narayanaswamy et al. 1999b; Boom et al. 1996; Pope et al. 2000). With this phenomenon properly accounted for, good agreement with field observations is obtained (Fig. 2.12).

In addition to liquid dropout, several other important phenomena can play an important role in determining the well productivity and need to be carefully evaluated. Because of the high flow rates of gas in the near-wellbore region, non-Darcy effects may be significant and may need to be accounted for (Narayanaswamy et al. 1999a; Wang and Mohanty 1999; Coles and Hartman 1998; Narayanaswamy 1999b). It has recently been documented that nonequilibrium effects may also play an important role in condensate accumulation in the near-wellbore region (Pope et al. 2000). The combination of non-Darcy flow, capillary-number-dependent relative permeability, and phase behavior makes the problem rather complex, and numerical simulations are needed to fully capture all of the physics of the problem. Clearly distinguishing the effects of liquid dropout from non-Darcy effects from production performance and pressure-transient tests can be challenging and may require compositional numerical models. Such models are widely available and have been used in estimating the productivity of gas wells, including condensate dropout.

The most direct method of reducing condensate buildup is to reduce the drawdown so that the bottomhole pressure remains above the dewpoint. In cases where this is not desirable, the impact of condensate formation can be reduced by increasing the inflow area and achieving linear flow rather than radial flow into the wellbore. This minimizes the impact of the reduced gas permeability in the near-wellbore region. Both of these benefits can be achieved by hydraulic fracturing.

Hydraulic-fracture stimulation is the most common method used to remedy condensate-buildup problems. The creation of a fracture results in a significant decrease in the drawdown needed to produce the well. In addition, the buildup of a liquid-hydrocarbon phase on the faces of the fracture does not impact the well productivity as significantly as in radial flow around the wellbore. Additional details of this are available in Kumar (2000).

Recently, the use of solvents and surfactants such as methanol has been suggested as a way to stimulate gas-condensate wells in which hydraulic fracturing is not the preferred option (Du et al. 2000; Al-Anazi et al. 2002). The use of methanol results in the removal of the condensate and water banks around a wellbore (Al-Anazi et al. 2005). This allows gas to flow unimpeded through the near-wellbore region, resulting in smaller drawdown and slower accumulation of condensate. Within certain ranges of temperature and pressure, the presence of a residual methanol phase in the near-wellbore region can also result in the inhibition of condensate formation for a period of time.

## 2.14 Formation Damage Resulting From Gas Breakout

In solution-gas-drive reservoirs, as the reservoir fluid pressure drops below the bubblepoint, a gas phase is formed. If this event occurs in the wellbore, gas bubbles formed help to lift the liquid hydrocarbons to the surface. However, if the bubblepoint is reached in the near-wellbore region, a significant gas saturation builds up around the wellbore resulting in a decrease in the oil relative permeability. Gas breakout may also induce precipitation of asphaltene in the near-wellbore region. As might be expected, this form of damage is more likely to occur later in the life of the reservoir as the average reservoir pressure is depleted below the bubblepoint.

This type of damage is relatively easy to diagnose if the production engineer has access to phase-behavior data. In many cases, however, lack of access to these data can result in an incorrect diagnosis of the reduction in productivity of the well. Such a misdiagnosis can lead to inaccurate recommendations for stimulation treatments.

In typical relative permeability curves, the change in the relative permeability to oil can be rather dramatic as the gas saturation increases. This decrease in oil permeability can have a dramatic effect on the well productivity. Oil-flow rates can decrease while gas-flow rates may increase rapidly over a relatively short duration.

The most common method to address gas-breakout problems is to hydraulically fracture the well in an attempt to reduce the drawdown needed to produce at a given rate. Repressurizing the reservoir is also an excellent alternative. The economics of reservoir repressurization need to be carefully evaluated in such applications. It should be noted that in cases where the average reservoir pressure drops below the bubblepoint throughout most of the reservoir, a gas cap may begin forming in the reservoir. This can, over a long time period, result in increased gas production into the wellbore from the gas cap.

## 2.15 Formation Damage Resulting From Water Blocks

If large volumes of water-based drilling or completion fluids are lost to a well, a region of high water saturation around the wellbore forms. In this region, the relative permeability to the hydrocarbon phases is decreased, resulting in a net loss in well productivity (Tannich 1975; Abrams and Vinegar 1985).

Regions of high water saturation, or water blocks, around the wellbore are expected to dissipate with time as the hydrocarbon fluids are produced. In general, when the viscous forces are significantly larger than the capillary forces, the water block will clear up rather rapidly. If, however, the capillary forces holding the water in place are larger than the viscous forces, (e.g., in tight gas reservoirs), water blocks may persist for a very long period of time. A capillary number, defined as the ratio of capillary to viscous forces, can be used to quantify this effect. When capillary forces are larger than or comparable to viscous forces, water blocks are hard to remove. On the other hand, when viscous forces dominate, water blocks will clear up in a matter of a few hours or days (Holditch 1979). Water blocks will generally be more troublesome for low-permeability, depleted gas wells in which the capillary number is significantly less than unity (Cimolai et al. 1993; Kamath and Laroche 2000).

There are three primary methods used to remove water blocks (McLeod and Coulter (1966) Mahadevan and Sharma (2003):

1. Surging or swabbing the wells to temporarily increase the capillary number.
2. Reducing surface tension through the addition of surfactants or solvents. This also has the net effect of increasing the capillary number by reducing the interfacial tension between the hydrocarbon and water phases so that the water block may be cleaned up during flowback.
3. The use of solvents or mutual solvents, such as alcohols, to solubilize the water and remove it through a change in phase behavior.

All of these three methods have been successfully applied in the field. The benefit of one method over another depends on the specific conditions of reservoir permeability, temperature, and pressure.

## 2.16 Formation Damage Resulting From Wettability Alteration

Converting a rock from water-wet to oil-wet results in a substantial reduction in the relative permeability to the hydrocarbon phase and an increase in relative permeability to the water (Fig. 2.12). Wettability alteration to less-water-wet conditions is, therefore, clearly undesirable.

The loss of surfactants in drilling and completion fluids (Sharma and Wunderlich 1987; Yan et al. 1993; Yan and Sharma 1989; Yan et al. 1993), loss off corrosion inhibitors and dispersants in stimulation fluids, and the use of resins for sand control can cause changes in wettability in the near-wellbore region. Care must be exercised when using oil-wetting surfactants in the wellbore to ensure that these fluids are not lost to the productive pay zone. Alteration of wettability in a region around the wellbore can result in an additional pressure drop because of the reduction in oil permeability. This additional pressure drop or skin is hard to distinguish from mechanical skin caused by physical plugging of pore throats. In effect, wettability alteration has the same net result as changing the effective permeability to the hydrocarbon phase in a region around the wellbore.

The use of solvents and water-wetting surfactants may be recommended in cases in which large volumes of oil-wetting surfactants such as oil-based muds have been lost to the formation.

### 2.17 Bacteria Plugging

Anaerobic bacteria are ubiquitously present in and around oil and gas wells (Carlson et al. 1961). Under most producing conditions, their growth is not stimulated because of the high-temperature and -pressure conditions. However, in some instances, the injection of water-based fluids can induce the growth of microbial populations and can result in significant declines in productivity or injectivity (Raleigh and Flock 1965). The growth of sulfur-reducing bacteria can also result in the generation of hydrogen sulfide gas and the fouling of flowlines and facilities.

The use of a bactericide (such as sodium hypochlorite or mixtures of other strong oxidizing agents and antibacterial agents) is sometimes an effective, though expensive, method of reducing this problem.

### 2.18 Conclusion

In this chapter, we have discussed methods to measure and quantify formation damage in oil and gas wells. Several different mechanisms responsible for causing formation damage were discussed. A better understanding of these mechanisms allows us to make recommendations for drilling, completion, and production operations that will reduce the extent of formation damage and maximize well productivity. Methods to remove the damage were also discussed. In many instances of near wellbore damage, acidizing is an effective method of removing the damage, while for some formation-damage mechanisms, other methods such as hydraulic fracturing may be more effective. It is, therefore, very important to correctly identify the damage mechanism before acidizing a well. The discussion in this chapter helps us to do that.

## References

Abrams, A. 1977. Mud Design to Minimize Rock Impairment Due to Particle Invasion. *J Pet Technol* **29** (5): 586–592. SPE-5713-PA. http://dx.doi.org/10.2118/5713-PA.

Abrams, A. and Vinegar, H. J. 1985. Impairment Mechanisms in Vicksburg Tight Gas Sands. Presented at the SPE/DOE Low Permeability Gas Reservoirs Symposium, Denver, 19–22 March. SPE-13883-MS. http://dx.doi.org/10.2118/13883-MS.

Afidick, D., Kaczorowski, N. J., and Bette, S. 1994. Production Performance of a Retrograde Gas Reservoir: A Case Study of the Arun Field. Presented at the SPE Asia Pacific Oil and Gas Conference, Melbourne, Australia, 7–10 November. SPE-28749-MS. http://dx.doi.org/10.2118/28749-MS.

Ali, S., Bailey, L., Ladva, H. et al. 1999. Alternative Methods Clean up Filter Cake. *Oil Gas J.* **97** (5): 54–55.

Allaga, D. A., Wu, G., Sharma, M. M. et al. 1992. Barium and Calcium Sulfate Precipitation and Migration inside Sandpacks. *SPE Form Eval* 7 (01): 79–86. SPE-19765-PA. http://dx.doi.org/10.2118/19765-PA.

Al-Anazi, H. A., Walker, J. G., Pope, G. A. et al. 2005. A Successful Methanol Treatment in a Gas/Condensate Reservoir: Field Application. *SPE Prod & Fac* **20** (1): 60–69. SPE-80901-PA. http://dx.doi.org/10.2118/80901-PA.

Al-Riyamy, K. and Sharma, M. M. 2002. Filtration Properties of Oil-in-Water Emulsions Containing Solids. Presented at the International Symposium and Exhibition on Formation Damage Control, Lafayette, Louisiana, USA, 20–21 February. SPE-73769-MS. http://dx.doi.org/10.2118/73769-MS.

Asar, H. and Handy, L. L. 1988. Influence of Interfacial Tension on Gas/Oil Relative Permeability in a Gas-Condensate System. *SPE Res Eng* **3** (03): 257–264. SPE-11740-PA. http://dx.doi.org/10.2118/11740-PA.

*ASTM D2500-11, Standard Test Method for Cloud Point of Petroleum Products.* 2011. West Conshohocken, Pennsylvania: ASTM International. http://dx.doi.org/10.1520/D2500-11.

*ASTM D97-12, Standard Test Method for Pour Point of Petroleum Products.* 2012. West Conshohocken, Pennsylvania: ASTM International. http://dx.doi.org/10.1520/D0097-12.

Azari, M. and Leimkuhler, J. M. 1990. Formation Permeability Damage Induced by Completion Brines. *J Pet Technol* **42** (04): 486–492. SPE-17149-PA. http://dx.doi.org/10.2118/17149-PA.

Bailey, L., Meeten, G., Way, P. et al. 1998. Filtercake Integrity and Reservoir Damage. Presented at the SPE Formation Damage Control Conference, Lafayette, Louisiana, USA, 18–19 February. http://dx.doi.org/10.2118/39429-MS.

Barkman, J. H. and Davidson, D. H. 1972. Measuring Water Quality and Predicting Well Impairment. *J Pet Technol* **24** (7): 865–873. SPE-3543-PA. http://dx.doi.org/10.2118/3543-PA.

Barnum, R. S., Brinkman, F. P., Richardson, T. W. et al. 1995. Gas Condensate Reservoir Behaviour: Productivity and Recovery Reduction Due to Condensation. Presented at the SPE Annual Technical Conference and Exhibition, Dallas, 22–25 October. SPE-30767-MS. http://dx.doi.org/10.2118/30767-MS.

Bartusiak, R., Behrmann, L. A., and Halleck, P. M. 1997. Experimental Investigation of Surge Flow Velocity and Volume Needed to Obtain Perforation Cleanup. *J Pet Sci Eng* **17** (1–2): 19–28. http://dx.doi.org/10.1016/S0920-4105(96)00053-8.

Behrmann, L. A. 1996. Underbalance Criteria for Minimum Perforation Damage. *SPE Drill & Compl* **11** (03): 173–177. SPE-30081-PA. http://dx.doi.org/10.2118/30081-PA.

Behrmann, L. A., Li, J. L., Venkitaraman, A. et al. 1997. Borehole Dynamics During Underbalanced Perforating. Presented at the SPE European Formation Damage Conference, The Hague, 2–3 June. SPE-38139-MS. http://dx.doi.org/10.2118/38139-MS.

Behrmann, L. A., Pucknell, J. K., and Bishop, S. R. 1992. Effects of Underbalance and Effective Stress on Perforation Damage in Weak Sandstone: Initial Results. Presented at the SPE Annual Technical Conference and Exhibition, Washington, D.C., 4–7 October. SPE-24770-MS. http://dx.doi.org/10.2118/24770-MS.

Behrmann, L. A., Pucknell, J. K., Bishop, S. R. et al. 1991. Measurement of Additional Skin Resulting from Perforation Damage. Presented at the SPE Annual Technical Conference and Exhibition, Dallas, 6–9 October. SPE-22809-MS. http://dx.doi.org/10.2118/22809-MS.

Bird, K. and Blok, R. H. J. 1996. Perforating in Tight Sandstones: Effect of Pore Fluid and Underbalance. Presented at the European Petroleum Conference, Milan, Italy, 22–24 October. SPE-36860-MS. http://dx.doi.org/10.2118/36860-MS.

Boom, W., Wit, K., Schulte, A. M. et al. 1995. Experimental Evidence for Improved Condensate Mobility at near Wellbore Conditions. Presented at the SPE Annual Technical Conference and Exhibition, Dallas, 22–25 October. SPE-30766-MS. http://dx.doi.org/10.2118/30766-MS.

Boom, W., Wit, K., Zeelenberg, J. P. W. et al. 1996. On the Use of Model Experiments for Assessing Improved Gas-Condensate Mobility Under Near-Wellbore Flow Conditions. Presented at the SPE Annual Technical Conference and Exhibition, Denver, 6–9 October. SPE-36714-MS. http://dx.doi.org/10.2118/36714-MS.

Borchardt, J. K. 1989. Cationic Organic Polymer Formation Damage Control Chemicals. In *Oil-Field Chemistry*, Vol. 396, Chap. 10, 204–221. ACS Symposium Series, American Chemical Society.

Borchardt, J. K., Roll, D. L., and Rayne, L. M. 1984. Use of a Mineral Fines Stabilizer in Well Completions. Presented at the SPE California Regional Meeting, Long Beach, California, USA, 11–13 April. SPE-12757-MS. http://dx.doi.org/10.2118/12757-MS.

Brooks, J. E., Yang, W., and Behrmann, L. A. 1998. Effect of Sand-Grain Size on Perforator Performance. Presented at the SPE Formation Damage Control Conference, Lafayette, Louisiana, USA, 18–19 February. SPE-39457-MS. http://dx.doi.org/10.2118/39457-MS.

Brown, K. E. and Beggs, H. D. 1977. *The Technology of Artificial Lift Methods*, I. Tulsa, Oklahoma: PennWell Publishing Inc.

Browne, S. V. and Smith, P. S. 1994. Mudcake Cleanup to Enhance Productivity of High-Angle Wells. Presented at the SPE Formation Damage Control Symposium, Lafayette, Louisiana, USA, 7–10 February. SPE-27350-MS. http://dx.doi.org/10.2118/27350-MS.

Browne, S. V. and Smith, P. S. 1994. Mudcake Cleanup to Enhance Productivity of High-Angle Wells. Presented at the SPE Formation Damage Control Symposium, Lafayette, Louisiana, USA, 7–10 February. SPE-27350-MS. http://dx.doi.org/10.2118/27350-MS.

Browne, S. V., Ryan, D. F., Chambers, B. D. et al. 1995. Simple Approach to the Cleanup of Horizontal Wells with Prepacked Screen Completions. *J Pet Technol* **47** (09): 794–800. SPE-30116-PA. http://dx.doi.org/10.2118/30116-PA.

Bruton, B. 1995. Estimate Formation Damage Effects on Horizontal Wells. *Petroleum Engineer International* 68 (8): 29–34.

Burnett, D. B. 1998. Wellbore Cleanup in Horizontal Wells: An Industry Funded Study of Drill-in Fluids and Cleanup Methods. Presented at the SPE Formation Damage Control Conference, Lafayette, Louisiana, USA, 18–19 February. SPE-39473-MS. http://dx.doi.org/10.2118/39473-MS.

Carlson, V., Bennett, E. O., and Rowe, J. A., Jr. 1961. Microbial Flora in a Number of Oilfield Water-Injection Systems. *SPE J.* **1** (2): 71–80. SPE-1553-G. http://dx.doi.org/10.2118/1553-G.

Chauvin, D. J., Jr. 1976. Selecting Packer Fluids: Here Is What to Consider. *World Oil* (June, 1976): 87–92.

Cimolai, M. P., Gies, R. M., Bennion, D. B. et al. 1993. Mitigating Horizontal Well Formation Damage in a Low-Permeability Conglomerate Gas Reservoir. Presented at the SPE Gas Technology Symposium Calgary, 28–30 June. SPE-26166-MS. http://dx.doi.org/10.2118/26166-MS.

Coles, M. E. and Hartman, K. J. 1998. Non-Darcy Measurements in Dry Core and the Effect of Immobile Liquid. Presented at the SPE Gas Technology Symposium, Calgary, 15–18 March. SPE-39977-MS. http://dx.doi.org/10.2118/39977-MS.

Colle, E. 1988. Increase Production with Underbalance Perforation. *Pet Eng Intnl* (July 1988): 39–42.

Coppel, C. P., Jennings, H. Y. Jr., and Reed, M. G. 1973. Field Results From Wells Treated With Hydroxy-Aluminum. *J Pet Technol* **25** (9): 1,108–1,112. SPE-3998-PA. http://dx.doi.org/10.2118/3998-PA.

Crawford, H. R. 1989. Underbalanced Perforating Design. Presented at the SPE Annual Technical Conference and Exhibition, San Antonio, Texas, USA, 8–11 October. SPE-19749-MS. http://dx.doi.org/10.2118/19749-MS.

Cunningham, W. C. and Smith, D. K. 1968. Effect of Salt Cement Filtrate on Subsurface Formations. *J Pet Technol* **20** (03): 259–264. SPE-1920-PA. http://dx.doi.org/10.2118/1920-PA.

Curtis, J. 2003. Environmentally Favorable Terpene Solvents Find Diverse Applications in Stimulation, Sand Control and Cementing Operations. Presented at the SPE Annual Technical Conference and Exhibition, Denver, 5–8 October SPE-84124-MS. http://dx.doi.org/10.2118/84124-MS.

Dake, L. P. 1978. *Fundamentals of Reservoir Engineering*. Amsterdam: Developments in Petroleum Science, No. 8, Elsevier Science BV.

Darley, H. C. H. 1972. Chalk Emulsion: A New Completion Fluid. *Pet Eng* **45** (7): 45–51.

Das, S. K., Sharma, M. M., and Schechter, R. S. 1995. Adhesion and Hydrodynamic Removal of Colloidal Particles from Sufaces. *Part Sci Technol* **13** (3–4): 227–247. http://dx.doi.org/10.1080/02726359508906680.

Detienne, J. -L., Creusol, M., Kessler, N. et al. 1998. Thermally Induced Fractures: A Field-Proven Analytical Model. *SPE Res Eval & Eng* **1** (1): 30–35. SPE-30777-PA. http://dx.doi.org/10.2118/30777-PA.

Du, L., Walker, J. G., Pope, G. A. et al. 2000. Use of Solvents to Improve the Productivity of Gas Condensate Wells. Presented at the SPE Annual Technical Conference and Exhibition, Dallas, 1–4 October. SPE-62935-MS. http://dx.doi.org/10.2118/62935-MS.

Eaton, B. A. and Smithey, M. 1971. Formation Damage from Workover and Completion Fluids. Presented at the SPE California Regional Meeting, Los Angeles, 4–5 November. SPE-3707-MS. http://dx.doi.org/10.2118/3707-MS.

Eylander, J. G. R. 1988. Suspended Solids Specifications for Water Injection from Coreflood Tests. *SPE Res Eng* **3** (04): 1287–1294. SPE-16256-PA. http://dx.doi.org/10.2118/16256-PA.

Ezeukwu, T., Thomas, R. L., and Gunneroed, T. 1998. Fines Migration Control in High-Water-Cut Nigerian Oil Wells: Problems and Solutions. Presented at the SPE Formation Damage Control Conference, Lafayette, Louisiana, 18–19 February. SPE-39482-MS. http://dx.doi.org/10.2118/39482-MS.

Fetkovich, M. J. 1973. The Isochronal Testing of Oil Wells. Presented at the Fall Meeting of the Society of Petroleum Engineers of AIME, Las Vegas, Nevada, USA, 30 September–3 October. SPE-4529-MS. http://dx.doi.org/10.2118/4529-MS.

Fetkovich, M. J. and Vienot, M. E. 1985. Shape Factor, $C_A$, Expressed as Skin, $S_{CA}$. *J Pet Technol* **37** (02): 321–322. SPE-13304-PA. http://dx.doi.org/10.2118/13304-PA.

Francis, P. 1997. Dominating Effects Controlling the Extent of Drilling-Induced Formation Damage. Presented at the SPE European Formation Damage Conference, The Hague, 2–3 June. SPE-38182-MS. http://dx.doi.org/10.2118/38182-MS.

Freitas, A. M. and Sharma, M. M. 1999. Effect of Surface Hydrophobicity on the Hydrodynamic Detachment of Particles from Surfaces. *Langmuir* **15** (7): 2466–2476. http://dx.doi.org/10.1021/la9807107.

Frick, T. P. and Economides, M. J. 1993. Horizontal Well Damage Characterization and Removal. *SPE Prod & Fac* **8** (01): 15–22. SPE-21795-PA. http://dx.doi.org/10.2118/21795-PA.

Gadde, P. B. and Sharma, M. M. 2001. Growing Injection Well Fractures and Their Impact on Waterflood Performance. Presented at the SPE Annual Technical Conference and Exhibition, New Orleans, 30 September–3 October. SPE-71614-MS. http://dx.doi.org/10.2118/71614-MS.

Glenn, E. E. and Slusser, M. L. 1957. Factors Affecting Well Productivity—II: Drilling Fluid Particle Invasion into Porous Media. *Petroleum Transactions* **210** (1957): 132–139. SPE-721-G.

Gray, D. H. and Rex, R. W. 1966. Formation Damage in Sandstones Caused by Clay Dispersion and Migration. *Clays Clay Miner* **14** (1): 355–366.

Gruesbeck, C. and Collins, R. E. 1982. Entrainment and Deposition of Fine Particles in Porous Media. *SPE J* **22** (6): 847–856. SPE-8430-PA. http://dx.doi.org/10.2118/8430-PA.

Halleck, P. M. 1997. Recent Advances in Understanding Perforator Penetration and Flow Performance. *SPE Drill & Compl* **12** (01): 19–26. SPE-27981-PA. http://dx.doi.org/10.2118/27981-PA.

Halleck, P. M., George, J., and Bast, M. 1999. The Character and Distribution of Damage around Perforations: Comparison of Balanced and Underbalanced Conditions. Presented at the SPE Eastern Regional Meeting, West Virginia, USA, 20–22 October.

Halleck, P. M., Poyol, E., and Santarelli, F. J. 1995. Estimating Perforation Flow from Variation in Indentation Hardness. *SPE Drill & Compl* **10** (04): 271–275. SPE-24769-PA. http://dx.doi.org/10.2118/24769-PA.

Hartman, K. J. and Cullick, A. S. 1994. Oil Recovery by Gas Displacement at Low Interfacial Tension. *J Pet Sci Eng* **10** (3): 197–210. http://dx.doi.org/10.1016/0920-4105(94)90081-7.

Henderson, G. D., Danesh, A., Tehrani, D. H. et al. 1998. Measurement and Correlation of Gas Condensate Relative Permeability by the Steady-State Method. *SPE Res Eval & Eng* **1** (2): 134–140. SPE-30770-PA. http://dx.doi.org/10.2118/30770-PA.

Hirschberg, A., deJong, L. N. J., Schipper, B. A. et al. 1984. Influence of Temperature and Pressure on Asphaltene Flocculation. Presented at the SPE Annual Technical Conference and Exhibition, New Orleans, 26–29 September. SPE-11202.

Hodge, R. M., Augustine, B. G., Burton, R. C. et al. 1997. Evaluation and Selection of Drill-in-Fluid Candidates to Minimize Formation Damage. *SPE Drill & Compl* **12** (03): 174–179. SPE-31082-PA. http://dx.doi.org/10.2118/31082-PA.

Holditch, S. A. 1979. Factors Affecting Water Blocking and Gas Flow from Hydraulically Fractured Gas Wells. *J Pet Technol* **31** (12): 1515–1524. SPE-7561-PA. http://dx.doi.org/10.2118/7561-PA.

Horner, D. R. 1951. Pressure Build-up in Wells. *Proc.*, 3rd World Petroleum Congress, The Hague, 28 May–6 June, WPC-4135, 503–521.

Houchin, L. R. and Hudson, L. M. 1986. The Prediction, Evaluation, and Treatment of Formation Damage Caused by Organic Deposition. Presented at the SPE Formation Damage Control Symposium, Lafayette, Louisiana, USA 26–27 February. SPE-14818-MS. http://dx.doi.org/10.2118/14818-MS.

Hsia, T. Y. and Behrmann, L. A. 1991. Perforating Skin as a Function of Rock Permeability and Underbalance. Presented at the SPE Annual Technical Conference and Exhibition, Dallas, 6–9 October. SPE-22810-MS. http://dx.doi.org/10.2118/22810-MS.

Hurst, W. 1953. Establishment of the Skin Effect and Its Impediment to Fluid Flow Into a Wellbore. *Pet. Eng.* **25** (October): B–6.

Israelachvili, J. N. 1992. *Intermolecular and Surface Forces*. London: Academic Press.

Jiao, D. and Sharma, M. M. 1992. Formation Damage Due to Static and Dynamic Filtration of Water-Based Muds. Presented at the SPE Formation Damage Control Symposium, Lafayette, Louisiana, USA, 26–27 February. SPE-23823-MS. http://dx.doi.org/10.2118/23823-MS.

Jiao, D. and Sharma, M. M. 1993a. Dynamic Filtration of Invert-Emulsion Muds. *SPE Drill & Compl* **8** (3): 165–169. SPE-24759-PA. http://dx.doi.org/10.2118/24759-PA.

Jiao, D. and Sharma, M. M. 1993b. Investigation of Dynamic Mud Cake Formation: The Concept of Minimum Overbalance Pressure. Presented at the SPE Annual Technical Conference and Exhibition, Houston, 3–6 October. SPE-26323-MS. http://dx.doi.org/10.2118/26323-MS.

Jiao, D. and Sharma, M. M. 1994. Mechanism of Cake Buildup in Crossflow Filtration of Colloidal Suspensions. *J. Colloid Interface Sci.* **162** (2): 454–462. http://dx.doi.org/10.1006/jcis.1994.1060.

Jiao, D. and Sharma, M. M. 1996. Mud Induced Formation Damage in Fractured Reservoirs. *SPE Drill & Compl* **11** (01): 11–16. SPE-30107-PA. http://dx.doi.org/10.2118/30107-PA.

Jones, F. O. Jr. 1964. Influence of Chemical Composition of Water on Clay Blocking of Permeability. *J Pet Technol* **16** (04): 441–446. SPE-631-PA. http://dx.doi.org/10.2118/631-PA.

Jones, L. G. and Watts, J. W. 1971. Estimating Skin Effect in a Partially Completed Damaged Well. *J Pet Technol* **23** (02): 249–252. SPE-2616-PA. http://dx.doi.org/10.2118/2616-PA.

Jones, L. G., Blount, E. M., and Glaze, O. H. 1976. Use of Short Term Multiple Rate Flow Tests to Predict Performance of Wells Having Turbulence. Presented at the SPE Annual Fall Technical Conference and Exhibition, New Orleans, 3–6 October. SPE-6133-MS. http://dx.doi.org/10.2118/6133-MS.

Jones, L. G., Blount, E. M., and Glaze, O. H. 1976. Use of Short Term Multiple Rate Flow Tests to Predict Performance of Wells Having Turbulence. Presented at the SPE Annual Fall Technical Conference and Exhibition, New Orleans, 3–6 October. SPE-6133-MS. http://dx.doi.org/10.2118/6133-MS.

Jones, R. R., Carpenter, R. B., and Conway, M. W. 1991. A Study of Formation Damage Potential During Cementing Operations. Presented at the SPE Annual Technical Conference and Exhibition, Dallas, 6–9 October. SPE-22777-MS. http://dx.doi.org/10.2118/22777-MS.

Kamath, J. and Laroche, C. 2000. Laboratory Based Evaluation of Gas Well Deliverability Loss Due to Waterblocking. Presented at the SPE Annual Technical Conference and Exhibition, Dallas, 1–4 October. SPE-63161-MS. http://dx.doi.org/10.2118/63161-MS.

Kawanaka, S., Park, S. J., and Mansoori, G. A. 1991. Organic Deposition from Reservoir Fluids: A Thermodynamic Predictive Technique. *SPE Res Eng* **6** (2): 185–192. SPE-17376-PA. http://dx.doi.org/10.2118/17376-PA.

Khilar, K. C. and Fogler, H. S. 1983. Water Sensitivity of Sandstones. *SPE J.* **23** (1): 55–64. SPE-10103-PA. http://dx.doi.org/10.2118/10103-PA.

King, G. E., Anderson, A., and Bingham, M. 1986. A Field Study of Underbalance Pressures Necessary to Obtain Clean Perforations Using Tubing-Conveyed Perforating. *J Pet Technol* **38** (6): 662–664. SPE-14321-PA. http://dx.doi.org/10.2118/14321-PA.

Klotz, J. A., Krueger, R. F., and Pye, D. S. 1974. Effect of Perforation Damage on Well Productivity. *J Pet Technol* **26** (11): 1033–1314. SPE-4654-PA. http://dx.doi.org/10.2118/4654-PA.

Kooijman, A. P., van den Hoek, P. J., de Bree, P. et al. 1996. Horizontal Wellbore Stability and Sand Production in Weakly Consolidated Sandstones. Presented at the SPE Annual Technical Conference and Exhibition, Denver, 6–9 October. SPE-36419-MS. http://dx.doi.org/10.2118/36419-MS.

Krueger, R. G. 1967. Effect of Pressure Drawdown on Clean-up of Clay- or Silt-Blocked Sandstone. *J Pet Technol* **19** (03): 397–403. SPE-1605-PA. http://dx.doi.org/10.2118/1605-PA.

Kumar, R. 2000. *Productivity Improvement of Gas-Condensate Wells by Fracturing*. MS thesis, The University of Texas at Austin, Austin, Texas, USA (August 2000).

Ladva, H. K. J., Tardy, P., Howard, P. R. et al. 2000. Multiphase Flow and Drilling Fluid Filtrate Effects on the Onset of Production. Presented at the SPE International Symposium on Formation Damage Control, Lafayette, Louisiana, USA, 23–24 February. SPE-58795-MS. http://dx.doi.org/10.2118/58795-MS.

Leontaritis, K. J. 1989. Asphaltene Deposition: A Comprehensive Description of Problem Manifestations and Modeling Approaches. Presented at the SPE Production Operations Symposium, Oklahoma City, Oklahoma, USA, 13–14 March. SPE-18892-MS. http://dx.doi.org/10.2118/18892-MS.

Leontaritis, K. J., Amaefule, J. O., and Charles, R. E. 1992. A Systematic Approach for the Prevention and Treatment of Formation Damage Caused by Asphaltene Deposition. Presented at the SPE Formation Damage Control Symposium, Lafayette, Louisiana, USA, 26–27 February. SPE-23810.

Longeron, D. G., Alfenore, J., Salehi, N. et al. 2000. Experimental Approach to Characterize Drilling Mud Invasion, Formation Damage and Cleanup Efficiency in Horizontal Wells with Openhole Completions. Presented at the SPE International Symposium on Formation Damage Control, Lafayette, Louisiana, USA, 23–24 February. http://dx.doi.org/10.2118/58737-MS.

Mahadevan, J. and Sharma, M. 2003. Clean-up of Water Blocks in Low Permeability Formations. Presented at the SPE Annual Technical Conference and Exhibition, Denver, 5–8 October. SPE-84216-MS. http://dx.doi.org/10.2118/84216-MS.

Mahajan, N. C. and Barron, B. M. 1980. Bridging Particle Size Distribution: A Key Factor in the Designing of Non-Damaging Completion Fluids. Presented at the SPE Formation Damage Symposium, Bakersfield, California, USA, 28–29 January. SPE-8792-MS. http://dx.doi.org/10.2118/8792-MS.

Mahmoud, M. A. 2014. Evaluating the Damage Caused by Calcium Sulfate Scale Precipitation During Low- and High-Salinity-Water Injection. *J Can Pet Technol* **53** (3): 141–150. SPE-164634-PA. http://dx.doi.org/10.2118/164634-PA.

Marek, B. F. 1979. Permeability Loss in Depletion of Reservoirs. Presented at the SPE Annual Technical Conference and Exhibition, Las Vegas, Nevada, USA, 23–26 September. SPE-8433-MS. http://dx.doi.org/10.2118/8433-MS.

Martins, J. P., Murray, L. R., Clifford, P. J. et al. 1995. Produced-Water Reinjection and Fracturing in Prudhoe Bay. *SPE Res Eng* **10** (03): 176–182. SPE-28936-PA. http://dx.doi.org/10.2118/28936-PA.

Matthews, C. S. and Russell, D. G. 1967. *Pressure Buildup and Flow Tests in Wells*, first edition, Monograph Vol. 1. Dallas: Henry L. Doherty Series, SPE of AIME.

Matthews, C. S., Brons, F., and Hazebroek, P. 1954. A Method for Determination of Average Pressure in a Bounded Reservoir. *Petroleum Transactions* **201** (1954): 182–191. SPE-296-G.

McAuliffe, C. D. 1973. Crude-Oil-Water Emulsions to Improve Fluid Flow in an Oil Reservoir. *J Pet Technol* **25** (06): 721–726. SPE-4370-PA. http://dx.doi.org/10.2118/4370-PA.

McClaflin, G. G. and Whitfill, D. L. 1984. Control of Paraffin Deposition in Production Operations. *J Pet Technol* **36** (11): 1965–1970. SPE-12204-PA. http://dx.doi.org/10.2118/12204-PA.

McKinney, L. K. and Azar, J. J. 1988. Formation Damage Due to Synthetic Oil Mud Filtrates at Elevated Temperatures and Pressures. Presented at the SPE Formation Damage Control Symposium, Bakersfield, California, USA, 8–9 February. SPE-17162-MS. http://dx.doi.org/10.2118/17162-MS.

McLeod, H. O. and Coulter, A. W. 1966. The Use of Alcohol in Gas Well Stimulation. Presented at the SPE Eastern Regional Meeting, Columbus, Ohio, USA, 10–11 November. SPE-1663-MS. http://dx.doi.org/10.2118/1663-MS.

McLeod, H. O., Jr. 1983. The Effect of Perforating Conditions on Well Performance. *J Pet Technol* **35** (01): 31–39. SPE-10649-PA. http://dx.doi.org/10.2118/10649-PA.

Mondshine, T. C. 1977. Completion Fluid Uses Salt for Bridging, Weighting. *Oil Gas J* (Aug. 22, 1977): 124–128.

Monger, T. G. and Fu, J. C. 1987. The Nature of $CO_2$-Induced Organic Deposition. Presented at the SPE Annual Technical Conference and Exhibition, Dallas, 27–30 September. SPE-16713-MS. http://dx.doi.org/10.2118/16713-MS.

Monger, T. G. and Trujillo, D. E. 1991. Organic Deposition During $CO_2$ and Rich-Gas Flooding. *SPE Res Eng* **6** (1): 17–24. SPE-18063-PA. http://dx.doi.org/10.2118/18063-PA.

Morgenthaler, L. N. 1986. Formation Damage Tests of High-Density Brine Completion Fluids. *SPE Prod Eng* **1** (06): 432–436. SPE-13811-PA. http://dx.doi.org/10.2118/13811-PA.

Muecke, T. W. 1979. Formation Fines and Factors Controlling Their Movement in Porous Media. *J Pet Technol* **31** (2): 144–150. SPE-7007-PA. http://dx.doi.org/10.2118/7007-PA.

Mungan, N. 1965. Permeability Reduction through Changes in Ph and Salinity. *J Pet Technol* **17** (12): 1449–1453. SPE-1283-PA. http://dx.doi.org/10.2118/1283-PA.

Narayanaswamy, G., Pope, G. A., Sharma, M. M. et al. 1999a. Predicting Gas Condensate Well Productivity Using Capillary Number and Non-Darcy Effects. Presented at the SPE Reservoir Simulation Symposium, Houston, 14–17 February. http://dx.doi.org/10.2118/51910-MS.

Narayanaswamy, G., Sharma, M. M., and Pope, G. A. 1999b. Effect of Heterogeneity on the Non-Darcy Flow Coefficient. *SPE J* **2** (03): 296–302. SPE-56881-PA. http://dx.doi.org/10.2118/56881-PA.

Newberry, M. E. and Barker, K. M. 1985. Formation Damage Prevention through the Control of Paraffin and Asphaltene Deposition. Presented at the SPE Production Operations Symposium, Oklahoma City, Oklahoma, USA, 10–12 March. SPE-13796-MS. http://dx.doi.org/10.2118/13796-MS.

Nowak, T. J. and Krueger, R. F. 1951. The Effect of Mud Filtrates and Mud Particles Upon the Permeabilities of Cores. Presented at the Drilling and Production Practice: Division of Production, Pacific Coast District Spring Meeting, Los Angeles, May. API-51-164.

Odeh, A. S. 1968. Steady-State Flow Capacity of Wells with Limited Entry to Flow. *SPE J.* **8** (01): 43–51. SPE-1797-PA. http://dx.doi.org/10.2118/1797-PA.

Paige, R. W. and Murray, L. R. 1994. Re-Injection of Produced Water—Field Experience and Current Understanding. Presented at the Rock Mechanics in Petroleum Engineering, Delft, Netherlands, 29–31 August. SPE-28121-MS. http://dx.doi.org/10.2118/28121-MS.

Pang, S. and Sharma, M. M. 1997. A Model for Predicting Injectivity Decline in Water-Injection Wells. *SPE Form Eval* **12** (3): 194–201. SPE-28489-PA. http://dx.doi.org/10.2118/28489-PA.

Patton, J. T. and Phelan, P. F. 1985. Well Damage Hazards Associated with Conventional Completion Fluids. Presented at the SPE Production Operations Symposium, Oklahoma City, Oklahoma, USA, 10–12 March. SPE-13800-MS. http://dx.doi.org/10.2118/13800-MS.

Pearson, J. R. A. and Zazovsky, A. F. 1997. A Model for the Transport of Sand Grains from a Perforation During Underbalance Surge. Presented at the SPE Annual Technical Conference and Exhibition, San Antonio, Texas, USA, 5–8 October. SPE-38634-MS. http://dx.doi.org/10.2118/38634-MS.

Peters, F. W. and Stout, C. M. 1977. Clay Stabilization During Fracturing Treatments With Hydrolyzable Zirconium Salts. *J Pet Technol* **29** (2): 187–194. SPE-5687-PA. http://dx.doi.org/10.2118/5687-PA.

Perkins, T. K. and Gonzalez, J. A. 1985. The Effect of Thermoelastic Stresses on Injection Well Fracturing. *SPE J* **25** (1): 78–88. SPE-11332-PA. http://dx.doi.org/10.2118/11332-PA.

Petitjean, L., Couet, B., Abel, J. C. et al. 1996. Well-Productivity Improvement by Use of Rapid Overpressured Perforation Extension: Case History. *J Pet Technol* **48** (02): 154–159. SPE-30527-JPT. http://dx.doi.org/10.2118/30527-JPT.

Pope, G. A., Wu, W., Narayanaswamy, G. et al. 2000. Modeling Relative Permeability Effects in Gas-Condensate Reservoirs with a New Trapping Model. *SPE Res Eval & Eng* **3** (2): 171–178. SPE-62497-PA. http://dx.doi.org/10.2118/62497-PA.

Priest, G. G. and Allen, T. O. 1958. Non-Plugging Emulsions Useful as Completion and Well-Servicing Fluids. *J Pet Technol* **10** (03): 11–14. SPE-951-G. http://dx.doi.org/10.2118/951-G.

Priest, G. G. and Morgan, B. E. 1957. Emulsions for Use as Non-Plugging Perforating Fluids. *Petroleum Transactions* **210** 177–182. SPE-682-G.

Pucknell, J. K. and Behrmann, L. A. 1991. An Investigation of the Damaged Zone Created by Perforating. Presented at the SPE Annual Technical Conference and Exhibition, Dallas, 6–9 October. SPE-22811-MS. http://dx.doi.org/10.2118/22811-MS.

Raleigh, J. T. and Flock, D. L. 1965. A Study of Formation Plugging with Bacteria. *J Pet Technol* **17** (2): 201–206. SPE-1009-PA. http://dx.doi.org/10.2118/1009-PA.

Read, P. A. and Ringen, J. K. 1982. The Use of Laboratory Tests To Evaluate Scaling Problems During Water Injection. Presented at the SPE Oilfield and Geothermal Chemistry Symposium, Dallas, 25–27 January. SPE-10593-MS. http://dx.doi.org/10.2118/10593-MS.

Roy, R. S. and Sharma, M. M. 2001. The Relative Importance of Solids and Filtrate Invasion on the Flow Initiation Pressure. Presented at the SPE European Formation Damage Conference, The Hague, 21–22 May. SPE-68949-MS. http://dx.doi.org/10.2118/68949-MS.

Ryan, D. F., Browne, S. V., and Burnham, M. P. 1995. Mud Clean-up in Horizontal Wells: A Major Joint Industry Study. Presented at the SPE Annual Technical Conference and Exhibition, Dallas, 22–25 October. SPE-30528-MS. http://dx.doi.org/10.2118/30528-MS.

Saripalli, K. P., Bryant, S. L., and Sharma, M. M. 1999. Role of Fracture Face and Formation Plugging in Injection Well Fracturing and Injectivity Decline. Presented at the SPE/EPA Exploration and Production Environmental Conference, Austin, Texas, USA, 21–3 March. SPE-52731-MS. http://dx.doi.org/10.2118/52731-MS.

Sarkar, A. K. and Sharma, M. M. 1990. Fines Migration in Two-Phase Flow. *J Pet Technol* **42** (5): 646–652. SPE-17437-PA. http://dx.doi.org/10.2118/17437-PA.

Schantz, S. S. and Stephenson, W. K. 1991. Asphaltene Deposition: Development and Application of Polymeric Asphaltene Dispersants. Presented at the SPE Annual Technical Conference and Exhibition, Dallas, 6–9 October. SPE-22783-MS. http://dx.doi.org/10.2118/22783-MS.

Schechter, R. S. 1992. *Oil Well Stimulation*, first edition. Englewood Cliffs, New Jersey, USA: Prentice Hall.

Sharma, B. G. and Sharma, M. M. 1994. Polymerizable Ultra-Thin Films: A New Technique for Fines Stabilization. Presented at the SPE Formation Damage Control Symposium, Lafayette, Louisiana, USA, 7–10 February. SPE-27345-MS. http://dx.doi.org/10.2118/27345-MS.

Sharma, M. M. 2000. The Nature of the Compacted Zone Around Perforation Tunnels. Presented at the SPE International Symposium on Formation Damage Control, 23–24 February. SPE-58720-MS. http://dx.doi.org/10.2118/58720-MS.

Sharma, M. M. and Wunderlich, R. W. 1987. The Alteration of Rock Properties Due to Interactions with Drilling-Fluid Components. *J Pet Sci Eng* **1** (2): 127–143. http://dx.doi.org/10.1016/0920-4105(87)90004-0.

Sharma, M. M. and Yortsos, Y. C. 1987. Fines Migration in Porous Media. *AIChE J* **33** (10): 16541662. http://dx.doi.org/10.1002/aic.690331009.

Sharma, M. M., Chamoun, H., Sita Rama Sarma, D. S. H. et al. 1992. Factors Controlling the Hydrodynamic Detachment of Particles from Surfaces. *J Colloid Interface Sci* **149** (1): 121–134. http://dx.doi.org/10.1016/0021-9797(92)90398-6.

Sharma, M. M., Pang, S., Wennberg, K. E. et al. 2000. Injectivity Decline in Water-Injection Wells: An Offshore Gulf of Mexico Case Study. *SPE Prod & Fac* **15** (1): 6–13. SPE-60901-PA. http://dx.doi.org/10.2118/60901-PA.

Sharma, M. M., Yortsos, Y. C., and Handy, L. L. 1985. Release and Deposition of Clays in Sandstones. Presented at the SPE Oilfield and Geothermal Chemistry Symposium, Phoenix, Arizona, USA, 9–11 March. SPE-13562-MS. http://dx.doi.org/10.2118/13562-MS.

Singh, T. and Sharma, M. M. 1997. Development of an Acid Degradable Drill-in Fluid for Fractured Reservoirs. Presented at the SPE European Formation Damage Conference, The Hague, 2–3 June. SPE-38153-MS. http://dx.doi.org/10.2118/38153-MS.

Sloan, J. P., Brooks, J. P., and Dear, S. F., III. 1975. A New, Nondamaging, Acid-Soluble Weighting Material. *J Pet Technol* **27** (01): 15–20. SPE-4782-PA. http://dx.doi.org/10.2118/4782-PA.

Soo, H. and Radke, C. J. 1984. Flow Mechanism of Dilute, Stable Emulsions in Porous Media. *Industrial & Engineering Chemistry Fundamentals* **23** (3): 342–347. http://dx.doi.org/10.1021/i100015a014.

Soo, H. and Radke, C. J. 1986. A Filtration Model for the Flow of Dilute, Stable Emulsions in Porous Media—I. Theory. *Chem Eng Sci* **41** (2): 263–272. http://dx.doi.org/10.1016/0009-2509(86)87007-5.

Standing, M. B. 1970. Inflow Performance Relationships for Damaged Wells Producing by Solution-Gas Drive. *J Pet Technol* **22** (11): 1,399–1,400. SPE-3237-PA. http://dx.doi.org/10.2118/3237-PA.

Suarez-Rivera, R., Stenebråten, J., Gadde, P. B. et al. 2002. An Experimental Investigation of Fracture Propagation During Water Injection. Presented at the International Symposium and Exhibition on Formation Damage Control, Lafayette, Louisiana, USA, 20–21 February. SPE-73740-MS. http://dx.doi.org/10.2118/73740-MS.

Suman, G. O., Jr. 1972. Perforations-a Prime Source of Well Performance Problems. *J Pet Technol* **24** (04): SPE-3445-PA. http://dx.doi.org/10.2118/3445-PA.

Suri, A. and Sharma, M. M. 2001. Strategies for Sizing Particles in Drilling and Completion Fluids. Presented at the SPE European Formation Damage Conference, The Hague, 21–22 May. SPE-68964-MS. http://dx.doi.org/10.2118/68964-MS.

Tambe, D. E. and Sharma, M. M. 1994. The Effect of Colloidal Particles on Fluid-Fluid Interfacial Properties and Emulsion Stability. *Adv Colloid Interface Sci* **52** (19 September 1994): 1–63. http://dx.doi.org/10.1016/0001-8686(94)80039-1.

Tannich, J. D. 1975. Liquid Removal from Hydraulically Fractured Gas Wells. *J Pet Technol* **27** (11): 1,309–1,317. SPE-5113-PA. http://dx.doi.org/10.2118/5113-PA.

Tariq, S. M. 1990. New, Generalized Criteria for Determining the Level of Underbalance for Obtaining Clean Perforations. Presented at the SPE Annual Technical Conference and Exhibition, New Orleans, 23–26 September SPE-20636-MS. http://dx.doi.org/10.2118/20636-MS.

Thomas, B. and Sharma, M. M. 1998. Distribution of Mud Induced Damage around Horizontal Wellbores. Presented at the SPE Formation Damage Control Conference, Lafayette, Louisiana, USA, 18–19 February. http://dx.doi.org/10.2118/39468-MS.

Thomas, D. C. 1988. Selection of Paraffin Control Products and Applications. Presented at the International Meeting on Petroleum Engineering, Tianjin, China, 1–4 November. SPE-17626-MS. http://dx.doi.org/10.2118/17626-MS.

Tuttle, R. N. and Barkman, J. H. 1974. New Nondamaging and Acid-Degradable Drilling and Completion Fluids. *J Pet Technol* **26** (11): 1221–1226. SPE-4791-PA. http://dx.doi.org/10.2118/4791-PA.

Valdya, R. N. and Fogler, H. S. 1992. Fines Migration and Formation Damage: Influence of pH and Ion Exchange. *SPE Prod Eng* **7** (04): 325–330. http://dx.doi.org/10.2118/19413-PA.

van den Hoek, P. J., Matsuura, T., de Kroon, M. et al. 1996. Simulation of Produced Water Re-Injection under Fracturing Conditions. Presented at the European Petroleum Conference, Milan, Italy, 22–24 October. SPE-36846-MS. http://dx.doi.org/10.2118/36846-MS.

van Oort, E., van Velzen, J. F. G., and Leerlooijer, K. 1993. Impairment by Suspended Solids Invasion: Testing and Prediction. *SPE Prod & Fac* **8** (03): 178–184. SPE-23822-PA. http://dx.doi.org/10.2118/23822-PA.

Venkitaraman, A., Behrmann, L. A., Blok, R. H. J. et al. 1997. Qualitative Analysis of Perforation-Induced Gravel-Pack Impairment Experiments. Presented at the SPE European Formation Damage Conference, The Hague, 2–3 June. SPE-38144-MS. http://dx.doi.org/10.2118/38144-MS.

Vogel, J. V. 1968. Inflow Performance Relationships for Solution-Gas Drive Wells. *J Pet Technol* **20** (01): 83–93. SPE-1476-PA. http://dx.doi.org/10.2118/1476-PA.

Wang, X. and Mohanty, K. K. 1999. Multiphase Non-Darcy Flow in Gas-Condensate Reservoirs. Presented at the SPE Annual Technical Conference and Exhibition, Houston, 3–6 October. SPE-56486-MS. http://dx.doi.org/10.2118/56486-MS.

Weeks, S. G. 1974. Formation Damage or Limited Perforating Penetration? Test-Well Shooting May Give a Clue. *J Pet Technol* **26** (09): 979–984. SPE-4794-PA. http://dx.doi.org/10.2118/4794-PA.

Wennberg, K. E. and Sharma, M. M. 1997. Determination of the Filtration Coefficient and the Transition Time for Water Injection. Presented at the SPE European Formation Damage Conference, The Hague, 2–3 June. SPE-38181-MS. http://dx.doi.org/10.2118/38181-MS.

Yan, J. and Sharma, M. M. 1989. Wettability Alteration and Restoration for Cores Contaminated with Oil-Based Muds. *J Pet Sci Eng* **2** (1): 63–76. http://dx.doi.org/10.1016/0920-4105(89)90051-X.

Yan, J. N., Monezes, J. L., and Sharma, M. M. 1993. Wettability Alteration Caused by Oil-Based Muds and Mud Components. *SPE Drill & Compl* **8** (01): 35–44. SPE-18162-PA. http://dx.doi.org/10.2118/18162-PA.

Yang, X. M. and Sharma, M. M. 1991. Formation Damage Caused by Cement Filtrates in Sandstone Cores. *SPE Prod Eng* **6** (04): 399–405. SPE-19305-PA. http://dx.doi.org/10.2118/19305-PA.

Yen, T. F. 1974. Structure of Petroleum Asphaltene and Its Significance. *Energy Sources* **1** (4): 447–463. http://dx.doi.org/10.1080/00908317408945937.

Zain, Z. 1999. Model Simplifies Filter Cake Lift-Off Pressure Determination. *Oil Gas J.* **97** (44): 70–75.

Zain, Z. M. and Sharma, M. M. 1999. Cleanup of Wall-Building Filter Cakes. Presented at the SPE Annual Technical Conference and Exhibition, Houston, 3–6 October. SPE-56635-MS. http://dx.doi.org/10.2118/56635-MS.

Zain, Z. M., Suri, A., and Sharma, M. M. 2000. Mechanisms of Mud Cake Removal During Flowback. Presented at the SPE International Symposium on Formation Damage Control, Lafayette, Louisiana, 23–24 February. SPE-58797-MS. http://dx.doi.org/10.2118/58797-MS.

Zhang, J., Rai, C. S., and Sondergeld, C. H. 1998. Mechanical Strength of Reservoir Materials: Key Information for Sand Prediction. Presented at the SPE Annual Technical Conference and Exhibition, New Orleans, 27–30 September. SPE-49134-MS. http://dx.doi.org/10.2118/49134-MS.

Chapter 3

# Acidizing Chemistry

**Murtaza Ziauddin,** Schlumberger

## 3.1 Introduction

This chapter focuses on the chemical and reaction engineering concepts common to both matrix and fracture acidizing treatments. Reactive fluids are injected in both treatments; however, the role played by these fluids in each treatment is different. In fracture acidizing, reactive fluids are injected to initiate and propagate a hydraulic fracture in the reservoir. The leakoff of the reactive fluid from the fracture is minimized, and the reactivity of fluid is tuned so that it is sufficient to etch the fracture face but not too rapid to spend entirely near the wellbore. In matrix acidizing treatments, the reactive fluid is injected at pressures below the fracture pressure. Matrix treatments in sandstone reservoirs require a careful balance of dissolution and precipitation reactions. They are designed to remove near-wellbore formation damage by dissolution. However, precipitation reactions often follow the dissolution reactions, and these have to be minimized for the treatment to be successful. Matrix treatments in carbonate reservoirs are designed so that the acid is placed at sufficient rates and volumes across all zones, creating wormholes (dissolution channels) that extend deep in the reservoir.

An engineering design for both matrix and fracturing acidizing treatment involves optimizing the types of reactive fluids, their injection rates, and volumes and sequences so that these functions are fulfilled and the stimulation effectiveness is maximized for a given cost. This chapter introduces concepts that are essential for understanding treatment design optimization presented in later chapters. In this chapter, first a brief introduction to chemical reactions is presented. This is followed by a description of some of the common dimensionless numbers used in assessing the relative significance of various physical and chemical processes occurring in the flow system. The reactions in acidizing predominantly occur between the treatment fluid and solid (rock) phase, and they depend on chemical properties of both phases. The origins and chemical compositions of rocks are described, followed by description of the chemistry of the main fluid types used in acidizing. The chemical interactions between the solid and fluid phase and laboratory techniques for measuring the reaction rate between the solid and fluids are discussed. The chapter is concluded by a description of some of the typical reaction types encountered in acidizing.

## 3.2 Introduction to Chemical Reactions

An acidizing treatment typically involves many different types of chemical reactions. A brief review of introductory concepts in reaction chemistry is presented here. Extended coverage of these topics can be found in chemical engineering textbooks (Fogler 2005; Kee et al. 2003).

**3.2.1 Reaction Kinetics.** A chemical reaction between two molecules takes place when they come in contact and have enough energy to overcome the activation-energy barrier. Reaction kinetics dictates how fast the chemical change or reaction will occur. Consider the reaction between molecules in $A$ and $B$ to form $C$.

$$A + B \rightarrow C$$

Assume that the rate of reaction is given by the rate law,

$$\text{rate of formation of } C = \frac{\mathrm{d}[C]}{\mathrm{d}t} = k[A][B],$$

where $k$ is the rate constant and [ ] indicates concentration of the species. The rate of reaction is proportional to the probability of collision between $A$ and $B$, which is proportional to the product of their concentrations. The rate constant $k$ is the number of collisions that lead to a reaction per unit time. Often, $k$ is described by the Arrhenius equation:

$$k = k_0 e^{\frac{-E_a}{RT}}$$

Where $k_0$, the pre-exponential factor, is the total number of collisions per unit time and the term $e^{\frac{-E_a}{RT}}$ is the probability that the collision will result in a reaction. $E_a$ is the activation energy for the reaction. If the activation energy is very low, then every collision successfully results in a reaction. The rate of reaction is then limited by the number of collisions. However, if the activation energy is high, only a small number of molecules in contact will react. The reaction rate is then limited by the energy barrier.

The rate laws used to describe reactions in matrix stimulation treatments are often empirically determined. The net reaction rate may not be first order with respect to concentration of both reactants as in the elementary reaction above. The effective reaction order can be either higher or lower than first order, and sometimes it can also be a fraction. This is because the underlyng reaction mechanism may have multiple steps. The slower steps (bottlenecks) in the reaction sequence determine the net reaction rate. These are called the rate-controlling steps, and they can change if the chemical or thermal environment for the reaction changes. Therefore, care must be exercised when empirically fitted reactions are used outside the environment they were calibrated in, because the rate-controlling steps in the application environment may be different.

**3.2.2 Chemical Equilibrium.** While reaction kinetics is concerned with the *rate* of chemical reaction, chemical equilibrium determines the *extent* of the reaction. Unlike kinetically controlled reactions, the concentrations of the reactants and products involved in an equilibrium-controlled reaction are independent of time and history. In matrix stimulation treatments, dissociation reactions of many inorganic and organic acids and dissolution/precipitation reactions of many minerals are typically faster than other chemical processes occurring locally, and they are generally assumed to be under equilibrium control. For a reaction under equilibrium control, equilibrium constants are used to relate the concentrations of reactants and products. Unlike constants in the kinetic rate law of the reaction, equilibrium constants are calculated from thermodynamic constants of the species involved in the reaction. Consider a simple ideal reaction under equilibrium control consisting of reactants $A$ and $B$, and products $C$ and $D$, for which the reaction stoichiometry is given by

$$aA + bB \leftrightarrow cC + dD \quad (3.1)$$

If the system is ideal, then the equilibrium constant for the system can be expressed in terms of concentrations as

$$K_{\mathrm{eq}} = \frac{[C]^c[D]^d}{[A]^a[B]^b} \quad (3.2)$$

where $K_{\mathrm{eq}}$ is the equilibrium constant and [ ] denotes concentration of the species. The equilibrium constant can also be expressed in terms of the free energy change for the reaction ($\Delta G°$) as

$$K_{eq} = \exp\left(\frac{-\Delta G^\circ}{RT}\right), \quad (3.3)$$

where $R$ is the gas constant and $T$ is the temperature. The $K_{eq}$ is a function of temperature and pressure only and does not depend on composition of the system. Therefore, if the composition of the system changes by subsequent reactions or addition of new reactants or products to the system, the same value of the equilibrium constant(s) can be used to calculate the equilibrium distribution in the new system provided the temperature and pressure remain constant and the equilibrium assumption is valid.

Equilibrium constants are not constants in the true sense because they depend on temperature and pressure. An increase in temperature may affect the forward and reverse reactions differently. The reaction that absorbs the most heat will increase that rate to a larger extent than the other reaction. A new equilibrium constant will now represent the new situation. The simple equation by van't Hoff can be used to compute the change in equilibrium constant resulting from the change in temperature. The equation can be expressed as

$$\frac{d \ln K_{eq}}{dT} = \frac{\Delta H^\circ}{RT^2}, \quad (3.4)$$

where $\Delta H^\circ$ is the standard enthalpy change of reaction, $T$ is the temperature, and $R$ is the gas constant. If the reaction is exothermic (i.e., if $\Delta H^\circ$ for the reaction is negative), then the equilibrium constant decreases as temperature increases. Conversely, $K_{eq}$ increases with temperature for an endothermic reaction.

If the standard enthalpy change of reaction is assumed to be independent of temperature, then integrating of Eq. 3.4 gives an even simpler result:

$$\ln \frac{K_{eq}}{K_{eq,ref}} = -\frac{\Delta H^\circ}{R}\left(\frac{1}{T} - \frac{1}{T_{ref}}\right) \quad (3.5)$$

Here $K_{eq,ref}$ is the value of the equilibrium constant at the reference temperature $T_{eq,ref}$. This approximate equation implies that a plot of ln $K_{eq}$ vs. the reciprocal temperature gives a straight line. This equation is helpful in interpolating and extrapolating equilibrium-constant data with reasonable accuracy.

The effect of pressure on equilibrium constants is typically smaller than the effect of temperature. However, for deep wells it can be significant and effect of pressure needs to be considered along with the change in temperature. The pressure dependence of the equilibrium constant can be calculated from

$$\left(\frac{\partial \ln K_{eq}}{\partial P}\right)_T = \frac{\Delta V^\circ}{RT}, \quad (3.6)$$

where $\Delta V^\circ$ is the molar volume change of the reaction with all reactants and products in their standard states (Langmuir 1997). If the molar volume change of the reaction is independent of pressure, then the integration of Eq. 3.6 yields

$$\ln \frac{K_{eq}}{K_{eq,ref}} = -\frac{\Delta V^\circ (P - P_{ref})}{RT}, \quad (3.7)$$

where $K_{eq}$ is the equilibrium constant at the desired pressure $P$. $K_{eq,ref}$ is the equilibrium constant at $P_{ref}$, which is typically 1 bar.

In very dilute aqueous solutions, the anions and cations behave in an ideal manner in which each ion will act as if it is independent of all other ions in the solution. In real solutions, especially where there are high concentrations of other ions, the ions are affected by the other ions in the solution and the fluid may not behave as if it has exactly the same number of ions as described by the concentration. The equilibrium constants in such nonideal systems are then expressed in terms of species activity. For example, if the chemical system considered previously of components *A, B, C,* and *D* is nonideal, then the equilibrium constant is given by

$$K_{eq} = \frac{\{C\}^c \{D\}^d}{\{A\}^a \{B\}^b} = \exp\left(\frac{-\Delta G^\circ}{RT}\right), \quad (3.8)$$

where { } denotes activity. The activity of the species can be thought of as an effective concentration of the species in solution. Even moderately concentrated solutions used in matrix treatments exhibit nonideal behavior. Strong acids and typical formation brines are far from ideal, and the nonideal formulation is required to represent them adequately.

**3.2.3 Homogeneous Reactions.** Homogeneous reactions are those reactions that occur in a single phase, and, therefore, they are independent of mass transfer of reactants and product across phase boundaries. In acidizing treatments, homogeneous reactions generally occur in the aqueous phase. Examples of a homogeneous reaction are dissociation of inorganic and organic acids in water. The dissociation of hydrochloric acid (HCl) can be expressed by the reaction

$$HCl\,(aq) + H_2O\,(aq) \rightarrow H_3O^+(aq) + Cl^-(aq). \quad (3.9)$$

The (aq) designation next to the species indicates that the species is present in the aqueous phase.

**3.2.4 Heterogeneous Reactions.** Heterogeneous reactions are those reactions in which the reactants are present in two or more phases, such as aqueous phase and the solid phase, or in which one or more reactants undergo a chemical change at an interface. Consider the heterogeneous reaction, which transforms the reactant *A* in the aqueous fluid phase and the mineral *M* in the solid phase into products *C* and *D* in the aqueous phase. The overall reaction can be written as

$$A(aq) + M(s) \rightarrow C(aq) + D(aq)$$

The (aq) designation next to the species indicates that the species is present in the aqueous phase, while the (s) designation indicates that the species is in the solid phase. A simplified sequence of steps involved in this reaction is depicted in **Fig. 3.1.**

1. Diffusion of reactant from bulk fluid phase to the mineral surface.
   The rate of this transport of A to the bulk phase is usually described by a mass-transfer coefficient, km, as

$$N_A = k_m\,([A] - [A_s]), \quad (3.10)$$

   where
   $N_A$ = mass flux to the solid surface, kmol/m²-s
   $k_m$ = mass-transfer coefficient, m³/m²-s
   $[A]$ = concentration of *A* in the bulk fluid phase, kmol/m³
   $[A_s]$ = concentration of *A* at the interface, kmol/m³.

   The difference in concentration represents the driving force for *A* to diffuse to the surface. Correlations for mass-transfer coefficients are often developed in terms of the Sherwood number, the dimensionless mass-transfer coefficient. For flow in porous media, an expression for Sherwood number proposed by Panga et al. (2005) is

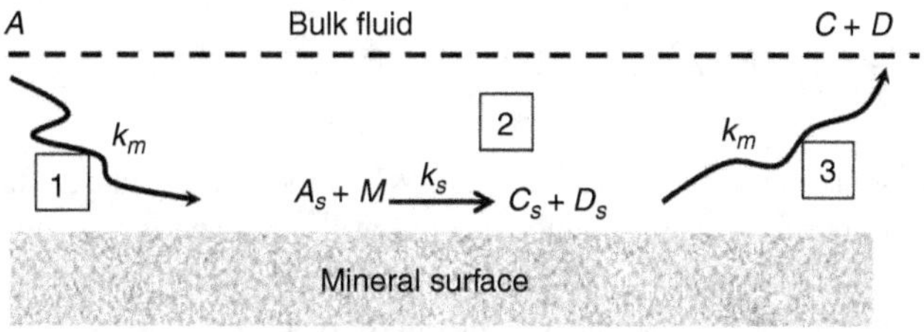


**Fig. 3.1—Steps involved in a typical heterogeneous reaction between a reactive species in the aqueous phase and the mineral in the solid phase.**

$$Sh = \frac{2k_m r_p}{D_m} = 2 + 0.6 Re_p^{1/2} Sc^{1/3},$$

where

$Sh$ = Sherwood number, dimensionless;
$r_p$ = pore radius, m;
$D_m$ = molecular diffusivity, m²/s;
$Re_p$ = pore Reynolds number$= \frac{u r_p}{\nu}$, dimensionless;
$Sc$ = Schmidt number $= \frac{\nu}{D_m}$, dimensionless;
$\nu$ = kinematic viscosity, m²/s;
$u$ = Darcy velocity, m/s;

2. Surface reaction of the mineral $M$ with $A$ to form products $C$ and $D$

$$A_s(\text{s}) + M(\text{s}) \rightarrow C(\text{s}) + D(\text{s}) \quad (3.11)$$

Assuming a first-order irreversible reaction, the rate law can be expressed as

$$r_{As} = k_s[A_s], \quad (3.12)$$

where

$r_{As}$ = rate of consumption of $A$ at surface, kmol/m³-s;
$k_s$ = rate coefficient for the reaction m³/m²-s;
$[A_s]$ = concentration of $A$ at the interface, kmol/m³.

The Arrhenius expression is often used to describe the dependence of $k_s$ on temperature.

The rate of mass transfer of reactant to the surface and the rate of consumption of the reactant by surface reaction are equal under steady state. This allows the unmeasured concentration of $A$ at the interface $[A_s]$ to be eliminated:

$$r_{As} = N_A = r_A \quad (3.13)$$

$$k_s[A_s] = k_m([A] - [A_s]) \quad (3.14)$$

$$[A_s] = \frac{k_m}{k_m + k_s}[A] \quad (3.15)$$

The surface reaction rate can then be expressed as

$$\dot{r}_{As} = r_A = \frac{k_m k_s}{k_m + k_s}[A] = k_0[A], \qquad (3.16)$$

where $k_0$ represents the overall rate constant. When the mass-transfer step is much faster than the surface reaction, the reactant concentration at the surface is the same as that measured in the bulk ($k_m >> k_s$). The observed reaction rate is controlled by the kinetics of the surface reaction, and this is termed a "kinetically controlled" reaction. However, if $k_s >> k_m$, the surface concentration of $A$ tends to zero, $[A_s] \approx 0$, and the reaction is under mass-transfer control.

3. Diffusion of products $C$ and $D$ back to the bulk fluid phase.
The last step in the reaction sequence is the diffusion of the products back to the bulk fluid phase:

$$C_s(\text{s}) \rightarrow C(\text{aq}) \qquad (3.17)$$

$$D_s(\text{s}) \rightarrow D(\text{aq}) \qquad (3.18)$$

If the direction of transport to the mineral surface is considered the positive direction, rates of transport of $C$ and $D$ are

$$N_C = k_m([C] - [C_s]) \qquad (3.19)$$

$$N_D = k_m([D] - [D_s]) \qquad (3.20)$$

**3.2.5 Key Dimensionless Numbers.** For flow of a reactive fluid through porous media, the location and extent of dissolution and precipitation reactions are determined by how the rate of these reactions relates to the rate of flow through the media. The time scale of these processes can vary widely, and it is useful to define dimensionless numbers to characterize these. Some common dimensionless variables used in assessing the relative importance of various physical and chemical processes occurring in the reaction system are defined below.

***Damköhler Number* (Da).** The Damköhler number (*Da*) is named after the German chemist Gerhard Damköhler. It is used in relating the reaction time scale to the time scales of other phenomena occurring in the system. For example, in a chemical system where both flow and reaction are occurring, it is defined as

$$Da = \frac{\text{characteristic fluid time}}{\text{characteristic chemical} - \text{reaction time}} \qquad (3.21)$$

In a flow system, the characteristic fluid time is often the time for convective mass transport, and then *Da* can be defined as

$$Da = \frac{\text{reaction rate}}{\text{convective mass-transport rate}} \qquad (3.22)$$

Note both the reaction rate and convective mass-transport rate have units of mass per unit of time.

***Thiele Modulus.*** The Thiele modulus is named after E. W. Thiele. It was developed to determine the effectiveness factor of a porous catalyst pellet by relating the rates of diffusion and surface reaction in the catalyst pellet. The Thiele modulus, $\theta$, is defined as

$$\theta = \sqrt{\frac{\text{"a" surface reaction rate}}{\text{"a" diffusion rate}}} \qquad (3.23)$$

From Eq. 3.23 relation, it is evident that when the Thiele modulus is large, the surface reaction rate is fast and the rate of diffusion limits the overall reaction; and when it is small, the surface reaction rate limits the overall rate.

The Thiele-modulus concept can be extended to acidizing applications. Here, the Thiele modulus is useful in comparing the rates of reaction of various minerals in the porous medium with chemical species in the bulk phase. Consider a first-order reaction with the rate law

$$rate = k_s [A_s], \qquad (3.24)$$

where $k_s$ is the rate constant for the surface reaction and $[A_s]$ represents the concentration of the reacting species at the mineral surface. The Thiele modulus can be defined as (Panga et al. 2005)

$$\theta = \sqrt{\frac{2k_s r_0}{D_m}}, \qquad (3.25)$$

where $r_0$ is average initial pore radius and $D_m$ is the diffusion coefficient. $D_m$ controls the supply of the fluid reactive species, while $k_s$ controls the rate of consumption of the species when it reaches the surface. For large values of $\theta$, the reaction between the mineral and fluid is diffusion limited, while for small values, it is surface reaction limited. For example, under typical acidizing conditions, the reaction rate of calcite with hydrochloric acid is much faster than diffusion of acid to the mineral surface; however, the reaction of quartz with HCl is generally surface reaction limited. For diffusion-limited reaction, the surface reaction kinetics can be ignored, and vice versa for kinetically limited reactions. Therefore, Thiele modulus allows a convenient means of determining the limiting parameters for a reaction scheme.

***Péclet Number.*** The Péclet number is named after the French physicist Jean-Claude Eugène Péclet. It is used in describing transport of mass or heat in a flow system. In transport of mass, it is defined as the ratio of the rate of advection of a chemical species to its rate of diffusion. In transport of heat, it is defined as the ratio of thermal energy transported by advection to the thermal energy transported by conduction.

***Acid Capacity Number.*** The acid capacity number, $N_{ac}$, describes the stoichiometry of the reaction between the mineral and the acid. It is defined as the molar ratio of acid present in the pore space to the acid required to dissolve the mineral accessible to the reaction (Fogler and McCune 1976):

$$N_{ac} = \frac{\phi_0 C_0}{\alpha(1-\phi_0)\Delta W}, \qquad (3.26)$$

where $\phi_0$ represents the initial porosity, $C_0$ represents the initial acid concentration, $\alpha$ represents the moles of acid required to dissolve 1 mole of the mineral, and $\Delta W$ is the molar concentration of the mineral accessible to the reaction.

## 3.3 Chemistry of Rocks and Minerals

Petroleum fluids are produced mainly from porous and permeable sedimentary rocks. Sedimentary rocks are formed by deposition of particles or sediments. Sandstones and carbonates are the two broad compositional groups of sedimentary rocks in which acidizing treatments are performed. The rock properties critical to design of acidizing treatments are porosity, permeability, and mineral composition and distribution. These properties are strongly influenced by the sediment source, as well as weathering, transport, deposition, and diagenesis of the sediments. The geology of sandstone and carbonates is adequately covered in the literature (Pettijohn et al. 1987; Chilingarian et al. 1992). A brief description of the origin and classification of sandstones and carbonates is presented here along with a discussion of rock properties critical to stimulation treatment design.

Sedimentary rocks are composed of detrital and authigenic constituents. Detrital constituents are those that are derived by mechanical/chemical disintegration of a parent rock. Authigenic constituents are those that form in place within sediments either soon after deposition, while sediment is still in an unconsolidated state, or during burial and digenesis. They can occur as cements, or crystallize in pore space as new minerals that do not act as cements, or form by replacement of original detrital minerals or rock fragments (Boggs 1992).

Porosity and permeability are central to the analysis of flow in porous media. The stimulation fluids access the minerals in the rock mostly through the porosity, and the type and distribution of porosity is required in defining the contact of the reactive fluid with rock. The pore space in the rock can be either connected or isolated, and, hence, two porosity types are generally defined. Total porosity, $\phi_t$, is defined as

$$\phi_t = \frac{\text{bulk volume} - \text{solid volume}}{\text{bulk volume}} \times 100 \quad (3.27)$$

Effective porosity, $\phi_e$, is defined as

$$\phi_e = \frac{\text{interconnected pore volume}}{\text{bulk volume}} \times 100 \quad (3.28)$$

Effective porosity is less than total porosity, and it is used in fluid-flow studies because it correlates better with permeability.

**3.3.1 Sandstones.** Sandstones are a form of clastic sedimentary rock because the sediments in sandstones are broken pieces or clasts of older rocks that are produced by weathering and erosion in a source area **(Fig. 3.2).** The sediments are transported from the source area to the place of deposition by water, wind, or ice. Sedimentation occurs when the particles settle out from suspension, such as when the water or wind slow down or when ice melts. The deposition can occur in a terrestrial or a marine environment. Examples of terrestrial environments are rivers, alluvial fans, glacial outwash, lakes, and deserts. Examples of marine environments are deltas, beaches, tidal flats, offshore bars, and turbidites (submarine channels and fans). Once deposited, the loose sediments are then compacted by pressure of overlying deposits and are cemented by the precipitation of minerals within the pore spaces. The most common cementing materials are silica and calcium carbonate, which are often derived from dissolution or alternation of the sediments after they are buried.

***Composition.*** Sandstones are generally composed of a much wider variety of minerals than carbonates. However, the mineral variety in sandstones is not as large as one would expect from the variety of minerals in the source area for sandstones. Theoretically, the total variety of minerals in the source area could be represented in the sandstone. However, this is usually not the case (Pettijohn et al. 1987; Boggs 1992). Most sandstones are composed of a few dominant minerals groups. This can be attributed to the processes of chemical weathering and physical and chemical attack during transport and deposition that tend to destroy or degrade chemically unstable and mechanically weak sand-sized grains. The framework grains of most sandstones are composed of stable minerals such as quartz, feldspars, and rock fragments. Detrital clay minerals are also found in some sandstones as matrix constituents. Coarse micas, such as muscovite, often make up a small percent of the framework grains in sandstones. Heavy minerals such as zircon, tourmaline, and rutile may constitute a small percentage of the detrital sandstone constituents.

***Silica.*** Silica is the most abundant mineral in sandstone and is generally found either as the framework mineral quartz or as a cement between sandstone grains. The chemical formula of silica is $SiO_2$. In the form of the mineral quartz, it is hard and chemically stable and its reactive surface area is generally much lower than that of clays. These properties allow it to survive multiple recycling events. The crystal system for $\alpha$-quartz, the most abundant form of quartz, is rhombohedral with a density of 2.648 g/cm$^3$. Silica is common cement for sandstone grains, and where it is present on quartz grains, it creates a rim around the quartz grain called an overgrowth. The overgrowth retains the same crystallographic continuity of the quartz grain; however, it may show increased reactivity because of a higher reactive surface area.

***Feldspars.*** Feldspars are the second-most-abundant mineral in sandstones after quartz. Feldspars of both detrital and authigenic origin are present in sandstones **(Fig. 3.3)** (Pettijohn et al. 1987).

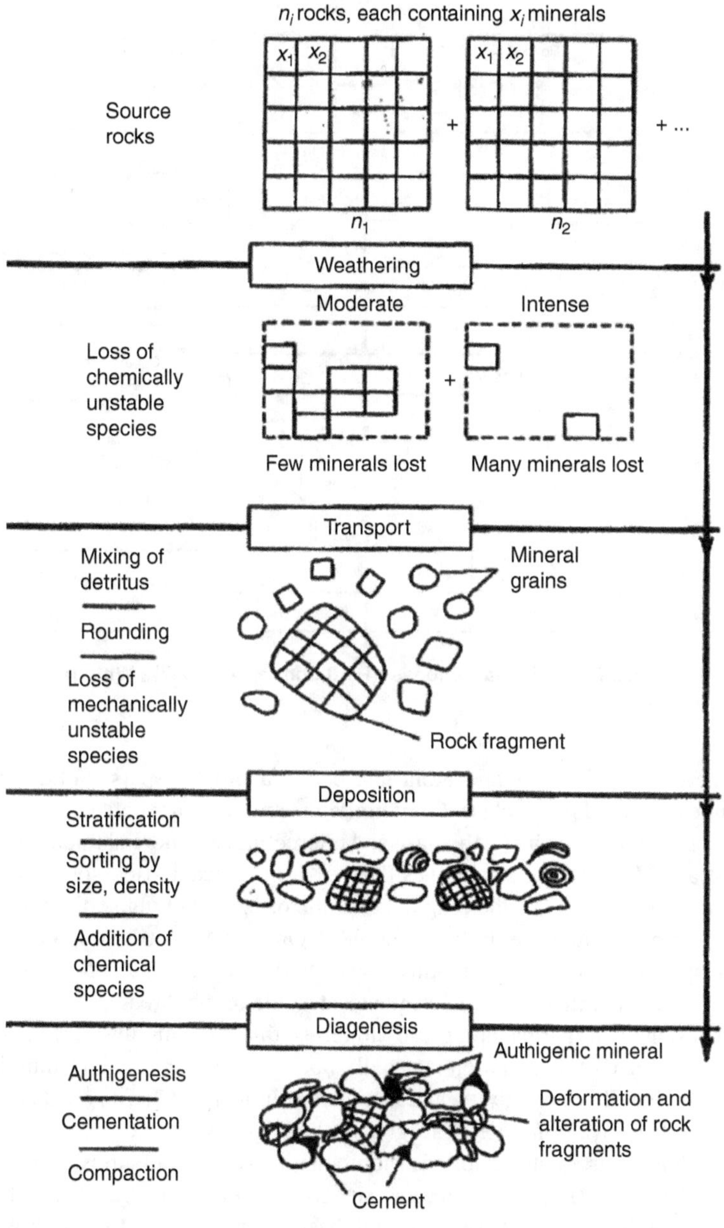


**Fig. 3.2—The sedimentary cycle of a sandstone. Though processes such as weathering and transport are shown as distinct, they overlap in nature (Pettijohn et al. 1987). With permission of Springer.**

Authigenic feldspars are common in sandstone, and the composition depends on the relative amounts of their components ($K^+$, $Na^+$, $Ca^{+}2$, and $SiO^2$) in solution and also the pH. Deer (1963) and Smith (1974a, 1974b) provide a detailed review of feldspar minerals.

The feldspar composition and type are generally expressed in terms of three end members (Fig. 3.3) (Pettijohn et al. 1987):

- Potassium-feldspar (K-feldspar), $KAlSi_3O_8$
- Albite $NaAlSi_3O_8$
- Anorthite $CaAl_2Si_2O_8$

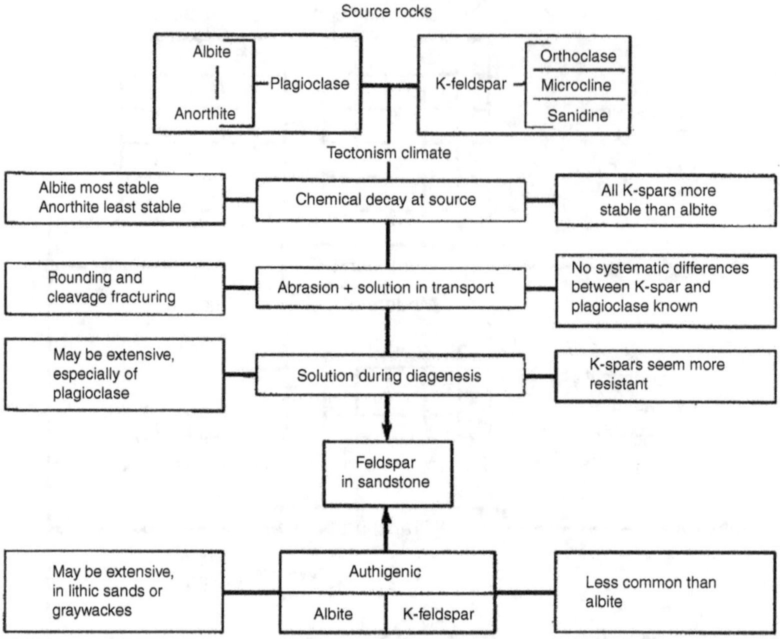


**Fig. 3.3—The origin of feldspars in sandstones (Pettijohn et al. 1987). With permission of Springer Science+Business Media.**

Solid solutions between K-feldspar and albite are called alkali feldspars. Solid solutions between albite and anorthite are called plagioclase. Only limited solution occurs between K-feldspar and anorthite. In general potassium-feldspars (orthoclase and microcline) are more abundant in sandstones than plagioclase feldspars. However, in sandstones derived from volcanic-rich source areas, plagioclase feldspar may be more dominant. Of the plagioclases, the ones containing sodium are more dominant than the ones containing calcium (Pettijohn et al. 1987). One of the contributing factors to the order of abundance of feldspars is their order of stability to weathering. K-feldspar is the most stable to weathering, followed by albite and then anorthite as the least stable of the three.

***Carbonates.*** Carbonate minerals are those minerals that contain the carbonate anion: $CO_3^{2-}$. In sedimentary rocks, the carbonate anion is usually bound to calcium, magnesium, iron, or sodium. Carbonates generally have high reactivity, and many are soluble in HCl. The most common carbonate minerals are calcite ($CaCO_3$), dolomite [$CaMg(CO_3)_2$], and siderite ($FeCO_3$). Aragonite is an orthorhombic form of calcium carbonate (calcite is rhombohedral) and occurs mostly in recent sediments.

***Clays Minerals.*** Kaolinite, illite, chlorite, and smectite are common clay minerals found in sandstone reservoirs. Clay minerals are classified on the basis of their ability to absorb water, their structure, and their chemical composition. Clay minerals that swell by absorbing water or other polar ions into their structure are called swelling clays or smectites, while clay minerals that do not are called nonswelling-type minerals. Most clay minerals have a phyllosilicate or sheet structure. There are basically two structure units involved in the atomic lattices of most of the clay minerals. One unit consists of a sheet with octahedral coordination composed of closely packed oxygen atoms in which alumina, iron, or magnesium atoms are embedded. The second unit is a sheet composed of silica tetrahedra. Clays can be classified according to the number and the type of sheets that combine to form a layer. For example, kaolinite is a 1:1 clay and is formed when one tetrahedral silica sheet combines with an octahedral alumina sheet. On the other hand, smectites and illites are 2:1 clays because an octahedral sheet is sandwiched between two silica tetrahedral sheets (Schechter 1990).

Clay minerals in sandstone reservoirs can be of both detrital and authigenic origin. Clay minerals of detrital origin generally occur as clasts or nodules in the matrix. The interaction of these clay types with stimulation fluids is generally not significant because the fluids preferentially flow through the much-higher-permeability matrix around the clays rather than through them. On the other hand, authigenic clay minerals generally occur as pore-filling components, are dispersed in the pore network, and are of prime importance to acidizing treatments. The total volume of these minerals may be low, but they are located in critical flow paths (choke points) of pore networks and are critical to the success of the treatment. These minerals can occur in pores as (a) discrete (not intergrown) particles, (b) intergrown crystal linings on pore walls, and (c) crystals bridging across pores **(Fig. 3.4).** These different clay morphologies significantly affect sandstone porosity and permeability. Sandstones of comparable grain size exhibit lower permeabilities if the clays are the pore-bridging

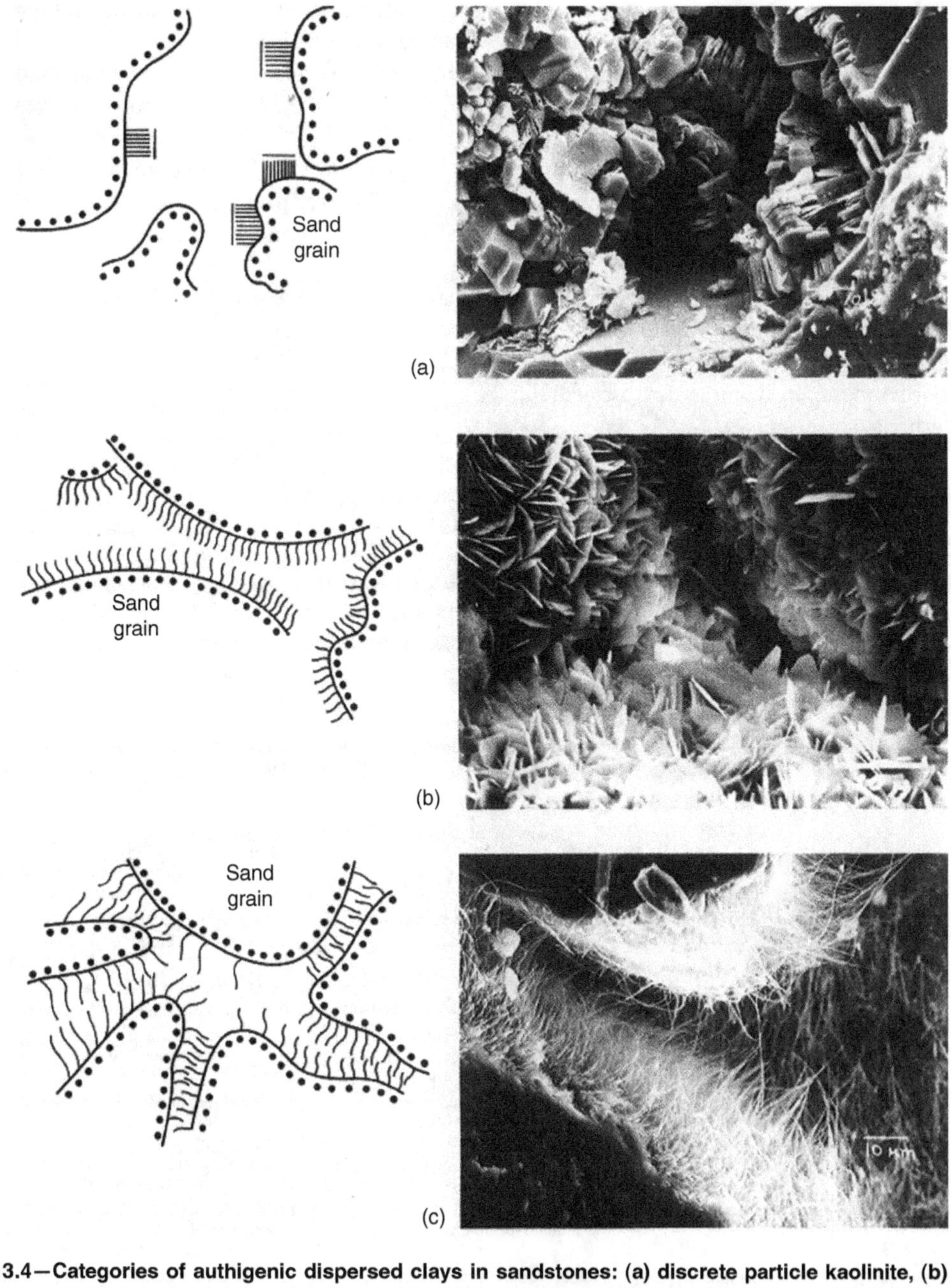


**Fig. 3.4—Categories of authigenic dispersed clays in sandstones: (a) discrete particle kaolinite, (b) pore-lining chlorite, (c) pore-bridging illite (Schechter 1990; Neasham 1977; da Motta et al. 1990). SCHECHTER, ROBERT S., OIL WELL STIMULATION, 1st ED. ©1990 Printed and electronically reproduced by permission of Pearson Education, Inc., New York, New York.**

type. Sandstones with significant quantities of both pore-lining and discrete clays can still have large permeabilities (Neasham 1977).

***Porosity and Permeability.*** The porosities of commercial conventional sandstone reservoirs typically range from 5 to 30%. Practical cutoff for conventional oil sandstone reservoirs is approximately 8%; however, for gas reservoirs, it is lower (Tiab and Donaldson 2012). The factors governing the magnitude of porosity in sandstone reservoirs are as follows (Tiab and Donaldson 2012):

(a) *Uniformity of grain size:* Uniformity or sorting in the gradation of grains. If small particles of silt or clay are mixed with larger sand grains, the effective porosity will be considerably reduced.
(b) *Degree of cementation or consolidation:* The highly cemented sand stones have low porosities, whereas the soft, unconsolidated rocks have high porosities.
(c) *Amount of compaction during and after deposition:* Compaction tends to lose voids and squeeze fluids out to bring the mineral particles close together. Generally, porosity is lower in deeper, older rocks.
(d) *Methods of packing:* With increasing overburden pressure, poorly sorted angular sand grains show a progressive change from random packing to a closer packing, thus reducing porosity. Crushing and deformation of the sand grains also leads to porosity reduction.

The key differences in porosity in sandstones and carbonates are compared in **Table 3.1** (Choquette and Pray 1970). Generally, sandstones are less heterogeneous than carbonate reservoirs. A 1-in. core plug may be representative of sandstone matrix porosity, but generally it is not adequate for carbonates. Furthermore, correlations for permeability and porosity are also better in sandstone reservoirs than carbonate reservoirs.

Comparison of Porosity in Sandstone and Carbonate Rocks

| Aspect | Sandstone | Carbonate |
|---|---|---|
| Amount of primary porosity in sediments | Commonly 25–40% | Commonly 40–70% |
| Amount of ultimate porosity in rocks | Commonly half or more of initial porosity; 15–30% common | Commonly none or only small fraction of initial porosity; 5–15% common in reservoir facies |
| Type(s) of primary porosity | Almost exclusively interparticle | Interparticle commonly predominates, but intraparticle and other types are important |
| Type(s) of ultimate porosity | Almost exclusively primary interparticle | Widely varied because of post-depositional modifications |
| Sizes of pores | Diameter and throat sizes closely related to sedimentary-particle size and sorting | Diameter and throat sizes commonly show little relation to sedimentary particle size or sorting |
| Shape of pores | Strong dependence on particle shape—a "negative" of particles | Greatly varied, ranges from strongly dependent "positive" or "negative" of particles to form completely independent of shapes of depositional or diagenetic components |
| Uniformity of size, shape, and distribution | Commonly fairly uniform within homogeneous body | Variable, ranging from fairly uniform to extremely heterogeneous, even within body made up of single rock type |
| Influence of diagenesis | Minor; usually minor reduction of primary porosity by compaction and cementation | Major; can create, obliterate, or completely modify porosity; cementation and solution important |
| Influence of fracturing | Generally not of major importance in reservoir properties | Of major importance in reservoir properties if present |
| Visual evaluation of porosity and permeability | Semiquantitative visual estimates commonly relatively easy | Variable; semiquantitative visual estimates range from easy to virtually impossible; instrument measurements of porosity, permeability, and capillary pressure commonly needed |
| Adequacy of core analysis for reservoir evaluation | Core plugs of 1-in. diameter commonly adequate for "matrix" porosity | Core plugs commonly inadequate; even whole cores (≈3-in. diameter) may be inadequate for large pores |
| Permeability/porosity interrelations | Relatively consistent; commonly dependent on particle size and sorting | Greatly varied; commonly independent of particle size and sorting |

Table 3.1—Comparison of porosity in sandstone and carbonate rocks (Choquette and Pray 1970). Republished with permission of AAPG; permission conveyed through Copyright Clearance Center, Inc.

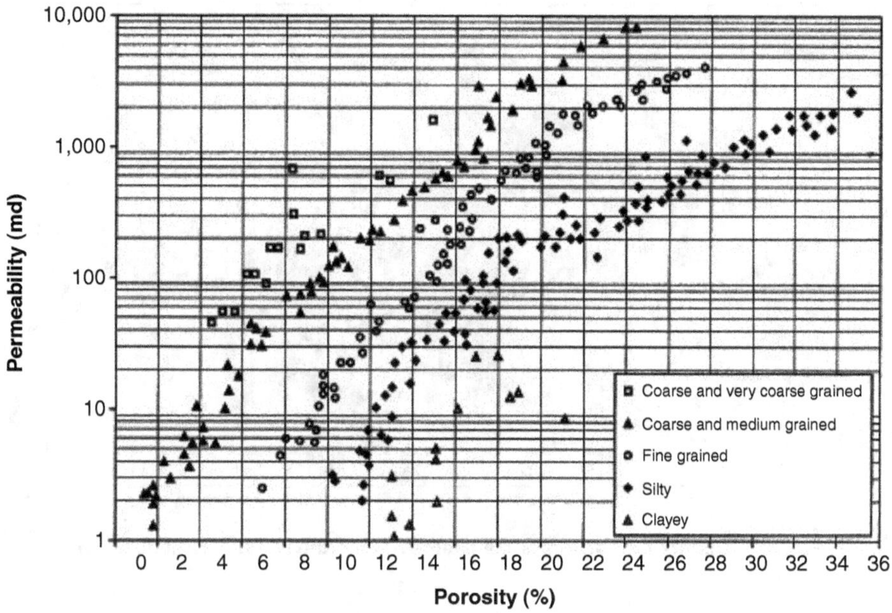


**Fig. 3.5—Influence of grain size on the relationship between permeability and porosity (Chilingarian and Wolf 1975). Reprinted with permission from Elsevier.**

Factors affecting the magnitude of permeability in sandstone reservoirs are (Tiab and Donaldson 2012)

(a) *Shape and size of sand grains:* If the rock is composed of large and flat grains uniformly arranged with the longest dimension horizontal, the horizontal permeability will be high and the vertical permeability will be medium to large. If the rock is composed of mostly large and rounded grains, its permeability will be considerably high and of the same magnitude in both directions. However, if the sand grains are small and irregular, the permeability of reservoir rock is generally low, especially in the vertical direction. **Fig. 3.5** shows the influence of grain size on the relationship between permeability and porosity (Chilingarian 1963). The figure shows permeability/porosity trends for (a) coarse-and very-coarse-grained, (b) coarse- and medium-grained, (c) fine-grained, (d) silty, and (e) clayey sandstones. The figure clearly shows that for a given porosity, the permeability is lower for smaller grain sizes.

(b) *Laminations:* Platy minerals such as muscovite and shale laminations act as barriers to vertical permeability. If these minerals are present in significant quantities, the horizontal permeability is generally 1.5 to 3 times larger than vertical permeability, and it may be approximately 10 times larger for some reservoir rocks.

(c) *Cementations:* Generally, the higher the degree of cementation, the lower the permeability. Permeability of unconsolidated sandstones is generally orders of magnitude higher than that of consolidated sandstones.

**3.3.2 Carbonates.** Carbonate rocks are a class of sedimentary rocks that are composed primarily of carbonate minerals. Carbonate minerals are salts of carbonic acid and contain the carbonate ion $CO_3^{2-}$. The two major types of carbonates are limestone, which is composed mainly of calcite or aragonite (different crystal forms of $CaCO_3$) and dolostone, which is composed primarily of the mineral dolomite [$CaMg(CO_3)_2$]. The carbonate sediments originate mainly through biological activity and to a lesser degree through inorganic precipitation (Wilson 1975; Reeckmann and Friedman 1982; Tucker and Wright 1990). Marine organisms, such as coral, mollusks, or foraminifera use materials dissolved in air or water to build their skeletons by secretions of aragonite or calcite. After the organism dies, the

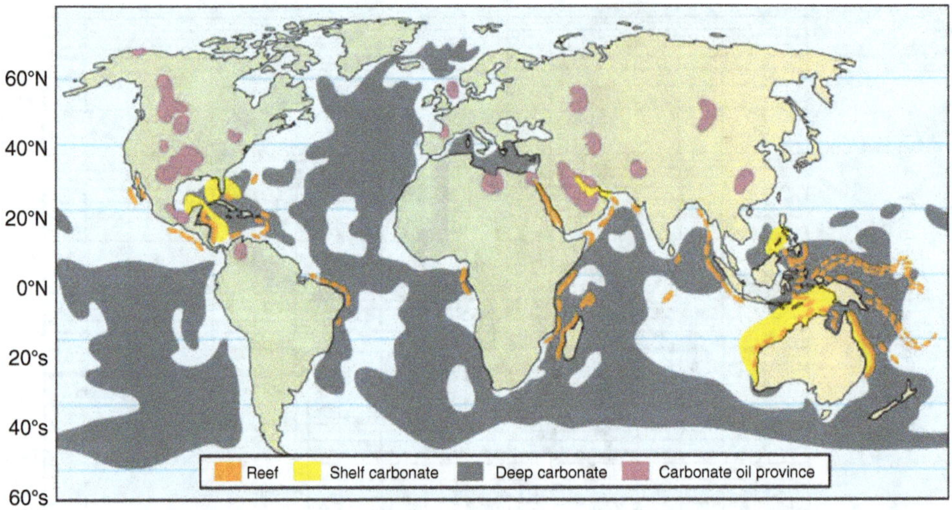


**Fig. 3.6—Distribution of carbonate basins around the world (Ahr et al. 2005).**

skeletal remains of these organisms form the grains in many carbonate rocks. The biological origins or carbonates limits them to specific water temperatures and other life-sustaining conditions and most carbonates are formed in an environment of warm, shallow, clear marine water **(Fig. 3.6).** Carbonates are also formed by chemical precipitation, such as when water evaporates from shallow onshore basins or when carbonate minerals precipitate from seawater. Carbonate rocks are also much more chemically reactive than sandstones and undergo substantial alternation, such as mineral dissolution and dolomitization. Dolomitization is the transformation of calcium carbonate to dolomite because of exposure to magnesium-rich water.

Unlike sandstones, the carbonate sediments typically undergo much less transport and many are deposited close to their point of origin. An exception is the less-abundant calcareous sandstones, which form when carbonate rocks are eroded, transported, and then deposited. Calcareous sandstones retain carbonate mineralogy and microporosity, but texturally they are similar to sandstones.

Two main limestone texture classification systems are in common use today, one by Folk (Folk 1974) and the second by Dunham (Dunham 1962). Folk's system recognizes four main limestone types **(Fig. 3.7).** Type I, designated by sparry allochemical rocks, consists chiefly of grains ("allochems") cemented by coarse calcite cement ("spar"). These rocks are similar in texture to well-sorted sandstones. Type II, designated by microcrystalline allochemical rocks, consists of allochems with significant presence of clay-sized calcite crystals or lime mud ("micrite") in the interstitial spaces. These rocks are texturally similar to clayey sandstones. Type III represents the opposite of Type I. It consists mostly of micrite with little or no allochems. Type IV represents rocks that are made of organic structures growing in situ and are referred to as "biolithite." The dominant organisms that make up the structure should be specified, such as coral biolithite or blue-green algal biolithite. If these rocks are broken up and redeposited, the resulting rock is either Type I or Type II, depending on the interstitial material.

After the main division of limestone in the aforementioned rock types, the Folk system further characterizes Types I and II rocks the basis of on the allochems of which they consist. Four types of allochems are defined.

1. *Fossils*: Whole, broken, or abraded fossils. They may range in size from gravel to fine sand.
2. *Oolites*: Spherical sand-sized particles that have a concentric or radial internal structure. They are formed when a small fragment of sediment acting as a seed gets washed around on the seabed where it accumulates layers of chemically precipitated calcite from the supersaturated water. The size of the ooids reflects the time they have been exposed to the water before they were covered later with sediment.

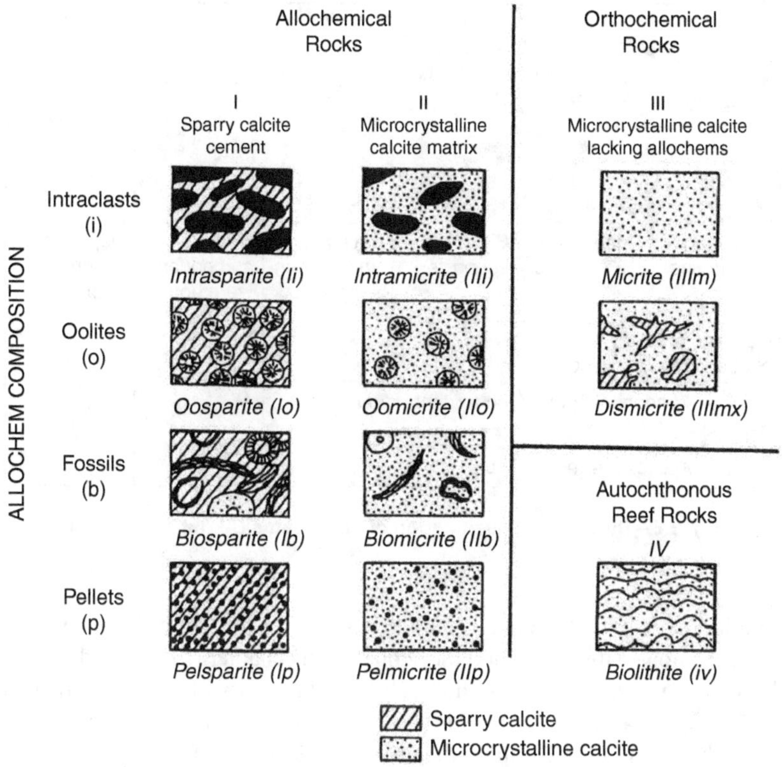


**Fig. 3.7—Classification of carbonate rocks of Folk (Folk 1962).**

3. *Peloids:* Spherical aggregates of microcrystalline calcite of coarse-silt to fine-sand size. Most peloids are fecal pellets produced by invertebrate animals.
4. *Interclasts*: Chunks of eroded limestone deposited as a conglomerate.

For example, a rock with fossils as an allochem in a micrite matrix is classified as biomicrite, whereas a rock with fossils in a spar matrix would be classified as biosparite (Fig. 3.7). Similarly a rock with oolites as the allochem in a micrite matrix would be oomicrite, and in spar matrix it would be oosparite. If the rock has more than one allochem or interstitial material, then they are listed in the order of least to most abundant in the classification. For example,

- oolites + fossils + spar matrix = oo bio sparite
- fossils + spar matrix + micrite matrix = bio spar micrite

Sometimes further refinements to the rock classification are added by defining the size of allochems.

The Dunham system is based on depositional texture. It classifies limestone primarily the basis of on whether the rock is grain-supported or mud-supported **(Fig. 3.8).** Grain-supported rocks are divided further into grainstones and packstones. Grainstones have little or no mud in their interstitial spaces, while packstones do. Mud-supported rocks with more than 10% of their volume occupied by larger grains are classified as wackestones, while those with less than 10% occupied by larger grains are termed mudstones. Boundstones are those rocks in which the original material provided support during deposition, such as reefs. Crystalline rocks are those that have lost their depositional fabric because of diagenetic recrystallization. Grainstones are similar to sparry allochemical rocks (Type I) in the Folk classification, while wackestone and packstone are similar to microcrystalline allochemical rocks (Type II), and mudstone is similar to microcrystalline calcite or micritic rocks (Type III). Boundstones

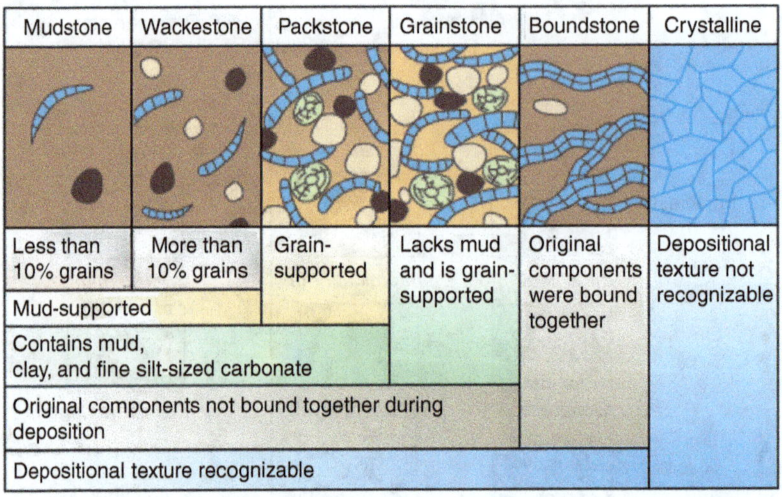


**Fig. 3.8—Dunham's classification of carbonates (Akbar et al. 1995).**

are similar to biolithites (Type IV). Embry and Klovan have further refined the boundstone/biolithite category, but a description of this is beyond the scope of this chapter (Embry and Klovan 1971).

Unlike the Folk scheme, Dunham deals with the original porosity of the rock. The Dunham scheme is useful for hand samples because it is based on texture, not the grains in the sample. A petrographic microscope is usually required to determine the nature of the allochems defined in the Folk scheme.

***Composition.*** Ancient carbonate rocks are mostly composed of calcite, dolomite, or combinations of these minerals. In contrast, many Pleistocene-to-Recent carbonate sediments and rocks are polymineralic and contain unstable mixtures of several carbonate polymorphs or species. At least 15 anhydrous carbonate minerals and several hydrous carbonate minerals exist in nature. However, compositionally carbonate rock types can be divided into two major rock types: limestone and dolostone. Limestone is composed mainly of the mineral calcite or aragonite, which are different crystal forms of $CaCO_3$. Dolostone is composed primarily of the mineral dolomite [$CaMg(CO_3)_2$]. Mixtures of calcite and dolomite are classified on the basis of the calcite or dolomite percentage **(Fig. 3.9).** Carbonates often contain variable amounts of silica or siliceous skeletal fragments and varying amounts of clay, silt, and sand carried by rivers. **Fig. 3.10** depicts the terminology used for limestone/clay mixtures. Similar

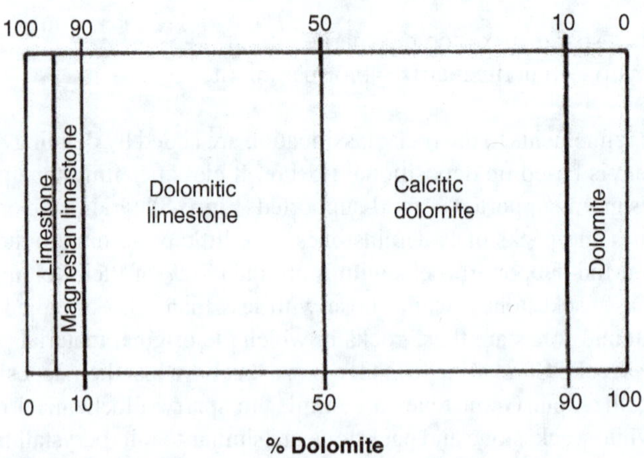


**Fig. 3.9—Classification of limestone/dolomite mixtures (Mazzullo et al. 1992).**

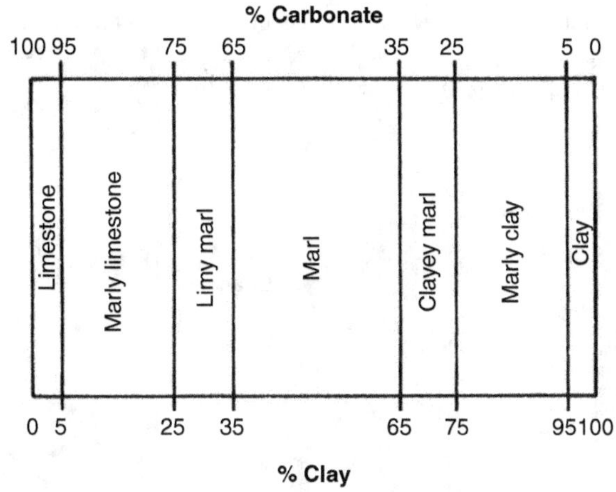


**Fig. 3.10—Terminology for limestone/clay mixtures. Similar terminology can be used for dolomites mixed with clay (Mazzullo et al. 1992).**

terminology can be used for dolomites mixed with clay (Mazzullo et al. 1992). Other minerals present in carbonate reservoirs are siderite ($FeCO_3$), anhydrite ($CaSO_4$), and pyrite ($FeS_2$).

Micrite or carbonate mud is an important constituent of many carbonate deposits. It consists of clay-sized crystals of calcium carbonate. It can be formed by either breakdown of calcareous algae skeletons or by the direct precipitation of calcite or aragonite from solution.

***Porosity and Permeability.*** Most carbonate reservoirs have porosity of approximately 5–15%, as compared to terrigenous sandstone reservoirs, which typically have porosities of 15–30% (Ahr 2008). The key differences in porosity in sandstones and carbonates are listed in Table 3.1. Unlike sandstones, the porosity/permeability crossplots for carbonate reservoirs commonly show large variability **(Fig. 3.11).** Therefore, porosity by itself is not a good indicator of permeability in carbonate rocks. Size and distribution of pore space must be included in models for permeability of carbonate rocks (Lucia 1999). The fluid/rock interaction occurs only through porosity; and thus the nature of the porosity is critical in determining the accessibility of the pore space and the reactive surface available for interaction. Four common carbonate-porosity classification schemes are described next.

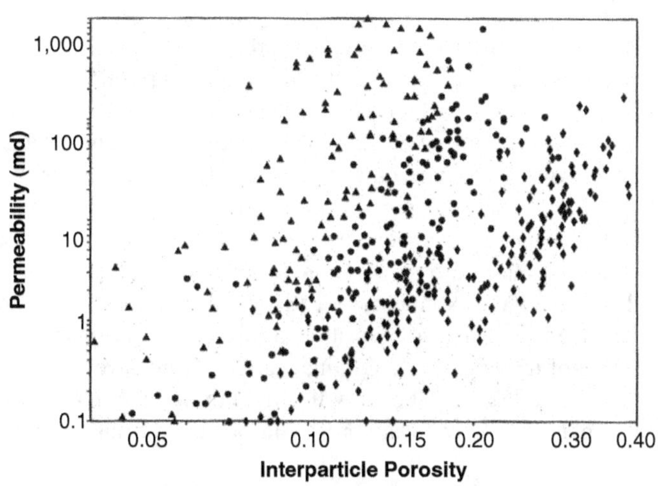


**Fig. 3.11—Plot of porosity and permeability for carbonate rocks, illustrating that there is no relationship between porosity and permeability in carbonate rocks without including pore-size distribution (Lucia 1999).**

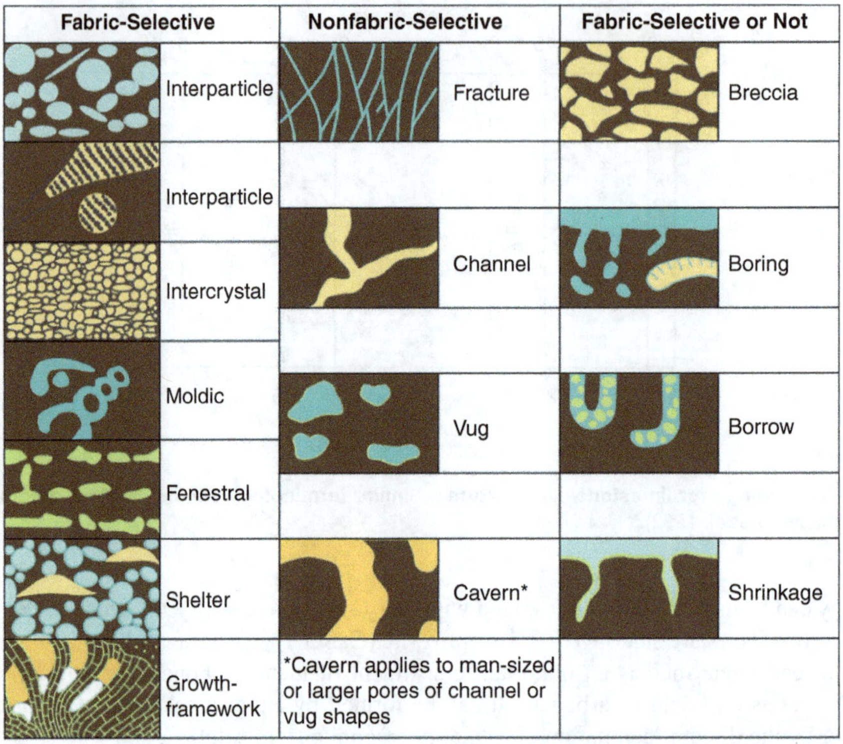


**Fig. 3.12—Choquette and Pray classification of carbonate porosity (Akbar et al. 1995).**

One of the first porosity-classification schemes was proposed by Archie (1952). His classification scheme consists of two parts: (a) the texture of the matrix and (b) the character of the visible pore structure. Archie in his pioneering work defined three textural categories (Types I through III) and four visible porosity classes (A through D) to characterize carbonate rocks. In Archie's classification, Type I texture is for compact crystalline matrix, Type II for chalky, and Type III for granular matrix. He classified the visible pore space on the basis of observations at 10X magnification. Pore sizes from 0 to 0.01 mm were classified as Class A, pore sizes from 0.01 to 0.1 mm as Class B and pore sizes > 0.1 mm as Class C; and Class D was reserved for samples that showed weathering, fracturing, or solution channels where the pore size was larger than the size of the cutting.

Choquette and Pray incorporated time and mode of origin in the classification of carbonate porosity (Choquette and Pray 1970). They divided porosity in carbonates into three main classes: (1) fabric-selective, (2) not fabric-selective, and (3) fabric-selective or not **(Fig. 3.12)** (Akbar et al. 1995). Seven different pore types were defined within the fabric-selective porosity class. These are interparticle porosity occurring between grains; intraparticle porosity occurring within skeletal grains; intercrystal porosity occurring within crystallized micrite and/or dolomite; moldic porosity resulting from dissolution of grains; large-scale framework porosity called fenestral porosity that usually results from dissolution of algal mat deposits; shelter porosity that describes pore space resulting from shelter offered by large overlaying grains; and growth framework porosity that is the natural outcome of organic processes such as coral reefs. Nonfabric-selective porosity includes fracture porosity; channel porosity caused by extensive leaching; vug porosity that results from extensive dissolution of materials and retains no evidence of the original host grain; and man-sized cavern porosity that results from highly extensive and prolonged leaching. Porosity that did not fit either class was grouped under the "fabric-selective or not" class. This includes porosity in breccia, boring, burrowing, and shrinkage (Akbar et al. 1995).

The Choquette-Pray classification provides a useful description of porosity because it allows one to resolve the response of carbonate rocks to reactive fluids in terms of 15 main pore types; however, a

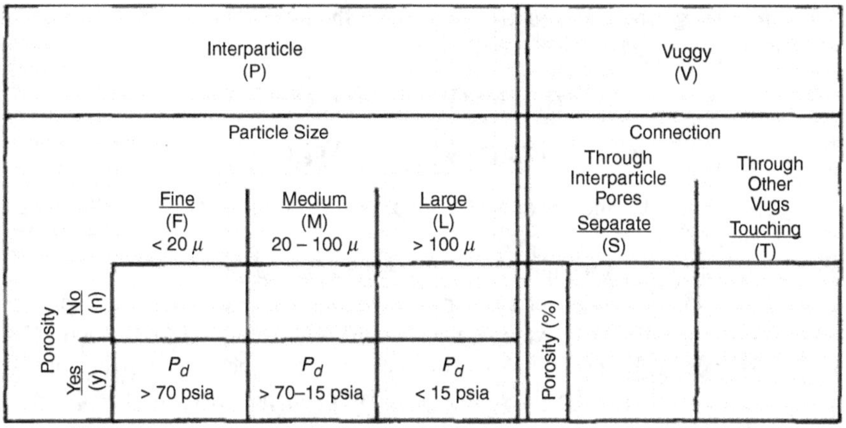


**Fig. 3.13—Lucia's classification of carbonate porosities (Lucia 1983).**

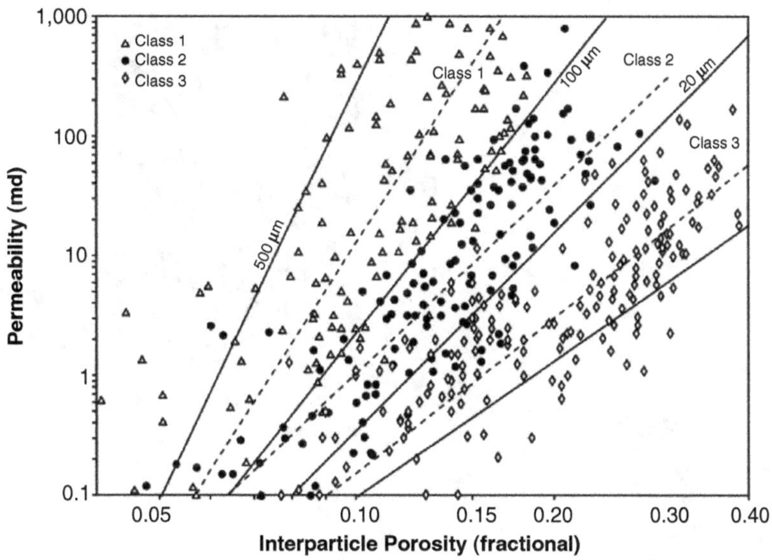


**Fig. 3.14—Porosity/air-permeability crossplot for nonvuggy limestones and dolostones (Lucia 1995).**

major limitation of their scheme is that it does not differentiate between connected and unconnected porosity, and, hence, it does not correlate directly with permeability.

Lucia expanded on Archie's work and related permeability, capillary properties, and *m* values of carbonate rocks to the particle size, amount of interparticle porosity, amount of separate vug porosity, and the presence or absence of touching vugs. His scheme is depicted in **Figs. 3.13 and Fig. 3.14.** The first-order division is made between interparticle and vuggy porosity. Interparticle porosity is subdivided further on the basis of particle size and porosity, with emphasis on particle size. The relationship between mercury/air extrapolated entry pressures and the average particle size is used to divide particle size into three groups. Rock with < 20 μm particles have high entry pressures, rock with 20 to 100 μm particles have intermediate entry pressures, and rocks with >100 μm particles have low entry pressures. Fig. 3.14 shows that these size groups, dubbed fine (< 20 μm), medium (20–100 μm), and large (>100 μm), give a reasonable partitioning to the porosity/permeability plot. Vuggy porosity is divided further into separate and touching vugs. An estimate of the percent of separate- and touching-vuggy porosity is necessary to complete the description. The classification is recorded as a composite symbol, the first part describing the interparticle pore space and the last part describing the vuggy pore space as shown here.

| (y,n)(F M L) | (S,%)(T) |
|---|---|
| Interparticle | Vuggy |

For example, "yLS$_5$T" would stand for apparent interparticle porosity, greater than 100 particle size, with 5% separate vugs and touching vugs present.

Other notable carbonate classification schemes are ones by Lønøy (2006) and Ahr et al. (2005). Lønøy combined the essential characteristics from the Choquette-Pray and the Lucia classification schemes with additional elements to develop a new classification system (Lønøy 2006). His classification includes 20 pore-type classes and allows reasonable correlations between porosity and permeability.

Ahr et al. (2005) proposed a genetic classification of carbonate porosity based on the mechanism for the origin of porosity. They recognized three independent mechanisms in their classification system.

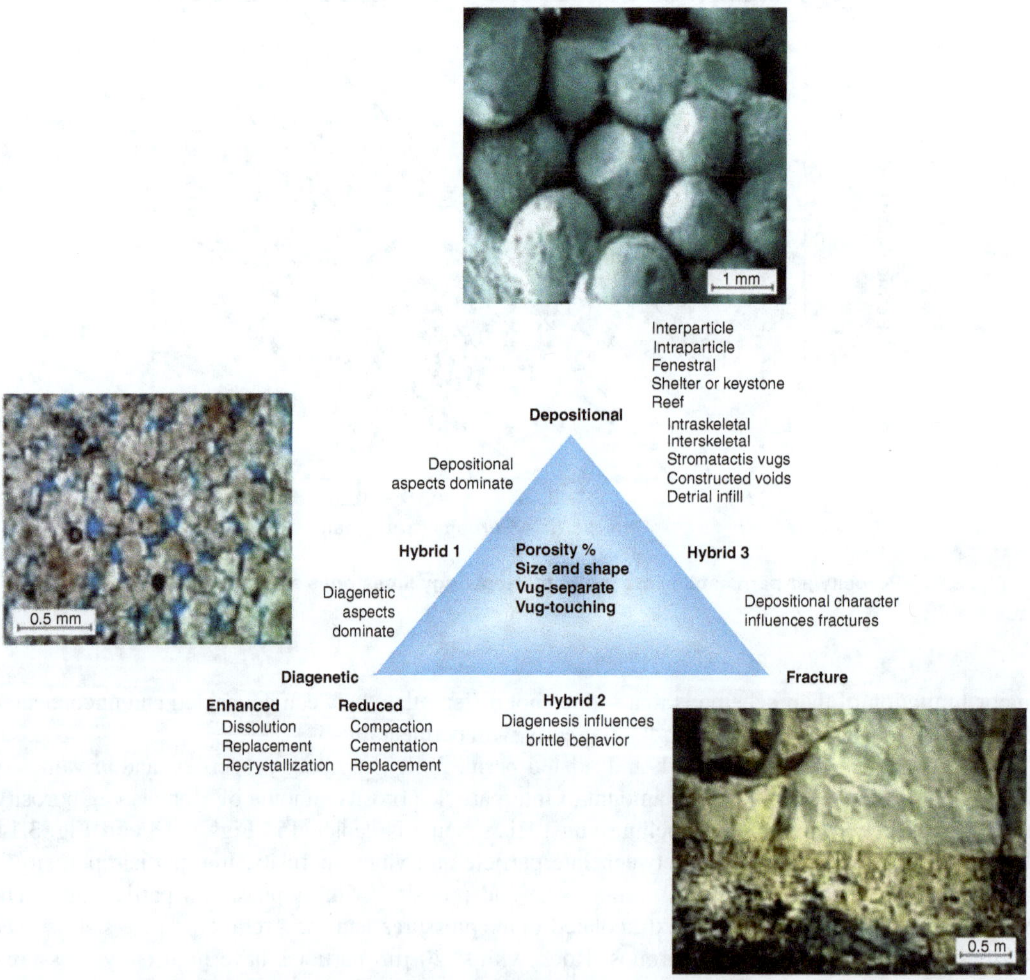


**Fig. 3.15—Genetic classification of carbonate porosity by Ahr et al. (2005). Depositional porosity represents the space that remains between grains, such as the ooids in the scanning electron micrograph (top), skeletal fragments, or other particles. Diagenetic porosity, filled with blue epoxy in a thin section of dolomitized limestone (left), can result from cementation, compaction, dissolution, recrystallization, or replacement processes. Fracture porosity can occur at more than one scale, as is shown by at least two fracture sets in the outcrop photograph (bottom right). An infinite variety of hybrid pores exists between the end members.**

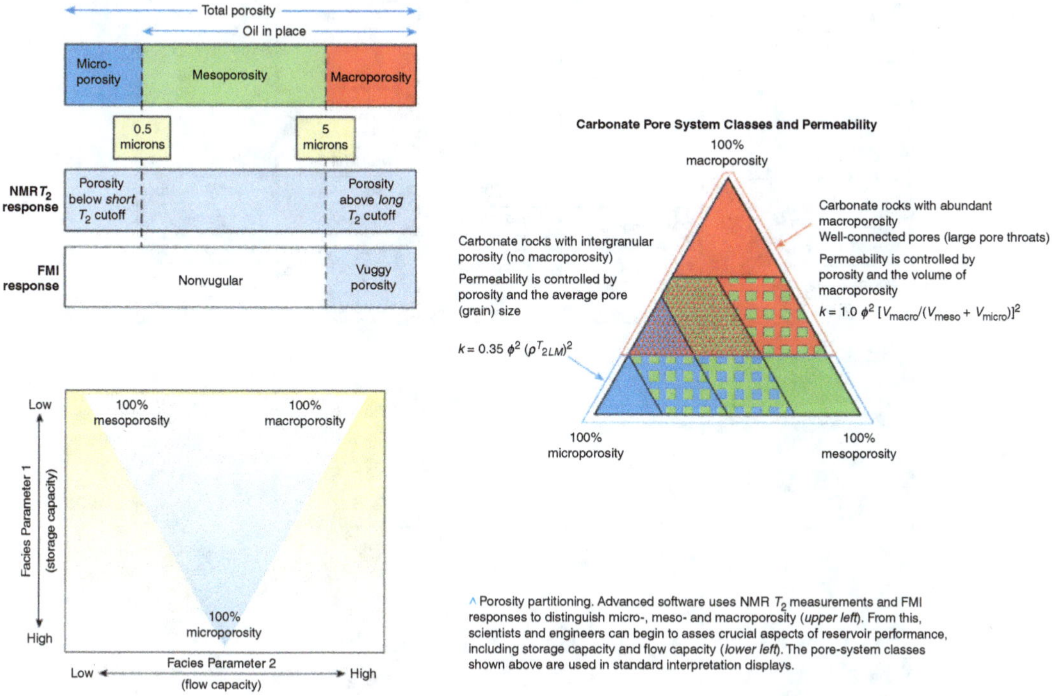


**Fig. 3.16—Carbonate pore-system and permeability classes based on NMR $T_2$ and FMI response (Ahr et al. 2005).**

These are (1) deposition, (2) diagenesis, and (3) fracturing (Ahr et al. 2005). These three mechanisms are plotted on a triangular diagram as end members **(Fig. 3.15).** The end-member processes are independent, but hybrid pore types exist between them and they are shown on the side. For example, depositional porosity altered by diagenesis, but with depositional texture, fabric, or bed forms still recognizable, is classified as a hybrid in which depositional attributes are dominant (Ahr 2008).

These porosity-classification systems can be implemented in software to assist with optimizing of acid stimulation treatments. One such attempt that is based primarily on Lucia's classification is described by (Ahr et al. 2005). Nuclear-magnetic-resonance (NMR) $T_2$ measurement and Formation Microimager (FMI) responses were used to distinguish between micro-, meso-, and macroporosity. These measurements were processed with the software to yield a log of micro-, meso-, and macroporosity in a giant field **(Fig. 3.16).** This distribution of pore sizes can then be used in acid stimulation models to predict the response of rock to acid injection. For example, the Wang et al. (1993) and Huang et al. (2000) have used pore size to estimate the optimum injection rates for different carbonate rock types (Wang et al. 1993; Huang et al. 2000).

A particularly useful classification of porosity for acidizing treatment design is by Zakaria et al. (2015a, 2015b). In this classification scheme, the acid accessible porosity is calculated by measuring the concentration of a nonreactive tracer in the effluent **(Fig. 3.17).** The accessible porosity is quantified by the flowing-fraction parameter, which is defined as the pore volumes of injected fluid at which half of the injected tracer concentration is observed in the effluent. Fig. 3.18 depicts the tracer concentration for Edwards Yellow limestone when a plain brine tracer and the brine tracer with a polymer base fluid are injected. The flowing fraction for the plain brine tracer is 0.9, while for the polymer tracer it is 0.7. The flowing fraction of the polymer tracer is lower because it is much more viscous, and it preferentially flows through the network of large pores. Zakaria et al. (2015a, 2015b) then used these flowing-fraction parameters to predict the response of plain and emulsified acids. They also found that these flowing fractions correlated well with NMR $T_2$ times. This suggests that it may be possible to

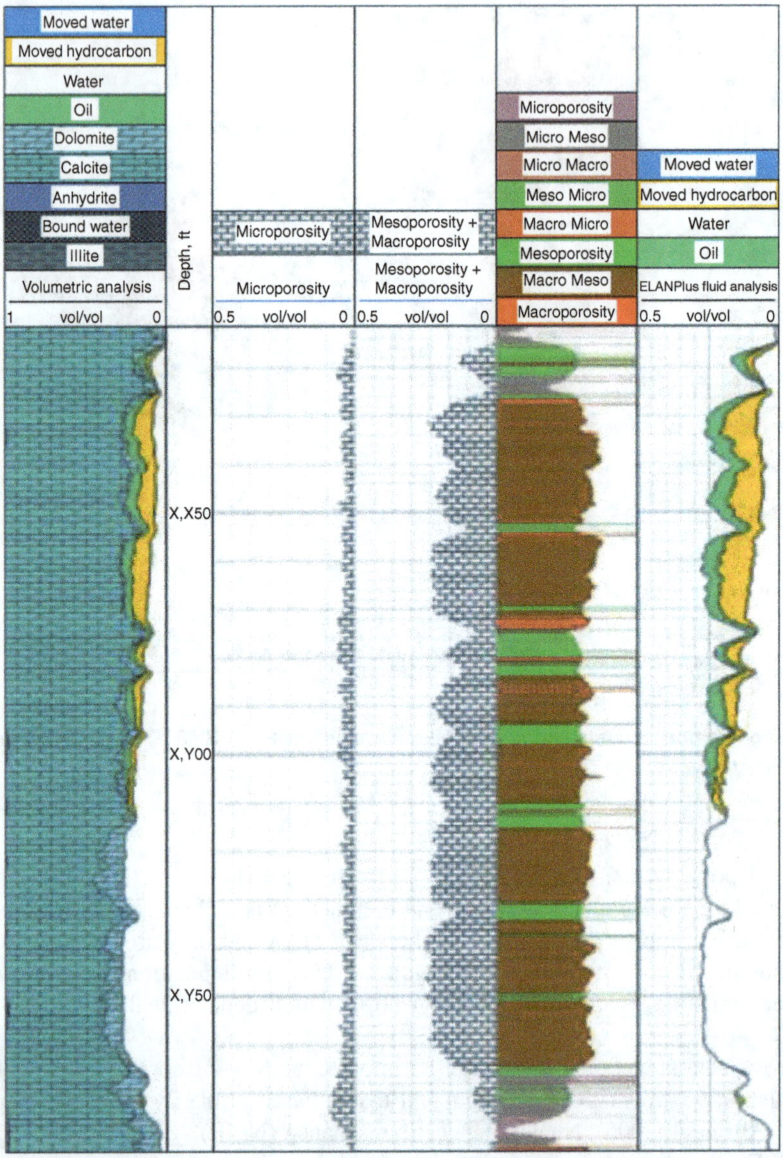


**Fig. 3.17—Porosity partitioning in a giant field. Pore types are shown in Tracks 2, 3, and 4. The final output from the software can be tailored to assist acidizing treatments (Ahr et al. 2005).**

create a wireline log of flowing fractions and display it on a track alongside permeability and other rock parameters required for treatment design.

## 3.4 Chemistry of Fluids

### 3.4.1 Common Constituents of Stimulation Fluids

***Hydrochloric Acid (HCl).*** HCl is by far one of the most commonly used acids for stimulation. It is supplied in concentrations of 32–36% and is normally diluted to 5–28% for field use.

1,000 gal of 15% HCl will dissolve 1,840 lbm (10.8 $ft^3$) of limestone ($CaCO_3$). Approximately 2,040 lbm of calcium chloride is produced in the reaction along with 40 gal of water and 6,640 $ft^3$ of carbon dioxide. The resulting spent acid is a 20% $CaCl_2$ brine solution. 1,000 gal of 28% HCl will dissolve 3,670 lbm (21.6 $ft^3$) of limestone. The resulting spent acid is approximately 35% $CaCl_2$ brine.

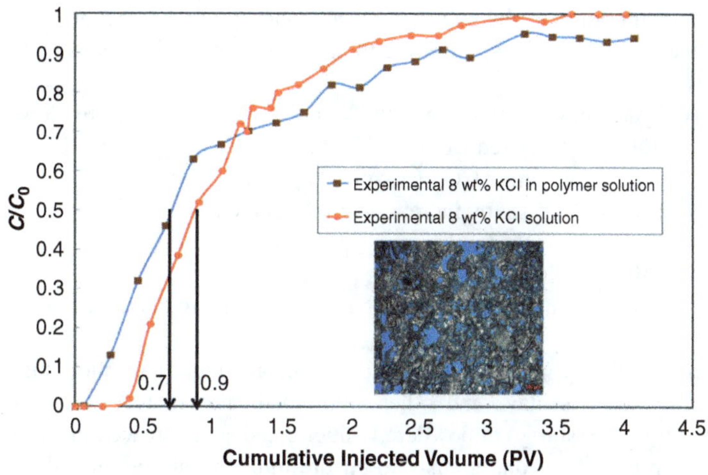


**Fig. 3.18—Normalized tracer concentration profile for Edwards Yellow limestone for plain and polymer tracer fluids (Zakaria et al. 2015a, 2015b). Copyright 2015 American Chemical Society.**

***Organic Acids.*** The main advantage of using organic acids is that they are less corrosive and easy to inhibit at high temperatures. Acetic acid ($CH_3COOH$) and formic acid (HCOOH) are two commonly used organic acids. These organic acids react at a slower rate with carbonate minerals than HCl and are used to replace HCl in situations where it is desirable to retard the reaction rate. They are also used to replace HCl for formations containing HCl-sensitive aluminosilicates.

***Mud Acids [Mixture of HCl and Hydrofluoric Acid (HF)].*** Mud acids are a mixture of HCl and HF, and they were originally used to remove formation damage caused by drilling mud. For many years, the standard mud-acid formulation consisted of 12% HCl and 3% HF and was commonly known as regular mud acid (RMA) or 12:3 mud acid. In recent years, the trend has been toward the use of lower-strength HF solutions such as 9% HCl and 1% HF, or 9:1 mud acid, because of the possibility of precipitation. For formations with HCl-sensitive clays, organic acids, such as formic and acetic acids, are commonly used to replace the HCl.

***Emulsions.*** An emulsion is a dispersion of two immiscible phases, such as oil in water or water in oil, stabilized with a surfactant. Emulsions occur throughout the petroleum industry and are usually detrimental to production because of their high viscosities. Solids play an extremely important role in both the formation and stability of emulsions encountered in the petroleum industry. However, many stimulation fluids are routinely emulsified to reduce the reactivity and leakoff of the fluid. These emulsions are moderately stable and are supposed to break up once their function is fulfilled. Formation of undesirable emulsions because of a stimulation treatment can be reduced by conducting compatibility tests between stimulation fluids and reservoir fluids.

***Foams.*** Like emulsions, foam is also a two-phase fluid stabilized by surfactants. In foams, the inner phase is a gas phase, and it is usually either $N_2$ or carbon dioxide ($CO_2$). The liquid phase is the external phase and is primarily saline water mixture with either a surfactant or a gellent, depending upon the required viscosity and stability. The liquid phase may also contain reactive fluids such as acids or hydrocarbon fluids such as diesel.

Foams are described by their quality:

$$\text{foam quality} = \frac{\text{gas volume}}{\text{foam volume}} \times 100 \quad (3.29)$$

The foam quality can range from 52–95%. Above 95%, the foam usually changes to a mist, with gas as the continuous phase. Below 52%, stable foam does not exist because there are no bubble/bubble interactions to provide resistance to flow or to gravity separation. Above 52% gas, the gas concentration is high enough that the bubble surfaces touch. Stable dispersions of gas in liquid can exist below 52% gas; it may not be appropriate to call them foams, but they can be used effectively as energized fluids.

The foams with $CO_2$ are denser than those with $N_2$ and, consequently, require lower surface-treating pressures. Lower treating pressures reduce pumping costs. However, $CO_2$ is much more soluble in oil and water than $N_2$ and, therefore, more $CO_2$ is required to saturate the liquid than $N_2$. Therefore, reductions in pumping costs may be offset by increases in material costs.

## 3.5 Measurement of Reaction Rates

Three main reactor geometries are used for characterizing acidizing reactions.

**3.5.1 Slurry Reactor.** Slurry reactors are used for measuring dissolution kinetics of naturally fine-grained minerals such as clays and other aluminosilicates in the surface-reaction-controlled regime. Typically, a mixture (slurry) of powdered mineral and acid is placed in a reaction vessel under pressure. The mixture is stirred continuously to ensure uniform mixing and adequate mass-transfer rates. Samples of the mixture are withdrawn from the reactor at various time intervals and filtered with syringes fitted with membrane filters. The filtrate samples are analyzed to monitor the progress of the reaction.

Kline and Fogler (1981a) describe a slurry reactor they used in measuring dissolution rates of several aluminosilicates in acid. The reactor they used in the study was a Hastelloy vessel of approximately 150 cm$^3$, with agitation provided by a magnetic stirring bar **(Fig. 3.19).** They maintained the reactor temperature with a surrounding water bath. For the small particles used in these experiments (less than 5 μm for clays and less than 20 μm for feldspar and quartz), the rate of mass transfer to the surfaces of these minerals was orders of magnitude faster than the dissolution rate. Therefore, the progress of the surface reaction could be monitored by directly measuring the concentration of the products in the bulk.

**3.5.2 Rotating Disk.** The rotating-disk apparatus consists of a disk of rock spinning in a volume of reactive fluid at a constant angular velocity $\omega$ (Levich 1962). The concentrations of the products or reactants in the bulk are measured at various time intervals while the disk is rotating at a fixed $\omega$. The global reaction rate at a given $\omega$ is computed from the change in concentration vs. time. The experiment is then repeated at different angular velocities to obtain the dissolution rate as a

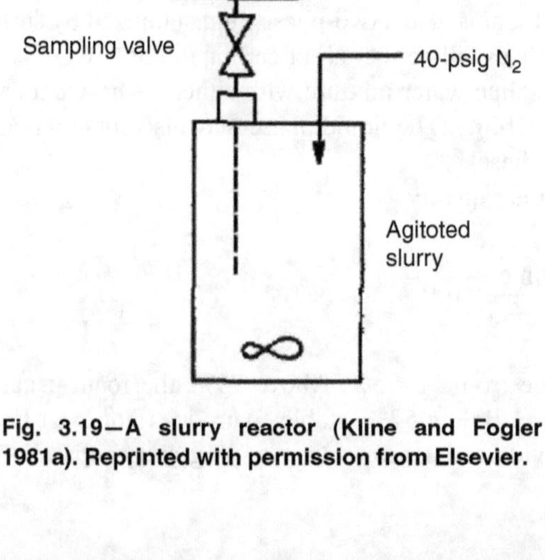


**Fig. 3.19—A slurry reactor (Kline and Fogler 1981a). Reprinted with permission from Elsevier.**

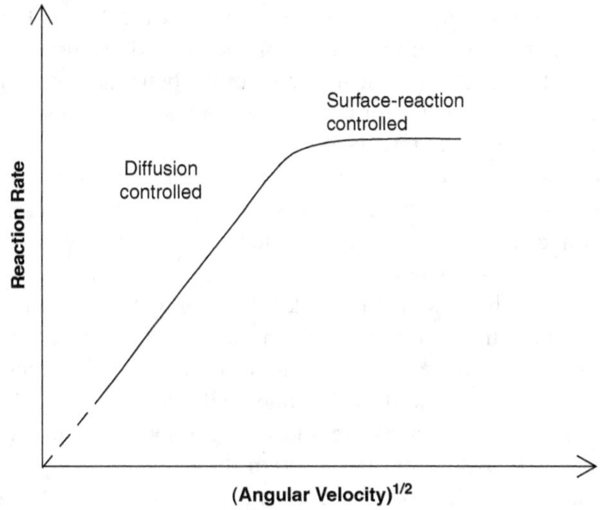


**Fig. 3.20—Reaction rate vs. square root of disk rotation speed for a reaction between a solid and solute in a rotating-disk apparatus. The diffusion-controlled and surface-reaction-controlled regions are indicated.**

function of $\omega$. The reaction-rate data are then plotted against $\sqrt{\omega}$ to indentify the reaction- and diffusion-limited regimes (**Fig. 3.20**). Additional experiments at different temperatures or different reactant concentrations are often required to obtain the full-rate law for the reaction system. The method has been successfully used to measure reaction kinetics of many reaction systems important to matrix stimulation treatments (Lund et al. 1973; Fogler and Lund 1975; Lund et al. 1975). There are several advantages of this system of studying reactions between fluid and solid surface over the traditional system of the fluid flowing over a flat plate (Litt and Serad 1964; Boomer et al. 1972):
A flow tunnel is not required because the fluid motion is induced by the disk.

- Large fluid volumes are not required.
- End effects are minor, whereas with a flat plate they can be a major factor.
- A disk rotating in an infinite fluid volume represents a 3D flow system for which there exists an exact solution to the Navier-Stokes equations.
- The heat- and mass-transfer coefficients are constant over the surface of the disk, whereas they decrease down the flat plate.

Levich (1962) provided an analytical solution for mass flux of a solute from a fluid to the surface of the rotating disk under laminar conditions as (Levich 1962; Boomer et al. 1972)

$$j = 0.62 D^{\frac{2}{3}} \nu^{\frac{-1}{6}} \omega^{\frac{1}{2}} C_0 \ \text{g mol/cm}^{2\text{-s}}, \quad (3.30)$$

where $D$ is the diffusion coefficient of the solute (cm²/s), $\nu$ is the fluid kinematic viscosity (cm²/s), $\omega$ is the angular velocity of the disk (radians/s), and $C_0$ is the solute concentration (g mol/cm³). This relation can be used to calculate mass-transfer coefficient $k_m$ and the diffusional boundary layer thickness, $\delta$, as follows:

$$k_m = 0.62 D^{\frac{2}{3}} \nu^{\frac{-1}{6}} \omega^{\frac{1}{2}} \ cm/s, \quad (3.31)$$

$$\delta \approx \frac{DC_0}{j} = 1.61 \left(\frac{D}{\nu}\right)^{\frac{1}{3}} \left(\frac{\nu}{\omega}\right)^{\frac{1}{2}} \ cm. \quad (3.32)$$

Eqs. 3.31 and 3.32 only apply when the flow in the vicinity of the disk is laminar. The theory also assumes an infinite disk spinning in an infinite fluid volume in the derivation of the equations. In practice, the disk diameter must be larger than the thickness of the boundary layer and the vessel diameter has to be at least twice the disk diameter for the observed transport rate to be almost independent of the vessel diameter (Gregory and Riddiford 1956).

Generally the reactions are studied at a high enough angular velocity, such that the rate of reaction is completely controlled by the surface reaction (Fig. 3.20). In this case, the surface reaction rate can be calculated directly from the overall reaction rate (Lund et al. 1973; Fogler et al. 1975). However, for many fast reactions, it may be difficult to achieve an angular velocity where the reaction is surface-reaction limited and the flow near the disk is laminar. For many fast reactions, the flow near the disk turns turbulent before the reaction becomes surface-reaction limited and the aforementioned relationship for mass transfer cannot be used. The rate of surface reaction generally slows down much more rapidly with temperature than the diffusion rate, and, therefore, in some cases, it is possible to measure the kinetics at a lower temperature and then extrapolate to the temperature of interest. For example, Lund et al. (1975) lowered the reaction temperature to measure the kinetics of the calcite-HCl system (Lund et al. 1975). However, for aqueous systems, the freezing point of the solution sets the limit to which the temperature can be lowered. If the reaction rate cannot be made to be completely under surface reaction control, an accurate model for diffusion coefficients may be necessary to estimate the contribution of the surface reaction (Lund et al. 1975).

**3.5.3 Parallel-Plate Reactor.** A parallel-plate reactor system, where reactive fluid flows between a set of parallel plates, with at least one of them having a reactive surface, has been used in several studies to determine reaction kinetics for fracture acidizing (Barron et al. 1962; Nierode and Williams 1971; Roberts 1974). The model equations for this reaction system are discussed in detail by Williams et al. (1979). In the derivation of the model equations, it is assumed that the mass transfer and chemical reaction have negligible effect on fluid velocity. The velocity profile is assumed to be laminar and neither influenced by the reaction taking place at the solid surface nor as a result of entry effects. **Fig. 3.21** shows a parallel-plate reactor system used by Nierode and Williams (1971). They used inert plates near the fluid entry and exit to reduce entrance effect. They also used an inert top plate to suppress natural convection effects because of higher density of the spent acid.

## 3.6 Reactions

Data on some of the common reactions encountered in matrix stimulation treatments are provided in this section.

**3.6.1 Dissociation of Acids in Aqueous Solutions.** An important property of acids is that they dissociate (ionize) in aqueous solutions by the reaction

$$HA(aq) + H_2O \leftrightarrow H_3O^+(aq) + A^-(aq), \quad (3.33)$$

where HA indicates a generic monoprotic acid that dissociates in presence of water to yield $A^-$, the conjugate base of the acid and the hydronium ion, $H_3O^+$. The equilibrium for the dissociation reaction is described by the equilibrium constant, $K_{eq}$,

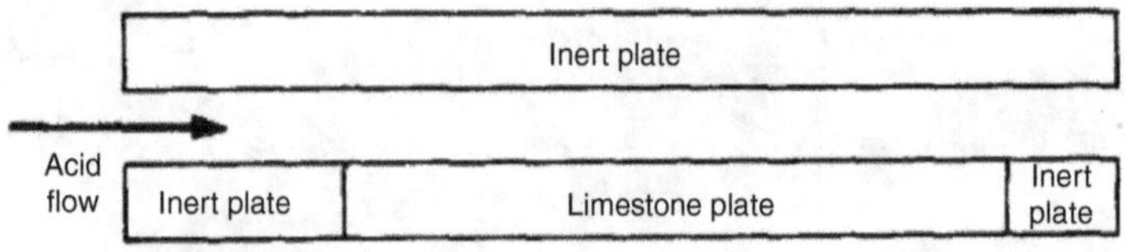


**Fig. 3.21—Parallel-plate reactor system of Nierode and Williams (1971) to determine reaction kinetics (Nierode and Williams 1971).**

$$K_{eq}=\frac{\{H_3O^+\}\{A^-\}}{\{HA\}\{H_2O\}}=\frac{[H_3O^+][A^-]}{[HA][H_2O]}\times\frac{\gamma_{A^-}\gamma_{H_3O^+}}{\gamma_{HA}\gamma_{H_2O}}, \quad (3.34)$$

where $\{X\}$ indicates the activity, $[X]$ the concentration, and $\gamma_X$ indicates the activity coefficient of the species $X$ ($\{X\}=[X]\gamma_X$). The temperature and pressure dependence of $K_{eq}$ can be calculated from the equations presented in section 3.2.2 Chemical Equilibrium.

Water is by far the dominant species in most dilute to moderate-strength acid solutions. Therefore, it is common to approximate the concentration of water $[H_2O]$ as a constant, equal to 55.5 mol/L. Furthermore, the dissociation equilibrium is often determined in solutions of high to medium ionic strength, and, therefore, the quantity $\frac{\gamma_{A^-}\gamma_{H_3O^+}}{\gamma_{HA}\gamma_{H_2O}}$ is also often assumed to be a constant (Rossotti and Rossotti 1961). Incorporating these constants and representing the hydronium ion concentration by the hydrogen ion concentration, $K_a$ (the dissociation constant of an acid) is defined as

$$K_a=\frac{[H^+][A^-]}{[HA]} \quad (3.35)$$

From this equation, $K_a$ appears to have dimensions of concentration, but, in fact, it is dimensionless because the concentration of water is assumed to be constant and has been incorporated into the value of $K_a$. Because the definition of $K_a$ is based on concentration and not activities (like a true equilibrium constant), $K_a$ depends on concentration as well as temperature and pressure. However, the concentration dependence is often assumed to be weak as described earlier. The values for $K_a$ span many orders of magnitude, and it is common practice to use the logarithmic measure, $pK_a$, to represent its value. $pK_a$ is defined simply as

$$pK_a=-\log_{10}K_a \quad (3.36)$$

$K_a$ and $pKa$ are both quantitative measures of the strength of an acid in solution. The larger the value of $K_a$ (or the smaller the value of $pK_a$), the stronger the acid.

Empirical dissociation constants for some common acids are listed in **Table 3.2** as a function of temperature.

Another common measure of the acidity of a solution is its pH value. pH is defined as

Empirical Constants of the Equation

$$-\log_{10}K_i=\frac{B_1}{T}-B_2+B_3T$$

| Equilibrium Constant | $B_1$ | $B_2$ | $B_3$ | Ref. |
|---|---|---|---|---|
| $K_1$ | –1,820* | –2.253* | – | Rossini et al. (1952); Garrels et al. (1960) |
| $K_2$ (formic) | 1,342.85 | 5.2743 | 0.015168 | Robinson and Stokes (1965) |
| $K_2$ (acetic) | 1,170.48 | 3.1649 | 0.013399 | Robinson and Stokes (1965) |
| $K_2$ (propionic) | 1,213.26 | 3.3860 | 0.014055 | Robinson and Stokes (1965) |
| $K_2$ (chloroacetic) | 1,229.13 | 6.1714 | 0.016486 | Robinson and Stokes (1965) |

*Computed from thermodynamic data.

Table 3.2—Empirical acid dissociation constants (Chatelain et al. 1976).

| Term | Symbol | Definition |
|---|---|---|
| Solubility | $S$ | Mass of a solid dissolved in a known weight (or volume) of fluid |
| Solubility product | $K_{sp}$ | $K_{sp} = \{M^{z+}\}^m \{X^{z-}\}^x$ |
| Saturation ratio | $SR$ | $SR = \{M^{z+}\}^m \{X^{z-}\}^x / K_{sp}$ |
| Saturation index | SI | $\log_{10}(SR)$ |

Table 3.3—Nomenclature.

$$\mathrm{pH} = -\log_{10}\left\{\mathrm{H}^+\right\} = -\log_{10}\left[\mathrm{H}^+\right]\gamma_{\mathrm{H}^+} \quad (3.37)$$

Note that the definition of pH is based on the activity of $H^+$. An activity-based definition is adopted because pH is measured in the laboratory by ion-selective electrodes that respond to activity.

**3.6.2 Solubility of Salts.** Equilibrium constants are very helpful in describing solubility of salts in water. The solubility (S) of a salt is the mass of the salt dissolved in a known weight (or volume) of fluid **(Table 3.3).** Consider for example the dissolution of sodium chloride (NaCl). The chemical equation

$$\mathrm{NaCl(s)} \leftrightarrow \mathrm{Na^+(aq)} + \mathrm{Cl^-(aq)}, \quad (3.38)$$

states that the dissolved sodium and chloride ions are in equilibrium with the solid NaCl salt. The forward reaction is the dissolution of NaCl(s) to produce one cation ($Na^+$) and one anion ($Cl^-$). The reverse reaction is the combination of the cation and anion to yield solid NaCl. When there are enough sodium and chloride ions present in the solution to exceed the solubility of the salt, a solid [NaCl(s)] forms.

For salts where there are different numbers of charges on the cations and anions, the generalized equation is

$$M_m X_x(\mathrm{s}) \leftrightarrow mM^{z+}(\mathrm{aq}) + xX^{z-}(\mathrm{aq}) \quad (3.39)$$

Here, $m$ is the number of the cations with a charge $z+$ and $x$ is the number of the anions with a charge of $z-$.

Like the dissociation constants of acids, the equilibrium constants for dissolution of salts is given special notation. For example, consider the equilibrium constant for the general dissolution reaction expressed in Eq. 3.39:

$$K_{\mathrm{eq}} = \frac{\{M^{z+}\}^m \{X^{z-}\}^x}{\{M_m X_x\}} \quad (3.40)$$

Since the activity of a pure solid is by definition equal to unity, Eq. 3.40 can be simplified to

$$K_{\mathrm{eq}} = K_{\mathrm{sp}} = \{M^{z+}\}^m \{X^{z-}\}^x, \quad (3.41)$$

where $K_{\mathrm{sp}}$ is the equilibrium constant for the dissolution of the salt, and is called the *solubility product* **(Table 3.4).** If the salt is dissolving into a dilute solution (i.e., into an ideal solution), then activities of the species are equal to their concentration and $K_{\mathrm{sp}}$ can then be expressed as

$$K_{\mathrm{sp}} = [M^{z+}]^m [X^{z-}]^x \quad (3.42)$$

| $M^{2+}$ | Radius ($10^{-8}$ cm) | Fluorides | Hydroxides | Carbonates | Sulfates |
|---|---|---|---|---|---|
| Mg | 0.31 | V.S.* | $1.2\times10^{-11}$ | $1.2\times10^{-3}$ | 2.4 |
| Ca | 0.65 | $2\times10^{-4}$ | $2.1\times10^{-2}$ | $1.5\times10^{-4}$ | $1.5\times10^{-2}$ |
| Sr | 1.13 | $1\times10^{-3}$ | $6.5\times10^{-2}$ | $7.0\times10^{-4}$ | $5.0\times10^{-3}$ |
| Ba | 1.35 | $1\times10^{-2}$ | $2.8\times10^{-1}$ | $1.0\times10^{-4}$ | $1.0\times10^{-5}$ |

Moles per 1000 g water at 1 atmosphere and saturated with $CO_2$ for carbonates
* Very soluble

Table 3.4—Solubility of alkaline earth salts (Becker 1998). Courtesy of PennWell Publishing Inc.

An extensive compilation of solubility data is provided by Linke (1958). $K_{sp}$ values for various ionic compounds can be obtained from the *CRC Handbook of Chemistry and Physics* (Lide 2006) or from the *Lange's Handbook of Chemistry* (Dean 1992). These data are helpful in assessing the scaling potential of oilfield brines.

In most brines encountered in the oil field, the ideal-solution approximation is not valid, and species activities must be used instead in equilibrium calculations. The activities of the dissolved ions at equilibrium are a function of temperature, pressure, and brine composition. Many models are available for calculating activities of species. These are discussed in textbooks on aqueous thermodynamics; for example, see Zemaitis et al. (1986). The various software programs can be used to automate these calculations (Frenier and Ziauddin 2008).

A further complicating factor is that many salt solutions can exist in a state of supersaturation (i.e., the concentrations of cations and anions in the solution are much larger than those predicted by chemical-equilibrium calculations). Supersaturated solutions can exist for a long time and may not produce a scale unless initiated, for example, by seed crystal or an appropriate surface. A useful quantity for such solutions is the saturation ratio (SR):

$$\mathrm{SR} = \{M^{z+}\}^m \{X^{z-}\}^x / K_{sp} \qquad (3.43)$$

SR for a salt measures the degree of its supersaturation. Precipitation can only occur when SR > 1.0. Much of the art and science of predication of scaling involves calculation of the real SR in complex brines. The log(SR) is called the saturation index (SI). Various correlations are available in the literature for estimating SI in terms of easily measurable quantities such as pH, alkalinity, and concentration of dissolved ions and ionic strength of the fluid. For example, the SI from Stiff and Davis (1952),

$$\mathrm{SI} = \mathrm{pH} - k - \mathrm{pCa} - \mathrm{pAlk}, \qquad (3.44)$$

is a relationship developed for predicting precipitation of calcium carbonate from brines. The term $k$ accounts for the ionic strength and species activities. Note that this is *not* the equilibrium coefficient. For downhole fluids, the SR values are usually calculated with geochemical models. Kan (2005) notes that calculations of solubilities in concentrated brines are frequently performed at standard temperatures and pressures, and this may lead to errors in estimating the SI values. The book on oilfield waters by Davies and Scott (2006) describes a number of saturation indices or scaling indices.

**3.6.3 Fluid-Mineral Reactions.** ***Carbonate-Acid Reaction.*** Lund et al. (1973a) measured the dissolution kinetics of dolomite in HCl from 25°C to 100°C with a rotating-disk system. They found that at 25°C, the dissolution of dolomite was surface-reaction-rate limited even at low disk-rotation speeds. At 100°C, the dissolution was mostly diffusion limited even at relatively high (500 rev/min) rotation speeds. The reaction parameters they determined are

$$\text{Dolomite-HCl Reaction: } 4HCl + CaMg(CO_3)_2 \rightarrow CaCl_2 + MgCl_2 + 2H_2O + 2CO_2, \qquad (3.45)$$

$$\text{Rate Law: } -r\text{HCl}\ \frac{\text{g mol HCl}}{\text{cm}^2/\text{s}} = 9.4\times10^{10}\, e^{-E/RT} C_{\text{H}^+}^{n}, \quad (3.46)$$

$$\text{where } E = 22.5\frac{\text{kcal}}{\text{g mol}},\ R = 1.987\times10^{-3}\frac{\text{kcal}}{\text{g mol}\cdot\text{K}}, \text{ and } n = \frac{6.2\times10^{-4}T}{\left(1-2\times10^{-3}T\right)} \quad (3.47)$$

They proposed a reaction mechanism with adsorption of hydrogen ion on the solid dolomite surface and subsequent reaction of the adsorbed hydrogen ion with the solid dolomite matrix. Lund et al. (1973) explained the unusual dependence of the reaction rate on a temperature-dependent fractional power of HCl by the Freundlich adsorption of $H^+$ onto the dolomite surface.

The rate of surface reaction of HCl with calcite is much faster than that with dolomite. Ideally, to measure the kinetics of the surface reaction, the experimental conditions have to be such that the surface reaction is the rate-limiting step. In particular, the rate of mass transfer to the surface has to be much faster than the surface reaction so that the overall reaction is independent of mass transfer. Lund et al. (1975) measured the kinetics of HCl-calcite reaction with a rotating-disk apparatus at 25°C and 800 psig and found it to be completely limited by mass transfer even at high rotation speeds. They then lowered the reaction temperature to –15.6°C in an attempt to make the system surface reaction limited. Sodium chloride was added to the acid solution to lower the freezing point. However, even at –15.6°C, the overall reaction rate was not completely independent of mass-transfer limitations. They used a multicomponent model to account for mass-transfer rates with reasonable accuracy and then provided an estimate for the surface-reaction rate.

The reaction parameters they determined are

$$\text{Calcite-HCl Reaction: } 2\text{HCl} + \text{CaCO}_3 \rightarrow \text{CaCl}_2 + \text{H}_2\text{O} + \text{CO}_2, \quad (3.48)$$

$$\text{Rate Law: } -r_{\text{HCl}}\ \frac{\text{g mol HCl}}{\text{cm}^2/\text{s}} = 7\times10^{6}\, e^{-E/RT} C_{\text{H}^+}^{0.63}, \quad (3.49)$$

$$\text{where } E = 15\frac{\text{kcal}}{\text{g mol}} \text{ and } R = 1.987\times10^{-3}\frac{\text{kcal}}{\text{g mol}\cdot\text{K}}. \quad (3.50)$$

Similar to the reaction mechanism of dolomite, Lund et al. (1975) proposed a mechanism with adsorption of the hydrogen ion onto the solid calcite surface and the subsequent reaction of the adsorbed hydrogen ion with the solid calcite matrix. Lund et al. (1975) explained the unusual dependence of the reaction rate on the fractional power of HCl by the Freundlich adsorption of $H^+$ onto the calcite surface. However, because of limited rate data in the surface-reaction-dominated regime, they were unable to determine the temperature dependence of the order of the reaction.

Under reservoir conditions, the reaction of calcite with HCl will almost always be mass-transfer limited and, therefore, the exact kinetics of the surface reaction are generally not required. Under mass-transfer-limited kinetics, the surface concentration of the $H^+$ is much smaller compared with the concentration in the bulk fluid. The expression for the reaction-rate expression then simplifies to

$$-r_{\text{HCl}}\ \frac{g\ mol\ HCl}{cm^2/s} = k_m C_{H^+}, \quad (3.51)$$

where $k_m$ is the mass-transfer coefficient. Note that the HCl-calcite reaction is first order in $C_{\text{H}^+}$ under these conditions. Highly concentrated acids are typically used in acidizing treatments, and the acid activity plays an important role. **Fig. 3.22** depicts the reaction rate as a function of acid concentration. Note the existence of a maximum reaction rate with respect to HCl concentration.

Schechter (1990) extended this HCl-calcite and HCl-dolomite rate law to weak acids, such as acetic and formic acids, by assuming that the reaction of HCl with carbonate minerals is actually a reaction

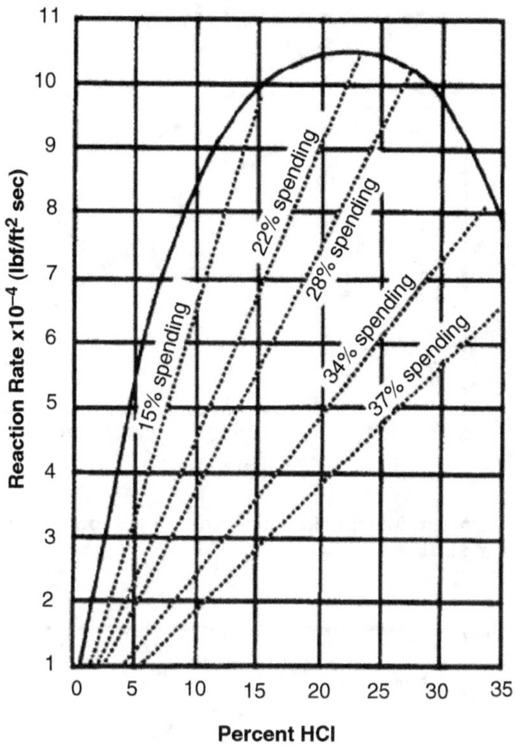


**Fig. 3.22—Effect of concentration of HCl on reaction rate and spending rate (Coulter et al. 1992).**

of H$^+$ with the mineral, and the dominant factor in determining the rate of a weak acid is the limited dissociation of the weak acid. He computed an effective H$^+$ concentration, and for weak acids he proposed the following rate law:

$$\text{Rate Law: } -r\text{weak acid } \frac{\text{gmol weak acid}}{\text{cm}^2/\text{s}} = k_{\text{HCl}} K_d^{\alpha/2} C_{\text{weak acid}}^{\alpha/2}, \quad (3.52)$$

where $k_{\text{HCl}}$ is the rate constant for the HCl-calcite or the HCl-dolomite reaction and $\alpha$ is the order of the reaction with respect to H$^+$ for the calcite or dolomite reaction. $K_d$ denotes the dissociation constant of the weak acid. At typical values of $K_d$ for formic acid, the surface-reaction rate is approximately two orders of magnitude slower than the reaction with HCl. The reaction rate of acetic acid is even slower than that of the formic acid.

Fredd and Fogler (1998b) studied the kinetics of acetic acid using a rotating-disk apparatus at 22–50°C and 800 psig and over a wide range of pH values. At pH values below 2.9, the dissolution is influenced by the rate of transport of reactants to the surface and the rate of transport of products away from the surface. The influence of the rate of transport of products is because of the overall reaction being reversible. The interplay between the two transport processes results in transport limitations that are much more significant than observed with either limitation independently. At pH values greater than about 3.7, the kinetics of the surface reaction also affected the rate of dissolution. They proposed a general model to account for the combined effects of the transport of reactants and products and the surface reaction.

A further complication with reaction of weak acids with calcite or dolomites is that the reaction does not usually go to completion under reservoir conditions because of the equilibrium limitations. This is because $CO_2$, a byproduct of the reaction, is held in the solution by reservoir pressure. The reaction does go to completion at low pressures in which $CO_2$ is allowed to escape. Chatelain et al. (1976) studied these equilibrium limitations of organic acids with carbonate minerals with a system

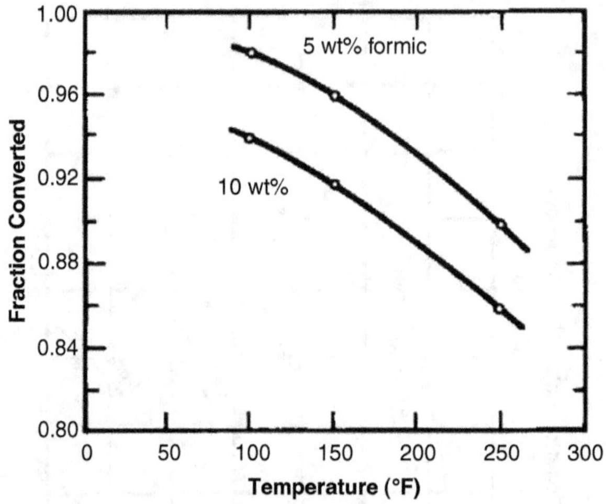


**Fig. 3.23—Effect of temperature and initial acid concentration on conversion in the formic acid/calcium carbonate system (Chatelain et al. 1976).**

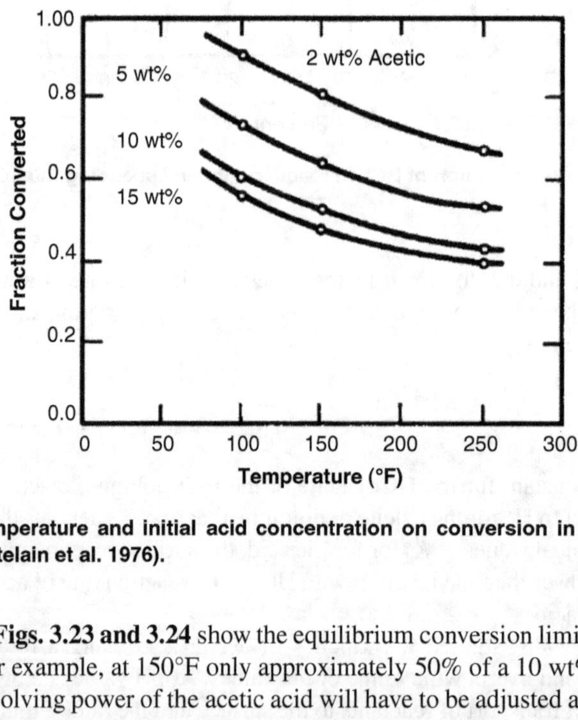


**Fig. 3.24—Effect of temperature and initial acid concentration on conversion in the acetic acid/calcium carbonate system (Chatelain et al. 1976).**

pressure of 1,000 psi. **Figs. 3.23 and 3.24** show the equilibrium conversion limits for formic and acetic acids, respectively. For example, at 150°F only approximately 50% of a 10 wt% acetic acid will react and, therefore, the dissolving power of the acetic acid will have to be adjusted accordingly. In practice, formic acid is generally not used above 10% concentration because of limited solubility of calcium formate and magnesium formate. Acetic acid is also generally not used above 10% because of the low conversion, even though calcium acetate is highly soluble in spent acid.

Williams et al. (1979) found that the equilibrium conversion of weak acids can be predicted approximately by the empirical equation

$$1.6\times10^4 K_a = \frac{C_{\mathrm{CaA_2}} C_{\mathrm{CO_2}}}{C_{\mathrm{HA}}}, \qquad (3.53)$$

where the $C_{\mathrm{CaA_2}}$ is the concentration of the salt of the weak acid HA. All concentrations in the equation are expressed in g mol/kg of water. Equilibrium conditions for dolomite are nearly the same as for

| Acid | | | Relative |
|---|---|---|---|
| Concentration (%) | Type | $CaCO_3$ Equivalent (lbm/1,000 gal acid) | Reaction Time* |
| 7.5 | HCl | 890 | 0.7 |
| 15 | HCl | 1,840 | 1.0 |
| 28 | HCl | 3,670 | 6.0 |
| 36 | HCl | 4,860 | 12.0 |
| 10 | Formic | 910 | 5.0 |
| 10 | Acetic | 710 | 12.0 |
| 15 | Acetic | 1,065 | 18.0 |
| 7.5<br>14 | Formic/<br>HCl mixture | 2,420 | 6.0 |
| 10<br>14 | Acetic/<br>HCl mixture | 2,380 | 12.0 |
| 8<br>14 | Formic/<br>Acetic mixture | 1,700 | 18.0 |

**Fig. 3.25—Reactions of different acidizing solutions with calcium carbonate (Coulter et al. 1992).**

limestone; therefore, a similar equation should also apply to dolomite (Chatelain et al. 1976). Besides the lower conversion of organic acids, the reaction rate is also lower. **Fig. 3.25** lists the relative reaction time of various acid blends with $CaCO_3$. The values are compared by using spending time of 15% HCl as unity.

Fredd and Fogler (1998a) studied the kinetics of calcite dissolution in the presence of calcium chelating agents at 21°C and 800 psig over a pH range of 3.3–12 using a rotating-disk apparatus. They found that the rate of dissolution increased significantly because of the presence of the chelating agent. The overall rate of dissolution is influenced by the combined effects of the hydrogen ion reaction, the chelating reactions, and the water reaction. The reaction that dominated the dissolution depends on the pH of the solution and the concentration and type of the chelating agent present. At low concentrations, the rate of dissolution is limited by the rate of transport of the chelating agent to the surface. An additional influence of the rate of transport of the calcium complex away from the surface is observed because of the competition for adsorption sites. The hydrogen ion reaction becomes significant at pH values below approximately 4. The water reaction was found to be negligible at conditions investigated in this study but is expected to be significant at extremely low chelating-agent concentrations and in the presence of weak complexing agents such as acetate. They found that the rate of dissolution varied considerably with pH, concentration, and type of chelating agent primarily because of the changes in the ionic species involved in the surface reaction. They proposed a surface chelating mechanism to describe these effects.

***Aluminosilicates-Hydrochloric Acid, -Formic Acid, -Acetic Acid Reactions.*** Unlike calcite, aluminosilicate minerals are not truly soluble in HCl; however, exposure to HCl does affect their structure. Typically, metal ions such as Al, Mg, Fe, Na, K, Ca are leached from the mineral structure leaving behind an amorphous silica residue. The extent of this reaction is often used as a measure of stability of aluminosilicates in acid. Weaker acids such as formic and acetic acids have a much lower concentration

| Acid Type and Strength | Percent Chloride Destroyed at 180°F<br>Exposure Time (hours) 1 | 2 | 4 | 10 | 20 | 50 | 120 |
|---|---|---|---|---|---|---|---|
| 15% HCl | 65 | 95 | 99+ | 100 | 100 | - | - |
| 7.5% HCl | 45 | 95 | 99+ | 100 | 100 | - | - |
| 3% HCl | 18 | 70 | 95 | 99+ | 100 | - | - |
| 10% Formic | 0 | 25 | 50 | 50 | 70 | 85 | 85 |
| 10% Acetic | 0 | 0 | 0 | 0 | 5 | 10 | 20 |

**Fig. 3.26—Stability of chlorite in acids at 180°F (Simon and Anderson 1990).**

Ions Leached From Chlorite by Various Acids
(120 hours exposure at 180°F)

| Fluid | Ionic Concentration (mpl) | | | | | Amorphous Material | Percent Chlorite Destroyed |
|---|---|---|---|---|---|---|---|
| | Mg | Al | Si | Ca | Fe | | |
| Deionized water | 1 | 1 | 7 | 6 | 1 | 0% | 0% |
| 10% acetic | 27 | 11 | 26 | 35 | 53 | 4% | 30% |
| 10% formic | 158 | 214 | 110 | 73 | 460 | 35% | 92% |
| 3% HCl | 158 | 265 | 125 | 65 | 565 | 55% | 100% |

**Fig. 3.27—Ions leached from chlorite by various acids at 180°F (Simon and Anderson 1990).**

of hydronium ion (higher pH), and stability of aluminosilicates is generally much higher in these acids. Stability of aluminosilicate minerals in other acids can be estimated from the $pK_a$ of the acid.

The leaching rates of metal ions from the mineral structure upon exposure to acid have been measured by several researchers. Ross (1969) studied the stability of eight chlorite clays in Si-saturated HCl solution. The rate of dissolution increased with increasing Fe and decreasing Mg content of the chlorite. The residue that remained after acid attack appeared to be an opaline amorphous hydrated silica. Simon and Anderson (1990) studied the stability of chlorite, illite, and kaolinite in hydrochloric, formic, and acetic acids. They also found that strong acids will leach Fe, Mg, and Al from chlorite clay by destroying the crystal structure and leaving behind an amorphous residue **(Fig. 3.26).** The stability was monitored by X-ray-diffraction. The clay slurries in each acid system and deionized water were filtered after 120 hours of exposure. Cations in solution were determined with atomic adsorption spectrometry **(Fig. 3.27).** The percent amorphous material was estimated by comparing the X-ray diffraction patterns to those of glass.

Hartman et al. (2006) studied the stability of kaolinite, analcime (a zeolite), chlorite, and illite in 15 wt% HCl at 25 and 100°C (77 and 212°F). They monitored the reaction progress by measuring the concentration of Al, Si, Na, or Mg ions in the aqueous phase **(Figs. 3.28 and 3.29).** Analcime was found to be the least-stable mineral, followed by chlorite, illite, and kaolinite. For analcime, the leaching rate of Al and Na were similar and stoichiometric; however, Si was first released from the structure but later precipitated (Fig. 3.29b). This suggests a multistep process comprising an initial dissolution followed by precipitation reactions. The reaction progress was modeled in a geochemical simulator using a combination of kinetics and equilibrium-controlled reactions. For example, for kaolinite the heterogeneous leaching of Al from the clay structure was modeled by the kinetics-controlled reaction,

$$12\,H^+_{(aq)} + Al_4Si_4O_{10}(OH)_{8(s)} \rightarrow 4\,Al^{+3}_{(aq)} + 2\,H_2O_{(aq)} + 4\,SiO_2 \times 2H_2O(s), \qquad (3.54)$$

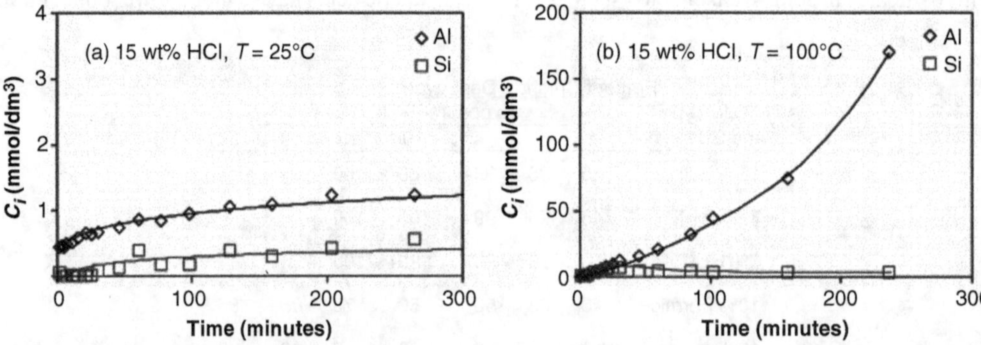


**Fig. 3.28—Elemental concentrations of Al and Si for the dissolution of kaolinite in 15 wt% HCl at 25 and 100°C (Hartman et al. 2006).**

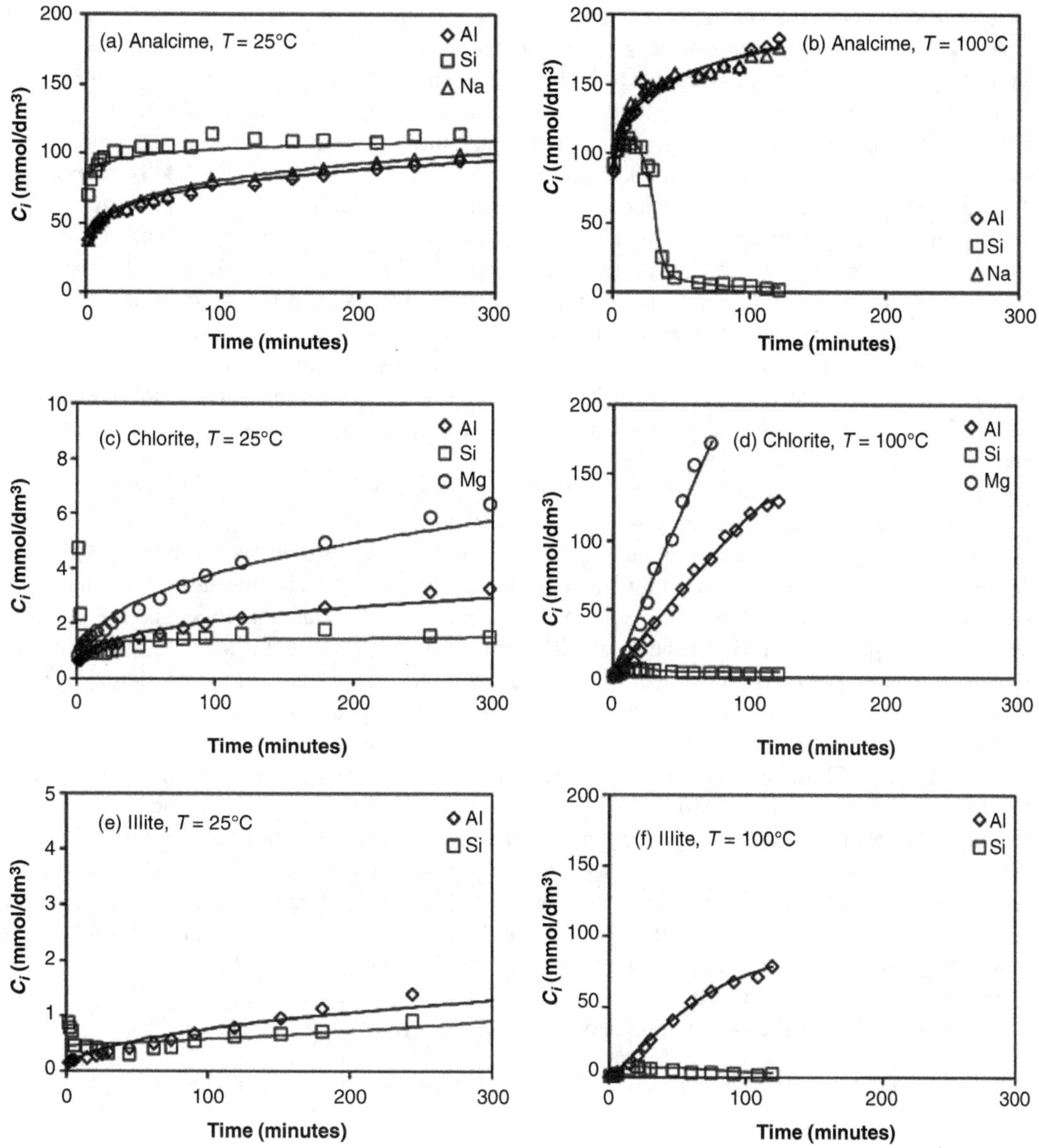


**Fig. 3.29—Elemental concentrations of Al, Si, Na, and Mg for the dissolution of analcime, chlorite, and illite in 15 wt% HCl at 25 and 100°C (Hartman et al. 2006).**

while the concentration of aqueous species concentrations such as for H+ was computed using equilibrium reactions.

***Quartz-HF.*** The reaction of quartz with HF is a first-order reaction with respect to HF (Hill et al. 1981; Blumberg and Stavrinou 1960):

$$6HF(aq) + SiO_2(s) \rightarrow H_2SiF_6\,(aq) + 2H_2O(aq), \quad (3.55)$$

$$\text{Rate} \propto k[HF]. \quad (3.56)$$

The reaction rate is much slower than the reaction of HF with feldspars and clays. However, it is not negligible and limits the depth surrounding the wellbore from which damage can be removed, and, therefore, it is critical in acid design (Hill et al. 1981; Schechter 1990).

***Aluminosilicate-Mud-Acid Reactions.*** The reaction of an aluminosilicate mineral with mud acid is typically limited by the kinetics of the surface reaction. It is a multistep process. Initially, HF reacts with the aluminosilicate mineral. Al and Si in the mineral are solubilized as complexes with F because of the affinity of the fluoride ion for these elements (Labrid 1975). This reaction of HF with the aluminosilicate mineral is known as the primary reaction. Other metals such as Na and K, which could be either adsorbed on the surface or substituted in the mineral structure, are leached or released as the reaction proceeds. Some of the salts of these metals, such as $K_2SiF_6$, $Na_2SiF_6$, and $Na_3AlF_6$, have low solubility and may precipitate out. These are known as primary precipitates. The extent of this precipitation can be estimated by solubility calculations described in Section 3.6.2. The formation of primary precipitates can be reduced by an ammonium chloride preflush before the mud-acid stage. The ammonium ion displaces the adsorbed metal ions from the surface. The precipitation is reduced as ammonium fluosilicate ($(NH_4)_2SiF_6$) and has a much higher solubility than the sodium and potassium fluosilicates. However, metal ions substituted in the aluminosilicate structure cannot be displaced by $NH_4Cl$ preflush, and careful selection of the mud-acid system is critical in avoiding excessive precipitation.

In excess acid, most of aluminosilicate eventually dissolves and the mass of the primary precipitates, if any, is smaller than mass of the original mineral. However, almost all acidizing treatments performed in the field have an excess of aluminosilicate, and Al-F and Si-F complexes react further with the aluminosilicate mineral, often leading to damaging precipitates. This is especially true as the spent acid moves deeper into the reservoir. The affinity of F to Al is higher than that to Si, and F is released from Si-F complexes first. The reaction of Si-F complex with the aluminosilicate is the secondary reaction (Gdanski 1999). The net effect of this reaction is that additional Al from the aluminosilicate is brought into solution (i.e., the Al-F ratio increases) and Si precipitates as silica gel. The reactions of Al-F complexes with aluminosilicates have been reported as tertiary reactions (Shuchart and Buster 1995; Gdanski 1998). In this reaction, F from the Al-F complex reacts with the aluminosilicate to yield an Al-F complex with a higher Al-F ratio, in addition to silica gel and other precipitates. The primary reaction is faster than the secondary reaction, which in turn is faster than the tertiary reaction.

***Primary Reactions.*** The rates of primary reaction of clays, feldspars, and zeolites with HF have been reported (Lund et al. 1973; Hekim and Fogler 1977; Kline and Fogler 1981a; Kline and Fogler 1981b; Hartman et al. 2006). A detailed description of the primary reaction of each of these minerals is outside of the scope of this chapter. The primary reaction of kaolinite and feldspar is described next to illustrate the principle involved.

*Kaolinite*: The primary dissolution reaction for kaolinite is (Hekim and Fogler 1977)

$$Al_4Si_4O_{10}(OH)_8(s) + xHF_{(aq)} + yH^+ \rightarrow 4\,AlF_n^{+(3n)}(aq) + 4\,SiF_m^{+(4-m)}(aq) + 18\,H_2O(aq) \quad (3.57)$$

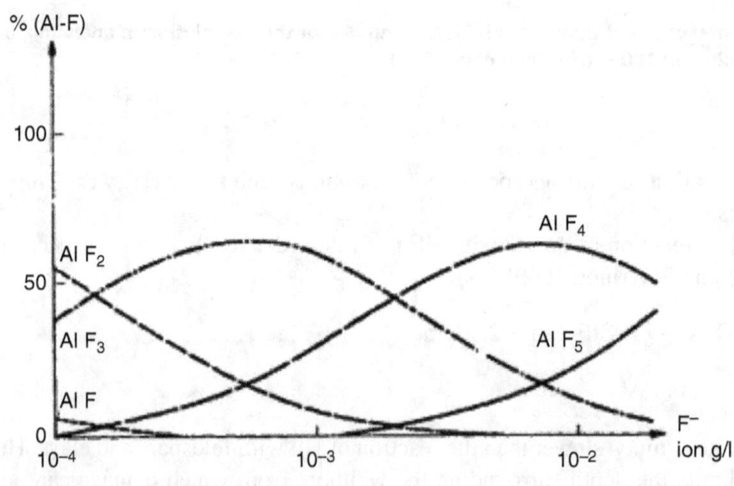


**Fig. 3.30—Theoretical Range of Al-F complexes with ionic fluoride concentration (Labrid 1975).**

Once in solution, Al and Si are distributed in various complexes (Hekim and Fogler 1977). For example, the total dissolved Al can be in several fluoride complexes:

$$[Al] = [Al^{3+}] + [AlF^{2+}] + [AlF_2^{+}] + [AlF3(aq)] + [AlF_4^{-}] + [AlF_5^{2-}] + [AlF_6^{3-}] + \ldots \quad (3.58)$$

The concentration of each complex depends on the total fluoride ion in solution **(Fig. 3.30).** Similarly, Si is also distributed in several complexes, depending on the fluoride concentration:

$$[Si] = [H_4SiO_4] + [SiF_4(aq)] + [SiF_5^{-}] + [SiF_6^{2-}]. \quad (3.59)$$

The determination of the reaction stoichiometry generally requires the use of a geochemical simulator (Hekim and Fogler 1977; Hartman et al. 2006).

*Feldspar:* Fogler et al. (1975) studied the primary dissolution reaction of potassium feldspar and sodium feldspar in HCl-HF acid mixtures with a rotating-disk apparatus at 25° and 100°C under a pressure of 40 psig. The composition of the acid mixture was varied from 0 to 2.5 g mol/L for HF and was 0–5 g mol/L for HCl. Under these conditions, the dissolution rate is limited by the rate of reaction of the acids with the solid surface and is not affected by mass-transfer limitations. For both feldspar types, they proposed the rate law of the form

$$-r_A = k\left(1 + KC_{HCl}^{a}\right)C_{HF}^{b}, \quad (3.60)$$

where the $-r_A$ represents the g mol of feldspar dissolved/cm$^2$-s, and $C_{HCl}$ and $C_{HF}$ indicate the concentrations of HCl and HF, respectively, in g mol/L.

For potassium feldspar, the parameters have the values

$$a = 0.4,\ b = 1.2,\ k = 1.25\times10^{-2} e^{\frac{-9.3}{RT}},\ \text{and}\ K = 5.7\times10^{-2} e^{\frac{1.9}{RT}}, \quad (3.61)$$

where $R$ is the gas constant equal to $1.987\times10^{-3}$ kcal/g mol·K, and the $T$ is the reaction temperature in K.

Similarly, for sodium feldspar the parameters have the values

$$a = 1,\ b = 1,\ k = 9.4\times10^{-4} e^{\frac{-7.8}{RT}},\ \text{and}\ K = 6.3\times10^{-2} e^{\frac{1.1}{RT}}. \quad (3.63)$$

The proposed mechanism suggests that the dissolution is limited in HF-HCl acid mixtures by the rate at which the hydrofluoric acid reacts with the solid surface. The strong effect of HCl on the dissolution rate is a result of adsorption of H$^+$ onto the mineral surface. The dissolution rates of the feldspars are only slightly different, as would be expected.

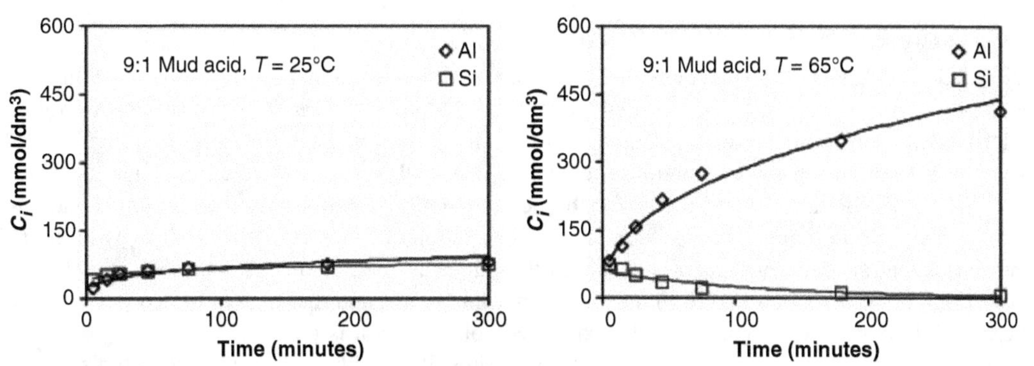


**Fig. 3.31—Elemental concentration of Al, Si for the dissolution of kaolinite clay in 9:1 mud acid (Hartman et al. 2006).**

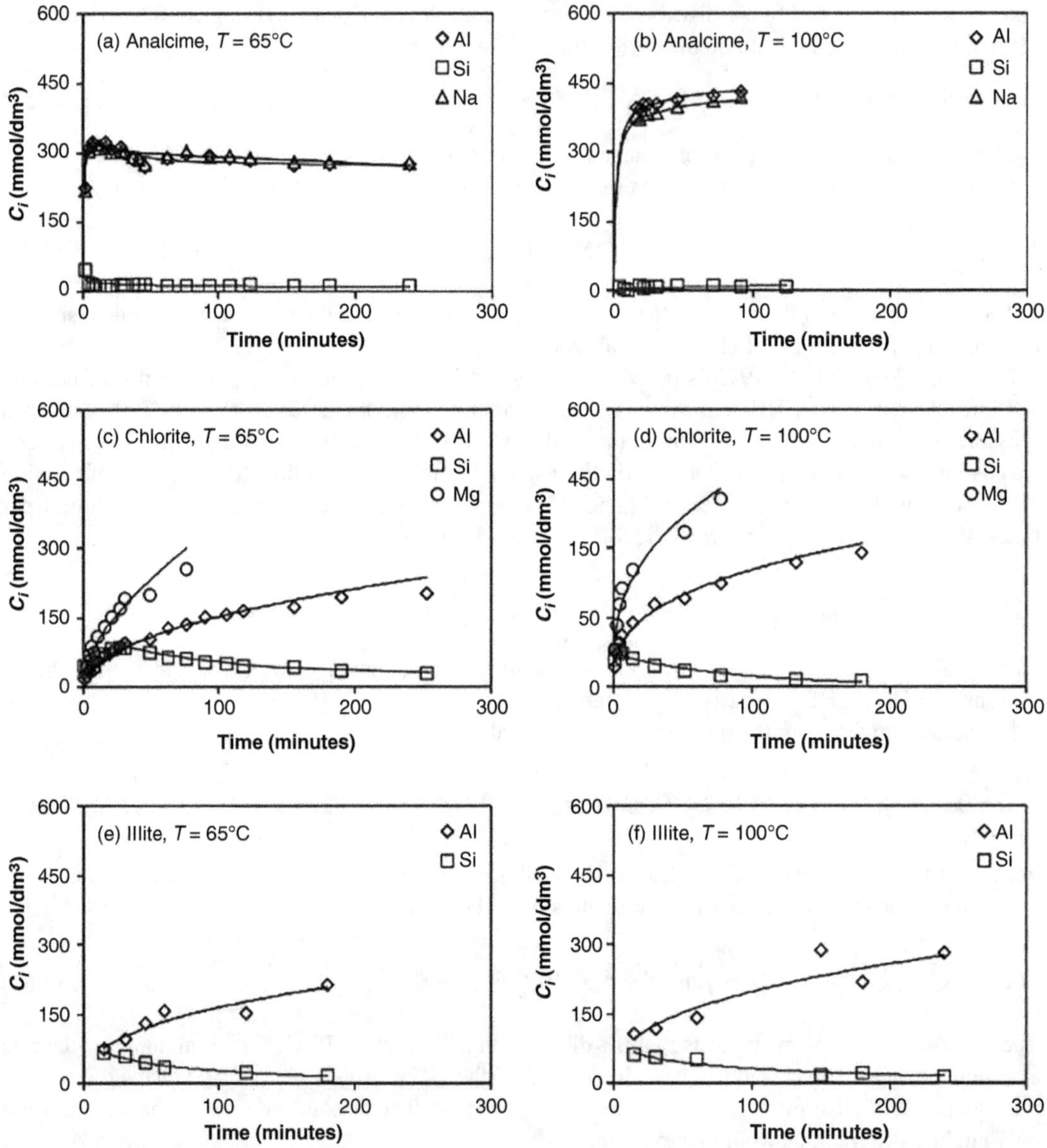


**Fig. 3.32—Elemental concentration of Al, Si, Na, and Mg for the dissolution of analcime, chlorite, and illite in 9:1 mud acid (Hartman et al. 2006).**

***Secondary Reaction.*** The reaction of Si-F complex with aluminosilicates is known as the secondary reaction. Gdanski (1999) modeled the secondary reaction as a kinetics-only reaction, with $HSiF_5$ as the effective reactive species. Hartman et al. (2006) studied the secondary reaction of kaolinite, chlorite, illite, and analcime with 9:1 mud acid in a slurry reactor. They followed the progress of the reaction by monitoring the concentration of Al, Si, Na, and Mg ions in solution. They used $SiF_6^{2-}$ or $H_2SiF_6$ as the main reactive species for the heterogeneous reaction with the aluminosilicate, and they used equilibrium models to calculate the concentration of $SiF_6^{2-}$ or $H_2SiF_6$ at any given time. Al and Si concentration during reaction of kaolinite with a 9:1 mud acid at 25 and 65°C are shown in **Fig. 3.31.** The molar ratio of Al:Si in unreacted kaolinite is 1:1, and the same ratio is observed during the dissolution at 25°C. No appreciable drop in Si concentration is observed during the experiment (≈5 hours). At 65°C, however, the concentration of Si peaks quickly and later drops as the secondary reaction progresses. The Al concentration continues to evolve because of a combination of primary and secondary reactions.

The reaction of analcime, chlorite, and illite with 9:1 mud acid at 65 and 100°C is shown in **Fig. 3.32** (Hartman et al. 2006). The concentration of Si peaked quickly because of the rapid primary reaction but then decayed as the secondary reaction progressed. However, the concentration of Al, Na, and Mg generally continued to increase as the secondary reaction progressed. These data from slurry reactor tests also provide the reaction time scales for primary and secondary reactions of these minerals. For example, at 25°C there is no appreciable secondary reaction of kaolinite until ≈300 minutes, and for chlorite clay, at 65°C it seems to be a in the tens of minutes. However, for analcime it seems to be very rapid, even at 65°C. In typical core-flow tests, the residence time of treatment fluids in the core is on the order of a few minutes and it may be difficult to quantify secondary reactions, especially if only short formation cores are available. Ziauddin et al. (2005) describe a method of combining slurry reactor data with data from short cores to quantify secondary reactions. In their method, the core-flow tests are only used to calculate the mineral surface area, which is then combined with reaction kinetics data from slurry reactors to predict performance of acidizing treatments.

***Tertiary Reactions.*** Shuchart and Buster (1995) identified tertiary reactions of Al-F complexes with aluminosilicates. They found that after silicon fluorides had reacted completely to produce silica gel (secondary reaction), aluminum fluorides continued to react with fresh aluminosilicates causing the Al/F ratio to increase and the acid concentration to decrease. Gdanski (1998) modeled this process as a kinetically controlled reaction of $AlF_2^+$ with the aluminosilicate. Gdanski and Shuchart (1998) successfully used these reactions in a fully kinetic geochemical model to simulate acidizing treatments. Geochemical simulators based solely on kinetics, require rate laws for all reactions of each mineral as well as reactions between aqueous species. These rate laws may not be readily available. Hartman et al. (2006) simulated tertiary reactions with a geochemical simulator using equilibrium and kinetically controlled reactions. In their model, the conversion of higher aluminum fluorides to lower aluminum fluorides and HF is modeled as an equilibrium process. The HF thus released reacted with the aluminosilicate through the kinetically controlled primary reaction. The equilibria between aluminum fluorides and other species are computed using thermodynamic constants, which are more readily available than kinetic constants. No separate reaction rate laws for the tertiary reactions were required.

## References

Ahr, W. M. 2008. Geology of Carbonate Reservoirs: The Indentification, Description, and Characterization of Hydrocarbon Reservoirs in Carbonate Rocks, first edition. Hoboken, New Jersey, USA: John Wiley & Sons, Inc. http://dx.doi.org/10.1002/9780470370650.

Ahr, W. M., Allen, D., Boyd, A. et al. 2005. Confronting the Carbonate Conundrum. *Oilfield Review* (Spring 2005): 18–29.

Akbar, M., Petricola, M., Watfa, M. et al. 1995. Classic Interpretation Problems: Evaluating Carbonates. *Oilfield Review* **7** (1): 38–57.

Archie, G. E. 1952. Classification of Carbonate Reservoir Rocks and Petrophysical Considerations. *AAPG Bull* **36** (2): 278–298.

Barron, A. N., Hendrikson, A. R., and Wieland, D. R. 1962. The Effect of Flow on Acid Reactivity in a Carbonate Fracture. *J Pet Technol* **14** (4): 409–415. SPE-134-PA. http://dx.doi.org/10.2118/134-PA.

Becker, J. R. 1998. *Corrosion and Scale Handbook*, first edition. Tulsa: PennWell Publishing Company.

Blumberg, A. A. and Stavrinou, S. C. 1960. Tabulated Functions for Heterogeneous Reaction Rates: The Attack of Vitreous Silica by Hydrofluoric Acid. *The Journal of Physical Chemistry* **64** (10): 1438–1442. http://dx.doi.org/10.1021/j100839a019.

Boggs, S. Jr. 1992. *Petrology of Sedimentary Rocks*, third edition. New York: Macmillan.

Boomer, D. R., McCune, C. C., and Fogler, H. S. 1972. Rotating Disk for Reaction Rate Studies in Corrosive Liquid Environments. *Rev Sci Instrum* **43** (2): 225–229. http://dx.doi.org/10.1063/1.1685599.

Chatelain, J. C., Silberberg, I. H., and Schechter, R. S. 1976. Thermodynamic Limitations in Organic-Acid/Carbonate Systems. *SPE J.* **16** (4): 189–195. SPE-5647-PA. http://dx.doi.org/10.2118/5647-PA.

Chilingarian, G. V. 1963. Relationship between Porosity, Permeability, and Grain Size Distribution of Sediments. *Proc.*, International Sedimentological Congress, Amsterdam, Summer 1963, 71–75.

Chilingarian, G. V. and Wolf, K. H. ed. 1975. *Compaction of Coarse-Grained Sediments, I*, Vol. 18A. Amsterdam: Developments in Sedimentology, Elsevier Scientific Publishing Company.

Chilingarian, G. V., Mazzullo, S. J., and Rieke, H. H. ed. 1992. *Carbonate Reservoir Characterization: A Geologic-Engineering Analysis, Part I*, first edition, Vol. 30 Amsterdam: Developments in Petroleum Science, Elsevier.

Choquette, P. W. and Pray, L. C. 1970. Geologic Nomenclature and Classification of Porosity in Sedimentary Carbonates. *AAPG Bull* **54** (2): 207–250.

Coulter, A. W. Jr., Hendrickson, A. R., and Martinez, S. J. 1992. Acidizing. In *Petroleum Engineering Handbook*. ed. H. B. Bradley. Chap. 54. Richardson, Texas: Society of Petroleum Engineers.

da Motta, E. P., Plavnik, B., and Schechter, R. S. 1990. Optimizing Sandstone Acidization. Presented at the SPE Formation Damage Control Symposium, Lafayette, Louisiana, USA, SPE-19426-MS.

Davies, M. and Scott, P. J. B. 2006. *Oilfield Water Technology*. Houston, Texas: NACE International.

Dean, J. A. ed. 1992. *Lange's Handbook of Chemistry*, 14th edition. New York: McGraw-Hill.

Deer, W. A. 1963. *Rock Forming Minerals: Framework Silicates*, Vol. 4. New York: John Wiley & Sons.

Dunham, R. J. 1962. Classification of Carbonate Rocks According to Depositional Textures. In *Classification of Carbonate Rocks—a Symposium*, edition. ed. W. E. Ham, 108–121. Tulsa: American Association of Petroleum Geologists.

Embry, A. F. and Klovan, J. E. 1971. A Late Devonian Reef Tract on Northeastern Banks Island, N.W.T. *Bull Can Petrol Geol* **19** (4): 730–781.

Fogler, H. S. 2005. *Elements of Chemical Reaction Engineering*, fourth edition. Upper Saddle River, New Jersey, USA: Prentice Hall International Series in the Physical and Chemical Engineering Sciences, Prentice Hall.

Fogler, H. S. and McCune, C. C. 1976. On the Extension of the Model of Matrix Acid Stimulation to Different Sandstones. *AIChE J* **22** (4): 799–805. http://dx.doi.org/10.1002/aic.690220426.

Fogler, H. S., Lund, K., and McCune, C. C. 1975. Acidization III—the Kinetics of the Dissolution of Sodium and Potassium Feldspar in HF/HCl Acid Mixtures. *Chem Eng Sci* **30** (11): 1,325–1,332. http://dx.doi.org/10.1016/0009-2509(75)85061-5.

Folk, R. L. 1962. Classification of Carbonate Rocks—a Symposium. In *Spectral Subdivision of Limestone Types*, first edition. ed. W. E. Ham, Vol. 1, Chap. 62–84. Tulsa: American Association of Petroleum Geologists.

Folk, R. L. 1974. *Petrology of Sedimentary Rocks*, second edition. Austin, Texas, USA: Hemphill Publishing Company.

Fredd, C. N. and Fogler, H. S. 1998a. The Influence of Chelating Agents on the Kinetics of Calcite Dissolution. *J Colloid Interface Sci* **204** (1): 187–197. http://dx.doi.org/10.1006/jcis.1998.5535.

Fredd, C. N. and Fogler, H. S. 1998b. The Kinetics of Calcite Dissolution in Acetic Acid Solutions. *Chem Eng Sci* **53** (22): 3,863–3,874. http://dx.doi.org/10.1016/S0009-2509(98)00192-4.

Frenier, W. W. and Ziauddin, M. 2008. *Formation, Removal, and Inhibition of Inorganic Scale in the Oilfield Environment*, first edition. Richardson, Texas, USA: Society of Petroleum Engineers.

Garrels, R. M., Thompson, M. E., and Siever, R. 1960. Stability of Some Carbonates at 250°C and One Atmosphere Total Pressure. *Am. J. Sci.* **258:** 402.

Gdanski, R. 1998. Kinetics of Tertiary Reactions of Hydrofluoric Acid on Aluminosilicates. *SPE Prod & Fac* **13** (2): 75–80. SPE-31076-PA. http://dx.doi.org/10.2118/31076-PA.

Gdanski, R. D. 1999. Kinetics of the Secondary Reaction of HF on Alumino-Silicates. *SPE Prod & Fac* **14** (4): 260–268. SPE-59094-PA. http://dx.doi.org/10.2118/59094-PA.

Gdanski, R. D. and Shuchart, C. E. 1998. Advanced Sandstone-Acidizing Designs with Improved Radial Models. *SPE Prod & Fac* **13** (4): 272–278. SPE-52397-PA. http://dx.doi.org/10.2118/52397-PA.

Gregory, D. P. and Riddiford, A. C. 1956. Transport to the Surface of a Rotating Disk. *J Chem Soc* **33** (0): 3756–3764. http://dx.doi.org/10.1039/JR9560001949.

Hartman, R. L., Lecerf, B., Frenier, W. W. et al. 2006. Acid-Sensitive Aluminosilicates: Dissolution Kinetics and Fluid Selection for Matrix-Stimulation Treatments. *SPE Prod & Oper* **21** (2): 194–204. SPE-82267-PA. http://dx.doi.org/10.2118/82267-PA.

Hekim, Y. and Fogler, H. S. 1977. Acidization—VI: On the Equilibrium Relationships and Stoichiometry of Reactions in Mud Acid. *Chem Eng Sci* **32** (1): 1–9. http://dx.doi.org/10.1016/0009-2509(77)80188-7.

Hill, A. D., Lindsay, D. M., Silberberg, I. H. et al. 1981. Theoretical and Experimental Studies of Sandstone Acidizing. *Society of Petroleum Engineers Journal* **21** (1): 30–42. http://dx.doi.org/10.2118/6607-pa.

Huang, T., Hill, A. D., and Schechter, R. S. 2000. Reaction Rate and Fluid Loss: The Keys to Wormhole Initiation and Propagation in Carbonate Acidizing. *SPE J* **5** (03): 287–292. SPE-65400-PA. http://dx.doi.org/10.2118/65400-PA.

Kan, A. T., Fu, G., and Tomson, M. B. 2005. Prediction of Scale Inhibitor Squeeze and Return in Calcite-Bearing Formation. Presented at the SPE International Symposium on Oilfield Chemistry, The Woodlands, Texas, USA, 2–4 February. SPE-93265-MS. http://dx.doi.org/10.2118/93265-MS.

Kee, R. J., Coltrin, M. J., and Glarborg, P. 2003. *Chemically Reacting Flow: Theory and Practice*, edition. Hoboken, New Jersey, USA: John Wiley & Sons, Inc.

Kline, W. E. and Fogler, H. S. 1981a. Dissolution Kinetics: Catalysis by Strong Acids. *J Colloid Interface Sci* **82** (1): 93–102. http://dx.doi.org/10.1016/0021-9797(81)90127-2.

Kline, W. E. and Fogler, H. S. 1981b. Dissolution Kinetics: The Nature of the Particle Attack of Layered Silicates in Hf. *Chem Eng Sci* **36** (5): 871–884. http://dx.doi.org/10.1016/0009-2509(81)85041-5.

Labrid, J. C. 1975. Thermodynamic and Kinetic Aspects of Argillaceous Sandstone Acidizing. *SPE J.* **15** (2): 117–128. SPE-5156-PA. http://dx.doi.org/10.2118/5156-PA.

Langmuir, D. 1997. *Aqueous Environmental Geochemistry.* Englewood Cliffs, New Jersey: Prentice Hall.

Levich, V. G. 1962. *Physiochemical Hydrodynamics*, second edition. Englewood Cliffs, New Jersey, USA: Prentice-Hall.

Lide, D. R., ed. 2006. *CRC Handbook of Chemistry and Physics*, 87$^{th}$ edition. Boca Raton, Florida: CRC Press.

Linke, W. F. 1958. *Solubilities: Inorganic and Metal-Organic Compounds.* Washington, DC: American Chemical Society.

Litt, M. and Serad, G. 1964. Chemical Reactions on a Rotating Disk. *Chem Eng Sci* **19** (11): 867–884. http://dx.doi.org/10.1016/0009-2509(64)85065-X.

Lønøy, A. 2006. Making Sense of Carbonate Pore Systems. *AAPG Bull* **90** (9): 1,381–1,405. http://dx.doi.org/10.1306/03130605104.

Lucia, F. J. 1983. Petrophysical Parameters Estimated From Visual Descriptions of Carbonate Rocks: A Field Classification of Carbonate Pore Space. *J Pet Technol* **35** (3): 629–637. SPE-10073-PA. http://dx.doi.org/10.2118/10073-PA.

Lucia, F. J. 1995. Rock-Fabric/Petrophysical Classification of Carbonate Pore Space for Reservoir Characterization. *AAPG Bull* **79** (9): 1275–1300.

Lucia, F. J. 1999. Petrophysical Rock Properties. In *Carbonate Reservoir Characterization*, first edition, Chap. 1, 1–22. Springer Berlin Heidelberg.

Lund, K., Fogler, H. S., and McCune, C. C. 1973a. Acidization—I. The Dissolution of Dolomite in Hydrochloric Acid. *Chem Eng Sci* **28** (3): 691–700. http://dx.doi.org/10.1016/0009-2509(77)80003-1.

Lund, K., Fogler, H. S., McCune, C. C. et al. 1973b. Kinetic Rate Expressions for Reactions of Selected Minerals with HCL and HF Mixtures. Presented at the SPE Oilfield Chemistry Symposium, Denver, 24–25 May. SPE-4348-MS. http://dx.doi.org/10.2118/4348-MS.

Lund, K., Fogler, H. S., McCune, C. C. et al. 1975. Acidization– II. The Dissolution of Calcite in Hydrochloric Acid. *Chem Eng Sci* **30** (8): 825–835. http://dx.doi.org/10.1016/0009-2509(75)80047-9.

Mazzullo, S. J., Chilingarian, G. V., and Bissell, H. J. 1992. Carbonate Rock Classifications. In *Carbonate Reservoir Characterization: A Geologic-Engineering Analysis, Part I*, first edition. ed. G. V. Chilingarian, S. J. Mazzullo, and H. H. Rieke, Vol. 30 Chap. 2. Amsterdam: Developments in Petroleum Science, Elsevier.

Neasham, J. W. 1977. The Morphology of Dispersed Clay in Sandstone Reservoirs and Its Effect on Sandstone Shaliness, Pore Space and Fluid Flow Properties. Presented at the SPE Annual Fall Technical Conference and Exhibition, Denver, 9–12 October. SPE-6858-MS. http://dx.doi.org/10.2118/6858-MS.

Nierode, D. E. and Williams, B. B. 1971. Characteristics of Acid Reaction in Limestone Formations. *SPE J.* **11** (4): 406–418. SPE-3101-PA. http://dx.doi.org/10.2118/3101-PA.

Panga, M. K. R., Ziauddin, M., and Balakotaiah, V. 2005. Two-Scale Continuum Model for Simulation of Wormholes in Carbonate Acidization. *AIChE J* **51** (12): 3231–3248. http://dx.doi.org/10.1002/aic.10574.

Pettijohn, F. J., Potter, P. E., and Siever, R. 1987. Mineral and Chemical Composition. In *Sand and Sandstone*, second edition, Chap. 2, 25–67. Springer New York.

Reeckmann, A. and Friedman, G. M. 1982. *Exploration for Carbonate Petroleum Reservoirs* edition. Hoboken, New Jersey, USA: John Wiley & Sons, Inc.

Roberts, L. D. 1974. The Effect of Surface Kinetics in Fracture Acidizing. *Society of Petroleum Engineers Journal* **144** (04): 385–395. SPE-4349-PA. http://dx.doi.org/10.2118/4349-PA.

Robinson, R. A. and Stokes, R. H. 1965. *Electrolyte Solutions*. London: Butterworths.

Ross, G. J. 1969. Acid Dissolution of Chlorites: Release of Magnesium, Iron, and Aluminum and Mode of Acid Attack. *Clays Clay Miner* **17** (6): 347–354.

Rossini, F. D., Wagman, D. D., Evans, W. H. et al. 1952. Selected Values of Chemical Thermodynamic Properties. Washington, DC: US Government Printing Office.

Rossotti, F. J. C. and Rossotti, H. 1961. Activity and Concentration Quotients. In *The Determination of Stability Constants and Other Equilibrium Constants in Solution*, edition. Chap. 2. New York: McGraw-Hill Series in Advanced Chemistry, McGraw-Hill Book Company, Inc.

Schechter, R. S. 1990. *Oil Well Stimulation*, first edition. Englewood Cliffs, New Jersey, USA: Prentice Hall.

Shuchart, C. E. and Buster, D. C. 1995. Determination of the Chemistry of HF Acidizing with the Use of F NMR Spectroscopy. Presented at the SPE International Symposium on Oilfield Chemistry, San Antonio, Texas, USA, 14–17 February. SPE-28975-MS. http://dx.doi.org/10.2118/28975-MS.

Simon, D. E. and Anderson, M. S. 1990. Stability of Clay Minerals in Acid. Presented at the SPE Formation Damage Control Symposium, Lafayette, Louisiana, USA, 22–23 February. SPE-19422-PA. http://dx.doi.org/10.2118/19422-MS.

Smith, J. V. 1974a. *Feldspar Minerals: Crystal Structure and Physical*, first edition, Vol. 1. New York: Springer-Verlag.

Smith, J. V. 1974b. *Feldspar Minerals: Chemical and Textural Properties*, first edition, Vol. 2. New York: Springer-Verlag.

Stiff, H. A. and Davis, L. E. 1952. A Method for Predicting the Tendency of Oil Field Waters to Deposit Calcium Carbonate, *Trans.*, AIME **195:** 213-216. SPE-952213-G.

Tiab, D. and Donaldson, E. C. 2012. *Petrophysics: Theory and Practices of Measuring Reservoir Rock and Fluid Transport Properties*, third edition. Oxford, UK: Gulf Professional Publishing.

Tucker, M. E. and Wright, V. P. 1990. *Carbonate Sedimentology*, first edition. Malden, Massachusetts, USA: Blackwell Publishing. http://dx.doi.org/10.1002/9781444314175.

Wang, Y., Hill, A. D., and Schechter, R. S. 1993. The Optimum Injection Rate for Matrix Acidizing of Carbonate Formations. Presented at the SPE Annual Technical Conference and Exhibition, Houston, 3–6 October. SPE-26578-MS. http://dx.doi.org/10.2118/26578-MS.

Williams, B. B., Gidley, J. L., and Schechter, R. S. 1979. *Acidizing Fundamentals*, second edition, Monograph Vol. 6. Dallas: Henry L Doherty Series, SPE of AIME.

Wilson, J. L. 1975. *Carbonate Facies in Geologic History*, first edition. New York: Springer Study Edition, Springer-Verlag. http://dx.doi.org/10.1007/978-1-4612-6383-8.

Zakaria, A. S., Nasr-El-Din, H. A., and Ziauddin, M. 2015a. Flow of Emulsified Acid in Carbonate Rocks. *Ind. Eng. Chem. Res.* **54** (16): 4,190–4,202. http://dx.doi.org/10.1021/ie504167y.

Zakaria, A. S., Nasr-El-Din, H. A., and Ziauddin, M. 2015b. Predicting the Performance of the Acid-Stimulation Treatments in Carbonate Reservoirs With Nondestructive Tracer Tests. *SPE J.* SPE-174084-PA (in press; posted 12 January 2015). http://dx.doi.org/10.2118/174084-PA.

Zemaitis, J. Jr., Clark, D. M., Rafal, M. and Scrivner, N. C. 1986. *Handbook of Aqueous Electrolyte Thermodynamics.* New York City: American Institute of Chemical Engineers.

Ziauddin, M. E., Gillard, M. R., Lecerf, B. et al. 2005. Method for Characterizing Secondary and Tertiary Reactions Using Short Reservoir Cores. *SPE Prod & Fac* **20** (2): 106–114. SPE-86520-PA. http://dx.doi.org/10.2118/86520-PA.

Chapter 4

# Carbonate Acidizing

**Frank F. Chang,** Saudi Aramco, and
**H. Scott Fogler,** University of Michigan

## 4.1 Introduction

Hydrocarbon productivity of a well drilled in carbonate reservoirs can be enhanced by injecting acids into the reservoir to dissolve part of the rock mass, thereby increasing the effective drainage radius of the well. The techniques of injection acid can be generally divided into two categories. The first category involves injecting fluids at injection rates high enough to result in the downhole pressure being above the fracturing pressure of the formation. This process, called fracture acidizing, creates cracks extending from the wellbore out into the formation with acid-etched patterns on the faces of the cracks. The second category involves lower acid-injection rates so that the resulting downhole pressure is lower than the fracture pressure of the formation. Acids are forced to percolate through the porous structure and dissolve the soluble portion of the rock. This process, called matrix acidizing, creates branched channels intersecting the wellbore to remove near-wellbore damage caused by drilling, completion, and previous production processes or to extend the effective wellbore radius to achieve negative skin. Fracture acidizing is discussed in another chapter of this monograph, so this chapter will focus only on carbonate matrix acidizing.

## 4.2 Carbonate Geological Considerations for Acidizing

Understanding the reservoir geology will help delineate the porosity and permeability distribution, the directional fluid-flow preference, the in-situ-stress orientation and magnitude, and the reservoir's interaction with injected acidizing fluids—all of which are critical to planning acid stimulation and predicting its outcome. The geology of sandstones composed of siliciclastic rocks have been studied and became better understood by the oil and gas industry because of their more-defined grain and pore structures. On the other hand, the carbonate rock structure and facies are much more complex. Characterizing them clearly for better understanding of hydrocarbon-flowing phenomena and for designing stimulation-treatment strategy remains a great challenge. The key relevance of carbonate geology to acid stimulation processes is the fluid-flow characteristics of the reservoir rock in various scales (i.e., porosity/permeability relationship). The complex and convoluted rock fabric in most carbonate rocks leads to a variety of pore types and, consequently, to heterogeneity in permeability. Even whole cores are not sufficient to derive representative porosity/permeability correlations (Akbar et al. 1995; Akbar et al. 2000).

**4.2.1 Rock Types.** On the basis of relative abundance of grain components and crystalline cement making up the rocks, carbonates can be categorized into six distinct classes (Dunham 1962; Nurmi and Frisinger 1983; Lucia 1983): grainstone, packstone, wackestone, mudstone, boundstone, and crystalline. **Fig. 4.1** illustrates the rock texture of these six carbonate rock classes.

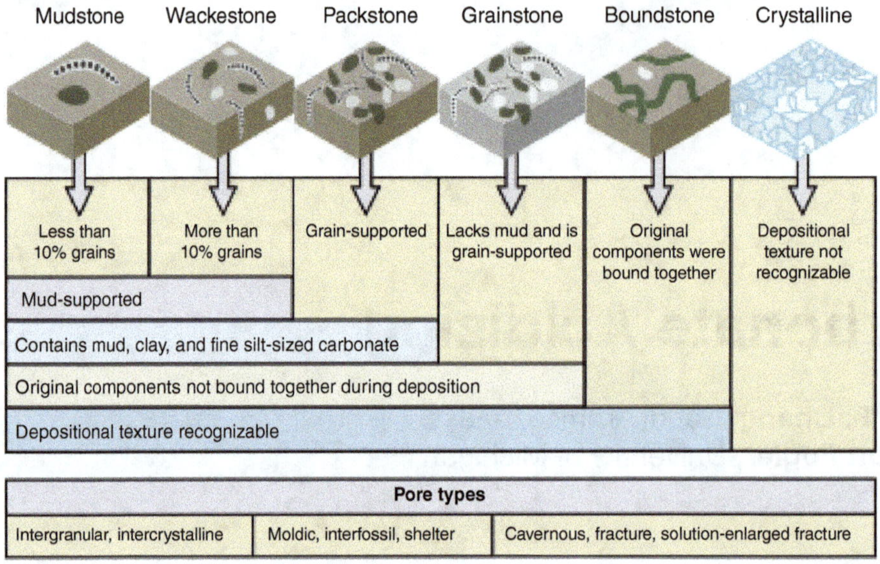


**Fig. 4.1—Carbonate rock classification based on depositional texture, grain types, rock fabric (Akbar et al. 2000).**

**4.2.2 Lithology and Heterogeneity.** While sandstone and shale contain a wide variety of silica and aluminosilicate minerals such as quartz, feldspar, plagioclase, and clays, carbonates mainly consist of just two minerals—calcite and dolomite. Unlike the sandstone grains transported over hundreds of miles from their sources to develop into sorted and well-defined grains with smooth surfaces, carbonate grains and matrix remain near their point of origin and deposition; they are chemically unstable and, hence, undergo major diagenetic processes such as chemical dissolution, recrystallization, reprecipitation, compaction and stylolitization, and fracturing (Akbar et al. 1995; Leet et al. 1978; Hoornveld 2009). The resulting carbonate pore systems can be intergranular, intragranular, vugular, cavernous, and naturally fractured, causing the flow network to be very complex. Dolomites and chalks are examples of carbonate rocks that have gone through various diagenetic alterations to form some of the important oil and gas reservoirs in the world (Cantrell et al. 2001; Scholle 1977).

**4.2.3 Porosity/Permeability Correlations.** Designing acid stimulation treatment in carbonate formations requires the knowledge of the rock porosity/permeability relationships. Though such data are often unavailable or unclear, the generalized porosity/permeability relationship for the various limestone depositional rock types presented in **Fig. 4.2** can provide some information. In this figure, the central line is the trend commonly found for oolite grainstones and coincides with the approximation for a 300-μm particle size (Nurmi and Frisinger 1983). Additional porosity/permeability trends for North Sea fine-grain chalk and some highly leached reef rock, or boundstone, are also shown to illustrate the great range of possibilities within limestone rocks.

## 4.3 Reaction Chemistry

The chemical nature of the carbonate rocks allow them to be readily dissolvable by fluids releasing hydrogen ions. The prevalent chemicals used in the oil and gas industry can be grouped into three different types: (1) strong acid, in particular hydrochloric acid; (2) organic acid, such as acetic and formic acid; and (3) chelating agent, such as ethylenediaminetetraacetic acid (EDTA). The detailed reaction chemistry between the carbonates and these fluids are different, but the reaction physics or kinetics can be generalized by considering these three steps: (1) the transport of reactants from the bulk solution to the surface of carbonate rock, (2) chemical reaction of the reactants on the rock surface with the rock composition, and (3) the transport of reaction products away from the carbonate surface into the bulk solution. These three steps occur in sequence and interfere with one another to dictate the overall rate of rock dissolved during the acidizing treatment process. The slowest step becomes the controlling step of the overall reaction rate.

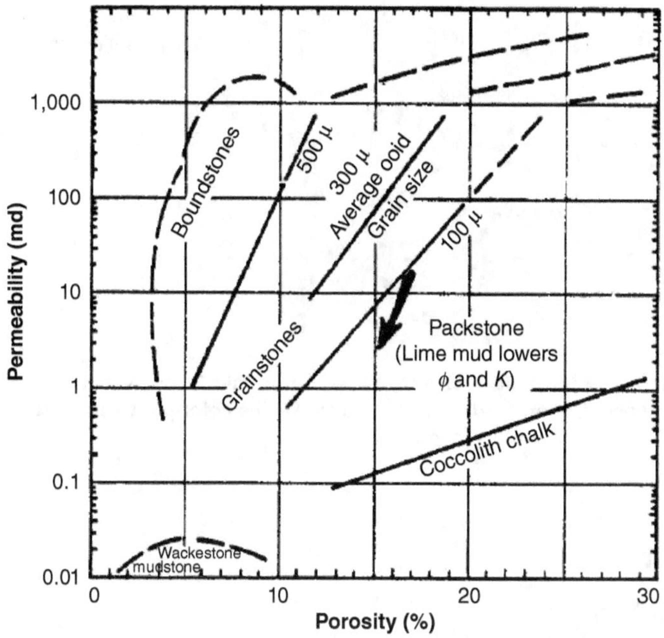


**Fig. 4.2—Porosity/permeability relationship for calcium carbonate rock types (Nurmi and Frisinger 1983).**

Intuitively, the reaction kinetics between the fluid and carbonates is heavily influenced by temperature. In addition, it is important to recognize that the transport of reactants and products to and from the rock surface, called mass transfer, is significantly impacted by convective fluid flow, fluid viscosity, and molecular and ionic diffusion. These factors allow the chemists and engineers to design fluid to alter fluid properties and pumping procedure to result in a desired dissolution pattern based on the reservoir properties. For example, acidizing fluid can be made more viscous or emulsified to reduce the diffusion coefficients of the reactants, leading to retardation in a mass-transfer-controlled reaction; or chemical additives can be added to adsorb on the rock surface and, hence, modify the rate of a surface-controlled reaction.

**4.3.1 Reaction of Strong Acid With Carbonates.** The most suitable and commonly used strong acid for carbonate acidizing is hydrochloric acid (HCl). Since strong acids are readily dissociated in water, the dissolution reactions of calcite and dolomite can be expressed as Eqs. 4.1 and 4.2, respectively:

$$CaCO_3 + 2H^+ \rightarrow Ca^{2+} + H_2O + CO_2 \quad (4.1)$$

$$CaMg(CO_3)_2 + 4H^+ \rightarrow Ca^{2+} + Mg^{2+} + 2CO_2 + 2H_2O. \quad (4.2)$$

The produced carbon dioxide ($CO_2$) can form carbonic acid ($H_2CO_3$), which will not affect the reaction chemically in the case of strong acid dissolving carbonates because it is not dissociated in such an environment that the solution is full of $H^+$ from the strong acid. However, should produced $CO_2$ be in a gaseous phase, the bubbling on the rock surface can significantly increase the mixing and mass transfer, leading to an increase in reaction rate.

The reaction between a strong acid and a carbonate solid surface can be viewed in **Fig. 4.3.** There exists a reactive boundary layer near the solid surface because of the reaction. Within the boundary layer there is a hydrogen-ion-activity gradient from the bulk to the solid surface. Ions are transported through the boundary layer to reach the solid surface by a diffusion process. The thickness of the boundary layer is a function of velocity of fluid flow past the solid surface. The faster the fluid velocity, the thinner the boundary layer.

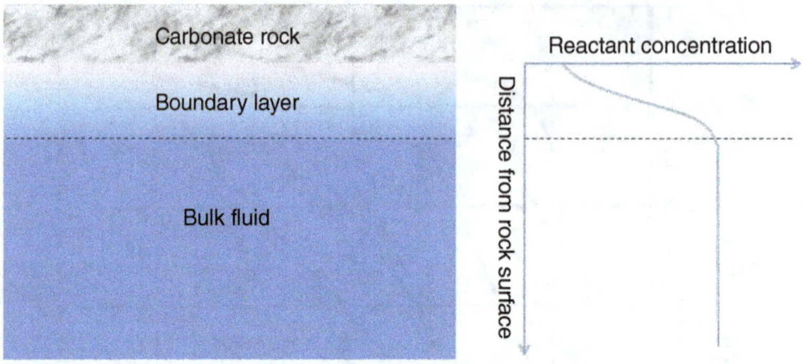


**Fig. 4.3—The reaction between fluid and rock generates a fluid layer near the solid surface. Within the layer, there exists a concentration gradient (represented by the color gradient). This fluid layer of changing concentration is the reaction boundary layer.**

$$J = K_{mt}\,(\bar{A}_b - \bar{A}_s), \qquad (4.3)$$

where

$J$ = reaction rate (mol/cm²/s)
$K_{mt}$ = mass-transfer coefficient (cm/s)
$\bar{A}_b$ = bulk acid activity
$\bar{A}_s$ = surface acid activity

Note that the activity is equal to the activity coefficient (α) multiplied by concentration:

$$\bar{A} = \alpha\, C.. \qquad (4.4)$$

In dilute solutions, the activity coefficient is equal to unity; therefore, activity can be replaced by concentration. High concentrations of acids are typically used in acidizing treatments. Strictly speaking, the activity of the ions should be used when computing the reaction kinetics. However, obtaining the activity coefficient is a tedious process; oftentimes, only concentrations are used. Readers are encouraged to consult advanced analytical-chemistry textbooks for calculation of activity coefficients of ions in a concentrated solution. In this chapter, concentrations will be used in the place of activity to make the illustration more intuitive for readers from a broader range of technical backgrounds. Eq. 4.3 can be rewritten in terms of concentration ($C$) as

$$J = K_{mt}\,(C_b - C_s).. \qquad (4.5)$$

The rate of surface reaction can be expressed by

$$R = k_r C_s^n, \qquad (4.6)$$

where

$C_s$ = surface concentration (mol/cm³),
$k_r$ = reaction-rate constant,
$n$ = reaction order, and
$R$= reaction rate (mol/cm³/s).

The mass-transfer rate of the reaction product from the rock surface to the bulk solution can be expressed similar to Eq. 4.5 as

$$J_p = K_{mtp}\,(C_{bp} - C_{sp}), \qquad (4.7)$$

where

$J_p$ = reaction rate (mol/cm²/s),
$K_{mtp}$ = mass-transfer coefficient (cm/s),
$C_{bp}$ = bulk concentration of reaction product (mol/cm³), and
$C_{sp}$ = surface concentration of reaction product (mol/cm³).

The concentration of hydrogen ions on the rock surface cannot be measured but can be mathematically calculated if the mass-transfer coefficient and the surface-reaction parameters are known. The evolution of surface concentration is determined by the balance between the rate of ions transported to the surface and the rate of ions consumed by the reaction:

$$\frac{\mathrm{d}C_s}{\mathrm{d}t} = JA - R, \qquad (4.8)$$

where

$A$ = the surface area of the solid rock (cm²),
$t$ = time (seconds), and
$J$ and $R$ are defined by Eqs. 4.5 and 4.6, respectively.

A rotating-disk apparatus is often used to study liquid/solid reaction kinetics in which the balance between the mass-transfer rate and surface reaction rate needs to be determined. The benefit of a rotating-disk reactor is that the boundary layer thickness across the disk surface is uniform. The mass transfer rate in a rotating disk was analytically solved by Levich (1962). The Levich solution for mass-transfer coefficient during the reaction of a Newtonian fluid with a solid on the rotating disk is

$$K_{mt} = 0.62 D^{2/3} \left(\frac{\mu}{\rho}\right)^{-1/6} \omega^{1/2}, \qquad (4.9)$$

where

$D$ = diffusion coefficient (cm²/s),
$\mu$ = fluid viscosity (g-cm/s),
$\rho$ = fluid density (g/cm³), and
$\omega$ = rotating speed (rad/sec).

**Fig. 4.4** illustrates the surface hydrogen ion concentration as the reaction progresses from the first contact between the acid and the rock. The figure was obtained from solving Eqs. 4.5, 4.6, and 4.8 simultaneously with a constant fluid velocity and constant rotating speed. When the surface reaction rate is significantly faster than the mass-transfer rate ($k_r/D$ = 10 000), the surface concentration rapidly depleted to zero as the surface reaction consumes every hydrogen ion diffused onto the surface. This reaction is considered to be mass-transfer controlled. As the mass transfer increases ($k_r/D$ = 1000), the resulting surface hydrogen concentration also increases because the surface reaction is not fast enough to consume all the ions. When equilibrium is established, there is a finite level of hydrogen on the surface. Further increase in the mass-transfer rate continues to increase the ion concentration on the surface until the surface reaction is slower than the mass transfer rate ($k_r/D < 1$); the surface concentration reaches the same level as the bulk concentration. This reaction will, therefore, be considered as surface-reaction controlled.

There are two things one can do to alter the mass-transfer rate. The first is to change the effective diffusion coefficient of the hydrogen through the boundary layer. This could be achieved by using higher acid concentration or changing the acid carrying liquid such as by emulsifying or increasing the viscosity of the acid. The second is to change the boundary layer thickness by changing the convective flow velocity of the bulk fluid. A higher fluid convective velocity brings the bulk concentration

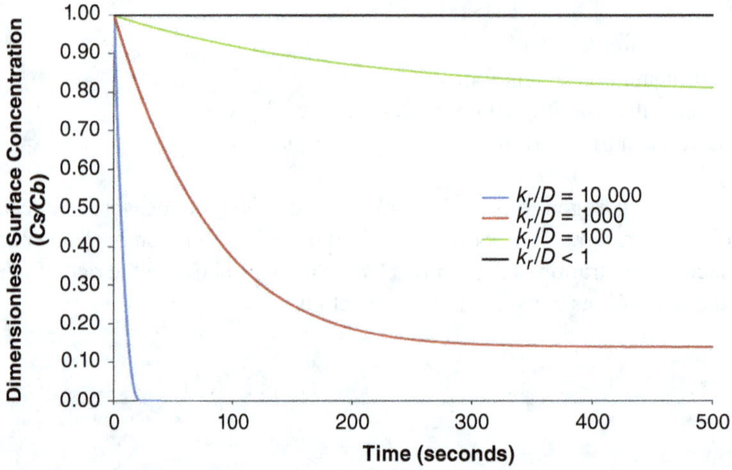


**Fig. 4.4—The surface hydrogen ion concentration during the acid-carbonate reaction as a function of time from the first contact between the acid and the rock.**

closer to the solid surface and effectively transports the reaction product away from the solid surface and therefore reduces the boundary layer. Lower fluid convective velocity decreases the mixing efficiency; therefore, the boundary layer thickness increases. **Fig. 4.5** illustrates the effect of increasing convective flow velocity on mass transfer in a rotating-disk reactor. At low rotating speed, the reaction is dominantly controlled by mass transfer (diffusion). As the rotating speed increases, the overall reaction rate increases because of the increase in mass transfer. Eventually, the surface cannot consume the delivered hydrogen ions, and further increase in mass transfer will not increase the overall reaction. The reaction therefore becomes surface-reaction controlled.

The reaction kinetics between HCl and limestone ($CaCO_3$) is considered to be mass-transfer controlled at any temperature above 25°C. The surface-reaction-rate constant ($k_r$) in Eq. 4.6 is found to be $1.34 \times 10^{-6}$ at –15.6°C with an activation energy of 15 kcal/mol, and the reaction order ($n$) is 0.63 (Lund et al. 1975). Above 25°C and above 1 mol/L of HCl concentration, the surface reaction rate is irrelevant if the reaction is fully mass-transfer controlled. The critical parameter to characterize such mass-transfer-controlled reactions will be the effective diffusion coefficient.

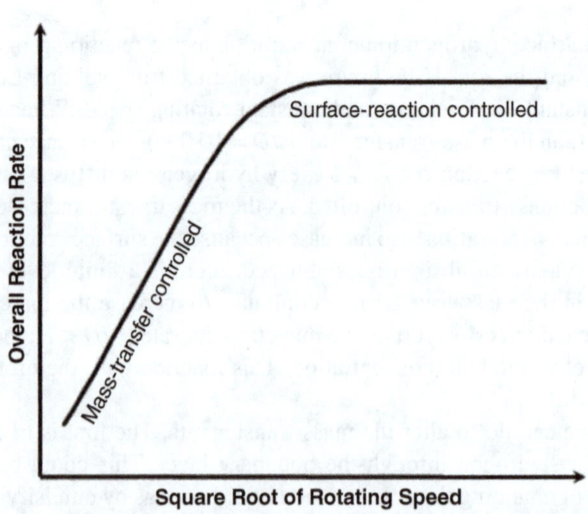


**Fig. 4.5—Reaction regime evolves from mass-transferr limited to surface-reaction limited as the convective transport increases by increasing rotating speed in a rotating-disk reactor (Williams et al. 1979).**

The reaction between HCl and dolomite is surface-reaction controlled until the temperature is above 100°C, in which the reaction begins to transition to mass-transfer controlled (Lund et al. 1973). The surface-reaction-rate expressions are as follows:

at 25°C, $R = 2.6 \times 10^{-6}\, C^{0.44}$;
at 50°C, $R = 6.6 \times 10^{-5}\, C^{0.61}$;
at 100°C, $R = 5.4 \times 10^{-3}\, C^{0.83}$.

In the preceding reaction-rate expressions, $C$ is the concentration in mol/cm$^3$. The activation energy for the reaction-rate constant ($k_r$) for HCl-dolomite reaction is therefore 22.5 kcal/mol.

To correctly characterize a mass-transfer-controlled reaction between acid and carbonate such as calcite, the diffusion coefficient needs to be measured under the specific temperature condition. Though experimental study of hydrogen ion diffusion exists (Vinograd and McBain 1941), the effective diffusion coefficient is more appropriate to be applied to the carbonate acidizing process because reaction products such as $Ca^{2+}$ and $Mg^{2+}$, and even a phase of $CO_2$, will significantly impact the real quantification of mass transfer. Rotating-disk and diaphragm cells have been used more realistically, though not perfectly by any means, to measure the effective diffusion coefficient in the carbonate acidizing process under reservoir conditions (de Rozieres 1994).

Using a rotating-disk apparatus to obtain effective diffusion coefficient of hydrogen ions is illustrated here for the HCl-calcite reaction (Eq. 4.1) (de Rozieres 1994). At steady state, every mole of $H^+$ ion transferred onto the surface of the marble disk will produce 2 moles of $Ca^{2+}$ ion on the surface. In a rotating-disk reaction-kinetics experiment, the $Ca^{2+}$ ion in the bulk fluid is actually measured instead of $H^+$ ion on the rock surface. Assuming the produced $Ca^{2+}$ on the rock surface is equal to that in the bulk fluid, since stoichimetrically the surface $H^+$ concentration must be twice the surface $Ca^{2+}$ concentration, the flux $J_H{}^+$ can therefore be expressed in the following equation:

$$J_{H^+} = 2J_{Ca^{2+}} \qquad (4.10)$$

Combining Eqs. 4.5, 4.9, and 4.10 yields

$$J_{Ca^{2+}} = 0.31 D^{2/3} \left(\frac{\mu}{\rho}\right)^{-1/6} \omega^{1/2} (C_b - C_s) \qquad (4.11)$$

The bulk and surface concentration terms in Eq. 4.11 are the $H^+$ concentrations. Since $C_s$ is zero when the reaction is fully controlled by mass transfer, Eq. 4.11 becomes

$$J_{Ca^{2+}} = 0.31 D^{2/3} \left(\frac{\mu}{\rho}\right)^{-1/6} \omega^{1/2} C_b \qquad (4.12)$$

All the parameters in Eq. 4.12 are known except for the diffusion coefficient ($D$). Using the measured $Ca^{2+}$ flux from the rotating-disk experiments, the $D$ can be solved by plotting $J_{Ca^{2+}} \Big/ 0.31\left(\frac{\mu}{\rho}\right)^{-1/6} C_b$ vs. $\omega^{1/2}$.

The slope will be $D^{2/3}$, and therefore the diffusion $D$ will be the slope raised to 3/2 power.

A significant factor that has been overlooked in carbonate acidizing practice is that when the acid penetrates into the formation, it is no longer the same fresh acid that is was near the wellbore. Therefore, the acid-carbonate reaction kinetics is continuously changing along the wormhole. This results in changing wormhole-penetration velocity as well as the final wormhole-penetration distance. The impact of reaction products on diffusivity should be accounted for to more accurately quantify the acid penetration into the reservoir. Diaphragm-cell experiments demonstrated that the diffusivity of the fresh acid is much higher than that of the spent acid because of the presence of counterions in the spent acid (Conway et al. 1999). Properly implementing the diffusion coefficient of spent acid in wormhole-penetration modeling will lead to a more accurate quantitative prediction.

Diaphragm cells can be used to directly measure the diffusion coefficient of acids containing various levels of reaction products. A diaphragm-cell device contains two chambers separated by a nonselective glass membrane. Fluid samples of different compositions are loaded in the two chambers. The ions in the two chambers are allowed to diffuse through the glass membrane because of their concentration gradient on the two sides of the membrane for a predetermined time. The diffusion coefficient of hydrogen ions under the influence of reaction products, $Ca^{2+}$ and $Mg^{2+}$, can therefore be determined. Using this method, Conway et al. (1999) published the following diffusion-coefficient correlation, which is useful for acidizing of carbonate under reservoir temperatures.

$$D_{H^+} = e^{-\frac{A}{T}+B\sqrt{\frac{[Ca^{2+}]}{[H^+]}}+C\sqrt{\frac{[Mg^{2+}]}{[H^+]}}+D[H^+]+E} \quad \text{.......... (4.13)}$$

In Eq. 4.13, the terms in square brackets are a concentration of the respective ions; the coefficients are as follows:

$A = -2918.54$,
$B = -0.589$,
$C = -0.789$,
$D = 0.0452$, and
$E = -4.995, -5.47, -7.99$ for straight, gelled, and emulsified acid, respectively.

The diffusion coefficients of $Ca^{2+}$ and $Mg^{2+}$ ions were also determined by Conway et al. (1999) using the diaphragm-diffusion-cell device:

$$D_{Ca^+} = e^{-\frac{A}{T}+B\sqrt{\frac{[Mg^{2+}]}{[Ca^{2+}]}}+C\sqrt{\frac{[H^+]}{[Ca^{2+}]}}+D[Ca^+]+E}, \quad \text{.......... (4.14)}$$

where

$A = 1476.11$,
$B = 0.501$,
$C = 0.0075$,
$D = 0.106$, and
$E = -8.38, -9.2963, -11.991$ for straight, gelled, and emulsified acid, respectively.

$$D_{Mg^+} = e^{-\frac{A}{T}+B\sqrt{\frac{[Ca^{2+}]}{[Mg^{2+}]}}+C\sqrt{\frac{[Ca^{2+}]}{[Mg^{2+}]}}+D[Mg^+]+E}, \quad \text{.......... (4.15)}$$

where

$A = -700.61$,
$B = 1.496$,
$C = 0.066$,
$D = 0.151$, and
$E = -11.89, -12.535, -15.225$ for straight, gelled, and emulsified acid, respectively

**4.3.2 Organic Acids.** Organic acids have been used in well stimulation when the reservoir temperature is high enough that excessive corrosion to well tubular goods is a concern (Harris 1961), or when the reservoir fluid is rich in asphaltenes, which tend to form sludge when encountering iron-containing spent HCl. Though organic acids react with carbonate at a much lower reaction rate than HCl, they should not be considered as good retarded acids for deep wormhole penetration because of their low carbonate-dissolving capacity. The other limitation of organic acid is that they cannot be used at high acid concentrations because of the limited solubility of their calcium salts. For instance, acetic and formic acids are typically used at concentrations less than 13 and 9 wt%, respectively, to avoid precipitation of calcium acetate and calcium formate (Robert and Crowe 2000). The cost of organic acid is significantly higher than that of HCl for equivalent mass of rock dissolved. The aforementioned factors prevent organic acids from being widely applied as the main carbonate stimulation treatment fluid. However, mixing HCl and organic acid can provide several benefits vs. using acid alone. Organic acids, such as formic acid, can be used as a corrosion-inhibitor intensifier for high-temperature applications (Dill and Keeney 1978; k and Nasr-El 2005; Katheeri et al. 2002). More interestingly, organic acids can be used to enhance acid penetration in the formation. When an organic acid is mixed with HCl, it does not dissociate to generate hydrogen ions because of its low dissociation constant. Therefore, the organic acid is preserved until HCl is nearly spent. As the hydrogen ions from HCl are depleted, the carboxylic groups start to dissociate, resulting in further carbonate rock dissolution from the tip of the acid front, and hence increasing acid penetration into the formation (Chang et al. 2008).

When a carbonate rock is dissolved by an organic acid, the reactant and products will form a weak acid salt and the solution becomes a buffer system, which significantly affects the reaction equilibrium and reaction rate (Lund et al. 1975). The low acid dissociation constant and the carbonic acid from produced $CO_2$ significantly limit the hydrogen available to dissolve the carbonate rock. The suppression hydrogen ion generation increases with increasing temperature (Chatelain et al. 1976). Therefore, under reservoir condition, the reaction kinetics between organic acids and carbonates is controlled by mass transfer of reactants and reaction products (Fredd and Fogler 1998) and the equilibrium of the acid dissociation while coexisting with $H_2CO_3$ (Buijse et al. 2004).

The acid attack of the carbonate rock described by Eq. 4.1 is applicable for organic acid. The amount of hydrogen ion, however, is driven by the acid dissociation:

$$HA \rightleftharpoons H^+ + A^-, \quad (4.16)$$

where

$A$ = organic ligands such as formate ($HCOO^-$) or acetate ($CH_3COO^-$).

It is noteworthy that the reactant in Eq. 4.1 is hydrogen ion ($H^+$) rather than the organic acid (HA). Unlike strong acids such as HCl in which the hydrogen ions are completely dissociated, the dissociation constant of the reaction in Eq. 4.15 is finite. Therefore, the reactant for carbonate dissolution ($H^+$) transferred to the rock surface from bulk solution is reduced. For example, at 25°C the dissociation constant ($K_a$) of acetic acid and formic acid is $1.74\times10^{-5}$ and $1.79\times10^{-4}$, respectively. The $H^+$ in a fresh 10% (≈1.67 M) acetic acid solution can be calculated by

$$K_a = \frac{[H^+][CH_3COO^-]}{[CH_3COOH]} \quad (4.17)$$

In Eq. 4.17, $[CH_3COO^-]=[H^+]$ and $[CH_3COOH]=1.67-[H^+]$. Substituting the acetate and acetic acid concentration in Eq. 4.17 yields

$$1.74\times10^{-5} = \frac{[H^+]^2}{1.67-[H^+]} \quad (4.18)$$

Hence, $[H^+]$ = 0.00538 M, which is equivalent to a pH 2.3 solution. Similarly, for a fresh 10% (2.17 M) formic acid solution with $K_a$= $1.79\times10^{-4}$, the hydrogen ion concentration $[H^+]$ = 0.0196 M, which is equivalent to a pH 1.7 solution. There is obviously not a sufficient amount of free hydrogen ions transferred to the rock surface even if the convective flow velocity is significantly increased. The experimentally determined calcite dissolution by acetic acid showed that actual hydrogen ions must have been significantly higher (Fredd and Fogler 1998). This equilibrium of acid dissociation (Eq. 4.16) on the rock surface accounts for the higher dissolution rate. The acid dissociation on the surface of the rock results in a concentration gradient, and therefore a boundary layer of HA in addition to $H^+$. Hence, as the dissolution reaction (Eq. 4.1 or 4.2) takes place, and the HA dissociation (Eq. 4.16) subsequently shifts to the right to provide additional $H^+$. In addition, the mass transfer of HA from bulk solution to the rock surface is taking place to continuously provide HA on the rock surface (Buijse et al. 2004).

As the dissolution reaction starts, there are two additional complications to the overall process. The first is the produced $CO_2$ that can further suppress organic-acid dissociation both on the rock surface and in the bulk solution by the formation of carbonic acid. In most reservoir-pressure conditions, the $CO_2$ is dissolved in the aqueous phase and generates carbonic acid:

$$H_2O + CO_2 \rightleftharpoons H_2CO_3 \quad (4.19)$$

$$H_2CO_3 \rightleftharpoons H^+ + HCO_3^- \quad (4.20)$$

$$HCO_3^- \rightleftharpoons H^+ + CO_3^{2-} \quad (4.21)$$

Carbonic acid equilibrium buffers the spent acid solution to a pH value of 4.5. The hydrogen ions from the carbonic acid dissociation limit the dissociation of the organic acid. **Figs. 4.6 and 4.7** show the formic acid and acetic acid dissociation as a function of pH, respectively. It can be seen that at pH 4.5, formic and acetic acids are not fully dissociated.

The second complication is the association of produced calcium ions and the ligands of the organic acid (Chang et al. 2008). The following reaction can also help shift the HA dissociation to the right to generate more $H^+$ by reducing the free $A^-$:

$$Ca^{2+} + A^- \rightleftharpoons CaA^+ \quad (4.22)$$

$$Ca^{2+} + 2A^- \rightleftharpoons CaA_2 \quad (4.23)$$

In summary, the organic-acid dissolution of carbonate rocks goes through the steps of mass transfer of reactant to the rock surface, surface reaction, and mass transfer of reaction products away from the rock surface. It is, however, complicated by the dissociation and association reaction described in

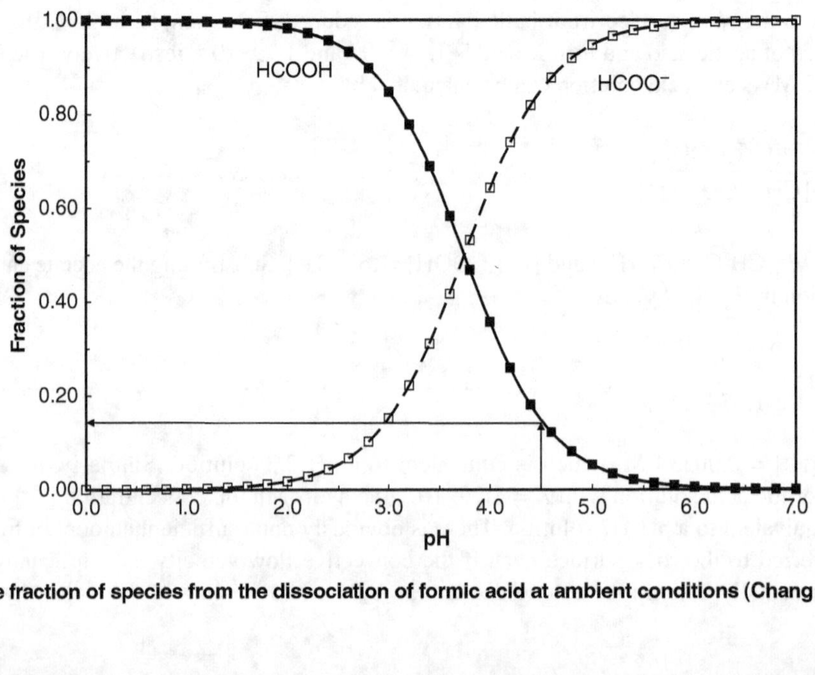


**Fig. 4.6—The fraction of species from the dissociation of formic acid at ambient conditions (Chang et al. 2008).**

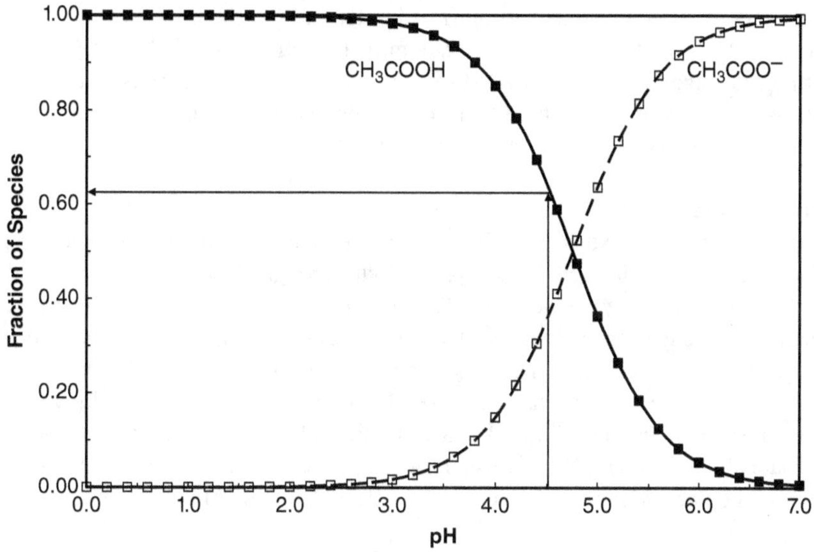


**Fig. 4.7—The fraction of species from the dissociation of acetic acid at ambient conditions (Chang et al. 2008).**

Eq. 4.15 and Eqs. 4.18 through 4.22. One noteworthy point is that the mechanism of surface reaction between the organic acid and carbonate rock is the same hydrogen attack as described by Eq. 4.1. Only the mechanisms of transporting the hydrogen ions to the surface and, subsequently, the reaction products away from the surface are different. It is safe to say that the mass transfer will be even slower and hence the reactions should be even more mass transfer-controlled. To rigorously determine the true reaction kinetics of carbonate dissolution by organic acids, not only the equilibrium expressions from Eq. 4.17 and Eqs. 4.19 through 4.22 should be used, but also, more importantly, the diffusion coefficient of reactants (HA, $H^+$) and reaction products ($Ca^{2+}$, $CaA_2$, and $CaA^+$) in the presence of all such counterions should be considered.

**4.3.3 Chelating Agents.** Chelating agents (chelants) can be alternative chemicals in stimulating carbonate reservoir where sludge and corrosion are of concern (Fredd and Fogler 1998). In the carbonate stimulation process, a chelating agent can be viewed as an organic acid and capable of extracting the metal ions from the rock by complexing them in the solution (Frenier et al. 2001). The reactions involved in carbonate dissolution by chelating agents include (a) a low-pH hydrogen attack (Eqs. 4.1 and 4.2) and (b) carbonate dissolution in water (Eq. 4.24 for calcite) and metal-ion complexation (Eqs. 4.25 and 4.26), which drives Eq. 4.24 to the right for continuous carbonate dissolution.

$$CaCO_3 \rightleftharpoons CO_3^{2-} + Ca^{2+} + H_2O \qquad (4.24)$$

$$H_{n-m}Y^{m-} \rightleftharpoons H^+ + H_{n-m-1}Y^{(m+1)-}, \qquad (4.25)$$

where

$m$ = state of dissociation of the chelant m=0, 1, 2, ... n−1 (higher $m$ at higher pH),
$n$ = total number of the carboxylate group in the chelant, and
$Y$ = fully deprotonated chelant.

$$H_{n-m-1}Y^{(m+1)-} + Ca^{2+} \rightleftharpoons CaY^{(m+1)-} + 2H^+ \qquad (4.26)$$

At low pH, both hydrogen attack and metal complexation take place with hydrogen attack being more dominant. The chelant at this condition is more effective in dissolving carbonate rock. At high pH, only metal complexation will take place, and the rock-dissolution capacity is decreased.

The most commonly used chelants for matrix stimulation of carbonates include ethylenediamine-tetraacetic acid (EDTA), hydroxyethyl ethylenediamine triacetic acid (HEDTA) (Frenier et al. 2001), and L-glutamic acid diacetic acid (GLDA) (LePage et al. 2011). The kinetics of chelants dissolving carbonate has not been extensively characterized, though chelants have been demonstrated to create wormholes in carbonate, particularly at high temperatures.

## 4.4 Wormhole Patterns

Matrix acidizing in carbonates extends a wellbore drainage radius by rapidly dissolving the rock to form wormholes intersecting the wellbore. The well productivity strongly depends on the length, diameter, and distribution of wormholes along the wellbore. The wormhole geometry, path, and distribution are controlled by the heterogeneity of the pore network and the reaction kinetics between the acid and the rock.

The heterogeneity of the rock pore network is determined by nature, but the reaction kinetics can be controlled by the engineers who practice the acidizing treatments. Fundamentally the engineering control will involve balancing the acid-rock reaction rate and the acid-injection rate. A dimensionless number, the Damköhler number ($N_{Da}$), is usually used to compare the reaction rate and the injection rate (Hoefner and Fogler 1988; McCune et al. 1975). The Damköhler number ($N_{Da}$) is defined as

$$N_{\mathrm{Da}} = \frac{\text{Rate of reaction}}{\text{Rate of fluid flow}} \quad (4.27)$$

The final dissolution pattern can be face dissolution, conical, dominant wormhole, ramified, or uniform dissolution when the acid is injected at higher and higher rates.

**Fig. 4.8** shows the wormhole morphology as a function of acid-injection rate during linear core-flow experiments (Fredd and Fogler 1999). When flow rate is low, acid is rapidly consumed near the acid entry point; therefore, the dissolution occurs only at the entrance and no live acid is available to dissolve carbonate into the core. The dissolution pattern will therefore look like the computed-tomography (CT) scan image in Fig. 4.8a. If the acid-injection rate is increased, the acid-consumption rate starts is overcome by the fluid propagation rate. Live acid becomes available at the leading edge of the dissolved cavity as the spent acid is pushed through the porous rock ahead of the cavity. The wormhole therefore starts to form and looks like the CT-scan image in Fig. 4.8b. At this condition, the wormhole is a conical shape with a wide base near the entrance and a narrow tip. If the

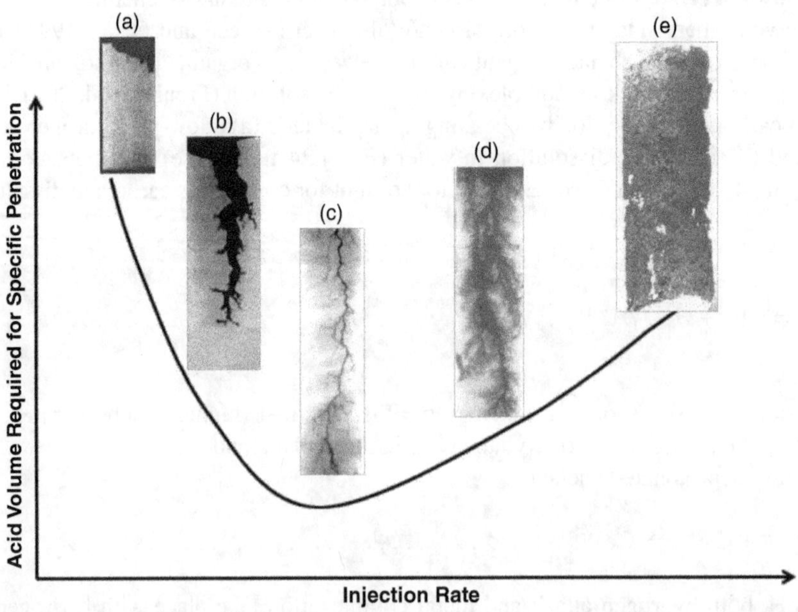


**Fig. 4.8—Wormhole morphology as a function of injection rate. [(a) Fredd and Fogler 1998; (b) Fredd and Fogler 1999; (c) Fredd and Fogler 1998; (d) Fredd and Fogler 1998; (e) Hoefner and Fogler 1988.]**

injection rate is further increased, it will lead to a balance between the acid consumption by reaction and the live-acid supply at the leading edge of the dissolution channel. Hence. a single and relatively-uniform-diameter wormhole is generated to penetrate through the core. The wormhole pattern looks like the CT-scan image in Fig. 4.8c. Further increase in flow rate causes the live acid to spread and seek small pores because the reaction can no longer keep up with the large amount of acid pumped at increasing pressure into the porous media. The acid will therefore leak off along the length of the wormhole and dissolve rock in an areal pattern. The dissolution channels ramify and look like the CT-scan image in Fig. 4.8d. Should the flow rate increase even further, the ramified channels will become uniform, resembling the acidizing in sandstone like the CT-scan image in Fig. 4.8e.

The curve on Fig. 4.8 also illustrates the amount of acid volume required to penetrate through a specific length of core. Obviously, the less the volume, the more efficient the acidizing process. Generating a dominant wormhole required the least amount of acid shown by Fig. 4.8c. Therefore, there exists an optimal injection rate when designing a matrix acidizing job to create the deepest wormholes with the minimum amount of acid.

The concept of pore volume to breakthrough (PVBT) has been used to interpret the acid efficiency. Frequently, the PVBT is also used to quantitatively scale up from linear core to the radial-flow geometry encountered in real reservoirs. An optimum injection rate is targeted for field execution by conducting linear core-flow acidizing experiments at multiple rates until a minimum PVBT is obtained at a certain injection rate. The cores commonly used in characterizing the PVBT are 3 to 6 in. Several carefully conducted experiments (Bazin et al. 1995; Chang 2012; Qiu et al. 2013) have shown that a dominant single wormhole pattern observed using short cores can turn into a conical wormhole pattern if the core used is longer. This is because the live acid delivered by convection to the tip of the wormhole decreases when core length increases. The acid concentration eventually becomes too low to ensure further propagation of the wormhole. Consequently, it is risky to quantitatively extrapolate the short-core experimental results into field-scale design.

Another common practice in core-flow experiments is using 1,000 psi as the system pressure. Experimentalists frequently claim 1,000 psi to be sufficiently high to keep the evolved $CO_2$ in solution during laboratory acidizing studies; hence, the results obtained using such pressure are applicable to real reservoir pressure, which is normally much higher than 1,000 psi. Thermodynamics studies and experiments by several researchers (Chang 2012; Mumallah 1991) indicate that 1,000 psi is far from being high enough to keep $CO_2$ in solution; the wormholing phenomena at 1,000 psi are by no means representative of those at high reservoir pressure. Assisted by the $CO_2$-bubble-forced convection, the mass-transfer-controlled acid-carbonate reaction rate at 1,000 psi is unrealistically higher than that in most of the real reservoir. It has been demonstrated that when raising the pressure to 3,000 psi, the wormhole-penetration velocity becomes constant along a 12-in. core, whereas the wormhole velocity reduces as it penetrates into the long core. Though the PVBT concept is good for explaining qualitatively the wormholing phenomena upon changing injection rate, one should refrain from applying it to quantitatively predict wormhole morphology in real reservoirs.

## 4.5 Wormholing Models

Different approaches have been taken by researchers to mathematically model the wormholing process at the core scale. The typical output of these models is a graphic image of the dissolution pattern and wormholing velocity through the core as a function acid volume and rate injected. The modeling results are often compared with the core-flow experiments to demonstrate the their ability to graphically reproduce the experimentally observed dissolution patterns to a certain degree of similarity (Hoefner and Fogler 1988; Daccord, et al. 1989) or to match experimentally obtained acid volume requirement for penetrating the entire core length, so called pore volume to breakthrough (PVBT). Additionally, efforts have been made to extrapolate the core-scale model to the field-scale model (Tardy et al. 2007; Buijse 2005).

A. Core-Scale Models

The core-scale models can be divided into three major categories: (1) fractal model (Daccord et al. 1993; Frick et al. 1994), (2) network model (Hoefner and Fogler 1989), and (3) mechanistic model (Panga et al. 2004).

1. Fractal Model
The pattern of carbonate dissolution by acid has been identified to be fractal, when observed from the linear and radial core-flow experiments using HCl dissolving carbonate rocks and water dissolving gypsum plaster (Daccord 1989). The analytical model can be used to describe an equivalent wormhole-penetration length, which is defined by the distance within which the pressure drop is zero. Therefore, in a linear core, the equivalent wormhole length is expressed as

$$\Delta P = \frac{\mu q (L - L_e)}{k_o \pi R^2}, \qquad (4.28)$$

where
$L$ = total core length,
$k_o$ = original core permeability,
$L_e$ = equivalent wormhole length,
$q$ = injection rate,
$R$ = radius of the core,
$\mu$ = viscosity, and
$\Delta P$ = differential pressure.

Many short linear core experiments showed that the differential pressure decreases linearly with time (Daccord et al. 1989; Wang 1993); hence, $L_e$ increases linearly with time.
Similarly for radial flow in a cylindrical core, the expression of an equivalent wormhole radius is

$$\Delta P = \frac{\mu q}{k_o 2\pi L} \ln \frac{R}{r_e}, \qquad (4.29)$$

where
$L$ = height of the cylindrical core,
$R$ = outer radius of the core, and
$r_e$ = equivalent wormhole radius.

Experiments showed that the $\Delta p$ and $r_e$ decrease proportionally to time raised to an inverse fractal dimension ($d_f$), as follows:

$$r_e = c_1 t^{\frac{1}{d_f}} \qquad (4.30)$$

In Eq. 4.30, $c_1$ is a constant, $t$ is time, and the fractal dimension ($d_f$) has a value from 1.5 to 1.7 on the basis of the radial-core-flow experimental data. It was also found that the wormhole-growth velocity ($v_e$) is linearly proportional to the rock solubility (Daccord et al. 1993), which is characterized by a dimensionless "acid capacity number" ($N_{ac}$). The $N_{ac}$ is defined as the ratio of the volume of the mineral dissolvable by a certain fluid volume to the volume of mineral in the porous media of equal volume to the fluid:

$$N_{ac} = \frac{CM}{\beta(1-\phi)\rho}, \qquad (4.31)$$

where
$C$ = concentration of the reactive fluid,
$M$ = molecular weight of the mineral,

$\beta$ = stoichometric coefficient in the reaction (mole of fluid to dissolved one more of mineral),
$\varphi$ = porosity of the porous medium, and
$\rho$ = density of the mineral.

Assuming an average velocity over the entire radius of the core sample in radial flow, the wormhole velocity up to breakthrough can be related to $N_{ac}$ and injection rate by

$$\frac{R}{t_{bt}} = c_2 N_{ac} q^{2/3}, \quad (4.32)$$

in which $c_2$ is a constant and $t_{bt}$ is the time required to break through the radial core sample with radius $R$. Combining Eqs. 4.30 and 4.32 yields

$$r_e = R^{\left(1-1/d_f\right)} c_2^{1/d_f} (N_{ac} q^{2/3} t)^{1/d_f} \quad (4.33)$$

Eq. 4.33 describes the dependence of equivalent wormhole penetration in a fractal pattern on injection rate and injected volume, which is the flow-rate-and-time multiplication. In a convection/diffusion process, the mass transfer of reactants toward the solid surface by diffusion impacts how much reactants are available to be delivered to the tip of the wormhole. Therefore, the Péclet number can be used to quantify the effect of reaction kinetics on acid dissolution at the tip of the wormholes. The Péclet number is defined by the ratio of the rate of transfer by convection to the rate of transfer by diffusion. In radial flow in porous media, it can be defined by (Daccord et al. 1993)

$$N_{\mathrm{Pe}} = \left(\frac{q}{\varnothing \pi L D}\right)\left(\frac{\sqrt{k_o}}{r}\right), \quad (4.34)$$

where $D$ is the diffusion coefficient; all other terms are defined previously.

Daccord et al. (1989) illustrated from their experiments that the wormhole velocity is proportional to flow rate raised to the (–1/3) power **(Fig. 4.9).**

Since the Péclet number is directional proportional to flow rate, it can therefore be related to wormhole velocity by the (–1/3) power as well. Applied in this way, Daccord et al. (1993) showed that Eq. 4.33 can be rewritten as

$$\frac{r_e}{R} = \left[\frac{c_3}{\pi^{\frac{1}{3}} R^{d_f - \frac{5}{3}}} \left(\frac{q}{\pi L D \varnothing} \frac{\sqrt{k_o}}{R}\right)^{-\frac{1}{3}} N_{ac} \left(\frac{qt}{LR^2}\right)\right]^{1/d_f} \quad (4.35)$$

The advantage of such an analytical model is that it can be easily extended from the laboratory scale to the field scale (Frick et al. 1994) to predict the skin-factor evolution during the acidizing treatment. Since the definition of skin is $s=(k_o/k-1)\ln(r/r_w)$, $r$ is now the equivalent radius $r_e$, within which there is no pressure drop [i.e., $k(r<r_e)=\infty$]; the skin is therefore equal to $-\ln(r_e/r_w)$.

2. Network Model
The network model for acidizing of carbonate involves constructing a 3D network of nodes connected by capillary tubes (bonds) to represent a core sample to be acidized (Hoefner and Fogler 1988). The fluid flows between nodes through the bonds connecting them and mixed in the nodes. The number of bonds coming in and going out of each node is defined by the

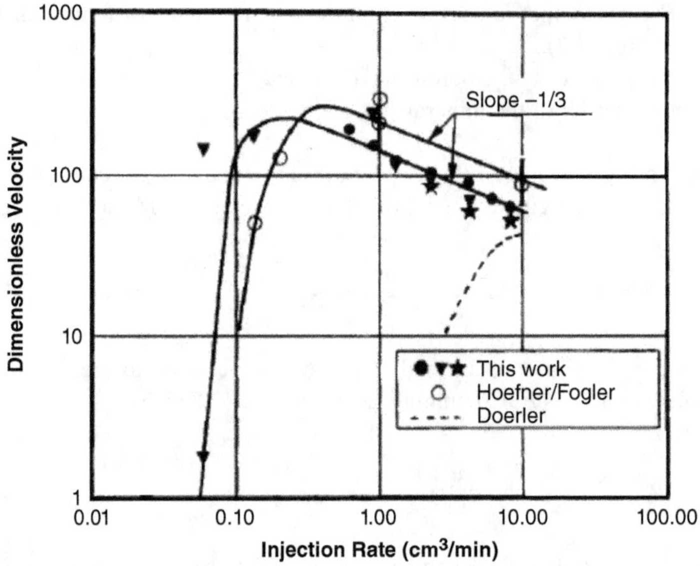


**Fig. 4.9—Core-flow experiments showing the wormhole velocity in proportional to flow rate raised to the (−1/3) power (Daccord et al. 1989).**

coordination number of the network. A simulation of a mercury-injection process through the bonds is matched to the experimentally obtained mercury-injection-porosimetry curve of the core. Once the match is successful, the size distribution is assigned to the bonds in the network. The flow through each bond in the network and the pressure at the node is calculated by simultaneously solving mass balances at each node for either constant-pressure or constant-flow-rate conditions by use of the Hagen-Poiseuille relationship for laminar flow in cylindrical tubes.

When dissolved by acid, the bonds will grow in diameter in proportion to the square root of the amount of the material dissolved from that bond. The rate of bond-diameter growth is therefore dictated by the reaction kinetics between the acid and the mineral. As discussed in the preceding section, for mass-transfer-limited reaction, the reaction rate is related to the acid concentration and diffusion coefficient. The acid concentration at a particular location in the network can be calculated by (Hoefner and Fogler 1989)

$$C = C_o e^{-a\left(\frac{D_e^{\frac{2}{3}} L}{q^{1-b} d^b v^{b-\frac{1}{3}}}\right)}, \qquad (4.36)$$

where

$a$ and $b$ = constants,
$C$ and $C_o$= concentration and initial acid concentration,
$d$ = bond diameter,
$D_e$ = diffusion coefficient,
$L$ = bond length, and
$q$ = flow rate.

The parenthetical term in Eq. 4.36 is the Damköhler number (Hoefner and Fogler 1988). And the diameter evolution of each individual bond diameter is expressed by (Hoefner and Fogler 1989)

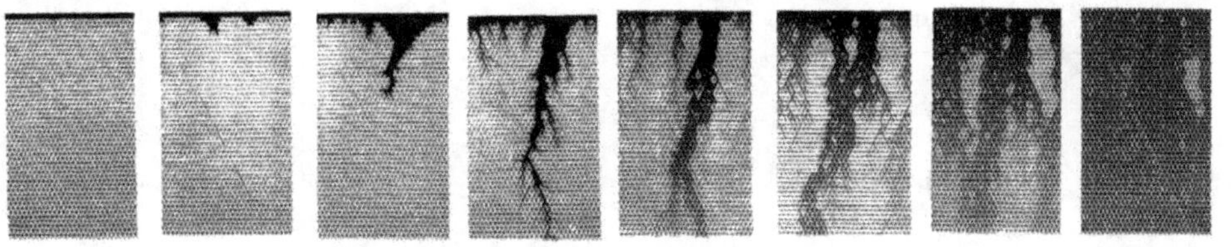

**Fig. 4.10—Wormholing pattern as a function of Damköhler number simulated by network model (Hoefner and Fogler 1988). The Damköhler number sequentially increases from left to right. Acid injected from top to bottom in each simulated porous medium.**

$$\frac{\Delta d}{d}=\left(\frac{C_o}{\beta}\right)\left(\frac{q\Delta t}{d^2L}\right)\left[1-e^{-a\left(\frac{D_e^{\frac{2}{3}}L}{q^{1-b}d^b v^{b-\frac{1}{3}}}\right)}\right], \quad (4.37)$$

in which $\beta$ is the stoichiometric constant of the acid-carbonate reaction.

With suitable boundary conditions, Eqs. 4.36 and 4.37 can be used to compute bond diameters and the pressure field across the core during the acid penetration. The contour of bond diameter can also be drawn to show the wormhole pattern as a function of permeability, flow rate, and reaction kinetics (Damköhler number). **Fig. 4.10** gives an example of network-model simulation to illustrate the impact of the Damköhler number on a wormholing pattern.

The advantage of the network model is that it simulates the fluid dynamics and chemical-reaction process at the microscopic level, making the scientific description of the process more fundamentally sound. Therefore, the rendered wormholing profile is more meaningful and realistic, though quantitatively it still requires match constants. The model is an excellent tool to demonstrate how individual parameters in the entire process impact the wormhole depth and diameter along the domain. The drawback of the network model is the requirement of large computational resources; this makes it not possible to extend it to the field level of studies.

3. Two-Scaled Continuum Models

   The two-scaled continuum model starts from the multicomponent mass- and momentum-balance equations to describe the fluid flow in porous media. This is defined at a Darcy scale, which is normally associated with core-level details. The Darcy-scaled formulation is then coupled with the mass-transfer-controlled reaction kinetics at a much smaller pore-scale level. This coupling makes the model capable of simulating reactant transport and pore-size evolution during the acidizing process (Kalia and Balakotaiah 2007). In a radial-flow ($r$–$\theta$) domain, the equations governing the acid flow and porosity evolution are

$$(u,v)=-\frac{K}{\mu}\left(\frac{\partial P}{\partial r},\frac{1}{r}\frac{\partial P}{\partial \theta}\right), \quad (4.38)$$

$$\frac{\partial \phi}{\partial t}+\frac{1}{r}\frac{\partial (ru)}{\partial r}+\frac{1}{r}\frac{\partial v}{\partial \theta}=0, \quad (4.39)$$

$$\frac{\partial(\phi C_b)}{\partial t}+\frac{1}{r}\frac{\partial(ruC_b)}{\partial r}+\frac{1}{r}\frac{\partial(vC_b)}{\partial \theta}=\frac{1}{r}\frac{\partial}{\partial r}\left(r\phi D_e\frac{\partial C_b}{\partial r}\right)+\frac{1}{r}\frac{\partial}{\partial \theta}\left(\frac{\phi D_e}{r}\frac{\partial C_b}{\partial \theta}\right)-k_{mt}a(C_b-C_s) \quad (4.40)$$

$$\frac{\partial \phi}{\partial t} = \frac{R(C_s)\beta a}{\rho_s}, \quad (4.41)$$

$$k_{mt}\left(C_b - C_s\right) = k_r(C_s) \quad (4.42)$$

In Eqs. 4.38 through 4.42,

$a$ = solid/liquid contact area;
$C_b$ and $C_s$ = bulk and surface acid concentrations, respectively;
$D_e$ = diffusion coefficient;
$K$ = permeability;
$k_{mt}$ and $k_r$ = mass-transfer coefficient and reaction rate constant, respectively;
$u$ and $v$ = velocity in radial and $q$ directions, respectively;
$\beta$ = dissolving power of the acid (mass of solid dissolved per mole of acid reacted); and
$\rho_s$ = density of the solid.

Eqs. 4.38 through 4.42 describe the transport of the acid species through the porous media and the porosity change as the reaction between the acid and the rock takes place. The model has to be completed by defining the new fluid-flow characteristics after the porosity field has been altered by the acid dissolution. This is achieved by the pore-scale modeling that correlates the pore size to porosity and then by correlating the porosity to the permeability. The following relationships are defined:

$$\frac{K}{K_o} = \frac{\phi}{\phi_0}\left[\frac{\phi(1-\phi_0)}{\phi_0(1-\phi)}\right]^{2c}, \quad (4.43)$$

$$\frac{r_p}{r_{po}} = \sqrt{\frac{K\phi_o}{K_o\phi}}, \quad (4.44)$$

$$\frac{a}{a_o} = \frac{\phi r_{po}}{\phi_o r_p} \quad (4.45)$$

In Eqs. 4.43 through 4.45, $r_p$ is average pore radius, and all other terms have been defined previously. The subscript "0" denotes the initial value of the respective properties. Acid dissolution of the rock creates new porosity and therefore new permeability; the new permeability is calculated by Eq. 4.43. With the new permeability, the new pore radius is calculated by Eq. 4.44. Finally, the new rock/fluid contact area is calculated by Eq. 4.45. Eqs. 4.38 through 4.45 are repeated using the newly updated values for all the parameters to continuously obtain new pore structure for the next timestep until the boundary is reached.

The two-scaled continuum model mechanistically simulates the physical and chemical processes of acid dissolution. It can be readily scaled up to the field scale with sufficient computing power. However, a key and difficult task is to be able to correctly correlate among pore size, porosity, contact surface, and permeability (Eqs. 4.43 through Eq. 4.45) for specific rock fabrics before the model can be quantitatively correct.

B. Field-Scale Models

The main purpose of a field-scale model is that it can be used to design acidizing treatment in the formation or to analyze the treatment results. It requires the model to accept a broad range of input geological and engineering parameters and still be able to execute rapidly.

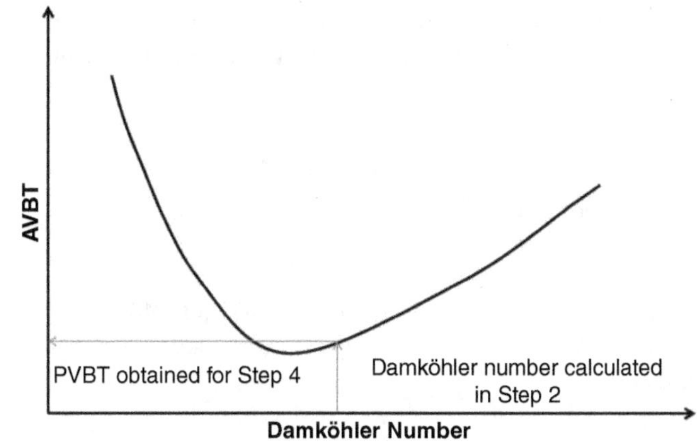


**Fig. 4.11—The PVBT vs. Damköhler number from acidizing experiments using linear cores at various injection rates.**

One field-scale model uses pore volume to breakthrough (PVBT) data measured from linear acidizing coreflood experiments to track the wormhole front propagation in radial flow (Tardy et al. 2007). The algorithm of the model can be summarized as follows:

1. The model first calculates an average Darcy velocity ($v_r$) at the front of the wormhole tip. In radial flow, the Darcy velocity is calculated by assuming the wellbore radius is equal to the wormhole penetration because there is no pressure drop from the wellbore to the tip of the wormhole.
2. From the Darcy velocity and known reaction kinetics, a corresponding Damköhler number can be calculated.
3. Construct a PVBT vs. Damköhler curve by conducting acidizing experiments on linear cores from the specific formation **(Fig. 4.11).** Multiple experiment runs are needed at various injection rates to obtain such a curve.
4. Obtain the PVBT for the Damköhler number calculated in Step 2 using the PVBT vs. Damköhler plot from Step 3. Note PVBT is not a volume but a velocity concept. The dimensionless wormhole velocity is equal to 1/PVBT.
5. The wormhole velocity from this radius will therefore be calculated by

$$v_{wh} = \frac{\lambda}{PVBT} \frac{v_r}{\phi_0}, \quad (4.46)$$

where

$v_r$ = the average Darcy velocity from Step 1,
$v_{wh}$ = the wormhole velocity, and
$\lambda$ = a fitting coefficient, that is approximately equal to 1.3.

The other field-scale model was developed by Buijse and Glasbergen (2005). The concept of modeling is similar to that described in Tardy et al. (2007). This model uses a simple wormhole velocity to preserve the dependence of wormhole velocity on injection rate raised to an exponent of 2/3, described by Daccord et al. (1989, 1993), and observations from linear short-core acidizing experiments (Bazin et al. 1995). The wormhole velocity is formulated as

$$v_{wh} = W_{\text{eff}} v_i^{\frac{2}{3}} B(v_i), \quad (4.47)$$

where

$v_i$ = interstitial velocity calculated at the current wormhole penetration,
$v_{wh}$ = wormhole velocity in radial flow,

$B$ = a function to describe the compact dissolution at low flow velocity, and
$W_{eff}$= wormhole efficiency factor that can be obtained experimentally.

In Eq. 4.47

$$v_i = \frac{q}{2\pi r_{wh} h\phi} \text{ and } B(v_i) = \left(1 - e^{-W_B v_i^2}\right)^2 \quad (4.48)$$

The matching parameters $W_{eff}$ and $W_B$ in Eqs. 4.47 and 4.48, respectively, can be obtained from linear or radial core-flow experiments, field-execution matches, or the following simple approximations:

$$W_{\text{eff}} = \frac{v_{i-\text{opt}}^{1/3}}{PV_{bt-\text{opt}}} \text{ and } W_B = \frac{4}{v_{i-\text{opt}}^2} \quad (4.49)$$

**4.5.1 Wormhole Density.** One may argue that the number of wormholes during acid dissolution of carbonate rock is dictated by the rock fabric and permeability heterogeneity. It can be shown, however, that the combined effect of reaction kinetics, acid dissolving capacity, and flow velocity plays a significant role in the wormhole branching process. These parameters are actually controllable by the treatment designer.

When two adjacent wormholes are too close to each other, they compete for the acid fluid. The result is that the smaller branch could cease to grow because the wider and/or the longer wormhole branch consumes most of the available acid by its higher hydraulic conductivity (Hoefner and Fogler 1988). On the other hand, the acid concentration at the tip of the longer wormhole may be significantly decreased because of acid spending. The acid in the shorter wormhole branch is fresher and can react with the rock more efficiently. This prevents the longer wormhole from penetrating unlimitedly, especially when the two branches are far apart. The wormholes eventually self-regulate to reach equilibrium (Chang 2012; Qiu et al. 2013). The wormholing pattern therefore becomes uniformly spread around the circumference and along the axis of the wellbore.

The network model can be used to calculate flow through wormholes. Simulation results predict that the difference in the lengths of two channels is very unlikely to exceed the distance between the channels. This concept is consistent with experimental observation that a branch from a dominant channel never grows much longer than the distance between the branch and the main channel. In a linear coreflood, the number of wormholes that are expected to propagate to a significant fraction of the total length of the core should be roughly equal to the ratio of the inlet area to the core length. In a radial coreflood, this means that no more than four or five channels will form in a plane (Hoefner and Fogler 1988).

Another way to look at the wormhole density is by a wormhole-initiation-pressure gradient (IPG) (Huang et al. 1999). The IPG is defined as the pressure gradient at which the optimal acid-injection rate (Wang 1993) for a single dominant wormhole is attained. When a wormhole is initiated at any given point along the wellbore, the local pressure distribution is instantaneously reduced. Within the region influenced by this pressure disturbance, no wormhole can be initiated because of the pressure gradient falling below IPG. Moving away from this region, the pressure disturbance vanishes and the local pressure gradient is maintained at the IPG, so another wormhole could therefore be initiated. Using the IPG concept, it can be estimated that in a homogeneous rock, there should be approximately four wormholes in a cross section of an openhole wellbore spreading approximately 4 in. apart along the wellbore axis (Huang and Hill 199).

## 4.6 Additives

Practically all acidizing systems require several different additives to be mixed with the acid solution. The additives are used to ensure compatibility of injected fluid with reservoir fluids, as well as to enhance the performance of the injected fluids so that they reach their intended functionality. Common components in an acidizing fluid system may include

1. Corrosion inhibitors to prevent corrosion of tubular goods and injection of iron into the formation to cause formation damage
2. Iron-ion stabilizers incorporating iron reducing and chelating agents to prevent iron-scale precipitation
3. Hydrogen sulfide ($H_2S$) scavenger to prevent iron sulfide (FeS) precipitation and the dangerous $H_2S$ gas once the fluid returns to surface
4. Nonemulsifier to prevent generation of emulsion when the acid and spent acid come in contact with reservoir crude
5. Surfactants to reduce surface tension between the injected and the reservoir fluids, suspend fines, or create foam (when foamed acid or diverting agent is used)
6. Mutual solvent to reduce the interfacial tension between the injected acid with the reservoir crude and to remove hydrocarbon from the rock surface so the aqueous acid fluid can readily react with the rock
7. Friction reducer to allow high-rate injection with available pumping capacity

There could be other additives, such as fluid-loss additives and viscosifying agents to fit the requirements according to specific reservoir-fluid and -rock properties and the type of acid used for the stimulation treatment. There is no universal acid composition because formations or formation fluids are never identical. Ensuring compatibility among all the additives, the main acid and reacted acid, and the reservoir fluid is critically important to prevent upset of the production flow path from the formation to the surface facility.

**4.6.1 Corrosion Inhibitors.** Corrosion is the loss of metal or alteration of metal as a result of chemical reactions. It causes serious damage to the pumping and storage equipment. More importantly, it weakens the well integrity and potentially causes loss of the entire well. Therefore, corrosion inhibitors are the most critical additive among all; they must always be used in acid treatments.

Corrosion during acidizing is caused by an electrochemical process: The metal (iron) becomes an anode (oxidation), and the hydrogen ions are reduced (cathode). Normally, the cathode requires another metal (such as platinum) to complete the electron flow, as shown in **Fig. 4.12;** however, the microscopic grainy structure and the imperfection of metal composition in the tubular provide the site for the anode/cathode reaction in the environment of the well.

Corrosion inhibitors are chemical components to protect metal from acids, brines, and even naturally occurring formation fluids. Their function is to interfere with the reactions on either the anode or the cathode. Therefore, corrosion inhibitors can be divided into anodic inhibitor and cathodic inhibitor. The corrosion inhibitor can make a physical barrier to the metal surface to retard diffusion of ions or to

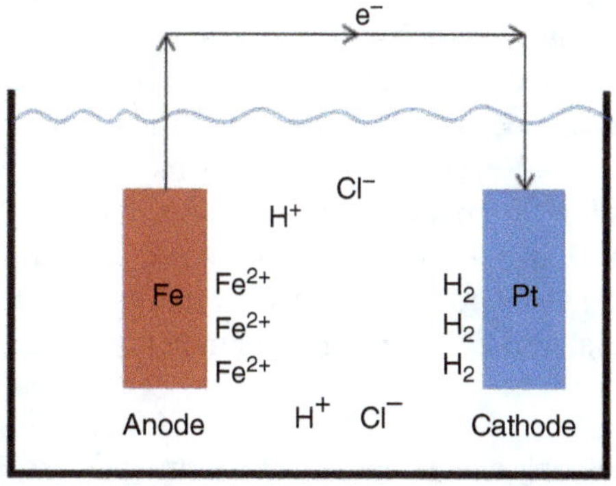


**Fig. 4.12—Redox reaction of iron in acid solution causing corrosion.**

block the anodic and/or cathodic reaction sites; it can interact with the intermediate corrosion reaction that adsorbed on the surface or it can affect the electrochemical reaction rate in the electrical double layer that developed around the metal/solution interface (Rostami 2009). The chemical nature of the inhibitor can be organic or inorganic.

Organic corrosion inhibitors usually consist of a polar group containing oxygen, sulfur, and nitrogen. The polar group is then attached to a long-chain hydrocarbon. The organic corrosion inhibitors may reduce cathodic and anodic reactions by adsorbing at the metal/metal-oxide surface. Adsorption occurs because of the polar or charge nature of organic molecules/ionic species first establishing a physisorbed surface film through van der Wals force; the film may then be further stabilized through chemisorptions. The most effective organic inhibitors are usually film-forming amines (Rostami 2009). Once adsorbed, the molecule of the polar group is attached to the metal surface, leaving the hydrocarbon chain stretching toward the fluid forming a hydrocarbon film that protects the metal from being closely contacted by the aqueous acid.

Inorganic anodic inhibitors are often sodium or potassium salts containing anions that form a precipitated salt layer on the anode. For example, sodium phosphate can form iron phosphate at the anode where iron dissolves. Inorganic cathodic inhibitors are often salts of zinc and nickel that form insoluble hydroxides at the cathode where hydroxyl ions are formed. Salts of arsenic and antimony can be also cathodic inhibitors by plating onto the cathode as free metal and hence limiting the hydrogen evolution.

Corrosion inhibitors function well at a temperature below 212°F (100°C). To improve the inhibition effect of corrosion-inhibitor molecules and to prolong the treating at high temperature, aids or intensifiers must be added to the acid. The most important and common intensifiers include aliphatic organic acids, iodide salts, formic acid, or formamide. Formic acid is one of the commonly used intensifiers. Formic acid requires a strong acid such as hydrochloric acid (HCl) and heat to decompose. The mechanism of reaction and decomposition of formic acid in HCl is suggested to happen by a series of reversible steps (Cassidy 2007).

**4.6.2 Iron Reducing Agents.** The sources of iron when acidizing a formation can include reservoir-rock mineral composition, tubing, casing, and storage and pumping equipment dissolved by the injected acid. The iron derived from the acid dissolution could be ferrous or ferric according to the following reactions:

$$2HCl + Fe \rightarrow FeCl_2 + H_2. \quad (4.50)$$

This reaction represents the HCl attack directly on the steel to produce ferrous ($Fe^{2+}$) ions. Often the steel could have patches of oxidized (rusted) spots. The rust is predominantly $Fe_2O_3$, which can be dissolved by HCl generating ferric ($Fe^{3+}$) ions as follows:

$$6HCl + Fe_2O_3 \rightarrow 2FeCl_3 + 3H_2O \quad (4.51)$$

Ferrous iron in solution is greenish, while ferric iron is orange in color. The more problematic iron species is the ferric ($Fe^{+3}$) ion because it can precipitate as $Fe(OH)_3$ when the pH of the fluid has reached 2–2.2, whereas $Fe^{+2}$ can precipitate as $Fe(OH)_2$ only when the pH is above 7–7.7 (Crowe 1985; Smith et al. 1969), which will not occur in a carbonate acidizing environment because of produced $CO_2$, as discussed in Section 4.3—Reaction Chemistry. A reducing agent is, therefore, warranted to convert the malign ferric ions ($Fe^{3+}$) into more-benign ferrous ions ($Fe^{2+}$).

A reducing agent is a chemical in a reduction-oxidation reaction that donates an electron to another chemical. Erythorbic acid ($C_6H_8O_6$) (Crowe 1985) and its sodium salt, hydrazine ($N_2H_4$), and hydroxylamine ($NH_2OH$) complex (Hall and Dill 1988) are common iron-reducing agents used in the oil field during acidizing treatments.

**4.6.3 Chelating Agents.** Chelating agents are used to form stable metal-chelate complexes. When the metal ions are surrounded by the large chelating agent to form a 3D structure **(Fig. 4.13),** they are essentially isolated from the other anions that can cause scale precipitation and other adverse reactions

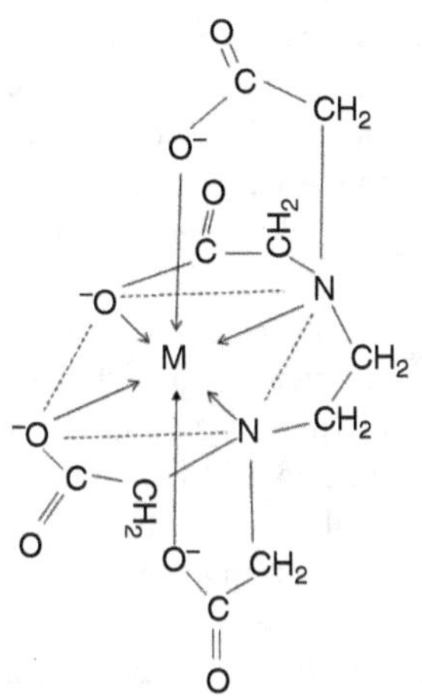

**Fig. 4.13—Metal-EDTA complex 3D structure.**

with the metal ions. The most important function of the chelating agent in acidizing applications is to prevent iron-related precipitations such as FeS, $Fe(OH)_3$, and sludge.

**4.6.4 $H_2S$ Scavengers.** Hydrogen sulfide ($H_2S$) is a gas formed by the action of sulfur-reducing bacteria, thermal alteration of organic sulfur, or thermal reduction of sulfate in formation fluids. It is known as "sour gas." $H_2S$ corrodes the iron goods by forming iron sulfide and hydrogen (Eq. 4.52). The precipitated FeS is a scale that causes severe formation damage.

$$H_2S + Fe \rightarrow FeS + H_2 \quad (4.52)$$

Acid could dissolve the FeS scale, thereby liberating $H_2S$. When there is ferric ion present because of acid dissolution of $Fe_2O_3$ mill scale, $Fe(OH)_3$ scale, or other ferric-ion-containing minerals such as hematite ($Fe_2O_3$), the $H_2S$, being a strong reducing agent, will react with the ferric ions ($Fe^{3+}$) to reduce them into ferrous ions ($Fe^{2+}$) and at the same time to precipitate sulfur (S) (Crowe 1985).

$$2Fe^{+3} + H_2S \rightarrow 2Fe^{+2} + S\downarrow + 2H^+ \quad (4.53)$$

To adequately prevent various precipitants encountered during acidization of sour wells, an additional additive, $H_2S$ scavenger, is needed in addition to the iron reducing and chelating agents. Many believe the reducing agent is not necessary because $H_2S$ itself is such a strong reducing agent that $Fe^{3+}$ ions will be nonexisting. Therefore, only chelating agents must be added to complex the produced ferrous iron to eliminate the reprecipitation of ferrous sulfide. There are many types of scavengers using various reaction mechanisms to remove $H_2S$ from the fluid composition (Foral and Al-Ubaidi 1995). Generally, the scavenger chemicals react rapidly with $H_2S$ to produce organic compounds of sulfur that are soluble in water. This modification of sulfur reduces the possibility of sulfide scale formation (Hall and Dill 1988).

**4.6.5 Antisludge Agents.** The asphaltenes in the crude oil are colloidal particles with diameter of 0.5 μm or smaller and with negative charge. They are suspended in crude oil by the natural surfactants in the oil. However, upon contacting positively charged ions such as $H^+$, $Fe^{2+}$, $Fe^{3+}$, and sometimes even $Ca^{2+}$, the colloids may flocculate and precipitate as sludge. Crude-oil sludge can plug flow channels and pore spaces to degrade the performance of acid treatments. Therefore, antisludge additives must be added to control the charge on the surface of the colloid to hold the particle in suspension during the acid treatment.

Antisludge chemicals are surfactants and a mutual solvent, plus iron control agents such as chelating agents.

**4.6.6 Friction Reducers.** Friction reducers are mostly water-soluble long-chain polymers. The polymer molecules stretch in the shear flow to suppress the turbulence while the fluid is injected at high rate. Though friction reducers are more frequently used in hydraulic-fracturing operations, they can be very relevant for acidizing in long horizontal wells where the injected fluid is exposed to a large surface area along the wellbore and high fluid velocity needs to be maintained for effective stimulation.

**4.6.7 Surfactants and Mutual Solvents.** Surfactants are widely used in acidizing fluids. They serve multiple functions to ensure the main treatment fluid performs the designed purposes of maximizing the productivity potential of the well. The functions of surfactants include minimizing interfacial tension between injected and formation fluids, controlling wettability, breaking or preventing emulsions and sludge, reducing water blocking after the stimulation job, and dispersing and suspending fines or other solids. Surfactants are also used as a foaming agent for diversion or emulsifying agent for reaction retardation.

Mutual solvents are mutually soluble in oil and aqueous phases. They serve functions similar to those of the surfactants, but they are more effective because unlike surfactants they do not get adsorbed on the rock surface; hence, they can penetrate deeper into the formation. And since they are truly soluble in both phases rather than relying on the surface interaction promoted by the surfactants, they are also more effective in wetting and emulsion-prevention purposes.

Surfactants and mutual solvents are sensitive in a mixture of many chemicals. They can destroy and be destroyed by other additives or among themselves. Therefore, the compatibility of all chemicals should be carefully checked before applying the final formulation to the stimulation treatments.

## 4.7 Treatment Design

**4.7.1 Skin Factors.** In sandstone acidizing, acid dissolves the pore-filling or pore-lining materials such as clays. The skeletal structure of the rock that is composed of quartz grains remains mostly intact. The objective is to restore the original rock permeability, so the skin factor returns to zero. Acidizing in carbonate differs from that in sandstone because acid dissolves the main structure of the rock matrix, which is made of calcite and dolomite. The result is many long and highly conductive channels, derived from complete dissolution of rock in that region, extending from the wellbore into the formation. Hence, the effective wellbore radius could be so large that the final skin becomes negative. Aside from several fundamental practices in preparing the wellbore and near-wellbore formations by pickling and preflushes, sandstone acidizing design focuses mainly on preventing reprecipitation of secondary and tertiary alumino- and fluosilicates, and carbonate acidizing focuses mainly on penetration and distribution of the wormholes.

**4.7.2 Treatment Considerations.** Generally, the considerations factored into acid treatment design in a carbonate formation include

1. Determine the wormhole-penetration requirement for the desired productivity increase after the acidizing treatment.
2. On the basis of bottomhole temperature and pressure, determine the reaction kinetics that dictates the wormhole penetration and select the main stimulation fluid.
3. Determine the fluid volume, number of stages, and diversion mechanisms according to reservoir geology.
4. Select additives on the basis of the reservoir-rock and -fluid chemical properties, as well as compatibility among the fluid components and with reservoir fluids.
5. Determine the maximum allowable injection rate and pressure without fracturing the formation during treatment.
6. Determine the fluid delivering mechanisms—bullheading or using coiled tubing.
7. Determine the shut-in times, flowback time, and the production mechanism—assisted lifting or natural production.
8. Determine the methods of post-stimulation assessment—quantify the treatment cost vs. productivity gain.

**4.7.3 Treatment Procedure and Volumes.** ***Pickling.*** The purpose of pickling is to prepare the flow path that the acidizing treatment fluid will travel by removing the undesirable materials from the tubing. The inner wall of tubular is often covered by greasy pipe dough, rust, fines, and steel shavings. The pickling fluid should be composed of acids, mutual solvents, surfactants, iron control agents, and a corrosion inhibitor to be able to dissolve, mix, and suspend these materials and carry them out of the wellbore by circulating them through the annulus to surface. Injecting the dirty fluid mixture into the formation could cause significant impairment to the reservoir and jeopardize the success of the stimulation job.

The volume of pickling fluid required depends on the length and diameter of the well trajectory and treating line. One-and-half to two total-flow-path volumes of pickling fluid is recommended.

***Preflush.*** The purpose of preflush is to condition the near-wellbore reservoir rock to become a become a receptive environment for the main acidizing treatment fluid to react with. The rock surface of oil reservoirs is covered by hydrocarbon that impedes the contact of acid. The preflush fluid will help wet the rock surface to increase its affinity to the aqueous phase so the rock/acid reaction can take place more readily. The other function of the preflush is to keep spacer between the main treatment fluid and the formation fluid. This is important when there is potential of forming emulsion and precipitants upon the contact of acid treatment fluid with formation oil and brine. The third function of preflush is to establish the injectivity to gain an idea of the allowable injected rate with the available pumping pressure capacity.

The formulation of the preflush fluid should contain mutual solvent and corrosion inhibitor carried by brine. Often, added chemicals in the preflush fluid may also include scale inhibitor, iron control agents, and nonemulsification surfactants. The volume of preflush should be designed on the basis of the intended penetration of main acid. In carbonate acidizing, this is very difficult to calculate because of the unsure nature of wormhole propagation. The stimulated radius may reach 25 ft away from the wellbore (McDuff et al. 2010), but it is unlikely that the preflush volume can be as such because preflush fluid invades the whole pore space as opposed to the main wormholing fluid that only flows in a small part of the pore space. A conventional rule recommends a preflush of 50 to 100 gal/ft of the treated section along the wellbore. In horizontal wells or long vertical wells, economy will play a determining role and 5 to 10 gal/ft is frequently practiced.

***Main Stimulation.*** The main stimulation fluid is injected after the preflush to either remove the damage near the wellbore or create deep penetrating wormholes. Depending on the reservoir temperature and pressure, the formation lithology, and the reservoir-fluid chemical and physical properties, the main stimulation fluid could be formulated in many different ways to provide the desired chemical and physical characteristics. The main fluid could be selected to be a simple HCl at various concentrations (mostly between 15 and 28 wt%), emulsified or viscosified HCl for reaction retardation, organic acids, or chelating agents. Compatible additives necessary to prevent corrosion, precipitation, and sludge should be included in the fluid formulation.

There has not been a technical specification on how much main stimulation should be pumped. Conventional recommendation calls for 50 to 100 gal/ft of the treated section along the wellbore. Similar to the preflush, economy may prohibit this in horizontal or long vertical wells. Often, 20 gal/ft or less is not an uncommon practice. Instead of relying on a rule-of-thumb approach, a better job design to decide the volume required should be based on the stimulation objectives (Chang and Abbad 2008). It is proposed that the design of the acid volume could be refined by modeling the productivity from a well intersected by wormholes against the expected production target from the stimulation job. Therefore, it requires a better quantification of wormhole geometry resulting from specific acid reacting with specific rock types under realistic reservoir conditions. Though carbonate acidizing studies have been around for nearly 50 years, a continuous quest toward the quantitative description of the wormholing phenomena should still be a worthwhile effort both in laboratory and field measurements.

***Diversion.*** The heterogeneous nature of carbonate rocks makes it impossible to have a uniform distribution of wormholes along the acidized sections. Even in more homogeneous formations, the region first dissolved by acid rapidly becomes a preferred flow path and consequently receives a large portion of the treatment fluids. This leads to uneven stimulation along the zones. Therefore, in addition to producing deep-penetration wormholes to extend the effective wellbore radius, placing the stimulation fluids to create wormholes in less easily drained reservoir sections is also a key factor to ensure the success a matrix acidizing job. Hence, diversion stages are incorporated in nearly every acidizing treatment design.

Understanding how each acid diversion technique works is the first step toward the optimized carbonate acidizing design. Diversion techniques developed for placing stimulation fluids generally utilize mechanisms such as flow dynamics, fluid viscosity, solids bridging, solid plugging, and mechanical tools. A combination of multiple mechanisms is often practiced to achieve the maximum efficiency. The following describes commonly used diverting materials and methodologies.

The strategy practiced most often in diversion is to pump at the maximum pressure differential and injection rates (MAPDIR) to force the acid into low-permeability zones (Paccaloni and Tambini 1993). This technique involves pumping the treatment fluids at the highest possible rate without fracturing. Results show that the MAPDIR technique allows fast total skin reduction in terms of pumping time, but at the expense of large acid volumes injected into the layer with high permeability. Furthermore, it does not allow treating the low-permeability layer fully. Nonetheless, the MAPDIR is a physically sound technique. It is always practiced when a chemical diverter is used.

Coiled tubing is a widely accepted and routinely prescribed tool for matrix acidizing. Spotting the treatment fluid with coiled tubing ensures the delivery of the treatment fluid against the target sections. When combined with chemical diverters and MAPDIR technique, coiled tubing provides a much better chance for success in acid coverage. The drawback of the coiled tubing is that the tubing diameter is much smaller than that of the drillpipe or production tubing used for bullheading treatments. Therefore, the injection rate is limited in situations in which sustained high rate and pressure are required. In addition, when treating long horizontal wells, extending coil tubing to the toe of the well requires tractors or vibrators (Nasr-El-Din et al. 2005; Bhalla 1995), significantly increasing the complexity of the treatment execution.

Mechanical zonal-isolation tools such as packers control the point of fluid entry. Though mechanical isolation tools are effective in accurately placing fluids, setting packers often requires multiple stages of operations that add time and cost to the treatment (Harrison 1972). In open hole, they could also be problem-prone if the borehole is not gauged.

Ball sealers are used in the cased-and-perforated completion to selectively block some perforations. Originally introduced in 1956, ball sealers are small spheres designed to seat in the perforations from the inside of the casing, thereby diverting injected fluids to other perforations (Economides et al. 1994). These balls are normally made of nylon, hard rubber, or biodegradable materials such as collagen (Bilden et al. 1998; Buffet et al. 2002). Ball sealers are only applicable to cased-and-perforated wells. In addition, sufficient differential pressure must be maintained across the perforations to keep the balls in place. The roundness and smoothness of the perforation holes and as the density difference between the balls and the treating fluid will all affect the performance of the ball sealers.

The preceding techniques could be considered mechanical means. Many chemical diverters are also used in matrix acidizing treatments, and they are summarized next. The fundamental difference between a chemical and a mechanical diversion is that a chemical diverting agent achieves diversion by increasing flow resistance inside the created channels, whereas a mechanical diversion process controls the fluid-entry point at the wellbore. Hence, the chemical diverting agents can be considered an internal diverting agent as opposed to the external mechanical diversion.

Foam is a dispersion of a gas in a liquid, with gas being the noncontinuous phase and liquid being the continuous phase. Its stability is largely affected by the nature of the fluid that it contacts. Foams have been used for acid diversion (Smith et al. 1969) most commonly in alternating slugs with acid. Foam is normally generated by commingled injection of a gas and a surfactant solution. As it flows through the porous medium, additional foam is generated. It resists the movement through the rock matrix therefore the ensuing acid can be diverted into the zones that have lower permeability. A case history (Ginest et al. 1993) showed that when treating many openhole water-injection wells completed in the Arab-D formation in Saudi Arabia, the injection pressure gradually rose throughout the treatment giving a positive indication of diversion. Foam blocking contains no solid material that could damage the formation. Foam expansion, while flowing the well back, helps clean the well at lower reservoir pressure.

A viscous-gel diverting agent uses its high viscosity or viscoelastic property to slow down the fluid penetration rate in the existing wormholes so that the acid following the diverter can have a better chance to contact the rock in which wormholes have not formed. The viscous gels use either a polymer solution or a viscoelastic surfactant (VES) (Chang et al. 1999). A polymeric diverting agent leaves residue after the gels break, potentially negating the stimulation by damaging the matrix or the surface of the wormholes, especially if the volume is designed incorrectly. Surfactant-based gel is claimed to be nondamaging; however, the potential of gel not breaking or forming an emulsion with the reservoir oil are drawbacks of this material.

In addition to alternating stages of acid and diverting agent, an improved technology of using a viscosity-control mechanism for acid diversion is an in-situ-gelled acid. The major advantage of an in-situ-gelled acid over alternating staged treatments is that a smaller number of stages is required to achieve diversion because the acid provides the stimulation and diversion at the same time. An in-situ-gelled acid uses unique polymer or surfactant chemistry to produce an acid system that has very low viscosity before reacting with the carbonate rocks. In-situ-gelled acid has been shown to improve acid placement in carbonate reservoirs (Saxon et al. 2000). The difference between an in-situ-gelled acid and external gelled acid is that the external gelled acids or diverting gels are already activated on the surface. Thus, when they reach the formation face, the viscosity is the same against the high-permeability and the low-permeability rock. The proportion of fluid entering each zone is simply determined by the permeability contrast, following Darcy's law. The in-situ-gelled acid offers the benefit of increasing viscosity inside the formation; therefore, when acid first enters the high-permeability zone and generates wormholes, its viscosity becomes higher than the acid still in the wellbore. This creates extra resistance in the already developed high-permeability channels to help alleviate the severity of the permeability contrast. The acid still in the wellbore, therefore, has a better chance to enter the low-permeability untreated zone.

The polymer-based in-situ-gelled acid is prepared by gelling the HCl acid with a polymer and adding into the system a crosslinker and an internal pH-activated breaker (Yeager and Shuchart 1997; Taylor and Nasr-El-Din 2003). When the acid spends upon dissolving carbonate, the increase in pH to near 2 causes the crosslinker to activate and the gelled acid becomes crosslinked. This crosslinked gel temporarily blocks the wormholes being created. Further increase in the pH to near 4 causes the internal breaker to activate and break the gel; the in-situ-gelled acid, hence, breaks down to nearly water-like viscosity to assist easy flowback. Since the system contains polymers, even though the bulk-fluid viscosity is dramatically reduced, there is still polymer residue, which can potentially plug the face of the wormholes (Mohamed et al. 1999).

In-situ-gelled acid system can also be formulated using a VES (Chang et al. 2001). The micelle structure dynamically changes as the pH electrolyte concentration changes upon the acid-carbonate reaction, resulting in viscosity and elasticity change. The gel breaks when hydrocarbons contact the fluid during flowback. The rheological property of VES is a complex function of the environment it is in; hence, it is relatively sensitive to the additives and contaminants (mainly $Fe^{3+}$). Special care must be taken when using this type of chemical.

Particle bridging is a common diverting technique used in carbonate formation. Particulate diverting agents are fine particles that form a medium having significantly lower permeability than the formation. This medium increases the flow resistance to divert acid to other parts of the formation. Materials are used that are not soluble in the original carrying fluid but later become soluble either by contacting hydrocarbon or a post-flush fluid. Rock salt in saturated solution and acid can be injected as solids and dissolved by a freshwater post-flush; benzoic acid flakes remain solid when added in acid, but they are soluble in both water and oil; solid organic acids that hydrolyze to release acid under reservoir temperature have been found applicable for diversion (Glasbergen et al. 2006). Hydrolysis reaction requires sufficient water to ensure full conversion of the solids into acid.

Recently, fibers have been added to viscous fluid to enhance the effect of diversion (Cohen et al. 2010). Fiber strands flocculate and entangle to form as highly resistant network against fluid flow. This is especially effective in blocking flow in high opening channels such as wormholes, fractures, and perforations.

Though the particulates are soluble in water or hydrocarbons, their removal requires contact with the solvent, post-flushed or produced. If the flowing stream and the solids do not effectively mix, the solids could remain as plugging material in the formation and negate the stimulation results.

***Post-Flush.*** The purpose of the post-flush stage is to push any insoluble material after the main acidizing processes far away from the wellbore hoping that the lower pressure gradient is not sufficient to mobilize it. More importantly, the post-flush is to prepare the near-wellbore region for hydrocarbon production. For example, the post-flush can contain solvents to break down the plugging material left by the diverting agent. The post-flush can also contain alcohol or mutual solvents to help more effectively flow back the water block formed by the large quantity of water injected during stimulation treatment. The volume of post-flush should again be customized on the basis of the size of the main stimulation and diversion. Often, post-flush and preflush of equal volumes has been practiced.

## References


Akbar, M., Petricola, M., Watfa, M. et al. 1995. Classic Interpretation Problems: Evaluating Carbonates. *Oilfield Review* **7** (1): 38–57.

Akbar, M., Vissapragada, B., Alghamdi, A. H. et al. 2000. A Snapshot of Carbonate Reservoir Evaluation. *Oilfield Review* **12** (4): 20–21.

Bazin, B., Bieker, H. P., Roque, C. et al. 1995. Improvement in the Characterization of the Acid Wormholing by "In Situ" X-Ray CT Visualizations. Presented at the SPE Formation Damage Control Symposium, Lafayatte, Louisiana, USA, 14–15 February.

Bhalla, K. 1995. Coiled Tubing Extended Reach Technology. Presented at the SPE Offshore Conference, Aberdeen, 5–8 September. SPE–30404–MS. http://dx.doi.org/10.2118/30404–MS.

Bilden, D. M., Lacy, L. L., Seiler, F. H. et al. 1998. New Water-Soluble Perforation Ball Sealers Provide Enhanced Diversion in Well Completions. Presented at the SPE Annual Technical Conference and Exhibition, New Orleans, 27–30 September. SPE-49099-MS. http://dx.doi.org/10.2118/49099–MS.

Buffet, M., Derbez, E., Leschi, P. et al. 2002. Pushing the Limits in Extra-Long Cased Horizontal Drains Acidification: Use of Double Trigger Perforation Technique to Maximize Productivity and Optimize Rig Time. Presented at the Abu Dhabi International Petroleum Exhibition and Conference, Abu Dhabi, 13–16 October. SPE-78543-MS. http://dx.doi.org/10.2118/78543–MS.

Buijse, M. A. and Glasbergen, G. 2005. A Semi-Empirical Model to Calculate Wormhole Growth in Carbonate Acidizing. Presented at the SPE Annual Technical Conference and Exhibition, Dallas, 9–12 October. SPE-96892-MS. http://dx.doi.org/10.2118/96892-MS.

Buijse, M., de Boer, P., Breukel, B. et al. 2004. Organic Acids in Carbonate Acidizing. *SPE Prod & Fac* **19** (3): 128–134. SPE-82211-PA. http://dx.doi.org/10.2118/82211-PA.

Cantrell, D. L., Swart, P. K., Handford, R. C. et al. 2001. Geology and Production Significance of Dolomite, Arab-D Reservoir, Ghawar Field, Saudi Arabia. *Geo Arabia* **6** (1): 45–60.

Cassidy, J. M., McNeil, R. I., and Kiser, C. 2007. Understanding Formic Acid Decomposition as a Corrosion Inhibitor Intensifier in Strong Acid Environments. Presented at the International Symposium on Oilfield Chemistry, Houston, 28 February–2 March. SPE-106185-MS. http://dx.doi.org/10.2118/106185-MS.

Chang, F. F. and Abbad, M. 2008. Optimizing Well Productivity by Controlling Acid Dissolution Pattern During Matrix Acidizing of Carbonate Reservoirs. Presented at the International Petroleum Technology Conference, Kuala Lumpur, 3–5 December. IPTC-12368-MS. http://dx.doi.org/10.2523/12368-MS.

Chang, F. F., Nasr-El-Din, H. A., Lindvig, T. et al. 2008. Matrix Acidizing of Carbonate Reservoir Using Organic Acids and Mixture of HCl and Organic Acids. Presented at the SPE Annual Technical Conference and Exhibition, Denver, 21–24 December. SPE-116601-MS. http://dx.doi.org/10.2118/116601-MS.

Chang, F. F. 2012. Revisiting Wormholing Phenomena During Carbonate Acidizing. Schlumberger Internal Research Report, OFSR/RN/2012/028/OTHER/C.

Chang, F., Qu, Q., and Frenier, W. 2001. A Novel Self-Diverting Acid Developed for Matrix Stimulation of Carbonate Reservoirs. Presented at the SPE International Symposium on Oilfield Chemistry, Houston, 13–16 February. SPE-65033-MS. http://dx.doi.org/10.2118/65033-MS.

Chang, F. F., Love, T., Affeld, C. J. et al. 1999. Case Study of a Novel Acid-Diversion Technique in Carbonate Reservoirs. Presented at the SPE Annual Technical Conference and Exhibition, Houston, 3–6 October. SPE-56529-MS. http://dx.doi.org/10.2118/56529-MS.

Chatelain, J. C., Silberberg, I. H., and Schechter, R. S. 1976. Thermodynamic Limitations in Organic-Acid/Carbonate Systems. *SPE J* **16** (04): 189–195. SPE-5647-PA. http://dx.doi.org/10.2118/5647-PA.

Cohen, C. E., Tardy, P. M. J., Lesko, T. M. et al. 2010. Understanding Diversion with a Novel Fiber-Laden Acid System for Matrix Acidizing of Carbonate Formations. Presented at the SPE Annual Technical Conference and Exhibition, Florence, Italy, 19–22 September. SPE-134495-MS. http://dx.doi.org/10.2118/134495-MS.

Conway, M. W., Asadi, M., Penny, G. S. et al. 1999. A Comparative Study of Straight/Gelled/Emulsified Hydrochloric Acid Diffusivity Coefficient Using Diaphragm Cell and Rotating Disk. Presented at the SPE Annual Technical Conference and Exhibition, Houston, 3–6 October. SPE-56532-MS. http://dx.doi.org/10.2118/56532-MS.

Crowe, C. W. 1985. Evaluation of Agents for Preventing Precipitation of Ferric Hydroxide from Spent Treating Acid. *J Pet Technol* **37** (4): 691–695. SPE-12497-PA. http://dx.doi.org/10.2118/12497-PA.

Daccord, G., Lenormand, R., and Liétard, O. 1993. Chemical Dissolution of a Porous Medium by a Reactive Fluid—I. Model for the "Wormholing" Phenomenon. *Chem Eng Sci* **48** (1): 169–178. http://dx.doi.org/10.1016/0009-2509(93)80293-Y.

Daccord, G., Touboul, E., and Lenormand, R. 1989. Carbonate Acidizing: Toward a Quantitative Model of the Wormholing Phenomenon. *SPE Prod Eng* **4** (01): 63–68. SPE-16887-PA. http://dx.doi.org/10.2118/16887-PA.

de Rozieres, J. 1994. Measuring Diffusion Coefficients in Acid Fracturing Fluids and Their Application to Gelled and Emulsified Acids. Presented at the SPE Annual Technical Conference and Exhibition, New Orleans, 25–28 September. SPE-28552-MS. http://dx.doi.org/10.2118/28552-MS.

Dill, R. W. and Keeney, B. R. 1978. Optimizing HCl-Formic Acid Mixtures for High Temperature Stimulation. Presented at the SPE Annual Fall Technical Conference and Exhibition, Houston, 1–3 October. SPE-7567-MS. http://dx.doi.org/10.2118/7567-MS.

Dunham, R. J. 1962. Classification of Carbonate Rocks According to Depositional Textures. In *Classification of Carbonate Rocks—a Symposium*. ed. W. E. Ham, Vol. Chap. 108–121. Tulsa: American Association of Petroleum Geologists.

Economides, M. J., Frick, T. P., and Nittman, J. 1994. Enhanced Visualization of Acid Carbonate Rock Interaction. *J Pet Technol* **46** (4): 822. SPE-27409-PA. http://dx.doi.org/10.2118/27409-PA.

Foral, A. J. and Al-Ubaidi, B. H. 1995. Evaluation of $H_2S$ Scavenger Technologies: Topical Report. M. W. Kellogg Company, GRI Report No. GRI–94/0197.

Frasch, H. 1896. Increasing the Flow of Oil Wells. US Patent No. 559,669.

Fredd, C. N. and Fogler, H. S. 1998. Alternative Stimulation Fluids and Their Impact on Carbonate Acidizing. *SPE J* **3** (01): 34–41. SPE-31074-PA. http://dx.doi.org/10.2118/31074-PA.

Fredd, C. N. and Fogler, H. S. 1999. Optimum Conditions for Wormhole Formation in Carbonate Porous Media: Influence of Transport and Reaction. *SPE J.* **4** (3): 196–205. SPE-56995-PA. http://dx.doi.org/10.2118/56995-PA.

Frenier, W. W., Fredd, C. N., and Chang, F. 2001. Hydroxyaminocarboxylic Acids Produce Superior Formulations for Matrix Stimulation of Carbonates at High Temperatures. Presented at the SPE Annual Technical Conference and Exhibition, New Orleans, 30 September–3 October. SPE-71696-MS. http://dx.doi.org/10.2118/71696-MS.

Frick, T. P., Kurmayr, M., and Economides, M. J. 1994. An Improved Modeling of Fractal Patterns in Matrix Acidizing and Their Impact on Well Performance. *SPE Prod & Fac* **9** (1): 61–68. SPE-23789-PA. http://dx.doi.org/10.2118/23789-PA.

Ginest, N. H., Phillips, J. E., and Al-Gamber, A. W. A. 1993. Field Evaluation of Acid Simulation Diverter Materials and Placement Methods in Arab-D Injection Wells with Openhole Completions.

Presented at the SPE Production Operationss Symposium, Oklahoma City, Oklahoma, USA, 21–23 March. SPE-25412-MS. http://dx.doi.org/10.2118/25412-MS.

Glasbergen, G., Todd, B. L., Van Domelen, M. S. et al. 2006. Design and Field Testing of a Truly Novel Diverting Agent. Presented at the SPE Annual Technical Conference and Exhibition, San Antonio, Texas, USA, 24–27 September. SPE-102606-MS. http://dx.doi.org/10.2118/102606-MS.

Hall, B. E. and Dill, W. R. 1988. Iron Control Additives for Limestone and Sandstone Acidizing of Sweet and Sour Wells. Presented at the SPE Formation Damage Control Symposium, Bakersfield, California, USA, 8–9 February. SPE-17157-MS. http://dx.doi.org/10.2118/17157-MS.

Harris, F. N. 1961. Applications of Acetic Acid to Well Completion, Stimulation and Reconditioning. *J Pet Technol* **13** (7): 637–639. SPE-63-PA. http://dx.doi.org/10.2118/63-PA.

Harrison, N. W. 1972. Diverting Agents-History and Application. *J Pet Technol* **24** (5): 593–598. SPE-3653-PA. http://dx.doi.org/10.2118/3653-PA.

Hoefner, M. L. and Fogler, H. S. 1989. Fluid-Velocity and Reaction-Rate Effects During Carbonate Acidizing: Application of Network Model. *SPE Prod Eng* **4** (1): 56–62. SPE-15573-PA. http://dx.doi.org/10.2118/15573-PA.

Hoefner, M. L. and Fogler, H. S. 1988. Pore Evolution and Channel Formation During Flow and Reaction in Porous Media. *AIChE J* **34** (1): 45–54. http://dx.doi.org/10.1002/aic.690340107.

Hoornveld, N. 2009. *Deformation History of the Ballino-Garda Line in the Southern Alps, Italy*. MS thesis, VRIJE Univrsiteit Amsterdam, Amsterdam (May 2009).

Huang, T., Zhu, D., and Hill, A. D. 1999. Prediction of Wormhole Population Density in Carbonate Matrix Acidizing. Presented at the SPE European Formation Damage Conference, The Hague, 31 May–1 June. SPE-54723-MS. http://dx.doi.org/10.2118/54723-MS.

Kalfayan, L. 2008. *Production Enhancement with Acid Stimulation*, second edition. Tulsa: PennWell Corporation.

Kalia, N. and Balakotaiah, V. 2007. Modeling and Analysis of Wormhole Formation in Reactive Dissolution of Carbonate Rocks. *Chem Eng Sci* **62** (4): 919–928. http://dx.doi.org/10.1016/j.ces.2006.10.021.

Al-Katheeri, M. I., Nasr-El-Din, H. A., Taylor, K. C. et al. 2002. Determination and Fate of Formic Acid in High Temperature Acid Stimulation Fluids. Presented at the International Symposium and Exhibition on Formation Damage Control, Lafayette, Louisiana, USA, 20–21 February. SPE-73749-MS. http://dx.doi.org/10.2118/73749-MS.

Leet, L. D. ed. 1978. *Physical Geology*, fifth edition. Englewood Cliffs, New Jersey, USA: Prentice Hall.

LePage, J., Wolf, C. D., Bemelaar, J. et al. 2011. An Environmentally Friendly Stimulation Fluid for High-Temperature Applications. *SPE J* **16** (1): 104–110. SPE-121709-PA. http://dx.doi.org/10.2118/121709-PA.

Levich, V. G. 1962. *Physiochemical Hydrodynamics*, second edition. Englewood Cliffs, New Jersey, USA: Prentice-Hall.

Lucia, F. J. 1983. Petrophysical Parameters Estimated from Visual Descriptions of Carbonate Rocks: A Field Classification of Carbonate Pore Space. *J Pet Technol* **35** (3): 629–637. SPE-10073-PA. http://dx.doi.org/10.2118/10073-PA.

Lund, K., Fogler, H. S., and McCune, C. C. 1973. Acidization—I. The Dissolution of Dolomite in Hydrochloric Acid. *Chem Eng Sci* **28** (3): 691–700. http://dx.doi.org/10.1016/0009-2509(77)80003-1.

Lund, K., Fogler, H. S., McCune, C. C. et al. 1975. Acidization—II. The Dissolution of Calcite in Hydrochloric Acid. *Chem Eng Sci* **30** (8): 825–835. http://dx.doi.org/10.1016/0009-2509(75)80047-9.

McCune, C. C., Fogler, H. S., and Cunningham, J. R. 1975. A New Model of the Physical and Chemical Changes in Sandstone During Acidizing. *Society of Petroleum Engineers Journal* **15** (5): 361–370. http://dx.doi.org/10.2118/5157-PA.

McDuff, D., Jackson, S., Shuchart, C. et al. 2010. Understanding Wormholes in Carbonates: Unprecedented Experimental Scale and 3D Visualization. *J Pet Technol* **62** (10): 78–81. SPE-129329-MS. http://dx.doi.org/10.2118/129329-MS.

Mohamed, S. K., Nasr-El-Din, H. A., and Al-Furaidan, Y. A. 1999. Acid Stimulation of Power Water Injectors and Saltwater Disposal Wells in a Carbonate Reservoir in Saudi Arabia: Laboratory Testing and Field Results. Presented at the SPE Annual Technical Conference and Exhibition, Houston, 3–6 October. SPE-56533-MS. http://dx.doi.org/10.2118/56533-MS.

Mumallah, N. A. 1991. Factors Influencing the Reaction Rate of Hydrochloric Acid and Carbonate Rock. Presented at the SPE International Symposium on Oilfield Chemistry, Anaheim, California, USA, 20–22 February. SPE-21036-MS. http://dx.doi.org/10.2118/21036-MS.

Nasr-El-Din, H. A., Al-Arnaout, I. H., Chesson, J. B. et al. 2005. Novel Techniques for Improved CT Access and Stimulation in an Extended-Reach Well. Presented at the SPE/ICoTA Coiled Tubing Conference and Exhibition, The Woodlands, Texas, USA, 12–13 April. SPE-94044-MS. http://dx.doi.org/10.2118/94044-MS.

Nurmi, R. D. and Frisinger, M. R. 1983. Synergy of Core Petrophysical Measurements, Log Data, and Rock Examination in Carbonate Reservoir Studies. Presented at the SPE Annual Technical Conference and Exhibition, San Francisco, 5–8 October. SPE-11969-MS. http://dx.doi.org/10.2118/11969-MS.

Paccaloni, G. and Tambini, M. 1993. Advances in Matrix Stimulation Technology. *J Pet Technol* **45** (3): 256–263. SPE-20623-PA. http://dx.doi.org/10.2118/20623-PA.

Panga, M. K. R., Ziauddin, M., Gandikota, R. et al. 2004. A New Model for Predicting Wormhole Structure and Formation in Acid Stimulation of Carbonates. Presented at the SPE International Symposium and Exhibition on Formation Damage Control, Lafayette, Louisiana, USA, 18–20 February. SPE-86517-MS. http://dx.doi.org/10.2118/86517-MS.

Qiu, X., Zhao, W., Chang, F. et al. 2013. Quantitative Modeling of Acid Wormholing in Carbonates—What Are the Gaps to Bridge. Presented at the SPE Middle East Oil and Gas Show and Conference, Manama, Bahrain, 10–13 March. SPE-164245-MS. http://dx.doi.org/10.2118/164245-MS.

Robert, J. A. and Crowe, C. W. 2000. Carbonate Acidizing Design. In *Reservoir Stimulation*, third edition. ed. M. J. Economides and K. G. Nolte, Chap. 17. New York: John Wiley & Sons, Inc.

Rostami, A. 2009. Review and Evaluation of Corrosion Inhibitors Used in Well Stimulation. Presented at the SPE International Symposium on Oilfield Chemistry, The Woodlands, Texas, USA, 20–22 April. SPE-121726-MS. http://dx.doi.org/10.2118/121726-MS.

Saxon, A., Chariag, B., and Rahman, M. R. A. 2000. An Effective Matrix Diversion Technique for Carbonate Formations. *SPE Drill & Compl* **15** (1): 57–62. SPE-62173-PA. http://dx.doi.org/10.2118/62173-PA.

Scholle, P. A. 1977. Chalk Diagenesis and Its Relation to Petroleum Exploration; Oil from Chalks, a Modern Miracle? *AAPG Bull* **61** (7): 982–1009.

Smith, C. F., Crowe, C. W., and Nolan, T. J., III. 1969. Secondary Deposition of Iron Compounds Following Acidizing Treatments. *J Pet Technol* **21** (9): 1,121–1,129. SPE-2358-PA. http://dx.doi.org/10.2118/2358-PA.

Smith, C. L., Anderson, J. L., and Roberts, P. G. 1969. New Diverting Techniques for Acidizing and Fracturing. Presented at the SPE California Regional Meeting, San Francisco, 6–7 November. SPE-2751-MS. http://dx.doi.org/10.2118/2751-MS.

Tardy, P. M. J., Lecerf, B., and Christanti, Y. 2007. An Experimentally Validated Wormhole Model for Self-Diverting and Conventional Acids in Carbonate Rocks Under Radial Flow Conditions. Presented at the European Formation Damage Conference, Scheveningen, The Netherlands, 30 May–1 June. SPE-107854-MS. http://dx.doi.org/10.2118/107854-MS.

Taylor, K. C., Al-Katheeri, M., and Nasr-El-Din, H. A. 2005. Development and Field Application of a New Measurement Technique for Organic Acid Additives in Stimulation Fluids. *SPE J.* **10** (2): 152–160. SPE-85081-PA. http://dx.doi.org/10.2118/85081-PA.

Taylor, K. C. and Nasr-El-Din, H. A. 2003. Laboratory Evaluation of in-Situ Gelled Acids for Carbonate Reservoirs. *SPE J.* **8** (4): 426–434. SPE-87331-PA. http://dx.doi.org/10.2118/87331-PA.

Van Dyke, J. W. 1896. Increasing the Flow of Oil Wells. US Patent No. 559,651.

Vinograd, J. R. and McBain, J. W. 1941. Diffusion of Electrolytes and of the Ions in Their Mixtures. *J Am Chem Soc* **63** (7): 2,008–2,015. http://dx.doi.org/10.1021/ja01852a063.

Wang, Y. 1993. *Existence of an Optimum Injection Rate in Carbonate Acidizing and the Effect of Rock Heterogeneity on Wormhole Patterns*. PhD thesis, The University of Texas at Austin, Austin, Texas, USA.

Williams, B. B., Gidley, J. L., and Schechter, R. S. 1979. *Acidizing Fundamentals*, second edition, Monograph Vol. 6. Dallas: Henry L Doherty Series, SPE of AIME.

Yeager, V. and Shuchart, C. 1997. In-Situ Gels Improve Formation Acidizing. *Oil Gas J* **95** (3): 70–72.

Chapter 5

# Sandstone Acidizing

**Lee Morgenthaler,** Shell E&P Company, and
**Scott Fogler,** University of Michigan

## 5.1 Introduction

In practice, matrix stimulation design can be based on general guidelines and experience, semiquantitative models, or detailed modeling and simulation. An overview of sandstone stimulation and the fundamental concepts that underlie all three design approaches will be presented. All design approaches have an empirical element because of uncertainty about reservoir response and behavior. The technical sophistication that can be applied is governed by the economics (potential gains and risk vs. cost) and the availability of critical information about the reservoir. As proposed by many authors, it is a progression from less- to more-sophisticated design methods, focusing on critical success factors to maximize the probability of success.

**5.1.1 Process Overview.** Sandstone acidizing is applied in to stimulate well inflow performance in sandstone reservoirs. The working definition of a sandstone for this purpose is reservoirs containing less than 5 to 8% carbonate minerals, with a balance of siliceous minerals (e.g., quartz, clays, feldspars, silica). Most of these reservoirs consist of a matrix of quartz grains and finer clays, feldspar, or amorphous silica particles (Farley et al. 1970; McLeod 1984). The texture (size and spatial distribution) of these particles is a function of the depositional environment and diagenetic history. Mineralogy and texture both influence the expected response to acid stimulation and therefore the design of the acid treatment (Morgenthaler et al 2008).

**5.1.2 Why Stimulate Sandstone Reservoirs?** The simple answer to this question is to remove inflow impairment. Chapter 2 discusses the underlying causes of formation damage in detail. Broadly speaking, they can be broken down into damage that occurs during drilling and completion operations and damage that occurs during production. Sandstone matrix acidizing addresses both causes of damage with the fundamental objective of creating or restoring effective hydraulic communication between the wellbore and the reservoir. As stressed in Chapter 2, the perforations and the near-wellbore formation, roughly the first meter or so from the wellbore, are most likely to become damaged and, when damaged, will have the greatest impact on well productivity.

Sandstone reservoirs are stimulated during completion operations to remove drilling-induced damage. In perforated cased-hole completions of unconsolidated reservoirs where sand control (gravel packing or frac packing) is required, it is critical to establish effective fluid leakoff through as many perforations as possible to ensure gravel placement and slurry dehydration. In both cased- and openhole sand-controlled completions, post-gravel-packing stimulation may be required to remove residual gravel-pack or sand-face damage from the completion operations.

During production, sandstone reservoirs are prone to damage by scale, wax, or asphaltene deposition driven by the phase behavior of reservoir fluids. Scales include carbonates, oxides (corrosion products), sulfides, and sulfates. In general, inhibition is the preferred method for managing sulfate scales. Production wells in seawater floods are especially vulnerable to damage from barium sulfate scale formed by the mixture of seawater and reservoir brine. Barium sulfate removal is discussed briefly in the following. Halite (salt) can be deposited in gas wells because of the evaporation of produced water. Other scales are usually treated by remedial stimulation with acids and special additive packages.

Organic materials, waxes and asphaltenes, can precipitate from the produced oil and impair perforations and the near-wellbore reservoir. In addition, they can create a coating on inorganic-solid deposits that resists penetration by aqueous acids. Solvents, mutual solvents, and surfactants are often included in matrix stimulations to dissolve these materials and water-wet mineral surfaces to make them more susceptible to acid attack. Many wells respond to solvent treatments alone, indicating that in those cases the primary impairment mechanism is organic-solid deposition. Generally, asphaltenes are thought to be the cause of this type of damage far more often than waxes, so it is essential to understand the stability of reservoir oil with respect to asphaltenes when designing a stimulation treatment.

Many sandstone matrix stimulations are performed to address fines migration. As reservoir fluids approach the wellbore, they accelerate because of the reduced cross-sectional area for flow. It is often observed in laboratory tests that sandstones exhibit a critical velocity at which the finest-grained material (clay-sized particles) is mobilized. The indicators of fines migration are (1) gradually increasing skin or other evidence of near-wellbore impairment with production, (2) failure to remove some of this impairment with solvents or hydrochloric acid (HCl), and (3) success, often temporary, removing this impairment with formulations containing hydrofluoric acid (HF). Response to HF and HF alone the is the critical identifier of fines migration.

**5.1.3 Matrix Stimulation Definition.** Virtually all sandstone acidizing is designed to restore or slightly improve the matrix permeability. Unlike carbonate fracture acidizing described in Chapter 4, the acid treatment of a sandstone reservoir will not increase the effective wellbore inflow area by creating an etched fracture surface. Acid treatment of an undamaged unimpaired sandstone reservoir will not provide significant production increase and is therefore likely to be uneconomic. However, in the case where the well is impaired, usually indicated by high-damage skin, the production response can be significant and very profitable.

**5.1.4 Critical Success Factors.** The first, and some would argue the most critical, factor in successful sandstone acidizing is proper candidate selection. The candidate well must have significant near-wellbore impairment. As discussed in Chapters 1 and 2, near-wellbore impairment is identified by analysis of the flowing behavior of the well, usually by a combination pressure-transient analysis and inflow/outflow simulation and production profile logging.

The second critical success factor is identification of the cause of the impairment. Risk factors such as the particle-size distribution in an unconsolidated sand (fines migration), chemical composition of the oil (asphaltenes, paraffin), formation-water composition (scale), and drilling- and completion-fluid history (mud invasion, emulsions, sludges) must be evaluated. These can then be mapped against the behavior of the well, and the most likely damage can be identified. Rule-based advisor systems have been used to make this analysis more consistent and successful. Identifying the cause of impairment is necessary to select the proper stimulation chemistry and design the sequence and timing of stimulation operations.

The next critical success factor is effective placement of the chemicals to ensure contact with the impairment for sufficient time to effect dissolution. Broadly speaking, there are three ways to apply stimulation fluid systems: (1) coiled tubing (see Chapter 4), (2) jointed pipe, and (3) injection down the production tubing and/or casing (bullheading), when bullheading placement can be improved by the use of diverters, as described in Chapter 7.

The last three critical success factors are proper attention to execution, proper management of flowback to return the well to production in an uninterrupted manner, and proper data collection and review to learn from the results. As we will emphasize in this chapter, the matrix stimulation process is highly empirical, and high-quality field data are indispensable.

**5.1.5 Safety Issues With Sandstone Acidizing.** The issue of safety during well intervention is a significant topic in itself (see Chapter 12); however, there are several health and safety issues unique to sandstone acidizing that must be managed. Sandstone acidizing involves the use of some of the most hazardous chemicals commonly used in oilfield operations. HF presents unique health hazards to personnel working with it in the form of persistent chemical burns that may not be apparent upon first exposure but which pose a significant threat if not treated properly and immediately. Use of proper personal protective equipment (PPE) is essential with all chemicals, and the requirements for HF are more stringent than many materials commonly used in drilling, completion, and well-servicing applications. Other chemicals, including HCl and scale solvents, are highly corrosive and will cause chemical burns. Proper PPE is always required. Some solvents and additives also contain toxic materials that may be identified in local environmental regulations [e.g., "priority pollutants" under the toxic substances control act (TSCA) in the USA]. These materials should be eliminated when practical and must be used and disposed of properly. It is critical to be aware of the chemicals that are being used and to communicate the hazards and mitigation measures to personnel who will be exposed to the chemicals and to use proper methods to prevent release into the environment. The safety data sheet (SDS) contains the information required and should be reviewed for all chemicals used in any given well intervention.

**5.1.6 Description of the Chemistry—the Staged Approach.** Reservoirs contain a variety of minerals and fluids and thus can be susceptible to multiple forms of impairment. Although it is highly desirable to determine the exact damage mechanism, it is often impossible to eliminate all possibilities. In addition, it is often necessary to eliminate one or more mineral or fluid components from the area to be contacted by the stimulation treatment. In the case of sandstone matrix, acidizing elimination of calcium and calcium carbonate from the near wellbore is critical to reduce the risk of secondary precipitation. In addition, crude-oil components that wet the formation mineral surfaces must be eliminated to ensure effective damage removal by aqueous fluids. A staged-treatment approach is usually used to accomplish this approach.

**5.1.7 Typical Treatment Sequences—Simple to More Complex.** ***Preparing the Wellbore.*** In some stimulation treatments, coiled tubing or jointed pipe (drillpipe or work string) are used to place the treatment at the target depth by circulating the wellbore fluids (completion fluid or produced fluids) out of the well. In this case, the engineer has control over the first fluid to enter the target formation. The cleanliness of this tubing must be checked.

In so-called bullheading treatments, the fluid in the wellbore will be the first fluid to enter the formation. In a large-capacity deep well, this volume can be on the order of several hundred barrels. When "bullheading," it is wise to attempt to clean the tubing by pumping an acidic "pickling" solution to the end of the tubing and then flowing it back before attempting the initial injection. In carbon-steel tubing, the pickle should be mildly acidic (often a combination of citric and organic acids) to remove rust and mill scale; some argue this step can be skipped with stainless-steel and corrosion-resistant alloy (CRA) tubing.

***Types of Treatments.*** If the well is known to have wax or asphaltene deposits, a solvent flush should be used. Organic deposit removal is often accomplished with a single fluid stage consisting of a solvent and, in some cases, chemical agents to enhance penetration of the deposit. In long intervals or heterogeneous reservoirs, it may be advisable to consider diversion techniques discussed in Chapter 7.

Calcium carbonate scale is usually removed with HCl. Organic acids may be used at higher temperatures, as discussed in carbonate acidizing above. Iron scales are usually treated with a combination of HCl and iron-reducing or -complexing agents that enhance the solubility of iron and stabilize to prevent precipitation. These additives are also used in matrix acidizing treatments to prevent sludges and secondary precipitation of iron compounds. Sulfate (calcium, barium, and strontium) scales normally require specialized chelating agents. Oftentimes, the chemistry works best at alkaline pH and is therefore incompatible with acidizing treatments. In contrast to the reactions of carbonates and aluminosilicates, the dissolution reactions of sulfate deposits are slow, generally requiring on the order of 12 to 24 hours to achieve optimum performance. In general, these deposits are very difficult to remove, and preventive measures based on the application of threshold scale inhibitors are recommended.

Removal of aluminosilicate fines that impair the near-wellbore region and perforations in sandstone reservoirs requires a sequence of fluids. The design objectives are to mitigate potential post-treatment impairment that can be caused by unfavorable reactions between the acid or spent acid and components of the crude oil; to remove calcium salts that might cause precipitates with spent HF solutions; and to prepare the clay and feldspar mineral surfaces for effective HF attack. This is normally accomplished with a combination of solvent, brine with surfactants and mutual solvents, and HCl with additives to remove calcium carbonate minerals from the near-wellbore region.

The main acid that does the bulk of the work in fines removal is a combination of HF, HCl, and sometimes organic acids. These acids attack finely divided clay and feldspar aluminosilicates without significantly affecting quartz matrix grains. The chemistry of attack on aluminosilicate minerals is complex (see Chapter 3). Sandstone matrix stimulation formulations are designed to balance effective removal of aluminosilicate fines with the risk of precipitation of a number of problematic materials, including fluorides and silicates formed with the many cations that are in solution. Most of these materials are soluble at a low pH but precipitate as the pH rises (as the acids spend). Later in this chapter, we will discuss several methods of designing matrix stimulation treatments to achieve this balance. Conventional HCl/HF and even organic acid/HF react rapidly with formation minerals. Both experience and theoretical modeling demonstrate that with these acids, the radius of penetration is limited to within a meter or so of the wellbore with any practical acid program.

Attempting to achieve deeper penetration with conventional HCl/HF requires impractical volumes and increases the risk of post-treatment impairment because of the secondary precipitation of the acid-reaction products. Oftentimes, fines-migration damage occurs again within a few months of the treatment. In this case, it may be desirable to achieve deeper treatment to dissolve more of the materials and remove more of them from the near-wellbore region, perhaps out to a radius where fluid velocity is below that required to mobilize fines. Deeper penetration is achieved with special chemical systems based on either slowing the chemistry of the secondary dissolution reaction or delaying the formation of HF. These systems are also claimed, on the basis of laboratory studies by their inventors, to affect mineral surfaces to prevent secondary precipitation or eliminate mobility of fines. Among the systems that have been developed are self-generating HF (SGHF) systems (Lybarger and Gates 1978a; Lybarger and Gates 1978b) based on ester hydrolysis to generate acids in situ; Clay Acid™ (Thomas and Crowe 1981; Thomas and Crowe 1981), which uses fluoro-boric acid to generate HF in situ; Fines Control Acid™ (Gdanski 1985), which includes additives to slow the complex reactions involving aluminum; and Sandstone Acid™ (Kalfayan and Watkins 1990), which uses a combination of phosphonates to generate acid in situ and modify the precipitation behavior.

Whenever HFs are used, it is necessary to pump a post-flush that displaces the HF reaction products away from the wellbore to mitigate risk of damage from secondary precipitation. It is important to realize that the post-flush is an integral part of the treatment and must meet certain requirements. Ammonium chloride is the only common brine acceptable for an overflush for HF. Cations commonly found in completion brines including calcium, sodium, and potassium all form insoluble silicates. Some authors recommend adding 3% or so of acetic acid to these brines to buffer the pH. Another acceptable practice is to include an overflush of the same acid used in the preflush (HCl or formic acid), and then follow with ammonium chloride brine. In summary, an HF acid treatment consists of essential stages, a preflush to remove calcium and iron ions from the near-wellbore region, an HF-containing stage to dissolve clays and other alumino silicates, and a compatible post-flush to displace the reaction products away from the near-wellbore region.

***Acid Additives.*** Acid formulations generally contain additives to either improve effectiveness or mitigate unwanted effects. Probably the most critical of these is corrosion control. Acids are intrinsically highly corrosive, and corrosion inhibitors are essential to prevent unacceptable damage to well tubulars and equipment. Even with inhibitors, exposure times must be limited to avoid damage. These are discussed in detail in Chapter 10.

During acid stimulation, iron may enter the solution from corrosion, dissolution of iron-containing minerals (e.g., siderite), or cation exchange with chlorite clays. Iron solubility depends on pH (more soluble at lower pH) and oxidation state (more soluble at higher pH in its reduced state). Oxidized iron ($Fe\ 3^{+}$) reacts with polar components of crude oil (resins and asphaltenes) to form emulsions or sludges that create formation damage. This can be worsened by the presence of iron oxide or iron sulfide solids to form an intractable solid that some authors have labeled "schmoo." Sludges can be mitigated by adding mutual solvents, surfactants, and iron-control agents to the treatment.

Effectiveness of acid attack on minerals is strongly affected by mineral wettability. Mutual solvents or surfactants are often recommended to ensure that mineral surfaces are water-wet. They can also help with post-stimulation recovery. Additives to mitigate these risks are discussed in Chapter 9.

Given all this complexity, a single-stage system with the right chemical behavior to dissolve fine-sized aluminosilicate particles—while keeping calcium, iron, aluminum, and silicon reaction products in solution—is highly desirable. While there is research on such systems going on, practical formulations are not available as of this writing.

## 5.2 Design Issues

Matrix stimulation design includes preparation of the wellbore and formulation of the preflush, main acid, and post-flush. Volumes need to be specified for each of these. The injection rate or rate program needs to be specified. In addition, the need for special placement or diversion techniques needs to be assessed and the diversion program designed. Generally, designs are classified as single stage, consisting of one sequence of preflush, main acid, and post-flush; or multistage, consisting of more than one such sequence separated by a "diverter." Such treatments can be bullheaded, injected using coiled tubing or a jointed-pipe work string inside the production tubing or casing.

Diverters are used in layered reservoirs because without them most of the stimulation treatment will enter the less-damaged and higher-permeability zones, leaving other portions of the completion untreated. Even in reservoirs with relatively uniform damage and permeability profiles, the first section to achieve stimulation will become preferred for injection, again limiting the ability to stimulate the entire completion interval. Diversion techniques and design will be discussed in detail in Chapter 8.

When confronted with these design decisions, the engineer can follow guidelines or models based on engineering principles. Guidelines incorporate the learning from models. So, one is implicitly using modeling results, though only the most generalized concepts. Several computer programs are available for matrix stimulation design. These generally combine rules-based systems and engineering models for semiquantitative comparison of options. The most sophisticated models add specific laboratory data to improve the quantitative accuracy for a specific reservoir.

## 5.3 Guidelines

The earliest publications on sandstone (Farley et al. 1970; Labrid 1975) matrix acidizing presented the three-stage approach based on relatively simple laboratory experiments and advocated the use of relatively high concentration of HCl and HF pumped at whatever rate the available pump capacity would provide. Field experience (Lybarger 1977) indicated that pumping lower concentration acids at lower rates improved effectiveness, at least for some fields; however, the underlying reason was not understood at the time. Subsequent experimental work by Gdanski (1994) **(Table 5.1)** led to a deeper understanding of the mechanism and improved guidance for selecting acid formulations based on mineralogy. Paccaloni (1979) has described a method of optimizing rate to provide the best possible placement while avoiding fracturing the formation. His method advocates designing treatments on the basis of bottomhole-treating pressure rather than rate. This is practical in wells equipped with surface read-out downhole-pressure monitoring capability (either permanent or temporary) or in wells in which the stimulation formulations are entirely liquid, allowing reasonable extrapolation from surface-measured pressures. When using coiled tubing or a work string, monitoring the surface pressure of a liquid-filled annulus provides a good measure of bottomhole-treating pressure offset by the constant hydrostatic

| | |
|---|---|
| **Sodium Feldspars** | |
| $T > 175$ | 13.5% HCl/1.5% HF |
| $T < 175$ | 9% HCl/1% HF |
| **Potassium Feldspars** | |
| $T > 250$ | 13.5% HCl/1.5% HF |
| $250 > T > 200$ | 9%HCl/1% HF |
| $200 > T > 175$ | 7.0% HCl/0.75% HF |
| $175 > T > 125$ | 6% HCl/0.5% HF |
| $125 > T > 100$ F | 6% HCl/0.4% HF |
| **Illite** | |
| $T > 200$ | 13.5% HCl/1.5% HF |
| $200 > T > 125$ | 9% HCl/1% HF |
| $125 > T > 100$ | 7% HCl/0.75% HF |

Table 5.1—Selection of acids based on mineral composition and temperature (Gdanski 1994).

head of the annulus liquid. When pumping foam or using gas to energize the stimulation treatment, it is very difficult to calculate bottomhole pressure (BHP) from surface pressure. Macleod et al. (1983) published a comprehensive set of guidelines based on experience with stimulation in a range of areas (**Table 5.2**). They provide guidelines for relating fluid formulation to mineralogy, selecting treatment volumes, selecting diverters, and determining what additives are needed.

Applying the guidelines in a reservoir with 25% porosity results in a recommended preflush volume of approximately 35 gal per net foot of reservoir. The main acid volume will be more variable, depending on reservoir permeability, pressure, and damage skin (Paccaloni 1979). In practice, the main acid volume is typically between 75 and 250 gal per net foot of reservoir. The overflush volume in this case will be on the order of 80 (3-ft displacement) to 200 (5-ft displacement) gal per net ft of interval.

## 5.4 Experimental Results

Much of the early work on understanding the science of sandstone matrix acidizing was developed from coreflood testing, often referred to as acid response tests. Studies by Shaugnessy and Kunze (1983), Thomas and Crowe (1981), Fogler et al. (1976), and others demonstrated much of the salient

| Reservoir Rock Compostion | Recommendation | Preflush |
|---|---|---|
| HCl solubility > 10% | HCl only, no HF | – |
| High permeability (100 md plus) | | |
| High quartz (80%), low clay 5% | 12% HCl/3% HF | 15% HCl |
| High feldspar (>20%) | 13.5% HCl/1.5% HF | 15% HCl |
| High clay (>10%) | 6.5% HCl/1% HF | Sequestered 5% HCl |
| High-iron chlorite clay | 3% HCl/0.5% HF | Sequestered 5% HCl |
| Low permeability (10 md or less) | | |
| Low clay (<5%) | 6.5% HCl/1% HF | Preflush with 7.5% HCl or 10% acetic |
| High chlorite | 3% HCl/0.5% HF | Preflush with 5% acetic |
| Volumes | | |
| Preflush | Sufficient to dissolve all carbonates within 2 ft of the wellbore | |
| HF | Adjust rate and volume to provide 2 to 4 hours of acid exposure time | |
| Overflush | Sufficient to displace the treatment 3 to 5 ft from the wellbore | |

Table 5.2—General guidelines for acid selection (Macleod et al. 1983).

chemistry, especially dissolution, but also indications of the potential for reprecipitation. Rate constants measured in these coreflood tests are used in models discussed in the following. The use of these models allows us to scale the laboratory experimental results to the size and geometry of the near-wellbore reservoir. Most core samples tested in these experiments were from reservoir cores taken in the direction of the sand-bedding planes and were therefore limited to approximately 3 in. (8 cm) in length. In such short cores, only the most rapid dissolution reactions proceed to completion; secondary dissolution and reprecipitation reactions are only observed partly or not at all.

A detailed study of well return fluids by Gdanski and Peavy (1986) revealed evidence of chemistry that was not observed in short-core acid response tests in which the sample is exposed to multiple pore volumes of acid. The explanation for this discrepancy was revealed in fractional-pore-volume experiments by Kline and Fogler (1981) and Gdanski (1998) and others in packed columns and longer-core samples. In these porous media, the full suite of reactions described in Chapter 2 near completion. Chemical composition of core effluent from these experiments closely resembled the well-flowback compositions observed in the field.

Gdanski discovered that the release and subsequent stability of potassium and sodium ions was the dominant determinant for creating effective permanent increase in porosity and permeability and removal of aluminosilicate formation impairment. On the basis of these observations, acids were formulated according to overall reactivity (HF concentration) and HCl/HF ratio. The results have been captured in the guidelines in Table 5.1. The general trend is conventional 12% HCl/3% HF for high-permeability clean sandstones, 13.5% HCl/1.5% HF for moderately clean (low clay and/or feldspar content), and 9% HCl/1% HF for the highest clay and feldspar content sands.

These guidelines apply to reservoirs at temperatures below approximately 220°F. Above that temperature, organic acid/HF systems are recommended to avoid creating excessive amounts of potentially damaging reaction products because of the rapid decomposition of clays in HCl.

## 5.5 Sandstone Acidizing Models

Models have been developed on the basis of reactor engineering concepts that stress kinetics (Lund and Fogler 1976; Hekim and Fogler 1980; Hill Lindsay et al. 1977), geochemical local equilibrium models that stress thermodynamics (Davies et al. 1994), and partial local equilibrium concepts that combine the kinetics of the reaction engineering models and the thermodynamics of the geochemical models (Bryant 1991; Sevougian et al. 1995). A detailed description of the simplest reactor engineering model, the "lumped parameter model," will be given because it can be solved analytically, it illustrates the key concepts, and it can be used to understand the interaction of the most critical design parameters: acid concentration and injection rate. This model can be used to scale up laboratory "core-response tests" to field conditions.

Thermodynamic local equilibrium models provide insight into the optimization of acid formulations for specific mineralogies and the balance between desirable aluminosilicate-damage dissolution and undesirable precipitation of reaction products that reduces stimulation effectiveness. However, these models require specialized computer codes, which although widely available, are beyond the scope of this monograph. Partial local equilibrium models explicitly consider the full suite of reactions, both precipitation and dissolution, as well as transport of reactants and products. Kinetics of both types of reactions are included. The reader is encouraged to read the references on this model simply to get a feeling for the complexity of the chemistry of matrix acidizing.

**5.5.1 Reactor-Based Models.** The acidization of sandstone reservoirs falls under an important class of reactors found in the mining, geothermal, and petroleum industries called porous dissolution reactors. Single and multiple reaction fronts can develop in the porous media, depending on the relative rate of dissolution of the minerals from the quartz matrix. Models have been developed that successfully predict the shape and movement of these fronts and the distance between multiple fronts.

Laboratory "acid response tests" are carried out to measure parameters for these models. Here HF/HCl mixtures are flowed through sandstone cores in a Hassler cell, effluent composition is analyzed, and change in permeability response is recorded. These data are used in mathematical models of clay dissolution and to relate the mass of mineral removed to permeability increase. The results of linear core experiments can be used to predict the depth of clay removal in radial geometry and hence the depth of stimulation.

***Derivation of the Model.*** In the case of sandstone acidization, the acid is consumed as it dissolves the minerals at and near the inlet face. Reaction rate is usually quite rapid with respect to the other transport steps. Once minerals are removed, acid must travel farther through the inert quartz matrix to reach undissolved minerals. A schematic of the movement of a reaction front through a bed is shown in **Fig. 5.1.**

After the acid has been flowing for a while, $\tau_1$, all the accessible minerals up to location $X_1$ have been removed, and at this point fresh acid meets undissolved minerals and the acid concentration drops to zero on the downstream side of this interface. After a time, $\tau_3$, all the accessible minerals up to $X_3$ have been dissolved and there is unreacted acid (fresh acid) on the left-hand side of the front at $X_3$. A short time later, the minerals between $X_3$ and the end of the core are consumed and fresh acid breaks through the core and the time at which fresh acid reaches the end of the core is called the breakthrough time, $t_B$.

Here, the equations that describe flow and reactions involving the dissolution of two minerals from an inert matrix will be derived. Solutions will be presented for the concentration profiles for the lumped parameter systems. This will be followed by a discussion on the relationship between the increase in porosity and permeability.

***The Balance Equations.*** Consider a core in which two solid minerals, solid *N* and *P*, are being reacted with liquid *A* and removed from an inert porous matrix *D*. Solid species *N*(*s*) could represent sodium of potassium feldspar and or clays that dissolve relatively fast in HF, and *P*(*s*), slower-dissolving minerals such as quartz, mica, or iolite.

Dissolution reactions:

$$A(1)+N(s)\xrightarrow{k_1}\text{Product}(\text{liq}) \quad (5.1)$$

$$A(1)+P(s)\xrightarrow{k_2}\text{Product}(\text{liq}) \quad (5.2)$$

The rate law for each reaction is assumed to be elementary.

Carrying out a differential balance of the liquid reactant (acid) *A* in the differential section between $z$ and $z + \Delta z$ yields the balance on acid:

$$\begin{pmatrix}\text{Molar}\\ \text{Flow Rate}\\ \text{In}\end{pmatrix} - \begin{pmatrix}\text{Molar}\\ \text{Flow Rate}\\ \text{Out}\end{pmatrix} + \begin{pmatrix}\text{Molar}\\ \text{Rate of}\\ \text{Generation}\end{pmatrix} = \begin{pmatrix}\text{Molar}\\ \text{Rate of}\\ \text{Accumulation}\end{pmatrix} \quad (5.3)$$

$$M_A\big|_z \quad - \quad M_A\big|_{z+\Delta z} \quad + \quad R_A A\phi\Delta z \quad = \quad \frac{\partial}{\partial t}(C_A A\Delta z\phi)$$

where

$A$ = cross-sectional area (m$^2$)
$z$ = axial distance (m),
$\phi$ = porosity,
$A\phi\Delta z$ = volume of liquid in the segment $\Delta z$ (m$^3$),
$R_A$ = rate of generation of acid per volume of porous bed (mol/m$^3$/s),
$C_A$ = concentration of liquid reactant (mol/m$^3$),
$M_A$ = molar flow rate of acid (mol/s),
$\upsilon$ = volumetric flow rate of acid (m$^3$/s), and
$U$ = superficial acid velocity (m/s).

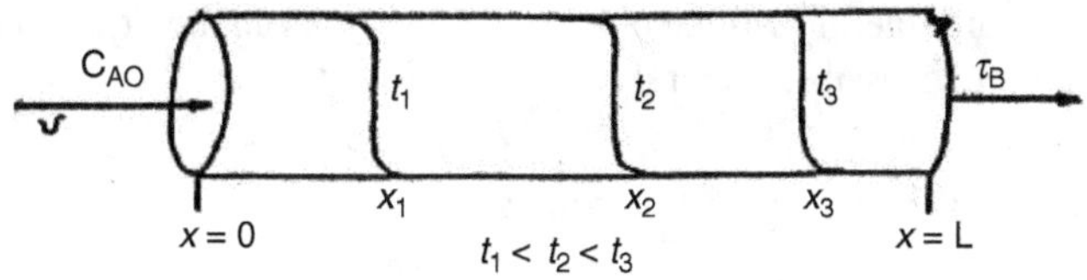


**Fig. 5.1—Movement of the reaction front (Fogler and Rege 1986). Reprinted by permission of Taylor & Francis Ltd.**

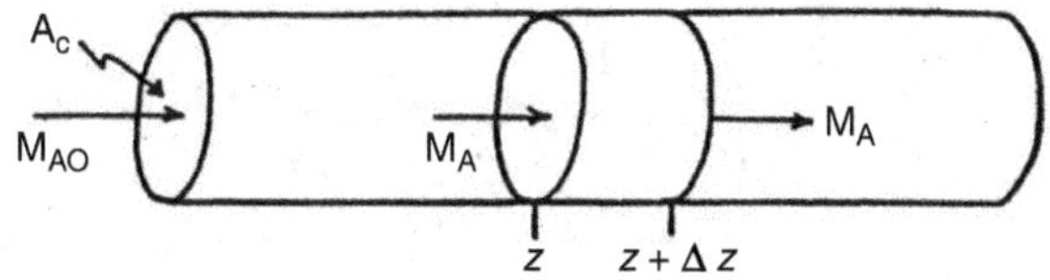


**Fig. 5.2—Differential section for acid balance (Fogler and Rege 1986). Reprinted by permission of Taylor & Francis Ltd.**

Dividing through by $A\Delta z$ and taking the limit as $\Delta z \to 0$, yields the partial-differential equation describing a reaction of acid,

$$-U\frac{\partial C_A}{\partial z} + R_A = \phi\frac{\partial C_A}{\partial t} \quad \text{(5.4)}$$

Since the dissolving material is not transported as a solid with the liquid flow, there is no inflow or outflow of the segment $\Delta z$. The mole balance on dissolving mineral $j$ is

$$\begin{pmatrix}\text{Rate}\\ \text{In}\end{pmatrix} - \begin{pmatrix}\text{Rate}\\ \text{Out}\end{pmatrix} + \begin{pmatrix}\text{Rate of}\\ \text{Generation}\end{pmatrix} = \begin{pmatrix}\text{Rate of}\\ \text{Accumulation}\end{pmatrix} \quad \text{(5.5)}$$

$$0 \quad - \quad 0 \quad + \quad r_j A\Delta z \quad = \quad \frac{\partial\left[W_j A\Delta z(1-\phi)\right]}{\partial t}$$

Dividing by $A\Delta z$ gives

$$(1-\phi)\frac{\partial W_j}{\partial t} = r_j, \quad \text{(5.6)}$$

where

$W_j$ = concentration of reacting solids (mol/m$^3$ of solid) and
$-r_j$ = rate of dissolution of solid mineral $j$ [(mol consumed)/m$^3$/s)]

In arriving at Eqs. 5.4 and 5.6, changes in φ with respect to the other terms in these equations have been neglected.

For $n$ different dissolving minerals, the rate of acid consumption is

$$-R_A = \sum_{j=1}^{n} \upsilon_j\left(-r_j\right) \quad \text{(5.7)},$$

where $\upsilon_j = \dfrac{\text{moles of liquid reactant consumed}}{\text{moles of solid } j \text{ consumed}}$ is the stoichiometric coefficient.

The rate of dissolution of mineral *j* will be taken to be first order in the liquid-reactant concentration and in the concentration of accessible mineral *j*.

$$-r_j = k_j C_A W_j \text{,} \quad (5.8)$$

where $k_j$ is the specific reaction rate for species *j*.

Combining the balance equations with stoichiometric and rate law relationships gives

$$\text{Liquid } A \;\; \phi \frac{\partial C_A}{\partial t} + U \frac{\partial C_A}{\partial z} + \sum^{n} \upsilon_j k_j W_j C_A = 0 \quad (5.9)$$

$$\text{Solid } j \;\; (1-\phi) \frac{\partial W_j}{\partial t} = -k_j W_j C_A \quad (5.10)$$

***Applying the Initial and Boundary Conditions.***

1. Initially, there is no acid on the core:

$$t = 0, z > 0, CA = 0.. \quad (5.11)$$

2. At the entrance to the core, the entering concentration is constant at $C_{A0}$.

$$t > 0, z = 0, CA = CA0. \quad (5.12)$$

3. No dissolution takes place ahead of the reaction front.

Volume of fluid that has entered the core at time *t* is equal to *UAt*.
Volume of core filled with fluid *z* is equal to $Az\phi$.

$$\begin{aligned} &\phi A_C z > U A_C t, \quad W_j = W_{j0} \\ &\phi Z/U > t, \qquad W_j = W_{j0} \end{aligned} \quad (5.13)$$

These equations can be defined in dimensionless variables as in **Table 5.3** (Hekim and Fogler 1980). Neglecting small changes in the porosity (i.e., $\phi = \phi_0$), the balance equations become

$$\frac{\partial \psi}{\partial \Theta} = \frac{\partial \psi}{\partial \varepsilon} + \left[ \Sigma \upsilon_j k_j W_{jo} \left( \frac{L}{U} \right) \eta_i \right] \psi = 0 \quad (5.14)$$

$$\frac{\partial \eta_j}{\partial \Theta} = \frac{\upsilon_j k_j W_{jo} L}{U} \bullet \frac{C_{A0} \phi}{\upsilon (1-\phi) W} \eta_j \psi \quad (5.15)$$

| Dimensionless Parameter | Symbol | Linear | Radial |
|---|---|---|---|
| Acid concentration | $\psi$ | $C_A / C_{A0}$ | $C_A / C_{A0}$ |
| Mineral "concentration" | $\eta_j$ | $W_j / W_{j0}$ | $W_j / W_{j0}$ |
| Flow rate | $\tau$ | $\phi L/U$ | $\phi r_w/U$ |
| Time | ~ | $t/\tau$ | $t/\tau$ |
| Distance | $\varepsilon$ | $z/L$ | $r/r_w$ |

Table 5.3—Dimensionless variables for linear and radial geometries (Hekim and Fogler 1980).

The first term on the right side of Eq. 5.15 is the Damköhler number for mineral $j$, $Da_j$,

$$Da_j = \frac{\left(\upsilon_j k_j W_{jo} C_{Ao}\right)}{F_{Ao}} V = \frac{-R_{jo} V}{F_{Ao}} \qquad (5.16)$$

$$Da_j = \frac{\text{rate at which liquid reactant is consumed by mineral } j}{\text{rate of transport of liquid reactant by convection}} \qquad (5.17)$$

The second term on the right side of Eq. 5.15 is the acid capacity number for mineral $j$, $Ac_j$,

$$Ac_j = \frac{C_{Ao}\phi}{\upsilon_j (1-\phi)(\Delta W_{jo})} = \frac{(\text{moles of solid } j \text{ dissolved})/(\text{pore volume of liquid})}{\text{total moles of solid } j \text{ present}} \qquad (5.18)$$

The Damköhler number determines the shape of the front, while the acid capacity number determines how fast the reaction front moves through the core. The dimensionless balances on the liquid and solid reactants along with the dissociated boundary conditions are

$$\left(\frac{\delta\psi}{\delta\Theta}\right)_{\varepsilon} + \left(\frac{\delta\psi}{\delta\varepsilon}\right)_{\Theta} + \Sigma Da_j \eta_j \psi = 0 \qquad (5.19)$$

$$\frac{\delta\eta_j}{\delta\Theta} = -Da_j Ac_j \eta_j \psi \qquad (5.20)$$

These equations describe the change in chemical composition of the region affected by the injected acid. However, the impact of these chemical changes on porosity and ultimately permeability must be understood to model the impact of matrix stimulation on well productivity.

***The Lumped Parameter Model (Lund and Fogler 1976; Hekim and Fogler 1980; Labrid 1975).*** In this model, all the minerals are assumed to consume the liquid reactant at the same rate,

$$\upsilon_j k_j = \upsilon_i k_i = k, \qquad (5.21)$$

and in the same stoichiometric amounts (i.e., $\upsilon_i = \upsilon_j$).
The total amount of all the minerals being dissolved is

$$\eta = \eta_1 + \eta_1 + \ldots + \eta_n = \frac{W}{W_o} \qquad (5.22)$$

Then,

$$\frac{\partial\psi}{\partial\Theta} + \frac{\partial\psi}{\partial\varepsilon} + Da\eta a = 0, \qquad (5.23)$$

$$\frac{\partial\eta}{\partial\Theta} = -DaAc\psi a \text{, and} \qquad (5.24)$$

$$\begin{aligned} \psi &= 1, & \varepsilon &= 0, & \Theta &> 0, \\ \psi &= 0, & \Theta &= 0, & \varepsilon &> 0. \\ \psi &= 1, & \varepsilon &> 0, \end{aligned} \qquad (5.25)$$

Combining Eqs. 5.23 and 5.24,

$$\frac{\partial \psi}{\partial \Theta}+\frac{\partial \psi}{\partial \varepsilon}-\frac{1}{Ac}\frac{\partial \eta}{\partial \Theta}=0 \qquad (5.26)$$

Eq. 5.26 can be solved with the aid of the method of characteristics. Using the transform, $[y = (\Theta - \varepsilon) Ac]$ yields

$$\psi=\left(1+\exp\left\{\mathrm{DaAc}\left[\left(1+\frac{1}{\mathrm{Ac}}\right)\varepsilon-\Theta\right]-\exp\left[DaAc\left(\varepsilon-\Theta\right)\right]\right\}\right)^{-1} \qquad (5.27)$$

$$\eta=\left(1+\exp\left\{-DaAC\left[\left(1+\frac{1}{Ac}\right)\varepsilon-\Theta\right]-\exp\left[-Da\varepsilon\right]\right\}\right)^{-1} \qquad (5.28)$$

There are two fronts associated with the dissolution process:

1. The front (i.e., interface) between the reacting liquid and the water initially present in the core,

$$\varepsilon=\Theta \qquad (5.29)$$

2. The reaction front between the unreacted liquid and unreacted solid,

$$\varepsilon=\left[\frac{Ac}{(1+Ac)}\right]\Theta \qquad (5.30)$$

Or, in terms of the dimensional variables, the position of the reaction front as a function of time is

$$z_r=\frac{Ac}{Ac+1}\frac{U}{\phi_0}t=\frac{Ut}{\phi_0\left(1+\dfrac{1}{Ac}\right)} \qquad (5.31)$$

The time to move completely through a core of length *L* (i.e., the breakthrough time) is

$$t_B=\frac{L}{U}\frac{(Ac+1)}{Ac} \qquad (5.32)$$

Liquid-reactant concentrations are given in **Fig. 5.3** and solid reactant profiles in **Fig. 5.4.** The Damköhler number determines the steepness of the profile (the higher the *Da*, the steeper the profile), while the acid capacity number determines how fast the front moves through the core.

Rearrangement of Eq. 5.32 shows that the acid capacity number can be calculated from the breakthrough time $t_B$,

$$Ac=\frac{L\phi}{Ut_B\left(1-\dfrac{L\phi}{Ut_B}\right)} \qquad (5.33)$$

The quantity

$$\frac{Ut_B}{L\phi}=\Theta_B \qquad (5.34)$$

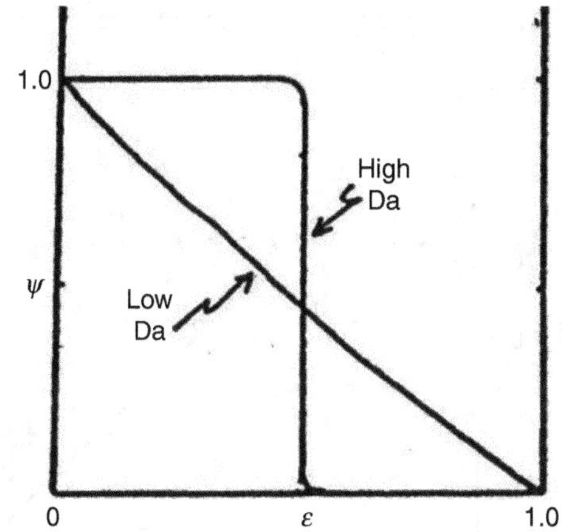


**Fig. 5.3—Effect of *Da* on ψ, the acid dissolution profile (Fogler and Rege 1986). Reprinted by permission of Taylor & Francis Ltd.**

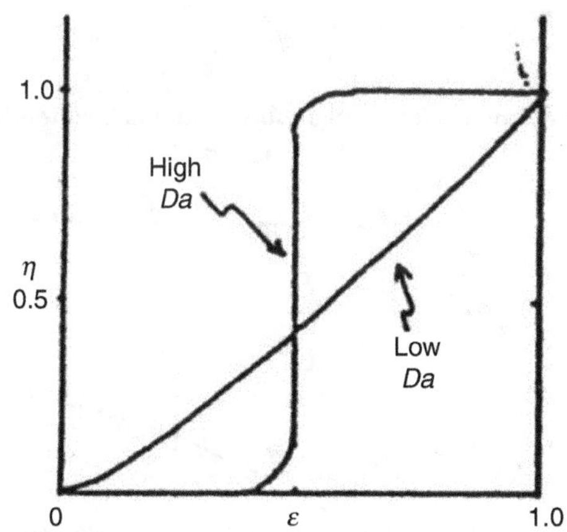


**Fig. 5.4—Effect of *Da* on η, the mineral dissolution profile (Fogler and Rege 1986). Reprinted by permission of Taylor & Francis Ltd.**

is the number of pore volumes of fluid that have passed through the core up to breakthrough

$$Ac = \frac{1}{\Theta_B\left(1-\frac{1}{\Theta_B}\right)} = \frac{1}{(\Theta_B - 1)} \quad (5.35)$$

The Damköhler number can be found by monitoring the exit concentration as a function of time. At the exit $\varepsilon = 1$ leading to

$$\ln\left[\left(\frac{1}{\psi_e}\right)-1\right] = -DaAc\,\Theta + \ln\left\{\exp\left[DaAc\left(1+\frac{1}{Ac}\right)\right]-\exp(DaAc)\right\} \quad (5.36)$$

These relationships are used to interpret the effluent acid concentration from an acid response test to determine the acid capacity and Damköhler numbers. The HF is flowed through a sandstone core in the axial direction. The concentration of HF is determined as a function of the pore volumes injected **(Fig. 5.5)** and plotted according to Eq. 5.36 **(Fig. 5.6).**

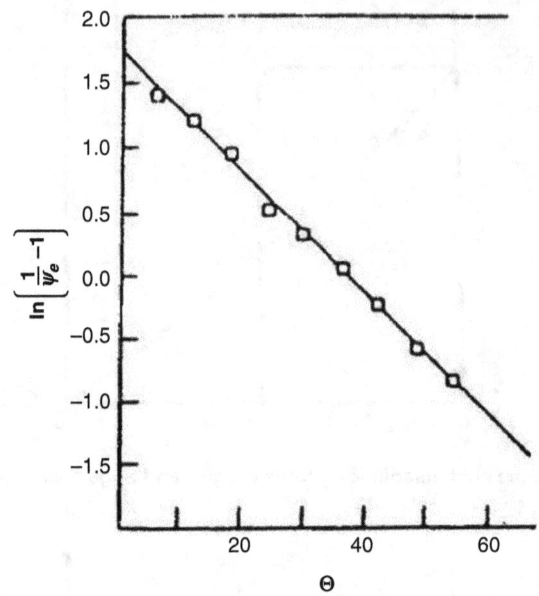


**Fig. 5.5—Effluent acid concentration as a function of pore volumes injected for Berea sandstone (Lund and Fogler 1976).**

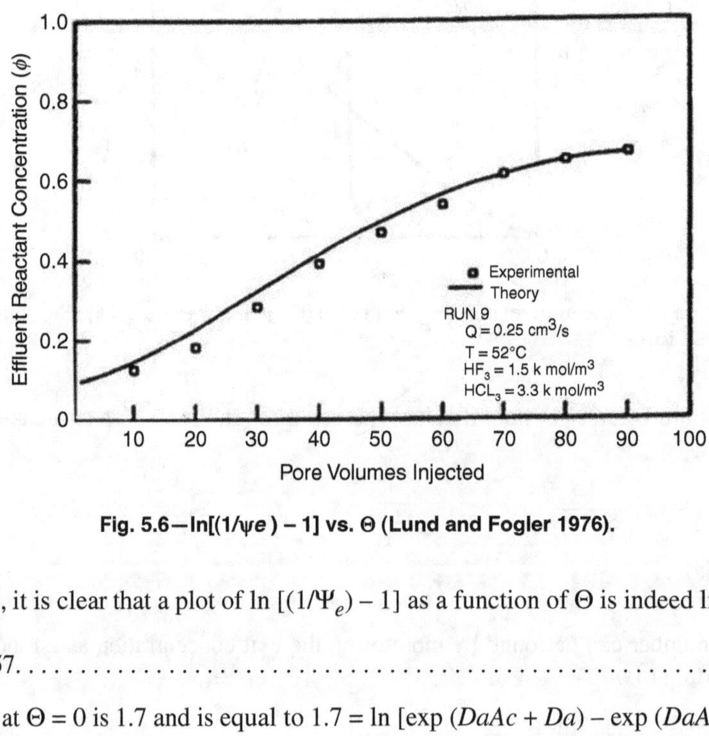


**Fig. 5.6—ln[(1/ψe) − 1] vs. Θ (Lund and Fogler 1976).**

From Fig. 5.5, it is clear that a plot of ln $[(1/\Psi_e) - 1]$ as a function of $\Theta$ is indeed linear, with slope

$$DaAc = 0.0457. \quad (5.37)$$

The intercept at $\Theta = 0$ is 1.7 and is equal to $1.7 = \ln[\exp(DaAc + Da) - \exp(DaAc)]$. ..... (5.38)

Eq. 5.37 is solved for *Da* and then Eq. 5.38 for *Ac*. For this example, the values are $Da = 1.83$ and $Ac = 0.025$. These values will then be used in the design of a well stimulation. Eqs. 5.23 and 5.24 in linear coordinates are transformed to cylindrical coordinates and use $Ac = 0.025$ and $Da = 1.83$ to determine acid concentration and mineral profile as one moves out into the formation from the wellbore.

***Estimating the Acid Capacity Number.*** The acid capacity number, which characterizes the acidization process, is a measure of how fast the reaction front moves through the porous medium and how much the acid-reaction front lags the liquid-injection front. In the absence of acid response coreflood testing, $Ac$ can be estimated from mineralogical data. Acid capacity number is given by

$$Ac = \frac{C_{A0}\phi_0}{\upsilon_j(1-\phi_0)(\Delta W_o)} = \frac{(\text{moles of solid } j \text{ dissolved})/(\text{pore volume of liquid})}{\text{total moles of solid } j \text{ present}}, \quad (5.18)$$

where $\upsilon$ = the average stoichiometric coefficient and $C_{A0}$ is the initial HF concentration (Lund and Fogler 1976).

The only unknown in Eq. 5.18 is the average stoichiometric coefficient $\upsilon$. In principle, it should be straightforward to calculate this number from mineralogical data. However, it is not because the chemical composition of aluminosilicates is highly variable, but rather that the reaction chemistry is complex and the spatial distribution of soluble minerals in the reservoir may render some soluble minerals inaccessible to acid flowing in the pores. Therefore, acid response experiments are recommended to measure the acid capacity number ($Ac$).

However, $Ac$ can be estimated for a typical sandstone using the following method. Minerals are assumed for simplicity to all have the same density; and hence, weight percent equals volume percent and the mineral compositions are idealized to some extent. The dissolvable minerals can be characterized as either feldspars, nonswelling clays (illite or kaolinite), or swelling clays (montmorillonite). Their stoichiometry for the primary reaction with HF and "concentration" are given in **Table 5.4.**

For a sandstone with 20% porosity and containing 10% feldspar, 5% kaolinite/illite, and 5% montmorillonite, and acidized with 1.5% (0.75 molar) HF,

$$\upsilon = (0.5 \times 14 + 0.25 \times 24 + 0.25 \times 40) = 23. \quad (5.39)$$

The "concentration" of dissolvable minerals

$$\Delta W = (0.10 \times 10.5 + 0.05 \times 5.4 + 0.05 \times 3.9) = 1.52. \quad (5.40)$$

$$\text{And } Ac = 0.2 \times 0.75/(1 - 0.2) \times 23 \times 1.52 = 0.0054. \quad (5.41)$$

Lund and Folger (1976) found that for many sandstones, approximately 40% of the total dissolvable minerals were accessible to acid. That is,

$$Ac = 0.2 \times 0.75/(1- 0.2) \times 23 \times 1.52 \times 0.4 = 0.0135. \quad (5.42)$$

Applying these two different results in Eq. 5.54 will result in significantly different results for acid-penetration distance, a critical design parameter. This illustrates the uncertainty introduced by the estimation technique and demonstrates the need for "acid response test" measurements.

It is important to bear in mind that these estimates capture only the primary reaction behavior. More-complex models that are not amenable to analytical solution are required to model the full complexity of the HCl-HF mineral reaction system in sandstone matrix acidizing. These models are reviewed in Section 5.8.

| Mineral | Stoichiometric Coefficient | "Concentration" (mol/L) |
|---|---|---|
| Feldspars | 14 | 10.5 |
| Kaolinite/illite | 14 | 5.4 |
| Montmorillonite | 40 | 3.9 |

Table 5.4—Stoiochimetric and "concentration" coefficients for aluminosilicate minerals (after Hill and Schechter 2000).

## 5.6 Relating Changes in Mineral Concentration to Changes in Porosity and Permeability

The porosity during dissolution is given by

$$\phi = \phi_o + (1-\phi_o)(W_o - W)M_F/\rho_s \quad (5.43)$$

The change in the porosity may also be expressed in dimensionless form by noting that the maximum possible change in the porosity is given by

$$(\Delta\phi)_{max} = \phi_{max} - \phi_o = (1-\phi_o)(W_o - W_l)M_F/\rho_s \quad (5.44)$$

Substituting this into Eq. 5.47 yields

$$\Delta\phi = (1-\eta)(\Delta\phi)_{max} \quad (5.45)$$

The change in porosity at any point and time in the core may be determined from Eqs. 5.27 and 5.45,

$$\frac{\Delta\phi}{(\Delta\phi)_{max}} = \frac{\exp\left[-DaAc\left(1+\frac{1}{Ac}\varepsilon-\Theta\right)\right]-\exp(-Da\varepsilon)}{1+\exp\left[-DaAc\left(1+\frac{1}{Ac}\varepsilon-\Theta\right)\right]-\exp(-Da\varepsilon)} \quad (5.46)$$

Eq. 5.46 predicts the change in porosity at any point and time, but not the change in the permeability. An experimental correlation between the permeability and the porosity has been determined independently by McCune and Fogler (1973) and correlated with results reported by Lund (1974):

$$\frac{k}{k_o} = \exp\left[\beta\frac{\Delta\phi}{(\Delta\phi)_{max}}\right], \quad (5.47)$$

where $k_o$ is the initial value of the permeability. In the quartzite core they tested, $(\Delta\phi)_{max}$ was approximately 0.08 and $\beta = 7.5$, so Eq. 5.51 becomes

$$\frac{k}{k_o} = \exp(95\,\Delta\phi) \quad (5.48)$$

The observed permeability did not necessarily reach the value predicted in Eq. 5.47. The maximum value of the permeability will be a function of the actual number of larger pores and the degree of collapse in the core. The relationship between permeability and amount of minerals dissolved is best determined experimentally from core-response tests with data fitted to equations of the form of Eq. 5.52.

Experimental corefloods are normally carried out on undamaged core samples; however, as discussed, impaired reservoirs are the best candidates for matrix acid stimulation. Reservoir impairment by fine aluminosilicate materials (clays and other similar-sized particles) can occur by shattering of grains during well perforating; filtration of fine materials entrained in drilling, completion, or injected fluids; or from mobilization and internal filtration because of the clay dispersion or mechanical disturbance by flow. Accumulation of these particles at pore throats leads to a loss in permeability, often without any change in bulk porosity. Put another way, particulate damage often shifts the porosity-permeability relationship.

The reactive flow described in the preceding is a self-reinforcing process. Acid is transported preferentially into the more-conductive pore networks where it removes the available impairment, allowing more acid and more improvement. Less-conductive pore networks receive less acid and clean up more slowly. The net result is that in porous media with a range of pore sizes, matrix stimulation creates a relatively few, relatively conductive pore networks through the damaged zone. **Figs. 5.7 and 5.8** illustrate this effect, showing the pore-size-distribution number-length parameter for an acidized core and an unacidized core vs. mean pore radius. Below 3 μm, the curves are nearly identical, while above 3 μm, the number-length parameter is greater for acidized cures. This indicates that it is primarily, if not solely, the larger pores that are responsible for the increase in permeability resulting from acidization. This increase can be accounted for in several ways—surface reaction at the pore wall increasing the radius of a pore, collision between two smaller pores forming one large pore, or removal of debris from a constriction in the pore.

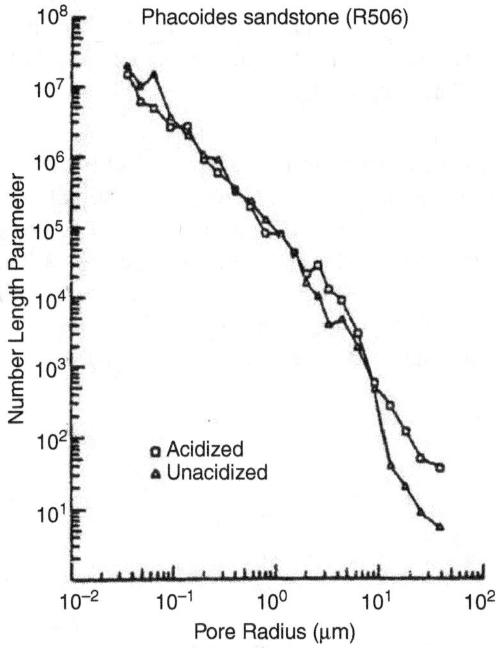


**Fig. 5.7—Relative pore numbers as a function of pore radius (Swift and Fogler 1977).**

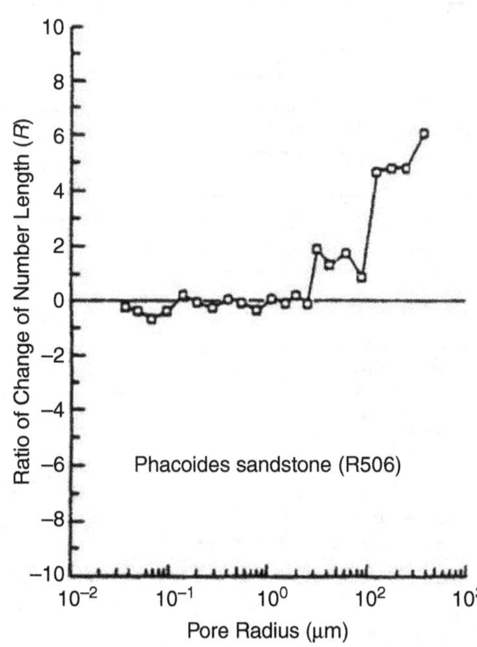


**Fig. 5.8—Relative change in pore numbers as a function of pore radium (Swift and Fogler 1977).**

Schecter and Gidley (1969), and McCune et al. (1975) have presented models that integrate the local porosity change over the pore-size distribution and have successfully matched experimental data (see **Fig. 5.9**). The original references can be consulted to apply these methods. More generally, an empirical relationship of the form of Eq. 5.52 is developed from acid response tests.

## 5.7 Measuring Critical Parameters for the Lumped Parameter Model—the Damköhler Number and Acid Capacity Number

Coreflood acid response tests (Farley et al. 1970) are used to determine the design parameters, β, *Da*, and *Ac* for a specific sandstone. Fig. 5.9 shows a schematic of the model and technique.

Eqs. 5.46 and 5.47 are combined with the measured permeability response to calculate the increase in overall permeability as a function of the pore volume of acid injected. **Fig. 5.10** shows a comparison of theory and experiment for Phacoides sandstone.

## 5.8 Scaling Up the Lumped Parameter Model to Radial Flow

Linear flow and reaction equations must be converted to cylindrical equations to apply the model to the stimulation of wells. As previously noted, the radial and linear solutions are equivalent if the appropriate dimensionless groups are applied (Hekim and Fogler 1980). The main design parameter of interest is how far the reactive acid front penetrates—how far from the wellbore will the acid dissolve aluminosilicate fines that can cause formation damage.

For radial flow, the differential mole balance on the acid flowing out from a wellbore is

$$\frac{\partial(\phi C)}{\partial t}+\frac{vr_w}{r}\frac{\partial C}{\partial r}-R_A=0, \quad (5.49)$$

where $v$ is the radial velocity in the porous formation.

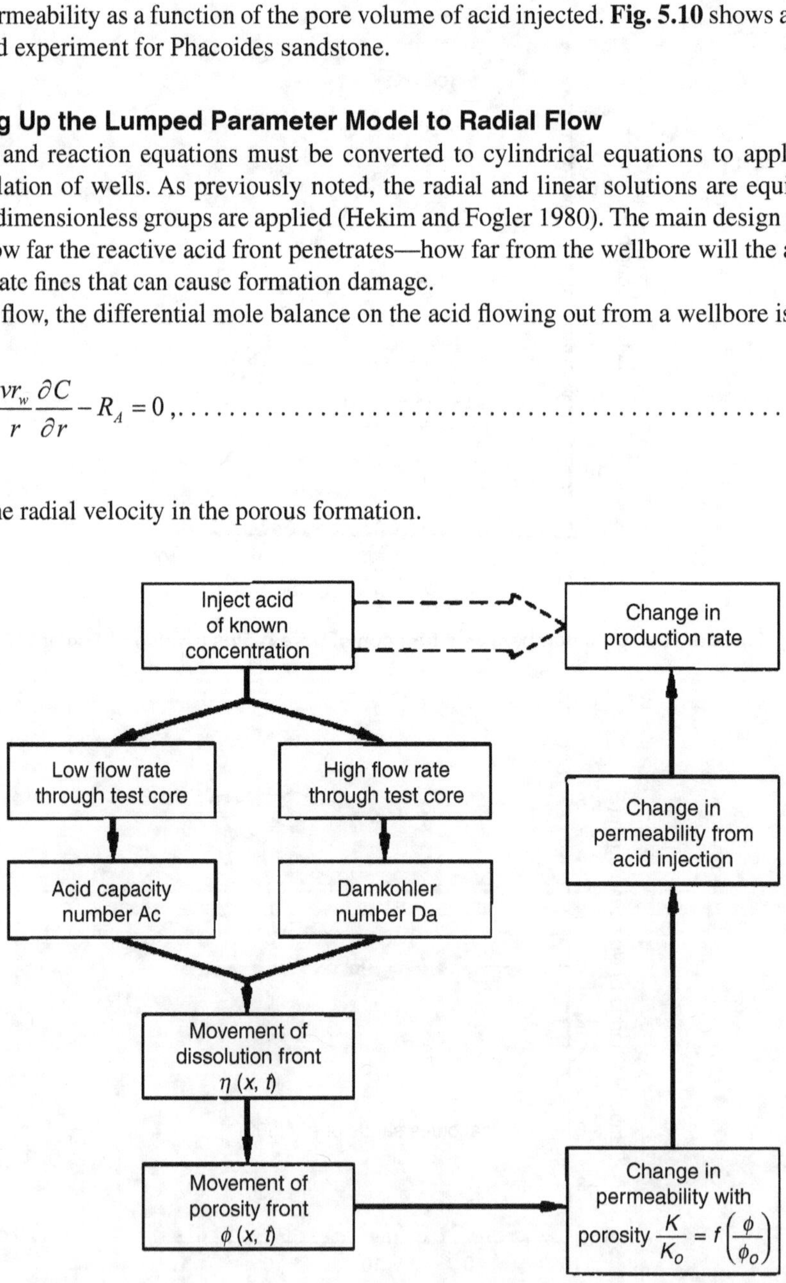


**Fig. 5.9—Prediction of production-rate changes following acidization (Fogler and Rege 1986). Reprinted by permission of Taylor & Francis Ltd.**

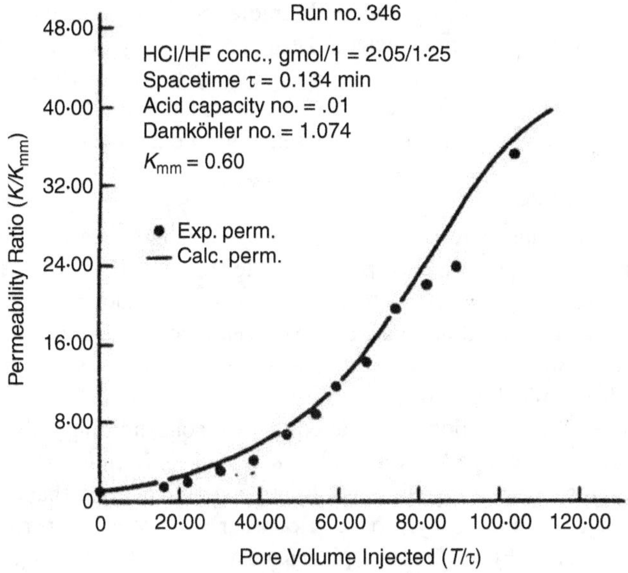


**Fig. 5.10—Permeability of a core during acidization vs. dimensionless time Run 346 (Swift and Fogler 1977).**

Eq. 5.49 is solved along with the mineral mole balance,

$$(1-\phi_0)\frac{\partial(W_o-W)}{\partial t}-r_A, \qquad (5.50)$$

the rate law

$$-R_A = kC_AW, \qquad (5.51)$$

and the stoichiometric relationships,

$$-R_A = \upsilon(-r_A), \qquad (5.52)$$

where $r_w$ is the radius of the wellbore. The mole-balance equation for radial flow is solved to obtain a relationship for the amount of acid necessary to stimulate a treatment radius $r_t$ into the formation,

$$r_t = \left[r_w^2 + \frac{V}{\pi\phi_0 h\left(1\frac{1}{Ac}\right)}\right], \qquad (5.53)$$

where $V$ is the total volume injected and $h$ is the height of the well openhole formation.

If (as is often the case) $Ac$ is much less than unity (0.005 to 0.025), Eq. 5.53 can be rearranged and simplified:

$$\left(r_t^{\,2}-r_w^2\right)=\frac{V\,Ac}{\pi\phi_0 h} \qquad (5.54)$$

and

$$\frac{\left(r_t^{\,2}-r_w^2\right)}{(r_R^2-r_w^2)} = Ac \qquad (5.55)$$

Eqs. 5.53 through 5.55 can be used to calculate the radial penetration of the primary acid-reaction front as a function of the volume of acid injected. The shape of the reaction front depends on the Damköhler number, Eq. 5.16. Data on reaction kinetics must be measured experimentally as described in the preceding. While the acid-penetration distance is a very useful result when designing matrix acid treatments, it must be remembered that the primary reactions are not the end of the story.

## 5.9 More-Complex Models

Experimental studies with short-core plugs (e.g., 2–3 in. in length) are dominated by the primary dissolution reactions and generally matched porous reactor models. However, field measurements of acid response and fluid returns demonstrate that the interaction is much more complex, as discussed in more detail in Chapter 2. Several approaches have been taken to address these complexities. All require the use of specialized computer codes that can be used to improve acid formulations to mitigate the risks of unfavorable chemical reactions.

Faber et al. (1994) neglected the flow and rate equations, concentrating only on Eqs. 5.4 and 5.5, modeling the acidization process as an array of "tanks" at equilibrium. They calculated total net reduction in solids volume, considering both dissolution and precipitation (but assuming all reactions occurred instantly to equilibrium). They produced contours of maximum permeability increase vs. HCl and HF concentration (see **Fig. 5.11**) for typical sandstone mineralogy and illustrated that there is an optimum HCl/HF system for any sand.

They use the relationship similar to Eq. 5.47:

$$\ln k = 80.57 \times \phi - 8.526, \quad (5.56)$$

where $k$ = the permeability and $\phi$ is the porosity expressed as a fraction.

They apply a Gibbs-free-energy minimization algorithm to the suite of dissolution-precipitation reactions in the mineral/HCl/HF system:

$$G(x^1, x^2, x^3 \ldots x^n) = \sum_{j=1}^{n} x_j \left[ g_j + \log c_j + \left( \log f_j \right) \right], \quad (5.57)$$

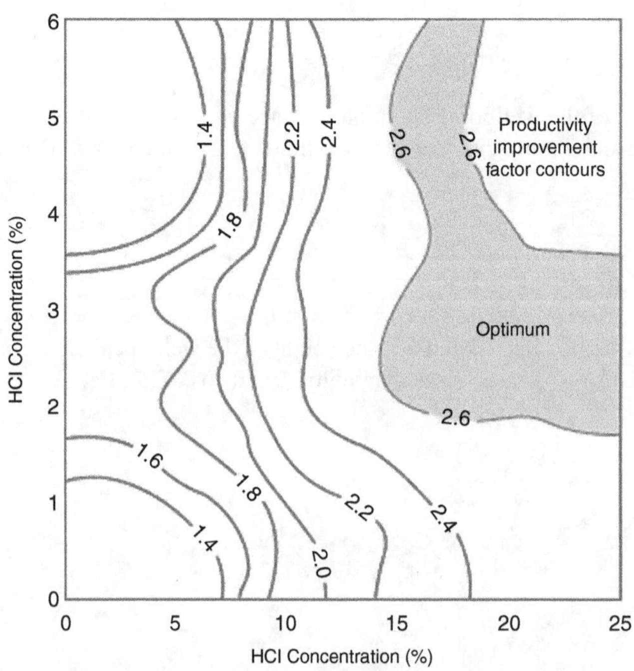


**Fig. 5.11—Productivity Index ($\kappa/\kappa_o$) contours illustrating acid formulation optimization (Davies et al. 1994).**

where

$G$ = the Gibbs free energy divided by $R$ (Boltzman's constant) $T$ (Temperature),

$G_j$ = standard Gibbs free energy of component $j$,

$c_j$ = the concentration of component $j$, and

$f_j$ = the activity coefficient of component $j$.

Walsh et al. (1982) applied a combination of the balance equations analogous to Eq. 5.9 and the thermodynamic equilibrium relationships analogous to Eq. 5.57. This reproduces the reaction fronts described by Fogler et al. (1976) and Hekim and Fogler (1980) but adds the result that precipitates occur at some of the reaction fronts (shaded region in **Fig. 5.12**). This is a result of the assumption that solids precipitate instantly when their $K_{sp}$ is exceeded. Walsh's model added the important finding that for any concentration of HF, there is a minimum concentration of HCl required to prevent precipitation.

Sevougian et al. (1992; 1993) relaxed the assumption of immediate precipitation. They considered the balance and flow equations (e.g., Eq. 5.7), the rate equations (e.g., Eq. 5.9), and thermodynamic equilibrium relationships (Eq. 5.58). They describe their model as partial local equilibrium. This result demonstrated that the precipitates are "smeared" behind the reaction front, as shown in **Fig. 5.13.**

In all of these models the porous medium is considered as a homogeneous assemblage of minerals. Porosity is assumed to be directly proportional to mineral mass within a volume element, and a simple relationship between porosity and permeability is assumed. The texture of the rock matrix, the spatial distribution of the mineral phases within it, is not explicitly considered. Its effect is captured in rate constants and in selection of which minerals to consider or neglect in the calculation.

## 5.10 Estimating the Maximum Injection Rate

In matrix acidizing, the objective is usually to distribute reactive acid as uniformly as possible in the impaired region around the wellbore and to stimulate as many perforations as possible. Unlike other stimulation techniques, fracturing is not desired. In wells with sand control, fracturing may disrupt the sand control and lead to failure of the completion. In perforated completions, fracturing will limit the ability to stimulate perforations not connected to the fracture. Fracturing is avoided by controlling

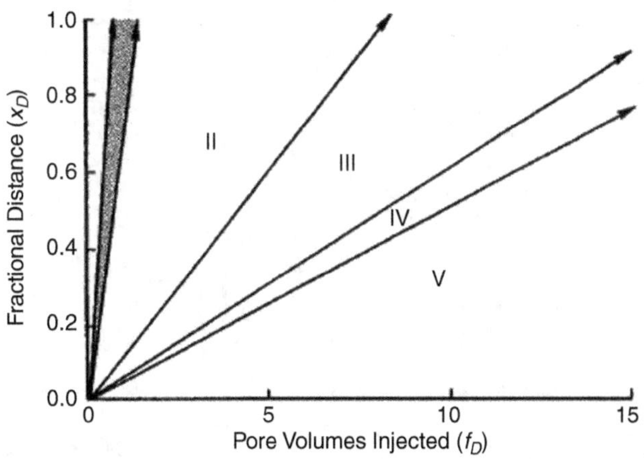


| Region | Available Quartz (kmol/m³) | Kaolinite (kmol/m³) | Calcite (kmol/m³) | Precipitated $Si(OH)_4$ (kmol/m³) | Precipitated $AlF_3$ (kmol/m³) | Precipitated $Al(OH)_3$ (kmol/m³) | pH | pHF |
|---|---|---|---|---|---|---|---|---|
| I | 0 | 2.5 | 3.0 | 0 | 0 | 0 | 6.2 | — |
| II | 0 | 2.5 | 3.0 | 0 | 0 | 0 | 3.7 | 5.4 |
| III | 0 | 0 | 0 | 8.2 | 4.9 | 0 | −0.6 | −0.3 |
| IV | 0 | 0 | 0 | 8.2 | 0 | 0 | −0.6 | −0.3 |
| V | 0 | 0 | 0 | 0 | 0 | 0 | −0.6 | −0.3 |

**Fig. 5.12—Calculated region of precipitation (shaded area) (Walsh et al. 1982).**

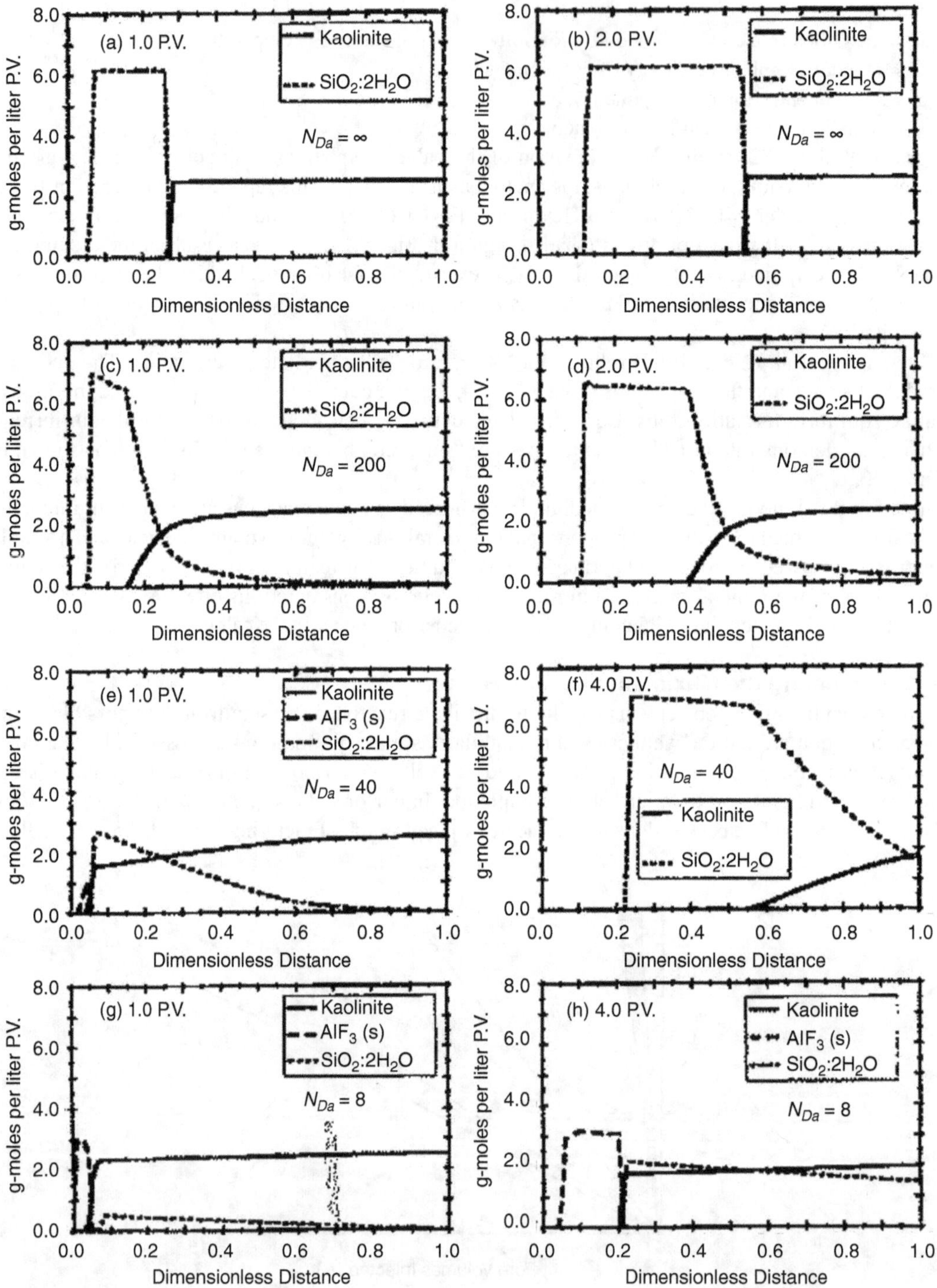


**Fig. 5.13—Reaction product deposition profiles from reactive transport simulator (Sevougian et al. 1995).**

the injection rate to maintain the injection pressure below the minimum in-situ stress of the reservoir being stimulated. At the same time, maintaining a high differential pressure across the damage zone is beneficial for injecting acid into the most-damaged zones. Paccaloni et al. (1995) and Paccaloni (1979) have presented a practical way to estimate the desired rates and pressures and use feedback from the injection pressure itself to monitor the progress of the stimulation.

The method is simple, practical, and field proven (Paccaloni 1979; Morgenthaler et al. 1995). The design is based on the simple steady-state radial flow equation.

The bottomhole-treating differential pressure is given by

$$P_i - P_e = (141.2q\mu_w/k_w h)(\ln r_b/r_w + S), \quad (5.58)$$

and the surface injection pressure is then

$$P_{si} = P_{iBH} - P_h + P_{fr}. \quad (5.59)$$

For the maximum-rate/maximum-pressure design, the maximum bottomhole-treating pressure is taken as

$$P_{iBHMAX} - P_e = 0.85\,(P_{FRAC} - P_e), \text{ where} \quad (5.60)$$

$$P_{siMAX} = P_{iBHMAX} - P_h + P_{fr} \quad (5.61)$$

$h$ = net pay thickness,
$q$ = injection rate,
$k_w$ = effective water (acid) permeability,
$P_{iBH}$ = bottomhole injection pressure,
$P_{si}$ = surface injection pressure,
$P_e$ = average reservoir pressure,
$P_h$ = hydrostatic head,
$P_{fr}$ = total friction pressure,
$P_{FRAC}$ = the bottomhole fracture pressure (often take as the minimum in-situ stress),
$\mu_w$ = acid viscosity,
$r_w$ = wellbore radius, and
$r_b$ = radius over which the pressure drop takes place (experience has demonstrated that using a value of 1.2 m for $r_b$ results is a reasonable analysis of well-pressure response during acidization), and
$S$ = skin.

A design plot is created showing the relationship between rate, skin, and surface pressure and is shown in **Fig. 5.14** as a series of curves (because of the dependence of $P_{fr}$ on $q$). An upper-limit curve corresponding to Eq. 5.61 shows the upper-limit of injection rate as a function of skin. As the well responds to stimulation, this rate limit increases.

This curve can be used in combination with knowledge of the skin to estimate an appropriate injection pressure and rate to initiate injection during the stimulation treatment. Operators can be advised to increase rate as the well responds and skin is reduced, to maintain near the maximum rate. Decisions about whether to continue or stop stimulation operations can also be made.

Method software evaluation tools have become available that calculate a real-time skin on the basis of these principles. They incorporate more-rigorous equations for pressure-transient analysis.

## 5.11 Computer Design Tools

The foregoing illustrates that effective design requires the integration of much information. Formation mineralogy, fluid properties, reservoir temperature, permeability, interval thickness, sand control, well metallurgy, and pressure all play a role understanding the damage mechanism in designing an effective treatment. The problem needs to be approached with a combination of rules and a variety of computational methods. Several authors have described "rule based, advisor systems" that facilitate balanced application of all these rules. Bartko et al. (1996) and Nitters et al. (1990) describe computer programs that combine a candidate-selection advisor, formation-damage advisor, and fluid-selection advisor with geochemical (equilibrium) and kinetic-based reaction models to facilitate the stimulation design process. These programs include computational capabilities to model reaction chemistry,

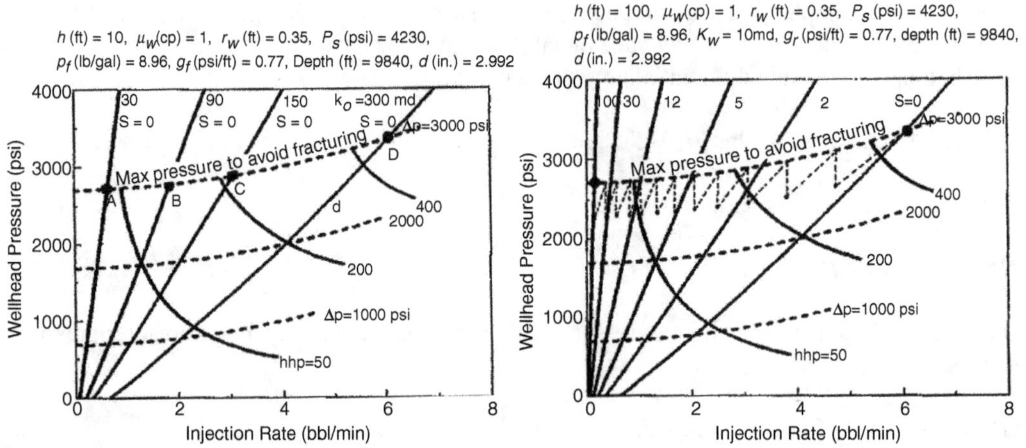


**Fig. 5.14—Changes in wellhead injection pressure with injection rate and skin during an acid treatment of a short interval and a long interval (Paccaloni 1995)**

pumping pressure response, and diversion effectiveness (see Chapter 8). Some authors have described programs to interpret core-response tests, as described in Section 5.7, directly into the design model (Samuel et al. 2007). Consistent use of these tools has often led to higher probability of success in stimulation campaigns. These tools are useful to the inexperienced engineer to ensure that critical factors are considered and provide the benefit of experience embedded in the expert system. For more-experienced engineers, the advanced computational tools enable a deeper understanding to the processes and enable exploring design alternatives.

### 5.12 Real-Time Monitoring

The development of distributed-temperature-sensing (DTS) technology based on laser interferometry in fiber-optic cables has enabled real-time measurement of temperature profiles during stimulation (Glasbergen, et al. 2009; Garcia et al. 2011). Interpretation of DTS during acidizing is still evolving. Interpretation models must take into account the effect of flow, which generally cools the treated interval along with the heating effect of the exothermic dissolution reactions. The best practice is to (1) obtain a baseline survey with an inert fluid such as brine or nitrogen, (2) continue taking data while pumping the stimulation, and (3) continue measuring the warm back of the well following stimulation. Integration of real-time data with design models can be expected to improve our understanding of matrix acidizing and lead to improved success.

## References

Bartko, K. M., Montgomery, C. T., Boney, C. L. et al. 1996. Development of a Stimulation Treatment Integrated Model. Presented at the Petroleum Computer Conference, Dallas, 2–5 June. SPE-35991-MS. http://dx.doi.org/10.2118/35991-MS.

Bryant, S. L. 1991. An Improved Model of Mud Acid/Sandstone Chemistry. Presented at the SPE Annual Technical Conference and Exhibition, Dallas, 6–9 October. SPE-22855-MS. http://dx.doi.org/10.2118/22855-MS.

Davies, D. R., Faber, R., Nitters, G. et al. 1994. A Novel Procedure To Increase Well Response to Matrix Acidising Treatments. *SPE Advanced Technology Series* **2** (1): 5–14. SPE-23621-PA. http://dx.doi.org/10.2118/23621-PA.

Farley, J. T., Miller, B. M., and Schoettle, V. 1970. Design Criteria for Matrix Stimulation with Hydrochloric-Hydrofluoric Acid. *J Pet Technol* **22** (04): 433–440. SPE-2621-PA. http://dx.doi.org/10.2118/2621-PA.

Fogler, H. S. and McCune, C. C. 1973. On the Extension of the Model of Matrix Acid Stimulation to Different Sandstones. Presented at the AIChE National Meeting, New Orleans, 11–15 March.

Fogler, H. S. and Rege, S. D. 1986. Porous Dissolution Reactors. *Chem. Eng. Commun.* **42** (4–6): 291–313. http://dx.doi.org/10.1080/00986448608911747.

Fogler, H. S., Lund, K., and McCune, C. C. 1976. Predicting the Flow and Reaction of HCL/HF Acid Mixtures in Porous Sandstone Cores. *SPE J.* **16** (5): 248–260. SPE-5646-PA. http://dx.doi.org/10.2118/5646-PA.

Garcia, S. J. R., Franco, E., Romandia, M. G. et al. 2011. Revolutionary Matrix Stimulation Process in Offshore Mexico Using Coiled Tubing Equipped with Optical Fibers (CT-EOF) and Distributed Temperature Survey (DTS). Presented at the SPE/ICOTA Coiled Tubing & Well Intervention Conference and Exhibition, The Woodlands, Texas, USA, 5–6 April. SPE-143318-MS. http://dx.doi.org/10.2118/143318-MS.

Gdanski, R. 1998. Kinetics of Tertiary Reactions of Hydrofluoric Acid on Aluminosilicates. *SPE Prod & Fac* **13** (2): 75–80. SPE-31076-PA. http://dx.doi.org/10.2118/31076-PA.

Gdanski, R. D. 1985. AlCl3 Retards HF Acid for More Effective Stimulations. *Oil Gas J* (October 28, 1985): 111–115.

Gdanski, R. D. 1994. Fluosilicate Solubilities Affect HF Acid Compositions. *SPE Prod & Fac* **9** (4): 225–229. SPE-27404-PA. http://dx.doi.org/10.2118/27404-PA.

Gdanski, R. D. and Peavy, M. A. 1986. Well Return Analysis Causes Re-Evaluation of HCL Theories. Presented at the SPE Formation Damage Control Symposium, Lafayette, Louisiana, USA, 26–27 February. SPE-14825-MS. http://dx.doi.org/10.2118/14825-MS.

Glasbergen, G., Gualtieri, D., Van Domelen, M. et al. 2009. Real-Time Fluid Distribution Determination in Matrix Treatments Using DTS. *SPE Prod & Oper* **24** (1): 135–146. SPE-107775-PA. http://dx.doi.org/10.2118/107775-PA.

Hekim, Y. and Fogler, H. S. 1980. On the Movement of Multiple Reaction Zones in Porous Media. *AIChE J* **26** (3): 403–411. http://dx.doi.org/10.1002/aic.690260312.

Hill A. D. and Schechter R. S. 2000. Fundamentals of Acid Stimulation. In *Reservoir Stimulation,* third edition, eds. M. J. Economides and K. G. Nolte. New York: John Wiley and Sons.

Hill, A. D., Lindsay, D. M., Silberberg, I. H. et al. 1977. Theoretical and Experimental Studies of Sandstone Acidizing. Presented at the SPE International Oilfield and Geothermal Chemistry Symposium, La Jolla, California, USA, 27–28 June. SPE-6607-PA.

Kalfayan, L. J. and Watkins, D. R. 1990. A New Method for Stabilizing Fines and Controlling Dissolution During Sandstone Acidizing. Presented at the SPE California Regional Meeting, Ventura, California, USA, 4–6 April. SPE-20076-MS. http://dx.doi.org/10.2118/20076-MS.

Kline, W. E. and Fogler, H. S. 1981. Dissolution of Silicate Minerals by Hydrofluoric Acid. *Industrial & Engineering Chemistry Fundamentals* **20** (2): 155–161. http://dx.doi.org/10.1021/i100002a008.

Kunze, K. R. and Shaughnessy, C. M. 1983. Acidizing Sandstone Formations with Fluoboric Acid. *SPE J.* **23** (1): 65–72. SPE-9387-PA. http://dx.doi.org/10.2118/9387-PA.

Labrid, J. C. 1975. Thermodynamic and Kinetic Aspects of Argillaceous Sandstone Acidizing. *SPE J.* **15** (2): 117–128. SPE-5156-PA. http://dx.doi.org/10.2118/5156-PA.

Labrid, J. C. 1975. Thermodynamic and Kinetic Aspects of Argillaceous Sandstone Acidizing. *SPE J* **15** (2): 117–128. SPE-5156-PA. http://dx.doi.org/10.2118/5156-PA.

Lund, K. 1974. *On the Acidization of Sandstone.* PhD thesis, University of Michigan, Ann Arbor, Michigan, USA.

Lund, K. and Fogler, H. S. 1976. Acidization—V: The Prediction of the Movement of Acid and Permeability Fronts in Sandstone. *Chem Eng Sci* **31** (5): 381–392. http://dx.doi.org/10.1016/0009-2509(76)80008-5.

Lybarger, J. H. 1977. Successful Well Work Demands Rigorous Quality Control. *Oil Gas J* **75** (21): 57–61.

Lybarger, J. H. and Gates, H. R. 1978a. SGMA Performance, Design, Part—1: New Acidizing System Yields Sustained Production Increase. *Oil Gas J* (October 16, 1978): 59–64.

Lybarger, J. H. and Gates, H. R. 1978b. SGMA Performance, Design, Part-2: Proper Reservoir Choice, Fluid Design Keys to SGMA Success. *Oil Gas J* (October 23, 1978): 137–142.

McCune, C. C., Fogler, H. S., and Cunningham, J. R. 1975. A New Model of the Physical and Chemical Changes in Sandstone During Acidizing. *SPE J.* SPE-5157-PA. http://dx.doi.org/10.2118/5157-PA.

McLeod, H. O. 1984. Matrix Acidizing. *J Pet Technol* **36** (12): 2055–2069. SPE-13752-PA. http://dx.doi.org/10.2118/13752-PA.

McLeod, H. O. Jr., Ledlow, L. B., and Till, M. V. 1983. The Planning, Execution, and Evaluation of Acid Treatments in Sandstone Formations. Presented at the SPE Annual Technical Conference and Exhibition, San Francisco, 5–8 October. SPE-11931-MS. http://dx.doi.org/10.2118/11931-MS.

Morgenthaler, L. N., Malochee, S. B., and Smith, C. D. 1995. Real Time Stimulation Evaluation in Deepwater Well Completions. Presented at the SPE Annual Technical Conference and Exhibition, Dallas, 22–25 October. SPE-30460-MS. http://dx.doi.org/10.2118/30460-MS.

Morgenthaler, L. N., Zhu, D., Mou, J. et al. 2008. Effect of Reservoir Mineralogy and Texture on Acid Response in Heterogeneous Sandstones. *SPE Prod & Oper* **23** (1): 39–48. SPE-102672-PA. http://dx.doi.org/10.2118/102672-PA.

Nitters, G. and Hagelaars, A. M. P. 1990. Careful Planning and Sophisticated Laboratory Support: The Key to Improved Acidisation Results. Presented at the European Petroleum Conference, The Hague, 21–24 October. SPE-20967-MS. http://dx.doi.org/10.2118/20967-MS.

Paccaloni, G. 1979. Matrix Stimulation Planning. 1. New Method Proves Value of Stimulation Planning. *Oil Gas J* **77** (47): 155–160.

Paccaloni, G. 1979. Matrix Stimulation Planning. 1. New Method Proves Value of Stimulation Planning. *Oil Gas J* **77** (47): 155–160.

Paccaloni, G. 1995. A New, Effective Matrix Stimulation Diversion Technique. *SPE Prod & Fac* **10** (3): 151–156. SPE-24781-PA. http://dx.doi.org/10.2118/24781-PA.

Samuel, E., Shaikh, A. K., Queiros, J. G. R. et al. 2007. Sandstone Matrix Stimulation Can Improve Brownfield Oil Production When the Chemistry and Procedures Are Correct. Presented at the International Symposium on Oilfield Chemistry, Houston, 28 February–2 March. SPE-106321-MS. http://dx.doi.org/10.2118/106321-MS.

Schechter, R. S. and Gidley, J. L. 1969. The Change in Pore Size Distribution From Surface Reactions in Porous Media. *AIChE J* **15** (3): 339–350. http://dx.doi.org/10.1002/aic.690150309.

Sevougian, S. D., Lake, L. W., and Schechter, R. S. 1995. A New Geochemical Simulator to Design More Effective Sandstone Acidizing Treatments. *SPE Prod & Fac* **10** (1): 13–19. SPE-24780-PA. http://dx.doi.org/10.2118/24780-PA.

Sevougian, S. D., Lake, L. W., and Schecter, R. S. 1992. A New Geochemical Simulator to Design More Effective Sandstone Acidizing Treatments. Presented at the SPE Annual Technical Conference and Exhibition, Washington, D. C., 4–7 October. SPE-24780-MS.

Sevougian, S. D., Schechter, R. S., and Lake, L. W. 1993. Effect of Partial Local Equilibrium on the Propagation of Precipitation/Dissolution Waves. *Ind Eng Chem Res* **32** (10): 2,281–2,304. http://dx.doi.org/10.1021/ie00022a013.

Swift, S. T. and Fogler, H. S. 1977. A Note on Capillary Model of Developments for Sandstone Acidization. *Chem Eng Sci* **32** (3): 339–342. http://dx.doi.org/10.1016/0009-2509(77)80218-2.

Thomas, R. L. and Crowe, C. W. 1981. Matrix Treatment Employs New Acid System for Stimulation and Control of Fines Migration in Sandstone Formations. *J Pet Technol* **33** (8): 1491–1500. SPE-7566-PA. http://dx.doi.org/10.2118/7566-PA.

Walsh, M. P., Lake, L. W., and Schechter, R. S. 1982. A Description of Chemical Precipitation Mechanisms and Their Role in Formation Damage During Stimulation by Hydrofluoric Acid. *J Pet Technol* **34** (9): 2087–2112. SPE-10625-PA. http://dx.doi.org/10.2118/10625-PA.

Chapter 6

# Acid Placement and Diversion

**Gerrit Nitters,** Nitters Petroleum Consultancy International,
**Valerie J. Yeager,** 3 Way Energy Consulting LLC, and
**W. R. Rossen,** Delft University of Technology, with contribution from
**John Misselbrook,** Consultant Associates Ltd.

## 6.1 Introduction

In any acid stimulation treatment, proper placement of the acid is critical to the success of the treatment. If the acid is not directed to the intended interval, the treatment objective, which is production or injection improvement, may be missed altogether. The delivery of stimulation fluids to the target destination is largely dictated by the completion design. So, for example, the challenge of proper placement is compounded in horizontal wells, which often contain a potentially producing (or injecting) interval covering several hundred feet or even thousands of feet.

Moreover, diversion in carbonates is generally more difficult than in sandstones because of the ability of the highly reactive hydrochloric acid (HCl) to dramatically increase permeability in carbonate rock. The acid reacts in the pore spaces and flow channels of the matrix, creating greater paths of least resistance that can counteract acid placement efforts.

However, the following are diversion techniques for both matrix acidizing and fracture acidizing of carbonates:

- Mechanical techniques
- Ball sealers (perforated completions; vertical or less-deviated wells)
- Particulates
- Viscosified fluids and emulsified acids
- Foams

The latter two are less applicable as diversion techniques in acid fracturing treatments.

In addition to the application of one of these techniques, for matrix acidizing treatments, injection at the highest possible matrix treating rate, not exceeding fracturing conditions [the maximum pressure differential and injection rates (MAPDIR) technique, discussed later in this chapter], has been promoted as the best means of ensuring sufficient acid placement in all zones.

Computer models have been developed for the simulation of various placement techniques. In this chapter, the effectiveness of the various techniques will be illustrated using one of these computer packages.

## 6.2 Diversion and Placement Philosophy

The success of matrix acid treatments depends to a large extent on the capability of getting the acid where you want it. First, we need to find out whether placement or diversion is actually needed. This depends largely on

1. The length of the interval to be treated. The longer the interval is, the more we need diversion; as a rule of thumb, any interval greater than 15 m (50 ft) will benefit from diversion.
2. Whether it is more or less homogeneous. In very homogeneous formations with little variation in permeability across the interval, formation damage is often also fairly uniformly distributed over the entire interval. Consequently, there is less need for application of a diversion agent. Simply pumping at a high rate tends to distribute the stimulation fluids effectively over the total interval, unless the interval is very long.
3. The presence of thief zones (i.e., large contrasts in permeability or skin damage across the interval) requires placement/diversion techniques for near-wellbore stimulation to be effectively used to "force" fluid into the low-permeability or most-damaged zones that will be bypassed if there are other paths with less resistance.
4. Presence of water-bearing intervals—if the oil and gas zones are damaged/lower permeability, then even small fluid characteristics such as higher relative permeability to the a damage-removal fluid that is typically aqueous based may result in a less-resistive path for fluid flow.
5. The completion type (e.g., open hole, perforated casing, slotted liner). Options for control of fluid flow by the application of the right diversion and placement method will be significantly impacted by the type of completion in the well. It will therefore affect the overall success of near-wellbore stimulation.

Ideally, a production-logging-tool (PLT) survey should be conducted to answer such questions. However, in practice it is not always possible to run a survey, in which case a calculated flow distribution based on permeability profile can be used.

## 6.3 Placement and Diversion

The terms placement and diversion are often used synonymously. However, in this chapter a distinct difference between placement and diversion is established. Placement refers to the ways in which the stimulation fluids are *introduced* in the well (e.g., pipe characteristics, pump rate). Diversion refers to the ways in which stimulation fluids are *distributed* across the target zone (e.g., by means fluid additives and fluid characteristics). So, in principle, a placement technique can be combined with a diversion method. Such a combination is often significantly beneficial in getting the acid where it is needed.

**6.3.1 Placement.** Placement strategy is an important step in the design of a matrix treatment, especially in horizontal wells. The goal is to achieve uniform penetration of the treating fluid throughout the entire target zone and/or into each natural-fracture system. If complete zonal coverage is not achieved, the full production (or injection) potential cannot be realized.

The importance of the placement strategy is magnified in a horizontal well because of the length of the interval. The placement strategy must address the type of tubing used to inject the fluid and diversion of the fluid from thief zones to the damaged sections. Although conventional tubing or casing can be used to place the treating fluid, this process can be time-consuming or inefficient. The following placement techniques can be considered:

1. Use of coiled tubing. Coiled tubing is perhaps the most common method to place acid at the desired location(s). Unlike conventional tubing, coiled tubing can be run in and out of the hole, and treating fluid and certain diverters can be injected during movement of the coiled tubing. Injection down the annulus of the coiled tubing and tubing/casing can help to further optimize fluid distribution over the entire/target interval.
2. MAPDIR (maximum pressure differential and injection rate). It is postulated that by maintaining maximum injection rate while always increasing it to maintain maximum allowable matrix injection pressure, the need for a diverter is removed. This technique is known as MAPDIR.

## 6.4 Diversion

Diversion must be achieved to ensure that the treating fluid is continuously removing damage, rather than simply being injected into a thief zone. Mechanical techniques (straddle packers or ball sealers) are not practical in many wells because they are completed openhole, with slotted liners. We will discuss the following options:

1. Mechanical methods
2. Ball sealers
3. Chemical methods
4. Particulates
5. Viscous fluids
6. Foams
7. Viscoelastic surfactants
8. Two-phase (relative permeability) effects

## 6.5 Decision Tree

Based on the philosophy above, **Fig. 6.1** presents a decision tree for assisting in the selection of the appropriate diversion technique.

In the remainder of this chapter, the effect of each technique on fluid distribution and skin reduction will be illustrated by use of a computer model. For the following example, a well penetrates a carbonate formation with four distinct layers (see **Fig. 6.2**). The general well and reservoir data are shown in the following and in **Tables 6.1–6.3.:**

The well and reservoir configuration are made up from real cases but do not represent an actual well or its treatment.

The base-case illustration **(Fig. 6.3)** represents the situation when neither diversion nor coiled tubing is used, but the acid treatment is just pumped through tubing (bullheading) into a carbonate formation.

Clearly, the most-damaged (Zone 3) does get the least amount of treating fluids.

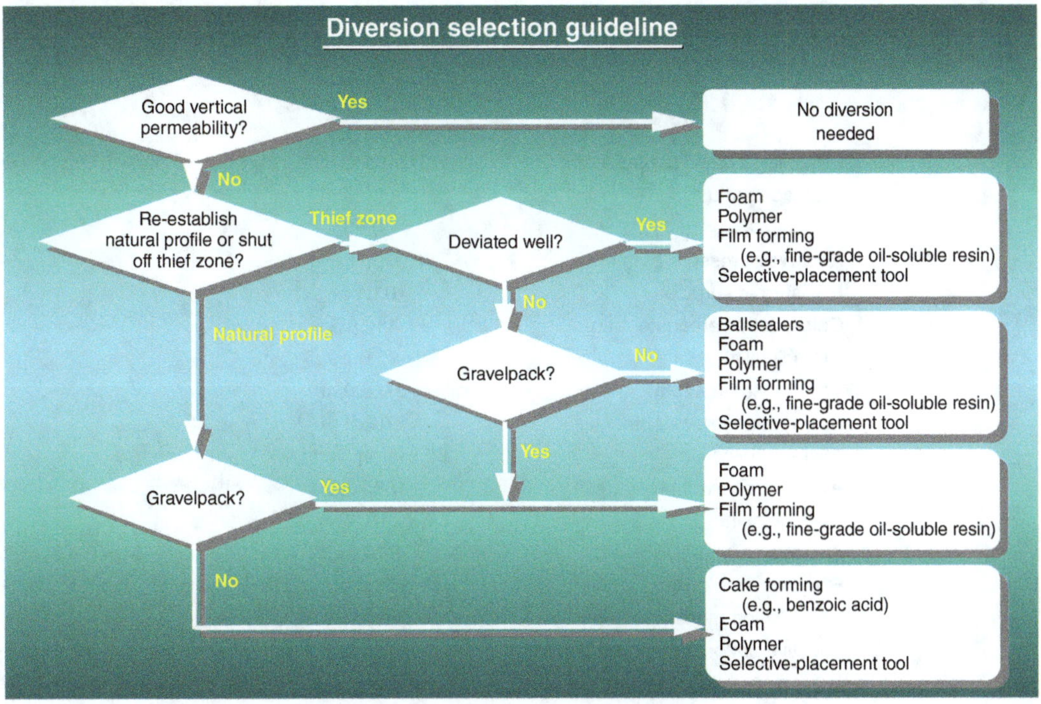


**Fig. 6.1—Diversion-selection decision tree.**

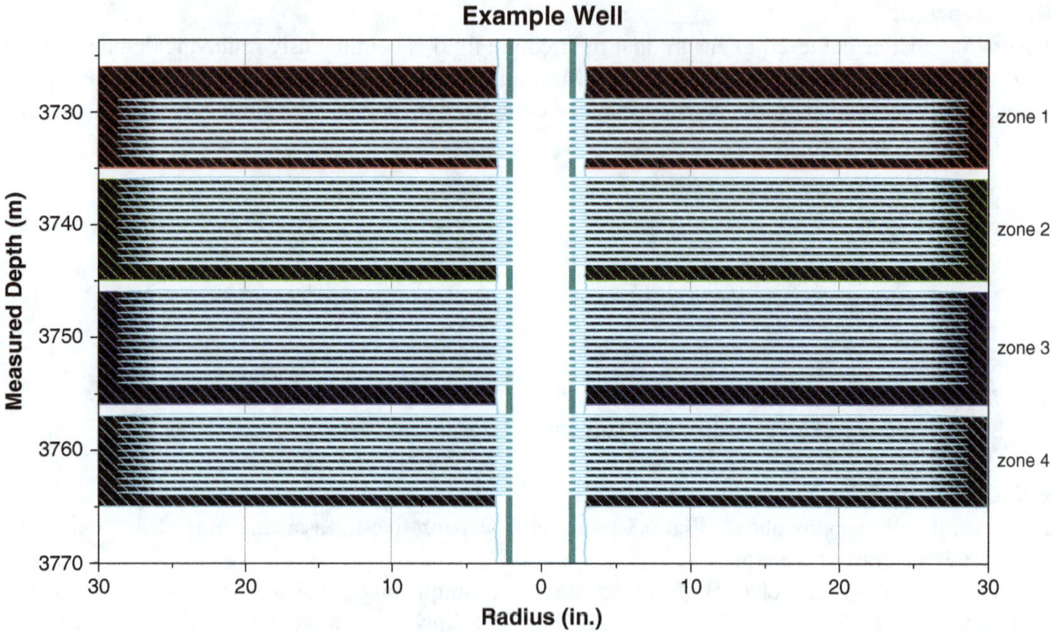


**Fig. 6.2—Example well schematic. Courtesy of Halliburton Energy Services.**

| Name | Top Measured Depth (m) | Bottom Measured Depth (m) | Permeability (md) | Porosity (%) | Pressure (bar) | Inflow (%) | Skin | Reservoir Type |
|---|---|---|---|---|---|---|---|---|
| Zone 1 | 3726.0 | 3735.0 | 50.0 | 11.00 | 188.2 | 28.00 | 4.08 | Carbonate |
| Zone 2 | 3736.0 | 3745.0 | 10.0 | 10.00 | 188.5 | 10.00 | –1.57 | Carbonate |
| Zone 3 | 3746.0 | 3756.0 | 100.0 | 10.00 | 188.7 | 2.00 | 390.46 | Carbonate |
| Zone 4 | 3757.0 | 3765.0 | 200.0 | 12.00 | 189.1 | 60.00 | 12.54 | Carbonate |

Table 6.1—Interval reservoir data.

| | |
|---|---|
| Bottomhole static temperature: | 91°C |
| Drainage radius: | 1000.0 m |
| $K_{vertical}/K_{horizontal}$: | 0.10 |
| Reservoir pressure gradient: | 0.0705 bar/m |
| Pressure at surface: | 0.0 bar |
| Calcite/total carbonate ratio: | 41.18% |
| HCl solubility: | 85.00% |
| Top measured depth: | 3726.0 m |
| Bottom measured depth: | 3765.0 m |
| Average deviation: | 65.5° from vertical |
| Average reservoir pressure: | 188.7 bar |
| Average skin: | 16.32 |
| Average permeability: | 87.2 md |
| Effective zone height: | 28.5 m |
| Effective top measured depth: | 3729.0 m |
| Effective bottom measured depth: | 3764.0 m |

Table 6.2—Interval reservoir data.

| Mineral | Zone 3 at 3752.0 m | Average % Content for Well* |
|---|---|---|
| Anhydrite | 15.0 | 15.0 |
| Calcite | 35.0 | 35.0 |
| Dolomite | 50.0 | 50.0 |

*These average content values are used in treatment design calculations.

Table 6.3—Mineralogy data.

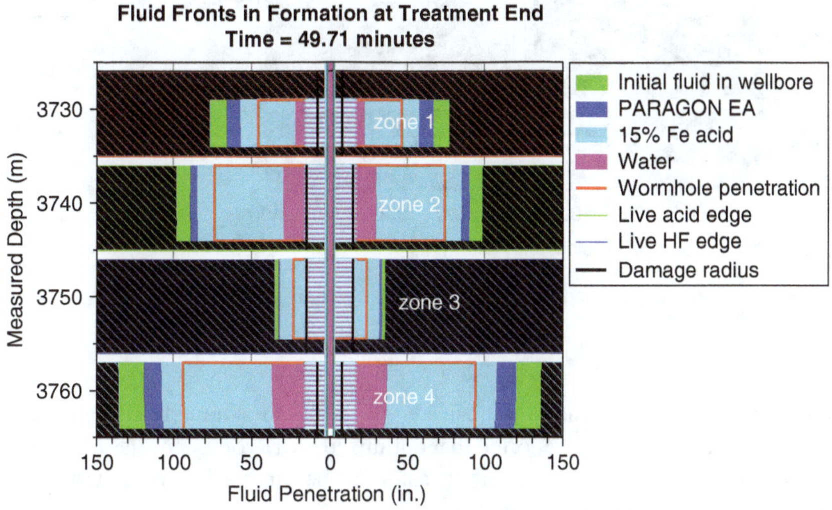


**Fig. 6.3— Base-case average rates. Courtesy of Halliburton Energy Services.**

## 6.6 Placement

**6.6.1 Coiled Tubing (CT).** The use of CT can be considered for the placement of matrix treating fluids, both in cased and openhole situations, especially when long intervals in horizontal wells are to be treated. However, as such, the use of CT does not necessarily change the penetration profile in the right direction. It is based on flow resistance along the well being at least significant compared to flow resistance entering the formation; otherwise, it would not matter where in the well the fluid is introduced, but it would flow along the well to the location where it is easiest to flow out into the formation.

**Fig. 6.4** shows that for the same well, just using CT placed at a fixed position some feet above the bottom of the well results in a greater amount of acid entering the lower-most zone.

There are a number of other ways in which CT can be used.

Acid can be spotted through the CT. It can be moved up or down (reciprocated) from the bottom toward the top of the interval to be treated. Other CT-based methods in matrix acidizing of horizontal wells involve either circulating up the CT/tubing annulus or bullheading acid or inert fluid along this annulus. Moreover, the use of CT with a jetting tool provides an excellent means of establishing direct communication of the acid with the formation face in long openhole sections (vertical or horizontal).

Stationary radial nozzles are simple and easy to manufacture locally but they do not provide complete circumferential coverage of the wellbore and, unless a nozzle lines up exactly with a perforation tunnel, the turbulence they create does not promote fluid swapping inside the perforation tunnel. Nozzles that are offset from the radial direction cause the fluid to swirl in the wellbore; this ensures full circumferential coverage, and, by optimizing the tangential component of the jet velocity, it increases

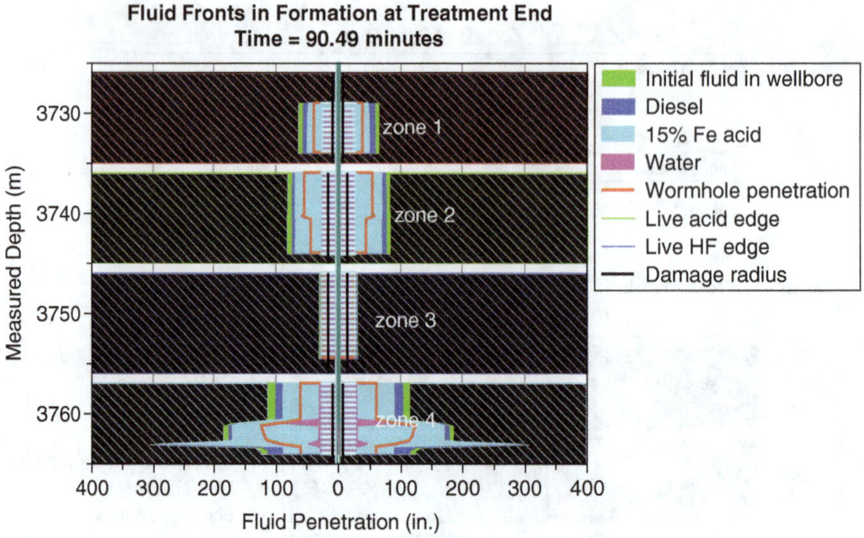


**Fig. 6.4—Same treatment with CT. Courtesy of Halliburton Energy Services.**

the potential for turbulent fluid exchange at the face of the perforation tunnel. Jetting tests with colored water on screens and sandpacks have reported that optimized swirl/vortex nozzles achieve full circumferential coverage of the screen and complete invasion of the sandpack, unlike stationary nozzles in which invasion efficiency is approximately 50%.

The combination of using CT and a (temporarily crosslinked) gelled acid system, or foam, as a diverting agent has proved effective as a diversion process in carbonate formations in horizontal wells.

## 6.7 Spotting Acid

A convenient way of placing the acid consists of circulating the acid down to the top of the perforated interval, provided that the completion allows circulating. For example, in the case of a straddle-pack completion, it might be possible to target the sleeve that is open if this technique is used in combination with CT. **Fig. 6.5** shows the fluid fronts in the formation with this technique.

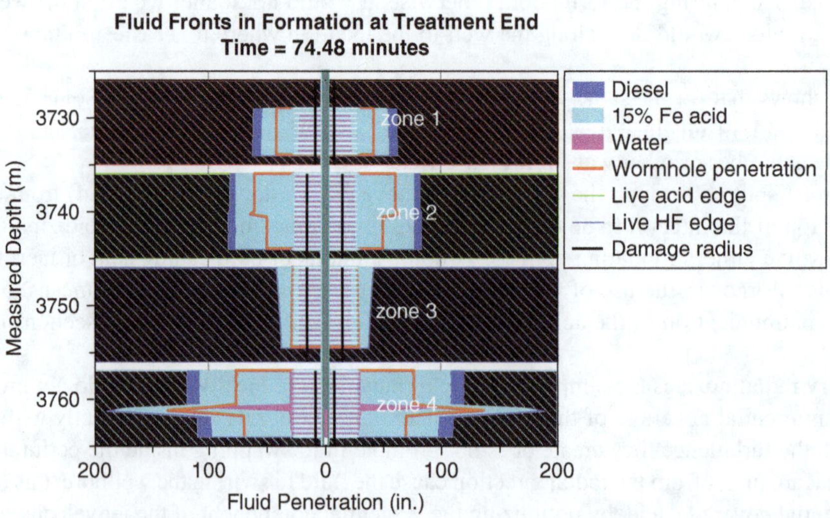


**Fig. 6.5—Same treatment: CT and spotting acid. Courtesy of Halliburton Energy Services.**

## 6.8 High Rate and Maximum Pressure

The maximum pressure differential and injection rates (MAPDIR) diversion technique has been promoted as the best means of ensuring sufficient acid placement in all zones during matrix stimulation treatments (Paccaloni 1995; Tambini 1992; Kalfayan and Martin 2009).

In practice, the method consists of injecting acid at the highest possible matrix treating rate at a constant bottomhole pressure slightly below the fracturing pressure. This requires continual increases in the injection rate as the acid removes damage and thereby increases the well injectivity. A high success rate has been reported with this technique, with success quantified as a large reduction in skin factor.

The MAPDIR technique does clearly result in the highest possible injection rate into low-injectivity intervals, while any of the other placement/diversion techniques may reduce injectivity into low-injectivity intervals. In this sense, MAPDIR achieves the goal of ensuring complete acid coverage of all zones without causing additional damage **(Fig. 6.6).**

The MAPDIR technique is especially suited for placement in matrix treatments of nonfractured carbonates. Here, the acid reaction rate is usually high, and mass transport limits the overall acid spending rate. The result of the acid/rock dissolution process therefore strongly depends on the acid-flow rate. When acid is pumped into a carbonate formation that contains zones with different degrees of damage, the zones with the least damage will accept the largest amount of acid, and these can be stimulated effectively. A lower volume of acid will flow into the more severely damaged zones. An increase in the total injection rate will result in higher flow rates into all zones. In particular, the zones with severe damage can benefit; the increased flow rate may cause a transition from a compact dissolution regime without much stimulation to a wormholing regime with effective stimulation.

A number of methods to monitor the decrease in skin factor during this matrix stimulation treatment from recorded wellhead pressure (or derived bottomhole pressure) and injection rate are available. The one presented by Paccaloni uses instantaneous pressure and rates from which the skin factor is derived during the treatment. For this purpose, the concept of damage ratio (DR) was introduced, which simply represents the ratio of productivity index with real skin (*S* at time zero) to that with zero skin. An example of the measured wellhead pressure superimposed on a graph of wellhead pressure vs. injection rate shows the progress of matrix treatment, or decreasing skin, in **Fig. 6.7.**

Fig. 6.6 demonstrates the effect of applying the maximum rates without fracturing on the distribution across the intervals. As the treatment progresses, the DR decreases to unity (i.e., zero damage or skin), indicating successful treatment. Even a slightly negative skin (stimulated beyond damage removal) can then be achieved.

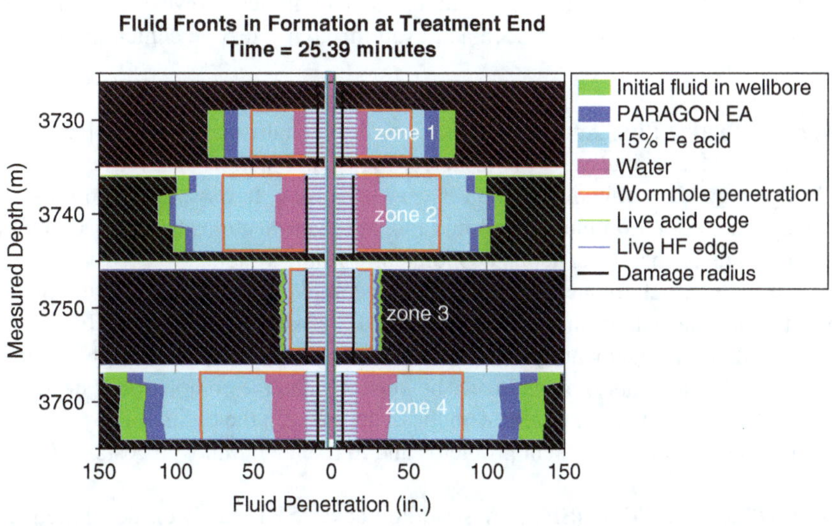


**Fig. 6.6—Placement profile bullheading simple HCl treatment in carbonates at maximum rates (MAPDIR). Courtesy of Halliburton Energy Services.**

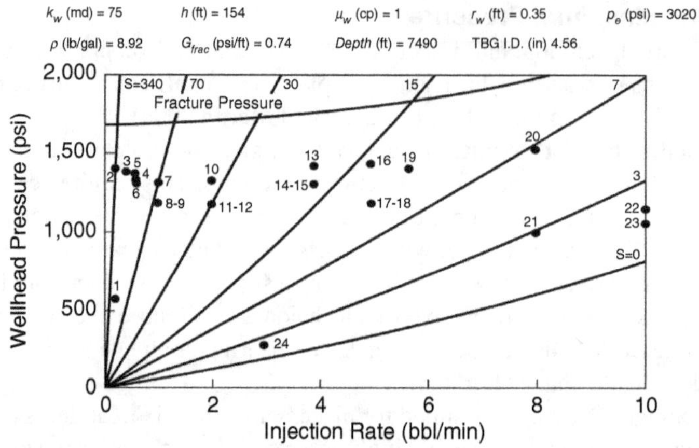


**Fig. 6.7—Job control graph during MAPDIR treatment. Courtesy of Halliburton Energy Services.**

On the basis of the analysis of more than 170 jobs, Paccaloni claimed that the MAPDIR technique is sufficient to obtain full coverage of the pay zone if the permeability contrast is less than 300 md and the pay zone does not exceed 70 m.

A drawback of the method is that it results in more acid than necessary being injected into high-injectivity intervals. In addition, the benefits of this method are reduced when the attainable bottomhole pressure is less than the desired value because of limitations in surface pressure or pumping capacity.

For better efficiency, Paccaloni recommended the use of CT for preacidizing activities wellbore cleanup and acid circulation for completion-string cleaning (pickling). Also recommended were spotting the acid in front of the pay zone by use of CT followed by simultaneous pumping through both the CT and tubing string to maximize the pump rate during matrix injection.

Paccaloni simply used the inflow equations for semisteady-state flow. In the next chapter, we will come back to this topic with a more refined method that evaluates the effect of diversion methods.

## 6.9 Diversion Methods

**6.9.1 Mechanical Techniques.** These techniques use mechanical means to isolate a particular zone for injection of stimulation fluids. Except for ball sealers (see below), mechanical diversion methods are usually more reliable than other diversion techniques. However, they are not effective in wells where zonal isolation has not been achieved through effective cementation of the production casing.

**6.9.2 Packer and Bridge-Plug Arrangements.** Complete zonal isolation is obtained by packing off a section of the completion interval—for instance, by a combination of a retrievable packer and a retrievable bridge plug. This effective method is very expensive, however, and requires a (workover) rig. Additionally, the perforation intervals should include blank sections to allow setting of the packers. In most cases, this is, therefore, not an economically viable system.

For horizontal wells, which frequently have openhole completions, or with a slotted/predrilled uncemented liner, this mechanical method is not suitable. External casing packers (ECPs) may help to improve wellbore fluid management by compartmentalizing the annulus, although the optimum ECP position for remedial treatments should already be known when the completion is designed. While in principle many ECPs could be used to deal with the uncertainty in the optimum position, their number is usually restricted to avoid deployment problems and to reduce initial well costs.

**6.9.3 Selective-Placement Tool (SPT).** A straddle-packer SPT has been developed that can run on CT, can pass through production tubing, and can be repeatedly set in casing. It is fitted with high-expansion straddle packers.

Nowsco UK Limited was the first company to bring such a tool on the market; several other companies followed. The current tool has the following limitations:

- Maximum operating temperature of 110ºC.
- The rubbers applied in the tool are sensitive to most solvents, particularly those with a high aromatic-hydrocarbon content.

A number of applications for the SPT have been identified. In the context of this document, the most relevant applications are

- Acid stimulation of specific intervals with optimum volumes of fluid
- Treatment of highly inhomogeneous carbonate formations

In many cases of chemical treatment, the SPT is the most economic means of ensuring 100% selectivity.

**6.9.4 Ball Sealers.** Ball sealers are small rubber-coated nylon or resin balls. They are available in a density range of 0.9–1.4 g/cm$^3$. They have often been used to divert fluids, both in matrix acidizing treatments and in acid fracturing treatments over long or multiple intervals. This diversion method has been successfully applied both in vertical perforated wells and, to a lesser extent, in slanted or horizontal wells. The absence of perforations in openhole completions means that ball sealers cannot be applied in that situation.

The action of ball sealers is based on the idea that the balls are carried along with the stimulation fluid toward a perforation, which they will subsequently shut off. In order for the balls to seat, the fluid velocity needs to be sufficiently high. As a guideline, the following values have been used: 3 bbl/min in a 4.5-in. casing, 4 bbl/min in 5.5-in. casing, and 7 bbl/min in 7-in. casing. The perforations should be of a consistent size, and be round and free from burrs. Ball sealers require a pressure differential of 100–200 psi to seat efficiently. This should be kept in mind when planning the number of perforations to be balled and the treatment pump rate. In order to compensate for nonseating balls, the number of balls that is being used is usually 10–20% higher than the number of perforations believed to be taking the fluid.

We see that the effect of the ball sealers in this case is rather limited **(Fig. 6.8).** This is because of the relatively low-injection rates that can be used to avoid fracturing. As a result, the seating efficiency of ball sealers is reduced.

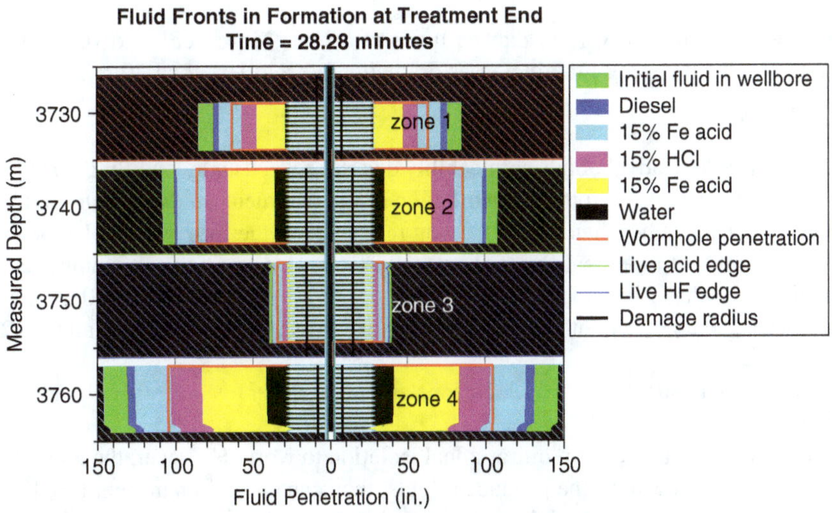


**Fig. 6.8—Use of ball sealers. Courtesy of Halliburton Energy Services.**

Buoyant balls tend to have a higher seating efficiency than nonbuoyant balls. However, buoyant balls have the disadvantage that they will be produced back to surface. As a compromise, the density of the balls could be chosen such that they will be buoyant in the treatment fluid, but have a density that is greater than the produced fluids. The balls will then sink to the sump, and can be left at the bottom of the well. However, during a workover, they can still pose problems. In deviated or horizontal wells, buoyant ball sealers tend to travel along the topside of the wellbore, thus reducing the chance of seating on perforations with other orientations than wellbore topside. Buoyant ball sealers may, therefore, be less applicable in deviated or horizontal wells, particularly with zero-degree phased perforations.

As to the quality control on density of commercial ball sealers, it has been observed that on average 20 to 30% of the ball sealers deviate more than the 5% scatter specified by the manufacturers. Especially for the 0.9- to1.0-g/cm$^3$-density ball sealers, such deviations cannot be tolerated. It is, therefore, suggested for field treatments to check the densities of the balls on-site by immersing them in fluids with the desired densities (upper and lower specification limit, such as methanol/water mixtures). Only ball sealers that float in the high-density fluid and that sink in the low-density fluid should be accepted.

Water-soluble balls, constructed from collagen, among other materials, have also been developed. This eliminates a range of potential difficulties, such as the possibility that the balls will remain downhole where they can make drilling out bridge plugs difficult or that they will fail to unplug during flowback in low-pressure reservoirs. Balls that are belatedly produced at surface can block chokes. The collagen balls vary more in diameter and show more imperfections on their surface than do balls made from rubber. The balls are stable in hydrocarbons and have a specific gravity of 1.2 g/cm$^3$. The solubility rate increases with increasing temperature and decreasing pH. Tests show that the balls can hold a pressure differential of 19.2 MPa at 50°C for 2 hours before failure. They have been used in up to 120°C for a short duration (20 minutes). This limits their use to relatively short stimulation jobs, and so their usefulness in treating long wells with multiple zones will be limited.

**6.9.5 Particulate Diverting Agents.** This technique uses chemical diverters, such as benzoic acid and its derivatives, to create a diverter cake either at the perforation or at the formation face.

Chemical diverters fall into two types:

1. Those forming a thin but very low-permeability film on the formation face. These can effectively plug off a high-injectivity or thief zone. However, there is the risk of almost total injectivity loss if too much diverter is used. Examples are fine-grained oil-soluble resins (particle size <10 μm) and liquid diverters.

**Fig. 6.9** shows that these diverters can be quite effective. More acid is diverted into Zone 3. However, note that the pumping time increases dramatically from less than 30 minutes to more than 14 hours.

2. Those creating a relatively high-permeability diverter cake, usually by filling up the perforation. Once the diverter cake is formed, distribution of the treatment fluids is controlled by the resistance of the cake rather than by the varying resistance of the formation. All zones will receive similar volumes of treatment fluids **(Fig. 6.10).** Examples are rock salt, commercial grades of benzoic acid, salts of benzoic acids, coarse grades of oil-soluble resins (>50 μm), wax particulates (beads, buttons), and other particulate solids soluble in oil or water (Kalfayan 2005).

**Fig. 6.11** shows that the flow rates into each zone are much closer to each other already after the first batch of diverting agent.

Neither type of chemical diverter requires zonal isolation to work. However, the main disadvantage is that if they are not removed by the produced fluids, they can cause impairment too. In most cases, cleanup is rather slow. Many cases of slow cleanup after diverted acid stimulation treatments may be attributed to the diverter not (completely) dissolving in the back-produced fluids.

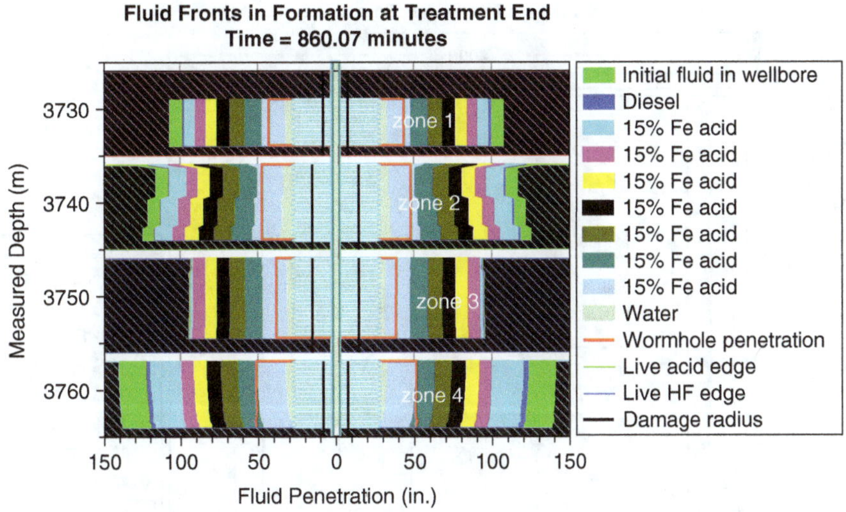


**Fig. 6.9—Diversion by film-forming agents. Courtesy of Halliburton Energy Services.**

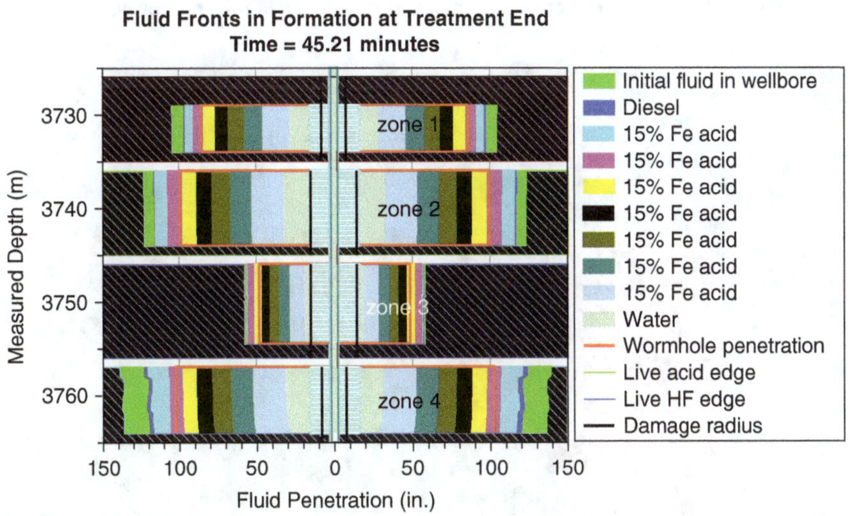


**Fig. 6.10—Cake diversion. Courtesy of Halliburton Energy Services.**

Nevertheless, chemical diverting agents, in particular benzoic acid, can be used successfully, provided a number of design rules are observed:

1. In general, use straight benzoic acid (rather than salts of benzoic acid), because it is soluble in formation water and oil, and it sublimates in gas.
2. The effective downhole particle size of benzoic acid is influenced by the type and order of addition of surfactants and inhibitors. The preferred order of mixing is surfactant, inhibitor, and finally the diverter. Acidic slurries prepared in this way show reasonably stable particles for 1 to 2 hours.
3. In order to maintain the low permeability of the diverter cake, continuous addition of benzoic acid throughout the treatment is preferred. **Fig. 6.12** shows the minimum concentration to be applied for effective diversion.
4. During acid fracturing treatments of uncased horizontal wells, the simplest method to isolate one created fracture from the next is to pump blocking material (rock salt, benzoic acid, or

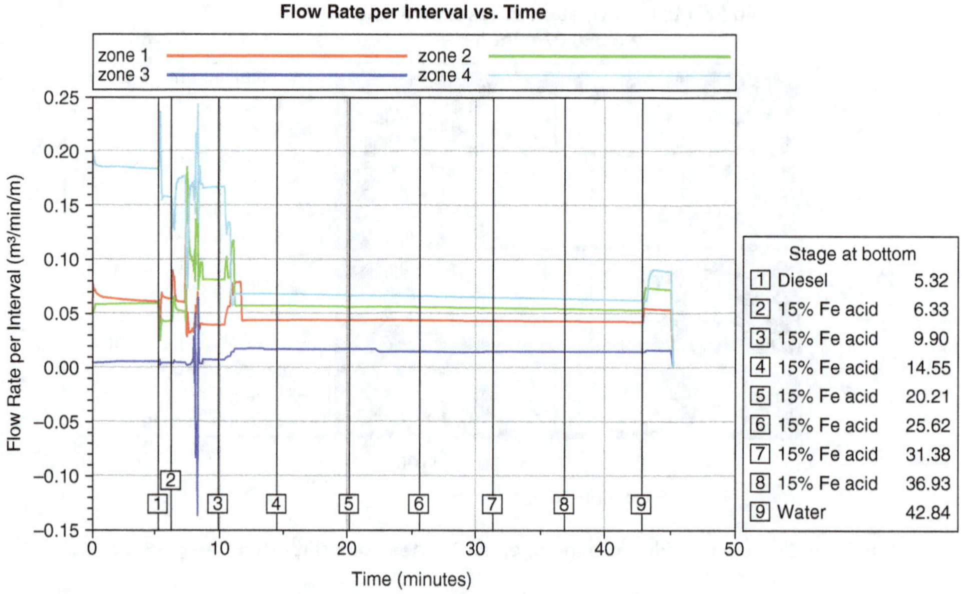


**Fig. 6.11—Flow rate per interval vs. time. Courtesy of Halliburton Energy Services.**

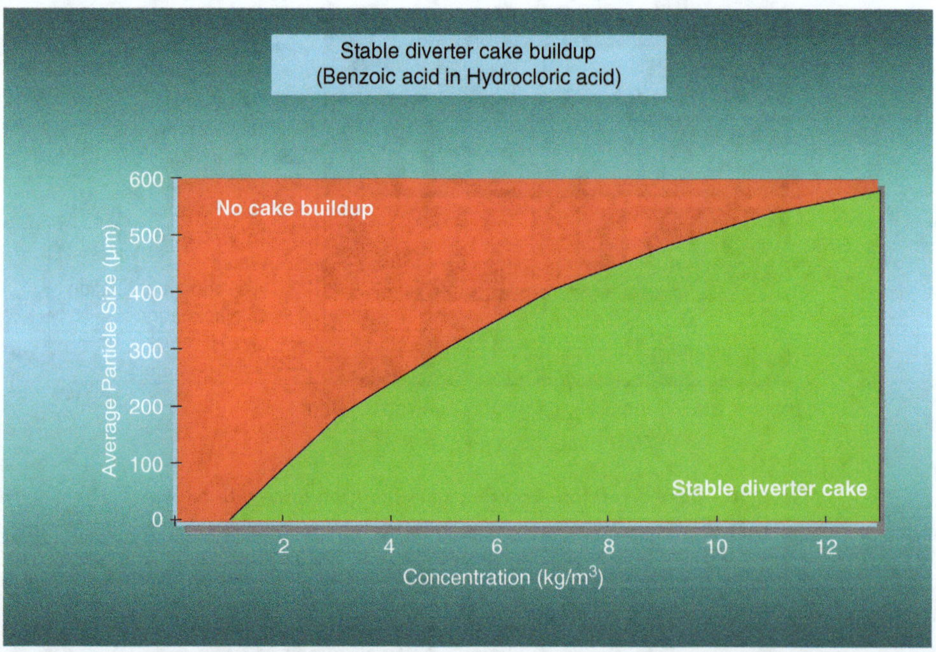


**Fig. 6.12—Stability of benzoic acid diverter cake in hydrochloric acid (Nitters and Davies 1989).**

mothballs) in batches between the stages of a continuous large fracture treatment. Unfortunately, this method does not provide much control over spacing of the fracture-initiation points along the wellbore.

5. Cake buildup is a function of concentration, inflow profile, length of perforated interval, shot density, injection rate, and other factors. Note that benzoic acid is slightly soluble in acid: At concentrations of less than 1 kg/m$^3$, no cake will be formed. This solubility increases with temperature. At temperatures above 90°C, the solubility of benzoic acid is so high that its use becomes impractical.

The use of particulate diversion material for matrix acidizing treatments in horizontal wells, which can be so long that cementing and perforating options are discarded because of their cost, would require so many diverter stages that such treatment would be prohibitively expensive. As a rough estimate, the diverting agent on its own would cost approximately USD 5,000 for every 100 ft of openhole section. In such a case, the preferred diverting technique consists of pumping viscous banks into sections of high fluid intake. This is discussed further in the next subsection.

**6.9.6 Viscosified Fluids.** Viscosified fluids, or acids viscosified by polymer (or by foam), are probably applicable to all completion types. They are of particular benefit in horizontal, openhole wells, in which many other diversion and selective-placement techniques cannot be applied. In the use of viscosified fluids, viscous pills (e.g., polymer solution) are injected between acid flushes. In order to work, a number of prerequisites should be fulfilled:

- An adequate volume of polymer must be placed in the unimpaired zone.
- The subsequent acid must have direct access to the impairment.
- The polymer must not cause impairment, and must be easily removed by the produced fluids.

In practice, these requirements, especially the latter, are difficult to meet. Currently available products have been used with limited success.

A better alternative is to gel the acid to be injected. The requirements for successful application of gelled acids are

- The acid gel should have sufficiently high viscosity during pumping at the treatment temperature.
- The acid gel should lose its viscosity after treatment to promote rapid cleanup.

With viscosified acid, the mechanism of diversion is through the increase in flow resistance in higher-permeability regions, which have been taking the treatment fluid first. Because gelled acid is a shear-thinning (non-Newtonian) fluid, its viscosity also increases farther away from the wellbore, thereby blocking the following acid and diverting it to other less-permeable zones. In such a way, the gelled acid acts as a self-diverting system. Biopolymer acid gels satisfy both of the preceding criteria and act as self-diverting acid systems. Moreover, such shear-thinning behavior also has the advantage of enabling the operator to pump the viscous acid at relatively high rates.

Some other biopolymers show a sharp drop in viscosity above a certain temperature. For some biopolymers this transition temperature at which the gel loses its viscosity is near 65°C, but this can be adjusted over a fairly wide range of temperature with various chemicals, including some acids. For instance, addition of 5% formic acid increases the transition temperature to near 75°C. Gelation of acid is sensitive to iron (III) ions. Therefore, a reducing agent (e.g., erythorbic acid or vitamin C) has to be added. Notably, an ordinary sequestering agent, such as citric acid, is not effective here. The acid gel loses its viscosity permanently because of the rapid chemical degradation of the polymer, which takes place above the transition temperature. These properties make a system with such biopolymers particularly useful for the removal of shallow damage, such as that caused by mud remnants.

Other systems that have been proposed and used for viscous acid diversion consist of acid mixed with a gelling agent and a pH-sensitive crosslinker. Initial spending of the acid during leakoff and wormholing will produce a rise in pH to a value of 2 to 3, which initiates a significant increase in viscosity because of crosslinking. The crosslinked gel will effectively stop any further fluid invasion and divert subsequent neat acid stages to different parts of the zone. The crosslink will break upon further spending, and at a pH value of above 4, the viscosity of the fluid will return to its original value. Such a pH-sensitive crosslinked gel is ideally suited for acid diversion in long intervals, and a number of treatments have been successfully performed in long horizontal openhole completions in the North Sea and the Middle East. Use of CT enhances fluid placement in treatment of such long intervals.

The following procedure for gelled acids has been applied successfully:

1. Run CT to below the perforated interval; if a large sump is present, a bridge plug may be placed to reduce the dead volume.
2. Inject the required volume of gelled acid by pumping through both the CT and the CT/tubing annulus, maintaining the same rate through both conduits.
3. Displace the acid with a suitable brine, diesel, or other fluid.
4. Close the well in for 12 to 24 hours to allow for sufficient reaction time (depending on the type of acid and damage) and proper breakdown of the viscosity.

**Fig. 6.13** shows a simulation of the placement profile using a pH-controlled gelled acid.

***Emulsified Acids.*** With emulsified acid, whereby the acid is present as the internal phase, the fluid system, which is a non-Newtonian fluid (shear-thinning behavior), has a significantly higher viscosity than the reservoir fluid. This will result in improved wellbore coverage, diverting more fluid to the low-permeability and/or damaged zones. The fluid velocity may still be low in these zones, but the retarded nature of the emulsified system still provides efficient wormholing, even at low rates.

***Foamed Fluids.*** Foam diversion is probably the least understood method. Therefore, we describe what is known about foam properties in more detail than the descriptions of other, better understood techniques. Foam offers several advantages as a diverting agent: It can act as a simple viscous fluid, but under some conditions, it can also spontaneously divert acid away from high-permeability layers. It is easy to clean up and introduces no permanent injectivity impairment. In cases where it spontaneously diverts flow away from high-permeability layers, no mechanical isolation is needed.

Although they share some similarities, there are important differences between foam for acid diversion and foams for drilling, fracturing, cementing, and liquid unloading of gas wells. In those applications, foam can be treated as a single-phase, though complex, fluid. When foam penetrates the geological formation it must be treated as a two-phase dispersion in which the flow of the gas is enormously reduced by the liquid films in the foam.

The behavior of foam in porous rock with no oil present is illustrated in **Fig. 6.14.** Foam behaves in two very different ways depending on foam quality (injected volume fraction of gas, usually expressed as vol%) (Alvarez et al. 2001). At low qualities, the pressure gradient is nearly independent of liquid-injection rate, while at fixed quality, flow is very shear-thinning. Represented as a power-law fluid, the power-law exponent can be as small as 0.4. Some studies find extremely shear-thinning behavior—virtually no increase in pressure gradient with increasing injection rate (Parlar et al. 1995; Zerhboub et al. 1994). At high foam qualities, pressure gradient is near independent of gas-injection rate, and at

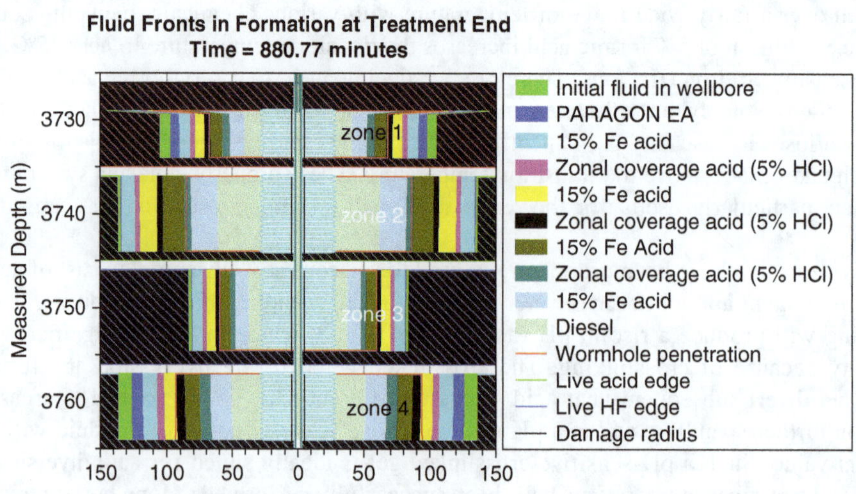


**Fig. 6.13—Example of fluid distribution using gelled acid. Courtesy of Halliburton Energy Services.**

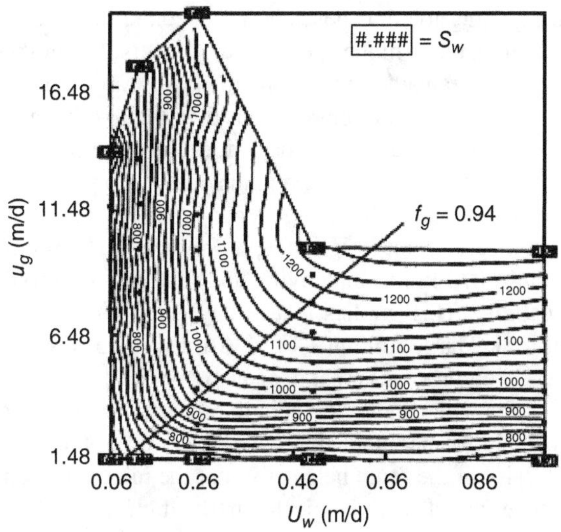


**Fig. 6.14—Two foam-flow regimes without oil. Pressure difference across a 2-ft sandpack as a function of gas ($U_g$) and water ($U_w$) superficial velocities (Osterloh and Jante 1992). The high-quality regime is at the upper left of this plot, and the low-quality regime to the lower right.**

fixed foam quality, apparent rheology can be either Newtonian or moderately shear-thinning or shear-thickening. The transition foam quality separating the low-quality and high-quality regimes varies with the application, depending on surfactant formulation, temperature, formation permeability, the gas in the foam, and other factors. The apparent viscosity of foam is at its maximum at this transition foam quality. In general, the transition foam quality increases with increasing formation permeability, decreasing temperature, increasing surfactant concentration, or any factor that makes the foam more stable (Alvarez et al. 2001). In Fig. 6.14, this transition between regimes happens at approximately 94% quality, but most formations with lower permeability would have this transition at a lower foam quality.

Most laboratory studies of foam use rock of permeabilities much higher than those typical of the field and represent the effect of damage using somewhat lower-permeability rock. Behenna (1995) and Thompson and Gdanski (1993) used thin wafers of very-low-permeability rock at the face of a core to represent a damaged region in sandstone and carbonate cores.

Foam in the high-quality regime is self-diverting; that is, foam shows a greater apparent viscosity in higher-permeability layers than lower-permeability layers. In some cases, this effect more than makes up for the difference in permeability, but more commonly it partially compensates for permeability differences. Though data are limited, it appears that there is little or no permeability effect in the low-quality regime.

Crude oils tend to destabilize most foam formulations; lighter crudes are thought to be more destructive of foam than heavier crudes. This could be an advantage or disadvantage, depending on the goals of the process. If the goal is to stimulate oil-bearing zones intermingled with watered-out zones, a foam that collapses in the presence of oil could be an advantage. It is also possible to formulate surfactants that are less sensitive to oil.

The choice of surfactant for a particular application of foam depends on the temperature, salinity, formation type, and oil present in the formation. It is, in general, difficult to find effective foaming surfactants at high formation temperatures. In addition, the foaming surfactant must be formulated for compatibility with the acid and with the corrosion inhibitors and other chemical additives to the acid. Foam cannot propagate faster than surfactant in the formation, and thus surfactant adsorption on the rock surfaces is an important consideration. A preflush of surfactant helps to satisfy adsorption ahead of the foam slug, and because most preflush goes to undamaged layers, this helps place more foam in those layers.

There are two ways of injecting foam for acid diversion: (1) in alternating slugs with the acid and (2) as a single slug of foamed acid. When injected in alternating slugs of foam and acid, the goal of foam is to place as much diverting foam as possible in the zone least in need of acid. It is *good* then if most foam enters the undamaged or higher-permeability layers. When the acid slug follows foam, most of the gas in the foam stays trapped in place, and the large trapped gas saturation greatly reduces the relative permeability to acid; as a result, more acid is diverted to other zones (Zeilinger et al. 1995; Zerhboub et al. 1994). In sandstone, the injected acid does not penetrate the foam evenly but fingers through it (Nguyen et al. 2009). The rock matrix remains intact (no wormholes are formed), but most of the damage would be removed within these fingers penetrated by acid. If most of the foam is in the layers not needing stimulation, this is not a big problem. Parlar et al. (1995) report that above a threshold $\nabla p$ there is virtually unlimited liquid injectivity, and that this threshold $\nabla p$ scales with roughly the (–½) power of permeability. This implies that at any given $\nabla p$, in a slug process, the highest-permeability zones would take large amounts of acid relative to lower-permeability zones. One advantage of the alternating-slug process is that once acid penetrates and stimulates a given zone, the next slug of foam can block that zone and divert acid to other zones in need of stimulation.

If foam is injected in a single foamed-acid slug, then diversion of the foam slug itself is crucial. In this case, the effective viscosity of the foam helps even out the injection profile, and the self-diverting nature of foam would be beneficial if foam is injected in the high-quality regime.

Acid injected into carbonates forms wormholes through the rock matrix (this is not the case in sandstones). Foam also alters the nature of the wormholing process in carbonates. Bernardiner et al. (1992) report that foam causes formation of longer, narrower wormholes with fewer branches, making acidization more efficient in the given layer. Thompson and Gdanski (1993) report that foam fills and temporarily blocks wormholes, diverting acid to layers with fewer wormholes, thus leading to a more-even distribution of acid between layers.

Several references give guidelines for design of a foam diversion process (Gdanski 1993; Zerhboub et al. 1994; Robert and Rossen 2000). As noted, foam can be injected as a spacer between acid stages (multistage process) or, alternatively, part or all of the treating acid can be injected as foam (one-stage process). A foamed-acid stimulation pumping schedule for shallow damage might read as follows:

1. Prepare the well and wellsite for the treatment by rigging up the CT unit and pumping equipment. Pressure test the surface lines, the blowout preventer, and tubulars to the required pressure.
2. Run the CT in the hole and clean the perforations with foam. Spot 0.12 $m^3/m$ (10 gal/ft) of foamed HCl while pulling the CT up at a velocity of 1 m/min. Maintain pump rate at 159 L/min (1 bbl/min).
3. Run the CT back to original depth. Spot 0.24 $m^3/m$ (20 gal/ft) of foamed mud acid. Pull the CT at 0.5 m/min while pumping at 159 L/min (1 bbl/min).
4. Inject 0.12 $m^3/m$ (10 gal/ft) of unfoamed 3% $NH_4Cl$ brine, while lowering the CT at 1 m/min and pumping at 159 L/min (1 bbl/min). An additional volume of $NH_4Cl$ brine should be injected after running in the CT to total depth.
5. Displace the tubing/CT annulus, while pulling the CT up the tubing, with filtered water, suitable brine, oil, etc. Keep the well full with displacement fluid to avoid swabbing acid foam into the wellbore while pulling.

## References


Alvarez, J. M., Rivas, H. J., and Rossen, W. R. 2001. A Unified Model for Steady-State Foam Behavior at High and Low Foam Qualities. *SPE J* **6** (3): 325–333. SPE-74141-PA. http://dx.doi.org/10.2118/74141-PA.

Behenna, F. R. 1995. Acid Diversion from an Undamaged to a Damaged Core Using Multiple Foam Slugs. Presented at the SPE European Formation Damage Conference, The Hague, 15–16 May. SPE-30121-MS. http://dx.doi.org/10.2118/30121-MS.

Bernadiner, M. G., Thompson, K. E., and Fogler, H. S. 1992. Effect of Foams Used During Carbonate Acidizing. *SPE Prod Eng* **7** (4): 350–356. SPE-21035-PA. http://dx.doi.org/10.2118/21035-PA.

Gdanski, R. D. 1993. Experience and Research Show Best Designs for Foam-Diverted Acidizing. *Oil Gas J* **91** (36): 85–89.

Kalfayan, L. J. 2005. The Art and Practice of Acid Placement and Diversion: History, Present State, and Future. Paper SPE 108802 presented as a Distinguished Lecture during the 2004–2005 season.

Kalfayan, L. J. and Martin, A. N. 2009. The Art and Practice of Acid Placement and Diversion: History, Present State and Future. Presented at the SPE Annual Technical Conference and Exhibition, New Orleans, 4–7 October. SPE-124141-MS. http://dx.doi.org/10.2118/124141-MS.

Nguyen, Q. P., Zitha, P. L. J., Currie, P. K. et al. 2009. CT Study of Liquid Diversion with Foam. *SPE Prod & Oper* **24** (1): 12–21. SPE-93949-PA. http://dx.doi.org/10.2118/93949-PA.

Nitters, G. and Davies, D. R. 1989. Granular Diverting Agents Selection, Design, and Performance. Presented at the SPE Production Operations Symposium, Oklahoma City, Oklahoma, USA, 13–14 March. SPE-18884-MS. http://dx.doi.org/10.2118/18884-MS.

Osterloh, W. T. and Jante Jr., M. J. 1992. Effects of Gas and Liquid Velocity on Steady-State Foam Flow at High Temperature. Presented at the SPE/DOE Enhanced Oil Recovery Symposium, Tulsa, 22–24 April. SPE-24179-MS. http://dx.doi.org/10.2118/24179-MS.

Paccaloni, G. 1995. A New, Effective Matrix Stimulation Diversion Technique. *SPE Prod & Fac* **10** (3): 151–156. SPE-24781-PA. http://dx.doi.org/10.2118/24781-PA.

Parlar, M., Parris, M. D., Jasinski, R. J. et al. 1995. An Experimental Study of Foam Flow through Berea Sandstone with Applications to Foam Diversion in Matrix Acidizing. Presented at the SPE Western Regional Meeting, Bakersfield, California, USA, 8–10 March. SPE-29678-MS. http://dx.doi.org/10.2118/29678-MS.

Robert, J. A. and Rossen, W. R. 2000. Fluid Placement and Pumping Strategy. In *Reservoir Stimulation*, third edition. ed. M. J. Economides and K. G. Nolte. Chap. 19. New York: John Wiley & Sons, Inc.

Tambini, M. 1992. An Effective Matrix Stimulation Technique for Horizontal Wells. Presented at the European Petroleum Conference, Cannes, France, 16–18 November. SPE-24993-MS. http://dx.doi.org/10.2118/24993-MS.

Thompson, K. E. and Gdanski, R. D. 1993. Laboratory Study Provides Guidelines for Diverting Acid with Foam. *SPE Prod & Fac* **8** (4): 285–290. SPE-23436-PA. http://dx.doi.org/10.2118/23436-PA.

Zeilinger, S. C., Wang, M., Kibodeaux, K. R. et al. 1995. Improved Prediction of Foam Diversion in Matrix Acidization. Presented at the SPE Production Operations Symposium, Oklahoma City, Oklahoma, USA, 2–4 April. SPE-29529-MS. http://dx.doi.org/10.2118/29529-MS.

Zerhboub, M., Touboul, E., Ben-Naceur, K. et al. 1994. Matrix Acidizing: A Novel Approach to Foam Diversion. *SPE Prod & Fac* **9** (2): 121–126. SPE-22854-PA. http://dx.doi.org/10.2118/22854-PA.

Chapter 7

# Treatment Evaluation and Real-Time Diagnosis

**Ding Zhu,** Texas A&M University

## 7.1 Introduction

Matrix stimulation treatments (most commonly using acid as a solvent) are aimed at overcoming the effect of near-wellbore damage by dissolving solids and, hence, increasing permeability. The skin factor, a parameter to measure altered flow conditions in the near-wellbore vicinity, is often used as a quantitative indicator of the level of damage. In the absence of completion effects (such as partial completion or poor perforations), a high positive skin factor indicates severe damage; and a negative skin factor is because of a higher effective permeability in the near-wellbore region after a successful stimulation. Selecting an optimal injection rate and appropriate acid system is a difficult process because of uncertainties about the damaged zone and the competing effects of mineral dissolution and reaction-product precipitation. Determining the evolution of skin factor during a matrix stimulation treatment is a means of measuring the effectiveness of acid or other solvent in overcoming formation damage. Treatment-evaluation procedure uses pressure and injection-rate data during a treatment to estimate skin-factor change as a function of pumping time. The procedure can be applied during a treatment as real-time evaluation, or after treatment, as post-treatment evaluation. It is a simple but valuable tool that helps engineers to optimize acid treatment design effectively. In this chapter, the theory of real-time monitoring of skin evolution and the procedure of stimulation evaluation are presented. Special conditions such as deviation from constant injection rate, gas-well treatment, and bottomhole-pressure convergence are also discussed. Field cases in this chapter illustrate the application of the method.

## 7.2 Diagnosis Method

Monitoring acid-stimulation treatment was suggested in the early 1980s (McLeod 1983) with watching the pressure and injection-rate responses during a treatment. The first theoretical approach of monitoring acid stimulation was introduced by Paccaloni et al. (1988). From the radial flow equation at stable condition, skin factor is calculated from bottomhole injection pressure and injection rate during a treatment. Later, the method was extended to use the theory of pressure behavior under transient flow conditions to directly calculate skin factor from treatment pressure and rate data. Continuous skin-factor evolution is readily obtained as a function of injection volume.

**7.2.1 Paccaloni Maximum $\Delta p$, Maximum-Rate Procedure.** Paccaloni's maximum $\Delta p$, maximum-rate procedure has two critical components for successful treatments: (1) acid should be injected at the maximum rate, and hence the maximum $p_{wf}$ possible without fracturing the formation; and (2) a treatment should be monitored in real time to ensure that the maximum-rate objective is being met and to determine when sufficient acid has been injected into the formation (Paccaloni et al. 1988).

To monitor the progress of a stimulation treatment, the Paccaloni method calculates a steady-state skin effect, according to Darcy's law, by

$$s = \frac{0.00708 kh \Delta p}{\mu q_i} - \ln\frac{r_b}{r_w}, \qquad (7.1)$$

where all of the variables are in standard oilfield units. In Eq. 7.1, $k$ is the permeability of the formation being treated, $h$ is the thickness of the pay zone, $r_w$ is the wellbore radius, and $r_b$ is the radius affected by acid injection. Paccaloni and Tambini (1993) suggested that a value of 4 ft be used. In the above equation, $\Delta p$ is the reservoir drawdown, defined as the difference between the reservoir pressure, $p_r$, and the buttonhole injection pressure $p_{wi}$. In terms of tubinghead injection pressure, $p_{ti}$, $p_{wi}$ can be expressed as

$$p_{ti} = p_{wi} - \Delta p_{pe} + \Delta p_F, \qquad (7.2)$$

where $\Delta p_{pe}$ is the hydrostatic pressure drop in the wellbore and $\Delta p_F$ is the frictional pressure drop in the wellbore. Eq. 7.1 relates $q_i$, and $p_{wi}$ (through $\Delta p$), to skin factor $s$. To monitor a stimulation treatment, Paccaloni constructs a graph of $p_{ti}$ vs. $q_i$, with skin factor as a parameter. Assuming that skin factor is the only parameter changing during an acid treatment, the graph shows a group of parallel lines with different skin-factor values. As a treatment proceeds, the wellhead pressure is plotted against the injection rate, and the evolving skin effect can be read from the graph. The fracture pressure vs. injection rate is also plotted on this graph, so that the rate can be held at the highest possible value without fracturing the formation. **Fig. 7.1** shows such a design chart for a matrix stimulation treatment.

The Paccaloni method is convenient to use, but it will tend to overestimate the skin effect because it neglects transient flow effects, as pointed out by Prouvost and Economides (1989) and confirmed by Paccaloni and Tambini (1993). This error can be severe when abrupt rate changes occur. However, in most treatments, it is not a serious problem because the error will be relatively constant and the evolution in skin effect is more important than its absolute value. Prouvost and Economides (1989) presented a procedure to calculate more accurately the changing skin effect during matrix acidizing. This technique is more applicable to a careful post-treatment analysis, when the injection-rate schedule from the treatment is known.

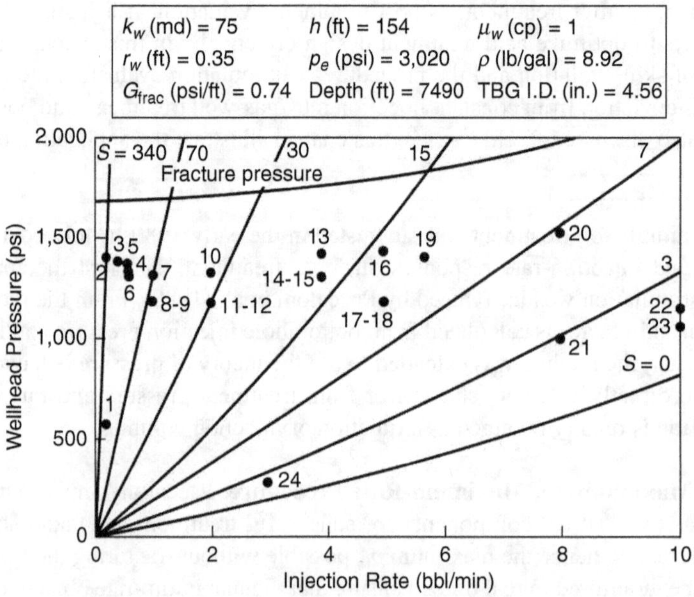


**Fig. 7.1—Paccaloni design chart for monitoring acidizing (Paccaloni and Tambini 1993).**

**7.2.2 Transient Pressure Response Method.** An alternative real-time monitoring procedure that accounts for transient effects and explicitly handles rate changes was presented by Zhu and Hill (1996). The equation for single-phase fluid and transient flow is used in their method. For an injection well, the injection pressure is related to the injection rate as (Earlougher 1977)

$$p_i - p_{wi} = \frac{162.6 q B \mu}{kh}\left[\log(t) + \log\left(\frac{k}{\phi \mu c_t r_w^2}\right) - 3.23 + 0.87s\right], \quad (7.3)$$

where $p_i$ is initial reservoir pressure and $t$ is time in hours. The rest of the parameters are listed in the Nomenclature at the end of the chapter. Eq. 7.3 can be simply rewritten as

$$\frac{p_{wi} - p_i}{q} = m\log(t) + b, \quad (7.4)$$

where

$$m = \frac{162.6 B \mu}{kh} \quad (7.5)$$

and

$$b = m\left[\log\left(\frac{k}{\varphi \mu c_t r_w^2}\right) - 3.23 + 0.87s\right] \quad (7.6)$$

In Eq. 7.4, the left-hand term is called inverse injectivity, which contains pressure and injection-rate data measured during a treatment as functions of injection time. Eq. 7.4 states that there exists a linear relationship between the inverse injectivity and the logarithm of time, with the slope $m$ and the intercept $b$ defined by Eqs. 7.5 and 7.6, respectively. At ideal conditions, $p_{wi}$ can be plotted against $\log(t)$, and the slope $m$ and the intercept $b$ can be estimated from the plot. Since during an acid treatment, the reservoir properties, such as permeability, formation volume factor, viscosity, and thickness of the reservoir, do not change, this leads to a constant slope $m$ from Eq. 7.5. Notice, from Eq. 7.6, that the intercept $b$ only changes with skin factor $s$. With $m$ and $b$ known, skin factor is readily calculated from Eq. 7.6:

$$s = \frac{1}{0.87}\left[\frac{b}{m} - \log\left(\frac{k}{\phi \mu c_t r_w^2}\right) + 3.23\right]. \quad (7.7)$$

To use this method, we first calculate the slope $m$ before a treatment from the properties of the reservoir and the fluid by Eq. 7.5. Then, we can either plot the inverse injectivity as a function of $\log(t)$ during the treatment, estimate the intercept $b$ from the plot, and calculate skin factor $s$ by Eq. 7.7, or calculate directly the intercept $b$ from the measured bottomhole pressure and injection rate by Eq. 7.4 and then calculate skin factor $s$ by Eq. 7.7. The method only applies to the treatments with constant injection rate.

---

**Example 7.1.** During an acid-stimulation job, the injection pressure at the bottom of the hole is measured at 4,000 psi at an injection rate of 2 bbl/min at 10 minutes of injection time. Calculate skin factor at this time. Reservoir pressure is 3,000 psi, thickness of pay zone is 200 ft, permeability is 50 md, viscosity is 2 cp, total compressibility is $10^{-6}$ psi$^{-1}$, formation volume factor is unity, and wellbore radius is 0.25 ft. From Eqs. 7.5 and 7.6,

$$m = \frac{(162.6)(1)(2)}{(50)(200)} = 0.0325 \quad (7.8)$$

and

$$b = \frac{4000 - 3000}{(2)(60)(24)} - 0.0325 \log(10/60) = 0.37 \quad (7.9)$$

Therefore, from Eq. 7.7,

$$s = \frac{1}{0.87}\left\{\frac{0.37}{0.0325} - \log\left[\frac{50}{(0.2)(2)(10^{-6})(0.25)^2}\right] + 3.23\right\} = 6.2 \quad (7.10)$$

Similarly, if at 15 minutes, the injection pressure decreased to 3,800 psi at the same injection rate of 2 bbl/min, then the skin factor is 3.5.

---

**7.2.3 Superposition Time Function for Unstable Injection Rate.** Practically, it is not possible to keep injection rate constant during an acid treatment. In case of changing injection rate, the principle of superposition can be applied to Eq. 7.4, which superposes the pressure responses at different rates. For a treatment with varying injection rates, the superposition principle leads to

$$\frac{p_i - p_{wf}}{q} = m\Delta t_{\text{sup}} + b \,, \quad (7.11)$$

where the superposition time function $\Delta t_{sup}$ in Eq. 7.11 is

$$\Delta t_{\text{sup}} = \sum_{j=1}^{N} \frac{q_j - q_{j-1}}{q_N} \log\left(t_N - t_{j-1}\right) \quad (7.12)$$

In Eq. 7.12, subscript $j$ denotes each timestep $j$ and $N$ denotes the current timestep.

At any time, the superposition time function can be calculated from the flow-rate history at the current time from Eq. 7.12, and then the intercept $b$ can be obtained from Eq. 7.11. Once $b$ is known, the skin at the current time is ready to be computed. Notice that the parameter $m$ (the slope of the plot of inverse injectivity vs. superposition time function) does not change as injection proceeds, and it can be obtained before treatment from the properties of the reservoir, as described in the constant-rate section (see Section 7.2.2). The skin factor for varying injection rate is calculated by

$$s = \frac{1}{0.87}\left(\frac{b}{m} - \Delta t_{\text{sup}} + 3.23\right) \quad (7.13)$$

With the superposition approach, the skin-factor deviation from the constant-rate model is corrected. We should observe no significant change on the skin curve when the injection rate changes.

---

**Example 7.2.** A treatment is carried out in a well with the following information:

$p_i$ = 2,250 psi, $B_o$ = 1.2 $\phi$ = 0.2, $c_t = 9 \times 10^{-5}$ psi$^{-1}$, $h$= 253 ft,
$k$ = 80 md, initial skin factor $s$ = 40, $r_w$ = 0.51 ft, and $\mu$ = 2.4 cp.

| Fluid at Perforations | Time (hours) | $q$ (bbl/min) | $p_{wf}$ (psi) |
|---|---|---|---|
| Water | 0.15 | 2 | 4,500 |
| Water | 0.30 | 2 | 4,489 |
| Water | 0.45 | 2 | 4,400 |
| Aromatic solvent | 0.60 | 2 | 4,390 |
| Aromatic solvent | 0.75 | 2.5 | 4,350 |
| Aromatic solvent | 0.90 | 2.4 | 4,270 |
| Aromatic solvent | 1.05 | 2.7 | 4,250 |
| 15% HCl | 1.20 | 2.5 | 4,200 |
| 15% HCl | 1.35 | 3.2 | 3,970 |
| 3% HF/12% HCl | 1.50 | 3.7 | 3,800 |
| 3% HF/12% HCl | 1.65 | 4.1 | 3,750 |
| 3% HF/12% HCl | 1.80 | 4.4 | 3,670 |
| 3% HF/12% HCl | 2.10 | 3.9 | 3,400 |
| 3% HF/12% HCl | 2.25 | 5.6 | 3,730 |
| 3% HF/12% HCl | 2.40 | 5.6 | 3,750 |
| 3% HF/12% HCl | 2.55 | 5.6 | 3,650 |
| 3% HF/12% HCl | 2.70 | 5.7 | 3,600 |
| 3% HF/12% HCl | 2.85 | 5.8 | 3,400 |

Table 7.1— Treatment records of pressure and rate (Economides et al. 2013). Economides, Michael J.; Hill, A. Daniel; Ehlig-Economides, Christine; Zhu, Ding, Petroleum Production Systems, 2nd ed. ©2013 Printed and electronically reproduced by permission of Pearson Education, Inc., New York, New York.

The injection schedule is shown in **Table 7.1.** To generate skin evolution during the treatment, we will show how the superposition time function is calculated, then the rest of the calculation is very similar to Example 7.1. For example, at the fifth record when the injection time is 0.75 hours (45 minutes), the superposition time function is

$$\begin{aligned}\Delta t_{\text{sup},5} &= \frac{q_1-q_0}{q_5}\log\left(t_5-t_0\right)+\frac{q_2-q_1}{q_5}\log\left(t_5-t_1\right)+\frac{q_3-q_2}{q_5}\log\left(t_5-t_2\right)\\&\quad+\frac{q_4-q_3}{q_5}\log\left(t_5-t_3\right)+\frac{q_5-q_4}{q_5}\log\left(t_5-t_4\right)\\&=\frac{2.5-2}{2.5}\log\left(0.75-0.6\right)=-0.165\end{aligned} \qquad (7.14)$$

Notice that for the first four data points, the injection rate is constant, and the superposition time function in this case can be simply reduced to log($t$). The calculation for $m$ and $b$ are the same as in Example 7.1. The slope $b$ the superposition time function $\Delta t_{sup}$, and the skin factor $s$ at each time recorded are listed in **Table 7.2.** The result shows that the treatment is successful, and the skin factor decreased from 38 to 2 at the end of stimulation.

---

**7.2.4 Horizontal Wells.** For horizontal wells, the transient flow models become more complicated compared with the vertical well models. We look for an analytical/semianalytical model that represents the relationship between superposition time function and inverse injectivity as a linear function so that the skin factor can be directly calculated following the procedures introduced before. The semi-infinite slab model developed by Goode and Thambynayagam (1987) is adopted here. The semi-infinite slab model defines the flow field as being bounded at the top and the bottom, and in the direction of wellbore length, but not in the reservoir width dimension. **Fig. 7.2** shows a sketch

| Fluid at Perforations | Time, $t$ (hours) | $q_i$ (bbl/min) | $p_{wf}$ (psi) | " $t_{sup}$ | $b$ | $s$ |
|---|---|---|---|---|---|---|
| Water | 0.15 | 2 | 4,500 | –0.824 | 0.799 | 38 |
| Water | 0.30 | 2 | 4,480 | –0.532 | 0.786 | 37 |
| Water | 0.45 | 2 | 4,400 | –0.347 | 0.754 | 36 |
| Aromatic solvent | 0.60 | 2 | 4,390 | –0.222 | 0.748 | 35 |
| Aromatic solvent | 0.75 | 2.5 | 4,350 | –0.265 | 0.589 | 27 |
| Aromatic solvent | 0.90 | 2.4 | 4,270 | –0.113 | 0.587 | 27 |
| Aromatic solvent | 1.05 | 2.7 | 4,250 | –0.121 | 0.517 | 23 |
| 15% HCl | 1.20 | 2.5 | 4,200 | 0.036 | 0.541 | 24 |
| 15% HCl | 1.35 | 3.2 | 3,970 | –0.111 | 0.376 | 15 |
| 3% HF/12% HCl | 1.50 | 3.7 | 3,800 | –0.117 | 0.293 | 11 |
| 3% HF/12% HCl | 1.65 | 4.1 | 3,750 | –0.092 | 0.256 | 9 |
| 3% HF/12% HCl | 1.80 | 4.4 | 3,670 | –0.051 | 0.225 | 7 |
| 3% HF/12% HCl | 2.10 | 3.9 | 3,400 | –0.041 | 0.206 | 6 |
| 3% HF/12% HCl | 2.25 | 5.6 | 3,730 | –0.186 | 0.187 | 5 |
| 3% HF/12% HCl | 2.40 | 5.6 | 3,750 | 0.207 | 0.182 | 5 |
| 3% HF/12% HCl | 2.55 | 5.6 | 3,650 | 0.292 | 0.167 | 4 |
| 3% HF/12% HCl | 2.70 | 5.7 | 3,600 | 0.348 | 0.157 | 3 |
| 3% HF/12% HCl | 2.85 | 5.8 | 3,400 | 0.404 | 0.129 | 2 |

Table 7.2—Real-time monitoring skin calculation (Economides et al. 2013). Economides, Michael J.; Hill, A. Daniel; Ehlig-Economides, Christine; Zhu, Ding, Petroleum Production Systems, 2nd ed. ©2013 Printed and electronically reproduced by permission of Pearson Education, Inc., New York, New York.

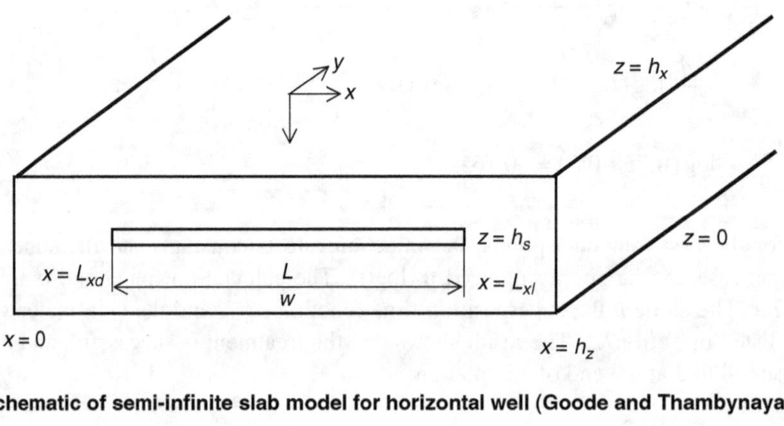


**Fig. 7.2—Schematic of semi-infinite slab model for horizontal well (Goode and Thambynayagam 1987).**

of the geometry for the model. The solution of the model calculates the slope $m$ and the intercept $b$ in Eq. 7.11 as

$$m = \frac{282.4 \mu B r_w'}{h_x h_z k_y}, \qquad (7.15)$$

$$\Delta t_{\text{sup}} = \sum_{j=1}^{i} \frac{q_j - q_{j-1}}{q_N} \sqrt{\pi \left( t_{D,N} - t_{D,j-1} \right)}, \qquad (7.16)$$

and

$$\begin{aligned} b = {} & m \frac{h_x^2}{\pi^2 v_x} \sum_{j=1}^{N} \frac{q_j - q_{j-1}}{q_N} \sum_{n=1}^{\infty} \left[ \frac{1}{n} \operatorname{erf}\left( v_x \pi n \sqrt{t_D - t_{D,j-1}} \right) \Xi_n^2 \right] \\ & + m \frac{h x h_z}{\left( L_{xl} - L_{xd} \right) v_z \pi} \sum_{j=1}^{N} \frac{q_j - q_{j-1}}{q_N} \sum_{l=1}^{\infty} \left\{ \frac{1}{m} \operatorname{erf}\left( v_z \pi l \sqrt{t_D - t_{D,j-1}} \right) \Xi_l \cos\left[ l\pi \frac{1}{h_z} \left( h_s + 1.47 r_w' \right) \right] \right\}, \qquad (7.17) \\ & + \frac{141.2 B\mu}{\sqrt{k_y k_z} L_w} s, \end{aligned}$$

where

$$\Xi_n = \frac{1}{n \left( L_{xl} - L_{xd} \right)} \left[ \sin\left( \frac{n\pi L_{xl}}{h_x} \right) - \sin\left( \frac{n\pi L_{xd}}{h_x} \right) \right], \qquad (7.18)$$

$$\Xi_l = \frac{1}{4 l r_w'} \left\{ \sin\left[ \frac{l\pi}{h_z} \left( h_s + 2 r_w' \right) \right] - \sin\left[ \frac{l\pi}{h_z} \left( h_s - 2 r_w' \right) \right] \right\}, \qquad (7.19)$$

$$t_D = \frac{0.000264 k_y t}{\phi \mu c_t r_w'}, \qquad (7.20)$$

$$v_x = \frac{r_w'}{h_x} \sqrt{\frac{k_x}{k_y}}, \qquad (7.21)$$

$$v_z = \frac{r_w'}{h_z} \sqrt{\frac{k_z}{k_y}}, \qquad (7.22)$$

and

$$r_w' = r_w \left( \frac{k_z}{k_y} \right)^{1/4} \qquad (7.23)$$

The geometry parameters $h_x$, $h_y$, $h_z$, $L_{xe,}$ and $L_{xd}$ are shown in **Fig. 7.2.**

The horizontal-well model introduced here is more complicated than the vertical-well model, and Eq. 7.17 is difficult for real-time applications because of the infinite-summation terms. For practical purposes, the infinite-summation terms are suggested to be approximated by the first 40 terms, which gives a satisfactorily stable result.

The model presented here for horizontal wells does not consider the effect of reservoir heterogeneity on skin factor from the measured pressure and flow rate. Specifically, this is an overall skin factor for the monitored well, and not a skin distribution along the horizontal wellbore. The distribution of permeability and skin along the horizontal lateral is lumped into three overall constant permeabilities, $k_x$, $k_y$, and $k_z$ when calculating the skin factor. The contact length of the wellbore to the reservoir in a horizontal well is much longer than in vertical wells (horizontal-section length vs. formation thickness), and unlikely, the permeability and degree of damage are uniformly distributed along a wellbore. Even though the overall skin factor may show a low value after acidizing, there may exist zones that still have unremoved damage. In a horizontal well, local skin distribution is hard to evaluate. For the treatments in which the wellbore is isolated into several sections, and the acid is pumped in stages over each isolated section, the monitored skin factor may reflect the local response of reservoir to acid injection.

**7.2.5 Monitoring Treatments in Gas Wells.** For treatments in gas wells, the single-phase-liquid-flow assumption made in the previous derivation is not valid, and we need a transient flow equation with a proper format (a linear relationship between inverse injectivity, $\Delta p/q$, and a time function) for gas reservoirs. When the reservoir pressure is high enough, gas viscosity $\mu$ and compressibility $Z$ are directly proportional to pressure, and the pressure transient flow equation for a constant flow rate can be simplified to (Zhu et al. 1998)

$$p_{wf} = p_i - 50{,}300 \frac{z_i \mu_{gi}}{2p_i} \frac{p_{sc}}{T_{sc}} \frac{qT}{kh} \left[ \log\left( \frac{tk}{\phi \mu c r_w^2} \right) + 0.87s - 3.23 \right] \quad (7.24)$$

Equation 7.24 can be written in the exact format of Eq. 7.11, but now the definition of $m$ is different:

$$m = 50{,}300 \frac{z_i \mu_{gi}}{2p_i} \frac{p_{sc}}{T_{sc}} \frac{T}{kh} \quad (7.25)$$

The same approach can be applied to obtain the skin factor for gas wells as for oil wells. The equations for skin calculation are Eq. 7.7 for constant injection rate or Eq. 7.13 for unstable injection rate.

**7.2.6 Skin Factor by Viscous Effects and Diversion.** When injecting a fluid with a different viscosity than the reservoir-fluid viscosity, the viscosity contrast affects the pressure response to injection. This effect on the injection pressure of a bank of fluid with one viscosity displacing a fluid with a different viscosity can be accounted for with an apparent viscous skin factor, $s_{vis}$, given by (Hill and Zhu 1996),

$$s_{\text{vis}} = \left( \frac{\mu_{\text{acid}}}{\mu_{\text{res}}} - 1 \right) \ln \frac{r_{\text{acid}}}{r_w}, \quad (7.26)$$

where $\mu_{acid}$ is the viscosity of injected fluid and $\mu_{res}$ is the viscosity of original reservoir fluid. Eq. 7.26 is analogous to the Hawkin's formula, which gives the skin factor for a region of altered permeability around a well. Assuming the acid penetrates uniformly into a formation of thickness $h$, and neglecting any residual saturation of the displaced fluid behind the acid front, the radial position of the acid is related to the injected-acid volume at any time by

$$r_{\text{acid}} = \sqrt{r_w^2 + \frac{V_{\text{acid}}}{\pi \phi h}} \quad (7.27)$$

Combining Eqs. 7.26 and 7.27, the apparent skin caused by the viscosity difference between the injected acid and the formation gas is given by

$$s_{\text{vis}} = \frac{1}{2} \left( \frac{\mu_{\text{acid}}}{\mu_{\text{res}}} - 1 \right) \ln \left( 1 + \frac{V_{\text{acid}}}{\pi \phi h r_w^2} \right), \quad (7.28)$$

and the damage skin factor, $s_d$, during acid injection is obtained by subtracting the viscous skin factor, $s_{vis}$, from the total apparent skin factor, $s_{app}$:

$$s_d = s_{\text{app}} - s_{\text{vis}} \quad (7.29)$$

Because the viscosity of the injected acid is typically many times larger than the viscosity of the gas being displaced, the apparent viscous skin factor can be quite significant. For example, using Eq. 7.28 for an acid with a viscosity of 1 cp displacing a gas with a viscosity of 0.02 cp, after the injection of 100 gal/ft of acid into a formation with a porosity of 0.2 and thickness of 100 ft from a 0.25-ft radius wellbore, the viscous skin factor is 32. The viscous skin factor is zero at the start of injection, increases rapidly during the early stages of injection, then more slowly later in time, according to the logarithmic relationship in Eq. 7.28.

On the other hand, when acidizing oil wells, because the viscosity of the oil is often similar to or larger than the viscosity of the acid, the viscous skin factor is small and will be negative when oil viscosity is greater than acid viscosity. For 100 gal/ft of injection of acid with a viscosity one-half that of the oil, the apparent viscous skin factor is –1.4, a minor correction to the skin factor obtained from real-time monitoring.

To correct the skin factor obtained from monitoring the injection rate and pressure, the viscous skin can be calculated as a function of injected volume by Eq. 7.28, and subtracted from the total apparent skin according to Eq. 7.29.

Diversion is commonly used in well stimulation to distribute injection fluids to desired locations. The purpose of diversion is to temporarily plug the high-permeability zones, which would receive more acid but need less treatment during acidizing, compared with the low-permeability zones (Economides et al. 2013). If diversion works effectively, acid will be directed to low-permeability zones. When this happens, the total skin effect (or apparent skin factor) will increase. Skin increased by diversion is removable by backflow after the treatment, and damaged skin would be lower when the well is put on production if the treatment is successful.

For a multiple-stage injection, $s_{div}$ can be estimated by (Nozaki and Hill 2009)

$$s_{\mathrm{div}} = \sum_{i=0}^{n-1} \frac{\mu_{i+1}}{\mu} \ln \frac{r_{i+1}}{r_i} - \ln \frac{r_n}{r_w} \quad (7.30)$$

For effective diversion, the monitored skin factor (the apparent skin) increases when diverters enter the formation. If skin factor decreases following diversion when pumping acid again, it indicates that, most likely, the diversion has succeeded and further stimulation is in process. By monitoring apparent skin factor, we can evaluate the effectiveness of diversion and stimulation.

## 7.3 Downhole Pressure Estimate

**7.3.1 Using Tubinghead Pressure.** It is common that, during an acid stimulation, pressure is measured at the surface ($p_{th}$). In such cases, the surface pressure needs to be converted to downhole injection pressure ($p_{wi}$) for skin calculation. For a single-stage treatment, this is straightforward (Economides et al. 2013):

$$p_{wi} = p_{th} + \frac{g}{g_c} \rho L \sin\alpha - \frac{32 f_f \rho q^2 L}{g_c \pi^2 D^5} \quad (7.31)$$

In Eq. 7.31, $L$ is the measured length of the well, $a$ is the deviation angle of the wellbore, $q$ is the injection rate, $D$ is the tubing inside diameter, and $f_f$ is the friction factor. The friction factor $f_f$ is a function of flow rate $q$ and can be obtained from Moody's diagram or calculated by the Chen equation when required in real time.

When a treatment has multiple stages, the fluids in the tubing could have different densities and viscosities. A more accurate way to calculate pressure drop is to track the length of each fluid traveling along the tubing, and use the corresponding density and viscosity in each section for each fluid. Assuming that injection fluids in different stages are immiscible, Eq. 7.31 becomes

$$p_{wi} = p_{th} + \frac{g \sin\alpha}{g_c} \sum_j \rho_j L_j - \frac{32 q^2}{g_c \pi^2 D^5} \sum_j \left(f_f\right)_j \rho_j L_j , \quad (7.32)$$

where $L_j$ is the column length of fluid $j$ and $r_j$ is the density of fluid $j$. Using field units, Eq. 7.32 becomes

$$p_{wi} = p_{th} + 0.433\sin\alpha \sum_j \gamma_j L_j - 7.335X10^{-7} \frac{q^2}{D^5} \sum_j \left(f_f\right)_j \rho_j L_j \quad (7.33)$$

In Eq. 7.33, $L$ is in ft, $r$ is in lbm/ft$^3$, $D$ is in in., and $q$ is in B/D. The result of calculated pressure $p_{wi}$ is in psi.

**7.3.2 Using Annular Pressure.** Annular pressure can be converted to bottomhole pressure, and can be used to confirm the calculation of pressure conversion by tubinghead pressure. An annulus is treated as a nonflowing column filled with liquid, and the calculation is much simpler compared with using tubing pressure. This method is very helpful when tubing is filled with compressible fluid, and pressure conversion from tubing head pressure becomes difficult. The equation to convert annular pressure to bottomhole pressure is

$$p_{wf} = p_{\text{ann}} + \frac{g}{g_c} \rho_{\text{ann}} L \sin\alpha , \quad (7.34)$$

where $p_{ann}$ is the annular pressure at surface, and $r_{ann}$ is the density of the annular fluid.

## 7.4 Data Required for Field Application

To apply the method of real-time monitoring of acid treatment, certain data are required. The reservoir and fluid properties are required to calculate the slope $m$, and the flow rate and pressure recorded during the treatment are used to calculate the intercept $b$. Then, the skin factor is estimated from $m$ and $b$ by Eq. 7.7 or Eq. 7.13. The treatment can be monitored and modified on the basis of the observation of skin-factor change.

**7.4.1 Pretreatment Reservoir Data**. Reservoir data required by the method include reservoir pressure, formation volume factor, porosity, total compressibility, viscosity of reservoir fluid, formation thickness, permeability, and initial skin factor. This information has to be obtained from other sources (e.g., geology, well logging, well testing, production history, reservoir simulation) before the acid treatment.

Since reservoir information obtained from other sources is often uncertain, sometimes errors in the reservoir data can result in nonpractical values of calculated skin factor (very high value or large negative numbers). To eliminate the effect of the errors in reservoir data, the reservoir data can be adjusted scientifically to yield meaningful skin factor. Some data in the group do not affect the skin calculation significantly in their reasonable ranges (for example, formation volume factor, porosity, and total compressibility), and some data are more reliable than others (formation thickness, porosity, and reservoir-fluid viscosity). The most sensitive parameter in the calculation is the reservoir pressure. It is recommended that reservoir pressure be adjusted first because it has a wider range to vary compared with other data.

In case the information of reservoir permeability and initial skin are not available, a simple buildup test can be applied to obtain the input data. In general, there will be an injection period before acid injection during which an inert fluid is pumped into the well to establish a stable injectivity. If the injection pressure and rate are recorded, the permeability and initial skin factor can be obtained from an injectivity test using the standard procedure of a buildup test. From Eq. 7.5, a slope, $m$, can be estimated from the inverse injectivity vs. log($t$) plot, and the permeability of the reservoir is

$$k = \frac{162.6 B\mu}{mh} \quad (7.35)$$

Also, from the intercept $b$ of the plot, the initial skin factor is

$$s = \frac{1}{0.87}\left[\frac{b}{m} - \log\left(\frac{k}{\phi \mu c_t r_w^2}\right) + 3.23\right] \qquad (7.36)$$

It is recommended to use the permeability and initial skin factor obtained from the pretreatment injectivity test because they should be closer to the well condition than the information obtained from other sources.

**7.4.2 Wellbore Data.** Wellbore data are used for pressure conversion from the surface value to the bottomhole value, which is required when bottomhole pressure is not measured. These data include wellbore radius, tubing diameter, deviation angle, and wellbore length. Deviation angle can be replaced by actual (measured) length and vertical length. Actual length is used for frictional pressure-drop calculation, and vertical length is for hydrostatic pressure drop. Wellbore-fluid properties (viscosity and density of tubing and annular fluid) before injection starts are also required for pressure conversion. Annular-fluid density is required only if the annular pressure is used to calculate bottomhole pressure.

**7.4.3 Injection Schedule.** Injection schedule is critical information for real-time monitoring. It is used for tracking different fluid-interface locations when injecting multiple stages so that the surface pressure can be correctly converted to the bottomhole condition from the surface. It is also used to calculate viscous skin effect. The actual schedule is tracked during a treatment for real-time monitoring, and the tracking record can be used for post-treatment evaluation. The exact injection schedule needs to be used in post-treatment evaluation, not the planned one.

## 7.5 Field Examples

Field applications of the real-time monitoring demonstrate the value of this technique in improving the efficiency of acid-stimulation design and execution. In this section, we will show some examples of field applications in vertical oil and gas wells, and in a horizontal well. The examples include both successful treatments and failures. Diversion effect is also evaluated in one of the examples.

**7.5.1 Optimizing Injection Volume by Monitoring of Skin Behavior.** The purpose of monitoring acid treatment is to determine if the reservoir is responding to the injection positively and if the volume of acid is appropriately designed. A good acidizing design should maximize the effectiveness of stimulation, avoiding excessive or insufficient acid volume injected. Injected-acid volume is designed on the basis of estimation of damaged-zone radius. The optimal design should be the one that covers the entire damaged zone. When the designed acid volume is insufficient, the treatment would leave an unacidized zone behind, and productivity cannot be fully recovered. On the other hand, if too much acid is injected, detrimental effects including additional damage to the formation can occur, especially in sandstone formations (precipitation or unconsolidated formation, for example), in addition to an increase in cost of treatment. Since the damaged-zone radius is hard to predict or measure, monitoring skin change can help us to modify and optimize the injection volume.

The first example is an acid treatment of an oil well with deep damage. The well is located in Placerita field in Bakersfield, California (Zhu and Hill 1996). The well is a shallow vertical well (a depth of 2,700 ft), with a permeability of 85 md, and severe damage in the near-wellbore region (an initial skin exceeding 100). The treatment is simply a two-stage injection, a hydrochloric acid preflush and a main stage of mud acid. The monitored skin evolution of this treatment is shown in **Fig. 7.3.** It is obvious that the formation responded well to both preflush and main acid injection. The absolute value of the skin factor may not be accurate because the final skin is approximately –10. This could be caused by the errors in the input data (e.g., in permeability). But the skin factor was reduced by 120 during the treatment, which indicates an effective stimulation. The well performance is dramatically improved by this stimulation. Assuming a value of $\ln(r_e/r_w)$ of 8, the productivity-index (PI) ratio before and after treatment would, in theory, be

$$\frac{(PI)_s}{(PI)_d} = \frac{\ln\left(\frac{r_e}{r_w}\right) + s_{\text{ini}}}{\ln\left(\frac{r_e}{r_w}\right)} = \frac{128}{8} = 16 \quad (7.37)$$

In reality, the ratio could be lower than 16, but the benefit from acid treatment is obvious.

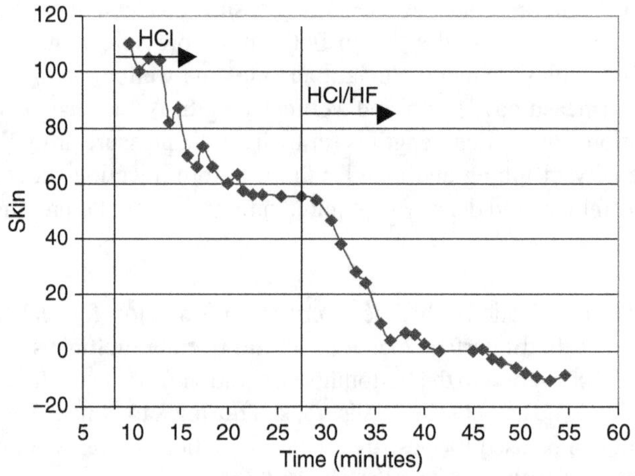


**Fig. 7.3—An example of successful acid treatment (Zhu and Hill 1996).**

The information obtained from on-site monitoring can be used to improve subsequent treatments in the same area. The next example shows a successful, but overdesigned treatment. The well is located in Canto do Amaro field in Brazil (da Motta et al. 1997). An acid treatment was designed for removing near-wellbore damage. The conventional treatment design in the area was 80 gal/ft, which was used in this treatment. **Fig. 7.4** shows the skin plot of the treatment. From Fig. 7.4, it can be seen that the treatment could be stopped at 80 minutes of pumping time. The skin factor decreased at a very slow rate after 80 minutes of injection, indicating that most likely the acid had covered the damaged region at approximately 80 minutes of injection. In following treatments, the acid volume was cut to 60 gal/ft, and then 40 gal/ft, and the results were satisfactory. Reduced acid volume decreased the cost of the treatments (injection fluid, operation, and manpower), and increased the efficiency of acidizing.

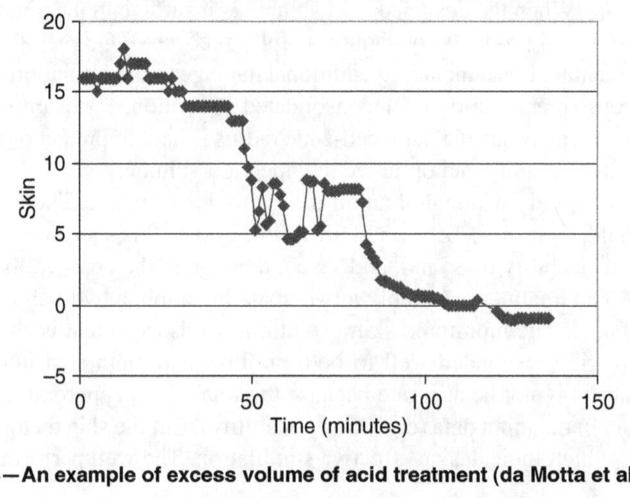


**Fig. 7.4—An example of excess volume of acid treatment (da Motta et al. 1997).**

Sometimes, excess acid not only increases the cost of stimulation, but also results in increasing skin factor. In more-extreme cases, excess acid could cause severe damage to formation. **Fig. 7.5** shows the skin plot for a failed acid treatment. The formation responded to the acid injection at the beginning. The skin factor decreased from 5 to close to zero. The treatment should have stopped after 20 minutes of injection because the next 30 minutes of injection did not add benefit to the stimulation. Then, the skin factor increased gradually during the next 20 minutes, and almost reached the initial skin. This is most likely because of the precipitation of reaction products. This could have been prevented if the injection had been stopped when the skin response to acid injection became flat.

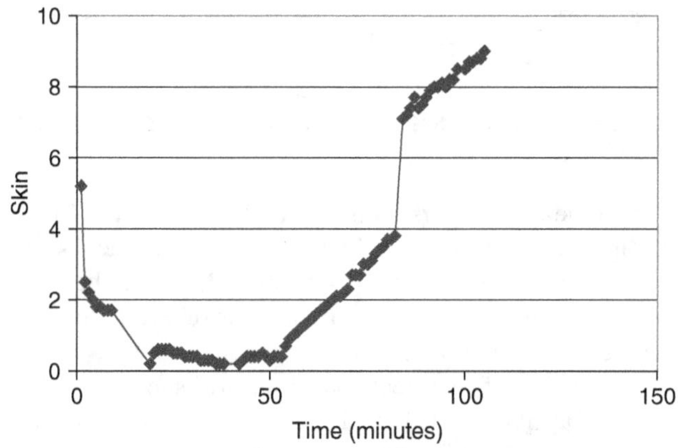


**Fig. 7.5—An example of severe damage caused by acid injection (da Motta et al. 1997).**

At 80 minutes of injection, severe damage apparently occurred, as the skin factor increased abruptly. The treatment was stopped because of high injection pressure, indicating plugging up of near-wellbore permeability.

The failed example could have been stopped on-site if the skin had been monitored in real time. Obviously, modifying treatments on-site is not easy, mainly because of operating issues. Sometimes the designed treatment cannot be altered, if, for example, the designed volume is not sufficient to remove the damage. A great benefit of real-time monitoring of acid treatment is in reviewing the treatment afterward and improving the design in the subsequent treatments, such as in the field case mentioned here.

**7.5.2 Evaluating Diversion.** Often, diversion is applied in acid treatments to evenly distribute the acid injected; the monitoring technique can evaluate the effectiveness of diversion. Since diversion adds resistance to flow on purpose, the apparent skin should increase if the formation is responding to diversion correctly, and the skin caused by diversion can be removed when the well is put on production.

The example shown here is to illustrate the application of diversion to cover multiple zones. **Fig. 7.6** shows the skin plot of a multiple-stage stimulation with diversion stages in between (Zhu and Hill 1996). The well is located in the Garza field near Post, Texas. The formation bears many thin zones. A multiple-stage stimulation was designed to treat as many thin zones as possible by alternating acid and diverting-agent injection. The skin plot shows that the apparent skin increased when the diverter was injected, and decreased when the acid was injected. The skin factor caused by diversion can be calculated by Eq. 7.30, and actual response of formation to acid injection is the difference between the apparent skin and diversion skin. In general, skin decreasing after diversion stage indicates further stimulation, ideally, in other zones than those that already have been stimulated. From the skin plot of Fig. 7.6, alternating successful stimulation and diversion can be seen. The production data after the treatment showed that several thin zones were treated, and the well performance was improved as expected.

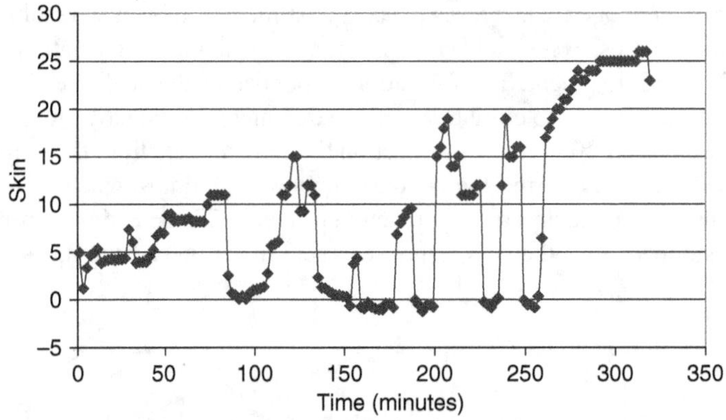


**Fig. 7.6—Using diversion to improve acid coverage (Zhu and Hill 1996).**

Another example of multiple-stage diversion is an oil producer in the Gulf of Mexico (Zhu et al. 1999b). The treatment consisted of four stages of acidizing, with each acid stage separated by a diversion with a stage spacer in between. The spacer used was $NH_4Cl$, and the diverter was HEC w/#650 sand. The initial skin was approximately 32. The formation is relatively thick (approximately 100 ft), with one major layer and several smaller layers.

**Fig. 7.7** plots the skin evolution of this treatment. It clearly shows that the first stage effectively reduced skin factor. The well also responded to each diversion stage positively (skin factor increases during diversion stages). Further skin decreasing is evidenced after each diversion stage when resuming acid injection, indicating a continuous stimulation in different zones. Even though the initial skin factor and the final skin factor have close values, considering that skin factor caused by diversion could be removed during flowback, the treatment design is reasonable, and the stimulation is successful. Post-treatment production confirms the evaluation.

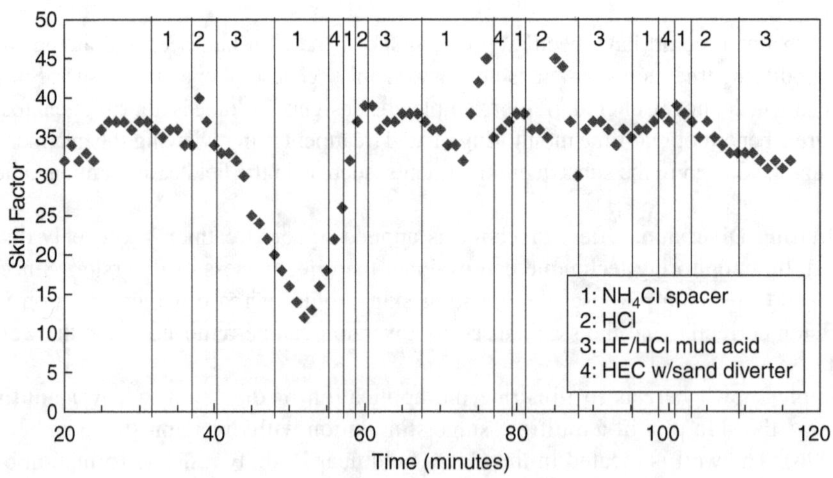


**Fig. 7.7—Skin response to diversion (Zhu et al. 1999b).**

**7.5.3 Gas Wells.** For gas wells, two modifications need to be made from the analysis for oil wells. First, the slope equation for gas wells, Eq. 7.25, is used; and the viscous skin factor caused by the viscosity difference between the reservoir fluid and injection fluid needs to be considered in the apparent skin that is calculated from the pressure and injection-rate data. We use an example of the gas wells in offshore Brazil to illustrate the procedure (Zhu et al. 1998). The Merluza field is located 180 km from the coast of São Paulo, Brazil, and has been commercially producing since May 1993. The six producer wells drain two gas reservoirs, and the total daily production before the acidizing campaign was 2 000 000 m$^3$ of gas and 900 m$^3$ of condensate. A production performance study showed that all the wells were damaged, probably by scaling, since the beginning of production and that acid treatments might be able to restore the productivity. Since these treatments had to be performed through the production tubing, which is made of 13-Cr steel, and because of the high reservoir temperatures, an organic-acid mixture was selected. The acid treatments were designed to remove scale damage with 25 gal/ft of 7% hydrofluoric acid or 5% acetic acid at the maximum possible injection rate. The skin was monitored in real time. The plots for two of the three wells acidized are shown in **Fig. 7.8** (Fig. 7.8a for Well MLZ-2 and Fig. 7.8b for Well MLZ-6). These plots show the apparent, viscous, and corrected damage skin curves during the treatment. Initially, during the treatment, as the pressure increased, the apparent skin also increased. Without discounting the effect of viscous skin, it appeared that the acid mixture was causing more damage, instead of removing it. However, the production of the MLZ-2 well increased from 187 000 to 731 000 m$^3$/d of gas, showing that the operation was, in fact, a huge success. Exactly the same feature was observed for the MLZ-6 well, though the gas-production increase was less impressive, from 144 000 to 255 000 m$^3$/d.

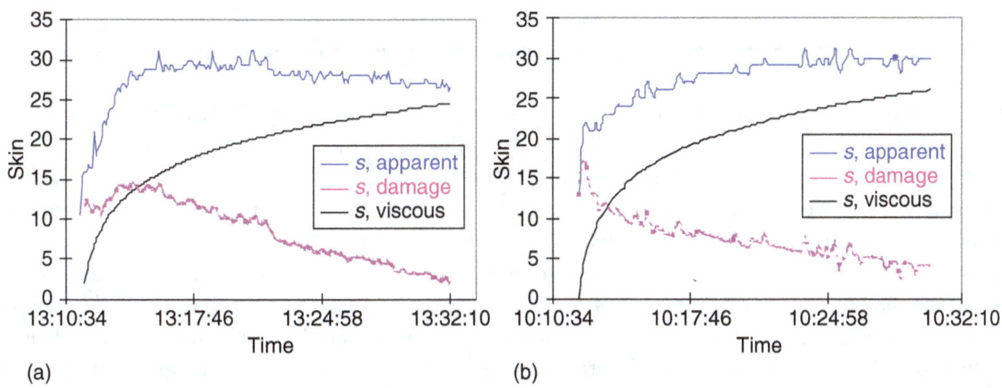


**Fig. 7.8—Skin response of offshore gas wells: (a) Monitored skin for well MLZ-2 and (b) monitored skin for well MLZ-6 (Zhu et al. 1998).**

After correcting for the effect of viscous skin, Fig. 7.8 shows that the damaged skin was actually reduced with the acid injection. The acid volume for Well MLZ-2 was enough to remove all the damage. The skin calculation predicted a tripled production after the treatment, which was confirmed by the production-history data. However, for Well MLZ-6, more than 25 gal/ft acid would probably have done a better job. There is room in MLZ-6 to further remove the residual damaged skin.

**7.5.4. Horizontal-Well Example.** Monitoring horizontal wells is much more of a challenge than monitoring vertical wells. As mentioned before, since single-point pressure data are used to calculate an overall skin response, the distribution of damage removal is hard to assess. The method of monitoring the treatment can provide a better understanding when combined with production history.

In general, matrix acidizing is a suitable stimulation method for horizontal wells in carbonate reservoirs. A critical issue in horizontal-well stimulation is acid placement. Because the contact length is much longer in horizontal wells compared with vertical wells, it is easy to have an injection rate much lower than the optimal injection rate for wormhole growth. The first example shows that when injection rate is lower than the optimal, the efficiency of stimulation slump.

The skin-analysis method is applied to a horizontal well in a carbonate formation in the Permian Basin of Texas (Zhu et al. 1999a). The permeability is 0.2–2 md, and the horizontal lateral length is near 900 ft. With an initial skin factor of near zero, the objective of this treatment was to create sufficiently long wormholes to give a post-treatment skin factor of –2 to –3. However, the injection rate was limited by the small coiled tubing used to convey the acid. The skin evolution determined by using the horizontal-well model is presented in **Fig. 7.9.** The analysis shows that the skin factor did not change notably throughout the entire treatment. This response was confirmed by the post-treatment productivity of the well, which was only 3% higher than the pretreatment productivity. Apparently, the low injection rate used in this treatment resulted in compact dissolution of the face of the carbonate formation, with no significant wormhole propagation. A subsequent acid fracture treatment successfully stimulated the well, confirming that deeper acid penetration was needed.

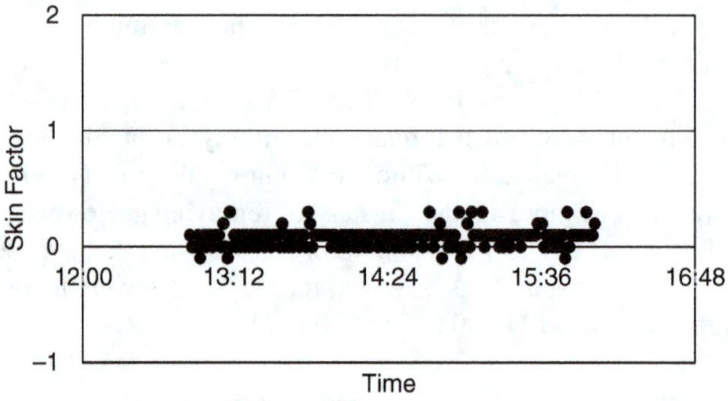


**Fig. 7.9—Insufficient injection rate resulting in unsuccessful treatment (Zhu et al.1999a).**

The second example in horizontal wells shows a successful case (Pandya 2012). The well was completed as a cased hole with perforations. The permeability is 10 md, and the pay zone is greater than 300 ft. Acid is pumped as bullheading injection over a 1,800-ft-long lateral. The surface pressure and injection-rate data were measured on-site during the acid job. **Fig. 7.10** plots the measured injection rate and the measured surface pressure recorded during the acidizing treatment. Converted bottomhole pressure is also shown on Fig. 7.10. Using the bottomhole pressure and injection-rate data, on the basis of the method presented in this chapter, skin factor is estimated and plotted in **Fig. 7.11.** The skin begins to decrease when the acid front enters into the formation, and the skin value decreases from 70 to 2 by the end of acid injection. It is inferred that the volume of acid injected was sufficient because the skin continues to decrease until the end of the injection. However, the skin drops more rapidly initially, compared with the later part of the treatment. This rapid decline in skin may be because of a high-permeability thief zone.

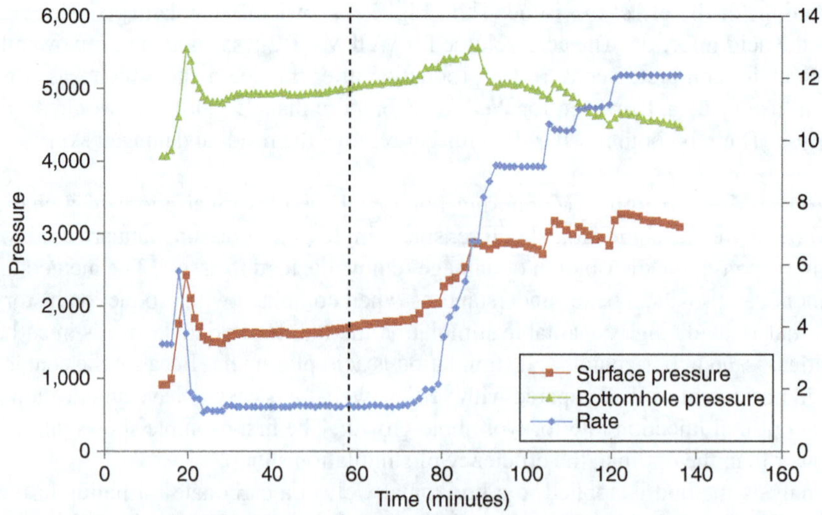


**Fig. 7.10—Treatment data for Case 2 (Pandya 2012).**

The PI ratio from the initial and final skin value was estimated to be 6. This was validated using the production test from before and after the stimulation process. The production rate of liquids from this

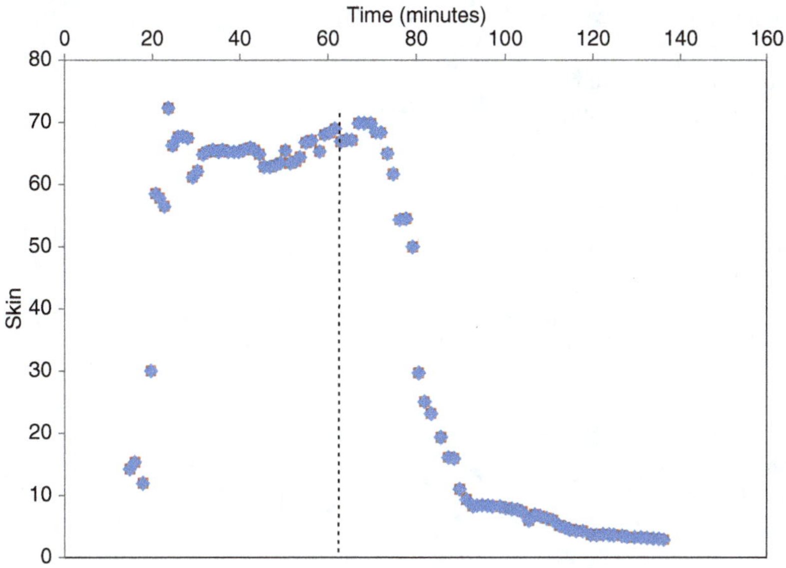


**Fig. 7.11—Skin evolution for Case 2 (Pandya 2012).**

well after stimulation increased from 805 to 4,754 B/D. Therefore, the production history confirms the skin evaluation, and it is clear that the acidizing treatment was successful in significantly reducing the near-wellbore damage.

## Nomenclature

$b$ = intercept
$B$ = formation volume factor
$c_t$ = total compressibility
$D$ = tubing diameter
$f_f$ = friction factor
$g$ = acceleration of gravity
$g_c$ = gravitational constant
$h$ = reservoir thickness
$h_s$ = coordinate of wellbore location in $z$-direction for horizontal model
$h_x$ = length of the flow field for horizontal model
$h_z$ = height of the flow field for horizontal model
$k$ = permeability
$k_x$ = permeability in $x$-direction
$k_y$ = permeability in $y$-direction
$k_z$ = permeability in $z$-direction
$L$ = length of tubing
$L_j$ = length of tubing occupied by fluid $j$
$L_w$ = length of horizontal well
$L_{xd}$ = left coordinate of wellbore in x-direction for horizontal model
$L_{xl}$ = right coordinate of wellbore in x-direction for horizontal model
$m$ = slope
$p_{ann}$ = annular pressure at surface
$PI$ = Productivity Index
$p_i$ = initial reservoir pressure
$p_r$ = reservoir pressure
$p_{sc}$ = standard pressure

$p_{ti}$ = surface tubing injection pressure
$p_{th}$ = surface pressure
$p_{wf}$ = bottomhole flowing pressure
$p_{wi}$ = bottomhole injection pressure
$q$ = flow rate
$q_i$ = injection rate
$r_{acid}$ = radial penetration of acid
$r_b$ = radius affected by acid injection
$r_e$ = drainage radius
$r_w$ = wellbore radius
$r_w'$= equivalent wellbore radius
$s$ = skin factor
$s_{app}$ = apparent skin factor
$s_d$ = damage skin factor
$s_{div}$ = diversion skin factor
$s_{ini}$ = initial skin factor
$s_{vis}$ = viscous skin factor
$t$ = time
$t_D$ = dimensionless time
$T$ = temperature
$T_{sc}$ = standard temperature
$V_{acid}$ = acid volume
$Z$ = compressibility
$\alpha$ = well inclination from vertical
$\Delta p$ = tubing pressure drop
$\Delta p_F$ = frictional pressure drop in wellbore
$\Delta p_{ppe}$ = hydrostatic pressure drop in wellbore
$\Delta t_{sup}$ = superposition time
$\phi$ = porosity
$\gamma$= specific gravity
$\mu$ = fluid viscosity
$\mu_{acid}$ = acid viscosity
$\mu_g$ = gas viscosity
$\mu_{gi}$ = gas viscosity at initial pressure
$\mu_{res}$ = original reservoir fluid viscosity
$\rho_{ann}$ = density of the annular fluid
$\rho_j$ = density of fluid $j$

## References


da Motta, E. P., dos Santos, J. A. C. M., Zhu, D. et al. 1997. Field Evaluation and Optimization of Matrix Acidizing Treatments. Presented at the SPE Production Operations Symposium, Oklahoma City, Oklahoma, USA, 9–11 March. SPE-37460-MS. http://dx.doi.org/10.2118/37460-MS.

Earlougher, R. C., Jr. 1977. *Advances in Well Test Analysis*, Vol. 5. Richardson, Texas: Monograph Series, SPE.

Economides, M. J., Hill, A. D., Ehlig-Economides, C., and Zhu, D. 2013. *Petroleum Production Systems*, second edition. Englewood Cliffs, New Jersey, USA: Prentice Hall.

Goode, P. A. and Thambynayagam, R. K. M. 1987. Pressure Drawdown and Buildup Analysis of Horizontal Wells in Anisotropic Media. *SPE Form Eval* **2** (4): 683–697. SPE-14250-PA. http://dx.doi.org/10.2118/14250-PA.

Hill, A. D. and Zhu, D. 1996. Real-Time Monitoring of Matrix Acidizing Including the Effects of Diverting Agents. *SPE Prod & Fac* **11** (2): 95–101. SPE-28548-PA. http://dx.doi.org/10.2118/28548-PA.

McLeod, H. O. Jr., Ledlow, L. B., and Till, M. V. 1983. The Planning, Execution, and Evaluation of Acid Treatments in Sandstone Formations. Presented at the SPE Annual Technical Conference and Exhibition, San Francisco, 5–8 October. SPE-11931-MS. http://dx.doi.org/10.2118/11931-MS.

Nozaki, M. and Hill, A. D. 2009. A Placement Model for Matrix Acidizing of Vertically Extensive, Heterogeneous Gas Reservoirs. Presented at the SPE Annual Technical Conference and Exhibition, New Orleans, 4–7 October. SPE-124881-MS. http://dx.doi.org/10.2118/124881-MS.

Paccaloni, G. and Tambini, M. 1993. Advances in Matrix Stimulation Technology. *J Pet Technol* **45** (3): 256–263. SPE-20623-PA. http://dx.doi.org/10.2118/20623-PA.

Paccaloni, G., Tambini, M., and Galoppini, M. 1988. Key Factors for Enhanced Results of Matrix Stimulation Treatments. Presented at the SPE Formation Damage Control Symposium, Bakersfield, California, USA, 8–9 February. SPE-17154-MS. http://dx.doi.org/10.2118/17154-MS.

Pandya, N. 2012. *A Systematic Study of Matrix Acidizing Treatments Using Skin Monitoring Method*. MS thesis, Texas A&M University, College Station, Texas. http://hdl.handle.net/1969.1/ETD-TAMU-2012-05-10763.

Prouvost, L. P., and Economides, M. J. 1989. Applications of Real-Time Matrix Acidizing Evaluation Method. *SPE Prod Eng* **4** (4): 401–407. SPE-17156-PA. http://dx.doi.org/10.2118/17156-PA.

Zhu, D. and Hill, A. D. 1996. Field Results Demonstrate Enhanced Matrix Acidizing Through Real-Time Monitoring. Presented at the Permian Basin Oil and Gas Recovery Conference, Midland, Texas, USA, 27–29 March. SPE-35197-MS. http://dx.doi.org/10.2118/35197-MS.

Zhu, D., Hill, A. D., and da Motta, E. P. 1998. On-Site Evaluation of Acidizing Treatment of a Gas Reservoir. Presented at the SPE Formation Damage Control Conference, Lafayette, Louisiana, USA, 18–19 February. SPE-39421-MS. http://dx.doi.org/10.2118/39421-MS.

Zhu, D., Hill, A. D., and Looney, M. D. 1999a. Evaluation of Acid Treatments in Horizontal Wells. Presented at the SPE Annual Technical Conference and Exhibition, Houston, 3–6 October. SPE-56782-MS. http://dx.doi.org/10.2118/56782-MS.

Zhu, D., Hill, A. D., and Morgenthaler, L. N. 1999b. Assessment of Matrix Acidizing Treatment Responses in Gulf of Mexico Wells. Presented at the SPE Mid-Continent Operations Symposium, Oklahoma City, Oklahoma, USA, 28–31 March. SPE-52166-MS. http://dx.doi.org/10.2118/52166-MS.

Chapter 8

# Acid Fracturing

**A. D. Hill,** Texas A&M University

## 8.1 Introduction

Acid fracturing is a stimulation technique in which acid is injected at pressures above the parting pressure of the formation, so that a hydraulic fracture is created. Usually, a viscous pad fluid is injected ahead of the acid to initiate the fracture, then plain acid, gelled acid, foamed acid, or an emulsion containing acid is injected. Fracture conductivity is created by the acid differentially etching the walls of the fracture; that is, the acid reacts nonuniformly with the fracture walls so that, after closure, the fracture props itself open, with the relatively undissolved regions acting as pillars that leave more dissolved regions as open channels. Thus, acid fracturing is an alternative to the use of proppants to create fracture conductivity after closure. The fracturing process itself is identical to that implemented with proppant fracturing.

The primary issues to be addressed in designing an acid fracturing treatment are the penetration distance of live acid down the fracture, the conductivity created by the acid (and its distribution along the fracture), and the resulting productivity of an acid-fractured well. Since acid fracturing should be viewed as an alternative means of creating fracture conductivity in a carbonate formation, a comparison with proppant fracturing should generally be made when planning a possible acid fracturing treatment.

## 8.2 Acid Transport and Fracture-Face Dissolution

To predict the distance to which fracture conductivity is created by acid and the final distribution of conductivity, the distribution of rock dissolution along the fracture is needed. This, in turn, requires a prediction of acid concentration along the fracture. The acid distribution along the fracture is obtained from an acid balance equation and appropriate boundary conditions. For linear flow down a fracture, with fluid leakoff and acid diffusion to the fracture walls, these are

$$\frac{\partial C}{\partial t}+\frac{\partial(u_x C)}{\partial x}-\frac{\partial(u_y C)}{\partial y}-\frac{\partial}{\partial y}\left(D_{\text{eff}}\frac{\partial C}{\partial y}\right)=0, \quad (8.1)$$

$$C(x,y,t=0)=0 , \quad (8.2)$$

$$C(x=0,y,t)=C_i(t), \quad (8.3)$$

and

$$Cu_y-C_L q_L-D_{\text{eff}}\frac{\partial C}{\partial y}=E_f C^n(1-\phi), \quad (8.4)$$

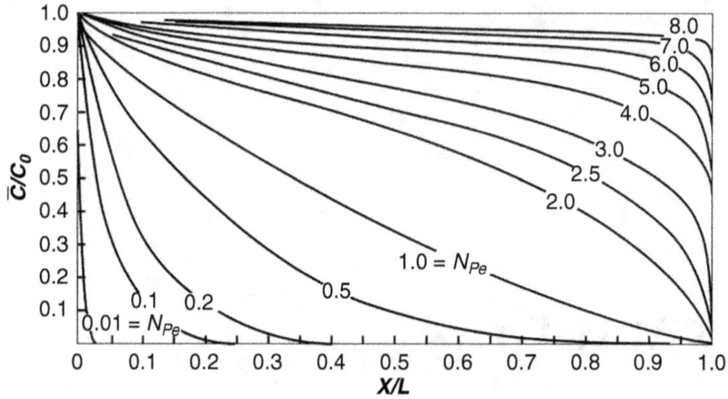


**Fig. 8.1—Acid-concentration profiles along a fracture (from Schechter 1992). SCHECHTER, ROBERT S., OIL WELL STIMULATION, 1st ED. ©1990 Printed and electronically reproduced by Permission of Pearson Education, Inc., New York, New York.**

where $C$ is the acid concentration, $u_x$ is the flux along the fracture, $u_y$ is the transverse flux because of fluid loss, $D_{eff}$ is an effective diffusion coefficient, $C_i$ is the injected-acid concentration, $E_f$ is the reaction-rate constant, $n$ is the order of the reaction, and $\phi$ is porosity. To solve Eq. 8.1, continuity and transport equations are needed as well as a model of the fluid-loss behavior. Complex numerical solutions of these equations that consider further complications such as the temperature distribution along the fracture, viscous fingering of low-viscosity acid through a viscous pad, the effect of the acid on leakoff behavior, and various fracture geometries have been presented by Ben-Naceur and Economides (1988), Lo and Dean (1989), and Settari (1993).

By assuming steady-state, laminar flow of a Newtonian fluid between parallel plates with constant fluid-loss flux along the fracture, Nierode and Williams (1971) presented a solution for the acid balance equation for flow in a fracture, adapted from the solution presented by Terrill (1965) for heat transfer between parallel porous plates. The solution for the concentration profile is presented in **Fig. 8.1** as a function of the leakoff Péclet number, defined as

$$N_{Pe} = \frac{\bar{u}_y w}{2D_{eff}}, \qquad (8.5)$$

where $\bar{u}_y$ is average leakoff flux and $w$ is the fracture width. The effective diffusion coefficient, $D_{eff}$, is generally higher than the molecular diffusion coefficient because of additional mixing caused by density gradients.

From Fig. 8.1, it is apparent that at low Péclet numbers, the acid concentration becomes very low before reaching the end of the fracture; at high Péclet numbers, high acid concentrations reach the end of the fracture. At low Péclet numbers, diffusion controls acid propagation because the leakoff rate is relatively low and the acid concentration is decreased as it moves down the fracture as reaction products diffusing back into the bulk solution dilute the acid. A low Péclet number is very unfavorable because it means that the created fracture is not all effectively contacted with acid. Thus, fracture fluids such as gelled or emulsified acids are needed to have long-etched lengths because the acid diffusion rate is lower in such fluids. Meanwhile at high Péclet numbers, fluid loss is the controlling factor—the acid concentration in the fracture is not diminished because of high fluid loss, but the bulk fluid transport in the fracture is reduced. With high fluid loss, and hence high Péclet number, the created fracture length is also low. An intermediate Péclet number of approximately 3 is the most favorable condition to create a long, etched acid fracture.

---

**Example 8.1. Acid Penetration Distance.** A 100-ft-high fracture has been created with acid injection at a rate of 40 bbl/min. When the fracture has grown to 300 ft in length, the total leakoff is equal to the total injection rate. For this time, assuming that the fracturing fluid is a gelled acid with $n' = 0.5$ and $K' = 0.2$, at what distance from the wellbore will the acid concentration be 50% of the injected concentration after 16 minutes of injection? And 10% of the injected concentration? Assume an effective diffusion coefficient of $10^{-4}$ cm²/s. The formation has a Young's modulus of $4 \times 10^6$ psi and a Poisson's ratio of 0.25.

***Answer.*** To answer this question, the fluid-loss Péclet number is needed. When the fluid-loss rate is equal to the injection rate, the average fluid-loss flux is the injection rate divided by the fracture-surface area:

$$\bar{u}_y = \frac{q_i}{4hx_f} \quad (8.6)$$

For a fracture height of 100 ft and a fracture length of 300 ft,

$$\bar{u}_y = \frac{(40 \text{ bbl/min})(5.615 \text{ ft}^3/\text{bbl})}{(4)(100 \text{ ft})(300 \text{ ft})} = 1.87 \times 10^{-3} \text{ ft/min} \quad (8.7)$$

Using the Perkins-Kern-Nordgren (PKN) fracture-geometry model, for a fracture length of 300 ft, the average fracture width is 0.25 in. (0.0208 ft). When the units are converted to ft²/min, the effective diffusion coefficient is $6.46 \times 10^{-6}$. From Eq. 8.5, the fluid-loss Péclet number is

$$N_{\text{Pe}} = \frac{(0.02083 \text{ ft})(1.87 \times 10^{-3} \text{ ft/min})}{(2)(6.46 \times 10^{-6} \text{ ft}^2/\text{min})} = 3.0 \quad (8.8)$$

Reading from Fig. 8.1 for a Péclet number of 3, the acid concentration will be 50% of the injected value at a dimensionless distance of 0.92, or at (0.92)(300) = 276 ft. A concentration of 10% of the injected concentration penetrates the full length of the 300-ft fracture.

---

**8.2.1 Acid Diffusion.** The effectiveness of acid fracturing is determined by several factors, including reaction between the acid and formation rocks, the transport of acid to the fracture surface, and the leakoff of the reaction fluids into the porous formation. The diffusional mass transfer of acid (hydronium ions) from the bulk acid solution to the walls of the fracture and the diffusion of calcium and magnesium ions back into the bulk solution are important parts of the acid-fracturing process, especially in fast-reaction cases such as hydrochloric acid (HCl) with calcium rocks. The higher the diffusion rate, the lower the concentration of acid along the fracture. The effective diffusion coefficient for straight HCl has been commonly estimated with the correlation of Williams and Nierode (1972) or that of Roberts and Guin (1974). The effective diffusion coefficients reported by Roberts and Guin (1974) are shown in **Fig. 8.2,** based on the Reynolds number in the fracture ($wu_x \rho/\mu$). However, on the basis of rotating-disk and linear-flow core experiments, Li et al. (1993) showed that effective diffusion coefficients for emulsified acids are more than an order of magnitude lower than those predicted by the Williams and Nierode correlation for straight acid.

Acid diffusion rates can be measured with a rotating-disk apparatus when the overall reaction rate is mass-transfer limited, a condition that is indicated by the overall reaction rate measured increasing linearly with the rotational speed of the disk. Rabie and Nasr-El-Din (2011) summarized diffusion coefficients from their work and those reported from several previous studies, as shown in **Table 8.1.**

Emulsified acids are often used in acid fracturing because of the low effective diffusion rates of acid dispersed in an oil-external emulsion. Effective diffusion coefficients for emulsified acids from several studies were summarized by Sayed et al. (2013) and are shown in **Table 8.2.**

**8.2.2 Role of Fluid Loss.** Leakoff or fluid loss from the fracture limits the distance to which live acid can propagate in an acid fracture. In fact, if the formation permeability is high enough, it may not be possible to keep a fracture open once significant dissolution of the formation begins

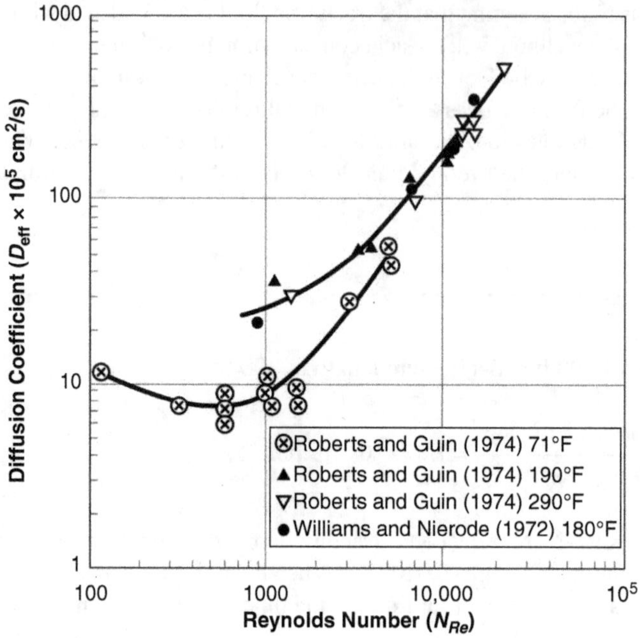


**Fig. 8.2—Effective acid diffusion coefficients (Roberts and Guin 1974).**

| Type of Acid | Concentration (M) | Temperature (°C) | Diffusivity ($cm^2/s$) | Reference |
|---|---|---|---|---|
| HCl | 0.50 | 25 | $3.6×10^{-5}$ | Kung (1998) |
| | 0.56 | 93.33 (200°F) | $8.3×10^{-5}$ | Conway (1999) |
| | 0.56 | 121 (250°F) | $1.1×10^{-4}$ | Conway (1999) |
| Formic | 0.5 | 25 | $3.0×10^{-5}$ | Kung (1998) |
| Acetic | 0.5 | 25 | $1.15×10^{-5}$ | Vitagliano and Lyons (1956) |
| Citric | 0.25 | 25 | $6.46×10^{-6}$ | Muller and Stockes (1957) |
| | 0.26 (5 wt%) | 40 | $6.8×10^{-6}$ | Akhaldi (2010) |
| | 0.39 (7.5 wt%) | 50 | $4.5×10^{-6}$ | Akhaldi (2010) |
| Lactic | 0.56 (5 wt%) | 26.67 (80°F) | $1.23×10^{-6}$ | Rabie and Nasr-El-Din (2011) |
| | 0.56 (5 wt%) | 93.33 (200°F) | $8.92×10^{-6}$ | Rabie and Nasr-El-Din (2011) |
| | 0.56 (5 wt%) | 121 (250°F) | $1.45×10^{-5}$ | Rabie and Nasr-El-Din (2011) |

Table 8.1—Diffusion coefficients for several acids.

| Authors | Diffusion Coefficient ($cm^2/s$) | Notes |
|---|---|---|
| Hoefner and Fogler (1989) | $10 × 10^{-8}$ | Microemulsion with calcite |
| de Rozieres et al. (1994) | $4.60 × 10^{-8}$ | 147°F with Carrara marble |
| Al-Mutairi et al. (2009) | $2.8187 × 10^{-8}$ | 185°F with calcite |
| Syed et al. (2015) | $1.413 ×10^{-8}$ to $8.367 × 10^{-8}$ | 230°F with dolomite rocks |

Table 8.2—Diffusion coefficients of emulsified acid systems used in previous and current studies.

(Furui et al. 2012). As acid flows into the matrix along the walls of the fracture, wormholes will form in the rock if the acid flux is sufficiently high. The wormholes in turn may increase the fluid-loss rate.

The effect of wormholes propagating into the matrix from the fracture walls can be accounted for in the standard Curtis leakoff coefficient. With acid injection, it is not expected that the acid solution will create a filter cake on the fracture face, even if a polymer-viscosified acid system is used. Thus, the resistance to fluid loss comprises only the compression of reservoir fluid and the pressure drop in the fracture fluid-invaded zone. Hill et al. (1995) showed that the presence of wormholes can be accounted for by the overall leakoff coefficient with wormhole, $C_{wh}$, as

$$C_{wh} = \frac{-\dfrac{1}{C_c} + \sqrt{\dfrac{1}{C_c^2} + 4\left(\dfrac{1}{C_{v,wh}^2} + \dfrac{1}{C_w^2}\right)}}{2\left(\dfrac{1}{C_{v,wh}^2} + \dfrac{1}{C_w^2}\right)}, \qquad (8.9)$$

where $C_c$ and $C_w$ are the compressibility leakoff coefficient and the wall-build leakoff coefficient, respectively, and $C_{v,wh}$ is the viscous leakoff coefficient with wormholes, which can be related to the viscous leakoff coefficient without wormhole, $C_v$, by a wormhole model. For example, if the volumetric model is used, then

$$C_{v,wh} = \sqrt{\frac{PV_{bt}}{PV_{bt} - 1}}\,C_v \qquad (8.10)$$

with $PV_{bt}$ as the pore volume required for acid to break through a core in a coreflood experiment.

The acid leakoff from a fracture may not follow the simple behavior described here if the formation is naturally fractured. Using a dual porosity model, Mou et al. (2012) showed that leakoff into intersected fracture networks are dramatically higher than leakoff into the matrix through the fracture faces.

**8.2.3 Analytical Solution for Acid Transport.** Schechter (1990) presented an analytical solution for the acid-concentration profile along an acid fracture that is useful for first-order approximations of acid transport and rock dissolution in acid fracturing. This solution requires that the fluid-loss Reynold's number $N_{Re*}$, defined below, be less than unity, and that the total leakoff from the fracture equal the injected volume (i.e., that the fluid have zero efficiency). The solution depends on the following dimensionless numbers:

$$N_{Re*} = \frac{\langle \bar{u}_N \rangle \bar{w} \rho}{2\mu} \qquad (8.11)$$

$$N_{Re} = \frac{u_A^0 \bar{w} \rho}{2\mu} = \frac{i\rho}{4h\mu} \qquad (8.12)$$

$$N_{Pe} = \frac{\langle \bar{u}_N \rangle \bar{w}}{2D_A} \qquad (8.13)$$

Next, the first six eigenvalues, $\lambda_n$, and constants, $G_n$, of an infinite-series solution are calculated by

$$\lambda_n = \sum_{i=0}^{3} g_{i,n} N_{Pe}^i + \sum_{i=1}^{2} h_{i,n} N_{Re*}^i, \quad (8.14)$$

$$G_n = \sum_{i=0}^{3} \bar{g}_{i,n} N_{Pe}^i + \sum_{i=1}^{2} \bar{h}_{i,n} N_{Re*}^i, \quad (8.15)$$

where the constants for these equations are given in **Table 8.3.** Then, the acid concentration at any position along a fracture of half-length $L$ is given by

$$\frac{\bar{c}}{c_i} = \sum_{n=0}^{\infty} G_n \left(1 - \frac{x}{L}\right)^{2\lambda_n^2/3N_{Pe}}, \quad (8.16)$$

Eq. 8.16 generates acid-concentration profiles like those shown in Fig. 8.1.

To predict the outcome of an acid-fracturing treatment, it is next necessary to calculate the amount of rock dissolution occurring along the fracture. This is generally characterized using a parameter called the ideal width, $w_i$. The ideal width is the width that would theoretically be created by the removal of the amount of rock that the injected acid could dissolve. Given the acid concentration at any point along the fracture and its dissolving power, this ideal width can be calculated. On the basis of the analytical model of acid concentration along a fracture just presented, Schechter (1990) showed that the ideal width profile is

$$w_i = \frac{\beta c_0 q t_A}{2\rho_f (1-\phi) h x_f} \sum_{n=0}^{\infty} G_n \left(1 + \frac{2\lambda_n^2}{3N_{Pe}}\right)\left(1 - \frac{x}{x_f}\right)^{2\lambda_n^2/3N_{Pe}}, \quad (8.17)$$

where the average ideal width, $\langle w_i \rangle$, for this equation is

$$\langle \bar{w}_i \rangle = \frac{\beta c_i w_i t_A}{2\rho_f (1-\phi) h x_f} = \frac{XV}{2(1-\phi) h x_f} \quad (8.18)$$

The ideal width calculated in this manner is used in correlations of acid-fracture conductivity.

Coefficients for Eq. 8.14 To Calculate $\lambda_n$

| $n$ | $g_{0,n}$ | $g_{1,n}\times10$ | $g_{2,n}\times10^3$ | $g_{3,n}\times10^4$ | $h_{1,n}\times10^3$ | $h_{2,n}\times10^3$ |
|---|---|---|---|---|---|---|
| 0 | 1.68231 | –2.26693 | 6.7544 | –1.8408 | 6.7593 | –4.6274 |
| 1 | 5.667053 | –0.69600 | 17.2931 | –2.9304 | 1.0032 | –3.4376 |
| 2 | 9.66842 | –0.39587 | 10.7545 | –0.5564 | –5.7028 | –0.4705 |
| 3 | 13.66772 | –0.27662 | 7.9375 | –0.1358 | –9.1500 | –0.5668 |
| 4 | 17.66740 | –0.21305 | 6.3431 | –0.0373 | –12.4496 | –0.71169 |

Coefficients for Eq. 8.15 To Calculate $G_n$

| $n$ | $\bar{g}_{0,n}\times10$ | $\bar{g}_{1,n}\times10^4$ | $\bar{g}_{2,n}\times10^4$ | $\bar{g}_{3,n}\times10^5$ | $\bar{h}_{1,n}\times10^4$ | $\bar{h}_{2,n}\times10^4$ |
|---|---|---|---|---|---|---|
| 0 | 9.10378 | –2.38279 | 14.9298 | –8.97017 | –7.08118 | –1.18392 |
| 1 | 0.53126 | 1.88909 | –12.5375 | 8.13482 | 4.01538 | 0.35148 |
| 2 | 0.15272 | 0.39035 | –1.6607 | 0.680785 | 1.03940 | 0.51540 |
| 3 | 0.06807 | 0.07330 | –0.4172 | 0.111312 | 0.58639 | 0.141225 |
| 4 | 0.03739 | 0.01901 | –0.1503 | 0.027559 | 0.35277 | 0.056322 |

Table 8.3—Coefficients for Eqs. 8.14 and 8.15.

**8.2.4 Numerical Modeling of Acid Transport and Rock Dissolution.** Acid fracturing is a complex process that includes the mechanics of hydraulic-fracture propagation, 3D reactive-fluid transport in the fracture, dissolution of rock along the fracture faces, and leakoff and subsequent reaction of reactive fluids in the rock matrix. Such a complex problem must be solved numerically to handle even parts of the overall complexity.

An acid-fracture treatment most commonly begins with the injection of a nonacidic pad gel to create the hydraulic fracture, and then the acid system (e.g., gelled acid, emulsified acid) follows. For this type of treatment, it is reasonable to decouple the hydraulic-fracture propagation problem from that of acid transport and dissolution by first generating the fracture domain with a hydraulic-fracture model, and then modeling acid transport and dissolution in this domain. We show results from Oeth (2013) to illustrate numerical-acid-fracture-model predictions obtained with this approach.

In Oeth's modeling approach, the flow field in the discretized 3D fracture domain generated by a commercial hydraulic-fracture simulator is obtained at every timestep by numerically solving the Navier-Stokes equations. Then, a numerical solution of Eqs. 8.1 through 8.4 yields the acid-concentration profile throughout the fracture domain at every timestep. From the amount of acid transported to the walls of the fracture by diffusion and convection, and knowing the acid dissolving power, the amount of rock dissolved is calculated throughout the fracture. The rock dissolution is then expressed as an ideal width so that the fracture-conductivity distribution can be calculated as described in the following sections. The reader is referred to Oeth (2013) for more details about this modeling procedure.

The data for an example acid-fracture simulation is given in **Table 8.4** and the results displayed in **Figs. 8.3–8.5** for fracture geometry at the start of the acid stage, the velocity field in the fracture, and the final etched width profile.

## 8.3 Acid-Fracture Fluids

An acid-fracturing treatment can be conducted using a simple HCl solution with no additives to increase the viscosity of the acid. However, it is not likely that a long conductive fracture can be created with straight acid because the amount of leakoff will be too great. Thus, the most common acid-fracturing fluids are viscosified acids, with viscosity created by adding polymers, by emulsifying the acid with oil, or by adding viscoelastic surfactants.

**8.3.1 Polymer-Viscosified Fluids.** Acids are most commonly viscosified for acid fracturing by adding polymers to the acid solution. Copolymers of polyacrylamide are the most commonly used polymers

| | |
|---|---|
| Acid type and concentration | 15 wt % hydrochloric acid (HCl) |
| Mineralogy | Limestone |
| Diffusion coefficient | $8\times10^{-6}$ cm$^2$/s |
| Injection temperature | 100°F |
| Pump rate | 40 bbl/min (constant),<br>20 bbl/min per halfwing |
| Injection time | 20 minutes |
| Young's modulus | 4.5 MMpsi |
| Closure stress | 2,000 psi |
| Pore volumes to breakthrough | 1.5 |
| Fraction of acid to react on surface before becoming leakoff | 0.3 |
| Power-law index | 0.65 |
| Consistency index | 0.05 Pa·s$^n$ |

Table 8.4—Example acid fracturing conditions.

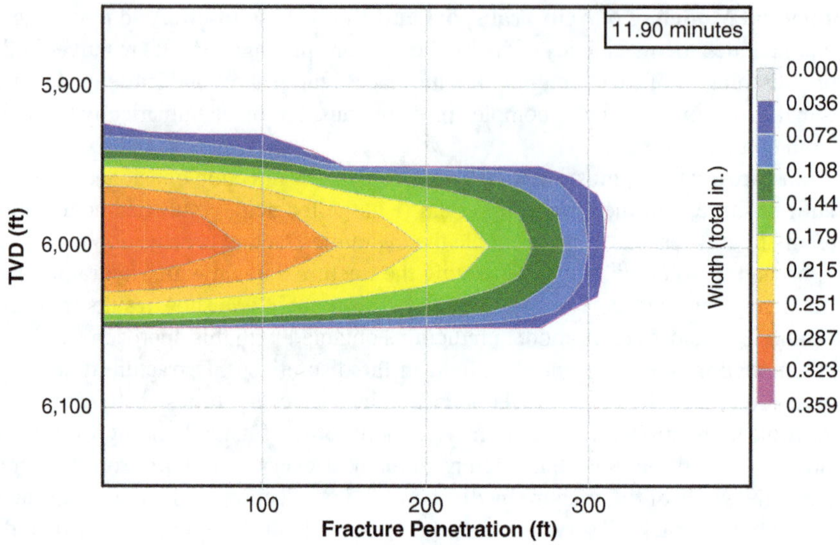


**Fig. 8.3—Fracture geometry from StimPlan at end of 30# crosslinked pad injection with stress barriers (Oeth et al. 2014).**

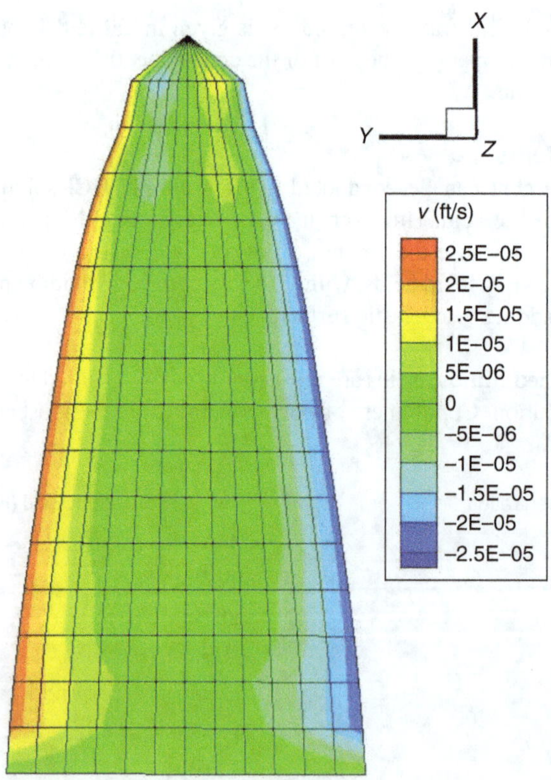


**Fig. 8.4—Velocity in the *y*-direction across fracture width for the gelled-acid case (Oeth et al. 2014).**

for viscosifying acid solutions. The resulting polymer-acid mixtures are non-Newtonian and are usually described by a power-law rheological model. The rheology of a polyacrylamide/HCl mixture presented by Gomaa et al. (2011) is shown in **Fig. 8.6.**

**8.3.2 Emulsions.** Oil-external emulsions have been used for many years as acid-fracturing fluids. An emulsion has a higher viscosity than either of the liquids used to create it, but in addition to the effect

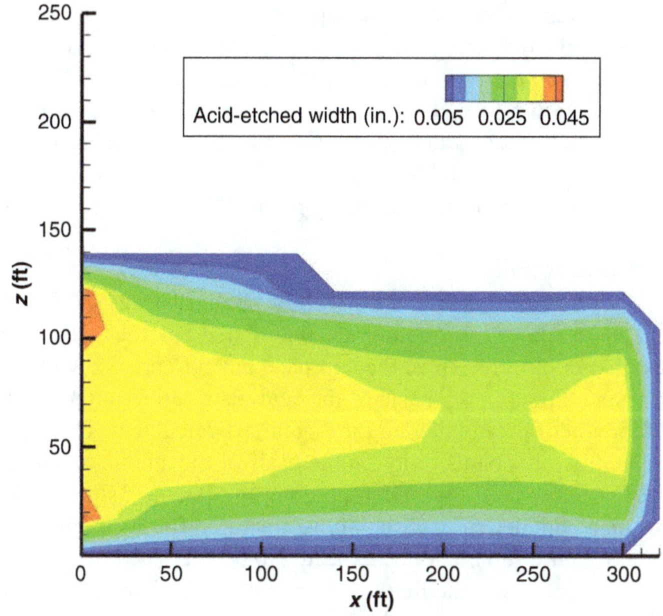


**Fig. 8.5—Acid-etched width for the gelled-acid case (Oeth et al. 2014).**

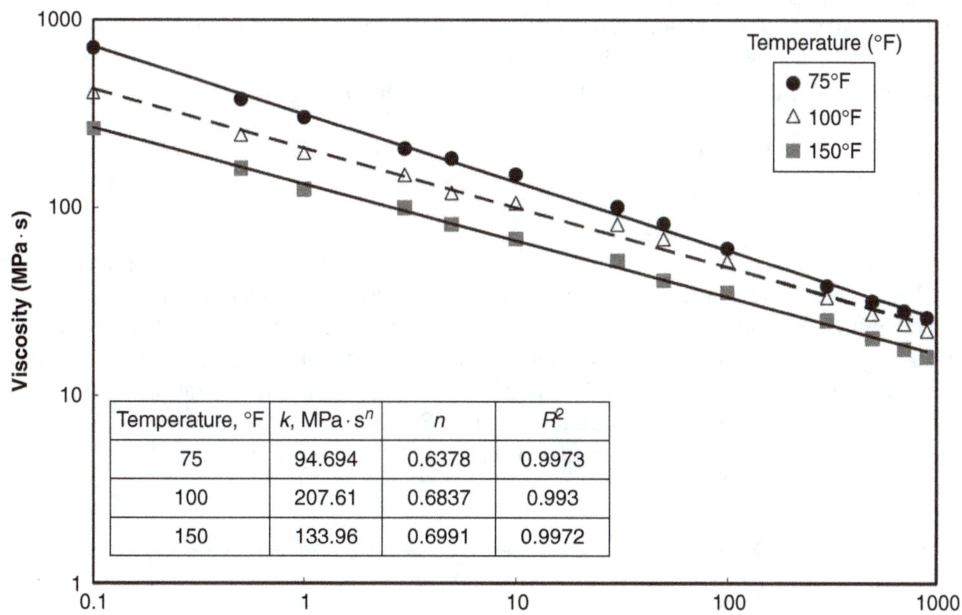


| Temperature, °F | $k$, MPa · s$^n$ | $n$ | $R^2$ |
|---|---|---|---|
| 75 | 94.694 | 0.6378 | 0.9973 |
| 100 | 207.61 | 0.6837 | 0.993 |
| 150 | 133.96 | 0.6991 | 0.9972 |

**Fig. 8.6—Rheology of a polymer-gelled acid.**

of high viscosity reducing the leakoff flux, the dispersion of the acid as fine droplets in a continuous oil phase reduces the diffusive flux of acid to the fracture walls. Li et al. (1993) reported the effective diffusion coefficient of a 70% acid/30% oil emulsion to be an order of magnitude lower than that of straight acid. Both the high viscosity and the lower diffusion coefficient allow emulsified acid to create longer etched-fracture length.

**8.3.3 Viscoelastic-Surfactant Fluids.** A class of surfactants generally referred to as viscoelastic surfactants can create high viscosity when a strong acid spends by reacting with carbonate minerals. As the pH increases and divalent-cation ($Ca^{++}$ and $Mg^{++}$) concentrations increase, the surfactant molecules arrange themselves in rod-like micelles that can create high viscosity just like long-chain polymer molecules. In an acid-fracturing operation using a viscoelastic-surfactant/acid mixture, this viscosification should occur in the rock matrix as acid leaks off into the formation, thus reducing the leakoff rate and allowing for a longer fracture to be created. In laboratory tests, Pournik et al. (2007) found that a viscoelastic-surfactant/acid fluid created considerably more conductivity than either polymer-gelled acid or emulsified acid, at least for small contact times (15 minutes). Some good field results have also been reported for this type of acid-fracturing fluid (Bartko et al. 2003).

## 8.4 Acid-Fracture Conductivity

The conductivity ($k_f w$) of an acid fracture is difficult to predict because it depends inherently on a stochastic process: if the walls of the fracture are not etched heterogeneously, very little fracture conductivity will result after closure. Thus, the approach taken to predict acid-fracture conductivity has been an empirical one. First, on the basis of the acid distribution in the fracture, the amount of rock dissolved as a function of position along the fracture is calculated. Then, an empirical correlation is used to calculate fracture conductivity on the basis of the amount of rock dissolved. Finally, since the conductivity usually varies significantly along the fracture, some averaging procedure is used to obtain an average conductivity for the entire fracture. The methods for predicting fracture conductivity of an acid fracture should not be expected to be very accurate. By measuring the effective conductivity of acid fractures in the field using pressure-transient testing, these techniques can be "field calibrated."

The amount of rock dissolved in an acid fracture is represented by a parameter called the ideal width, $w_i$, defined as the fracture width created by acid dissolution before fracture closure. If all the acid injected into a fracture dissolves rock on the fracture face (i.e., no live acid penetrates into the matrix or forms wormholes in the fracture walls), the average ideal width is simply the total volume of rock dissolved divided by the fracture area, or

$$\bar{w}_i = \frac{XV}{2(1-\phi)h_f x_f}, \qquad (8.19)$$

where $X$ is the volumetric dissolving power of the acid, $V$ is the total volume of acid injected, $h_f$ is the fracture height, and $x_f$ is the fracture half-length. For Péclet numbers greater than approximately 5, the acid concentration is almost equal to the injected concentration along most of the fracture length, and the actual ideal width is approximately equal to the mean ideal width. For lower Péclet numbers, more acid is spent near the wellbore than farther along the fracture, and the concentration distribution must be considered to calculate the ideal-width distribution. Schechter (1990) presented one such solution **(Fig. 8.7),** based on the concentration profiles of Fig. 8.1.

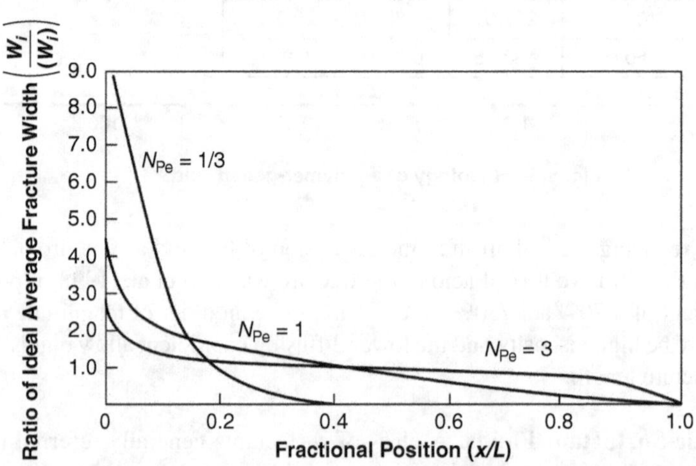


**Fig. 8.7—Ideal-fracture-width profile along a fracture (from Schechter 1992).**

**8.4.1 Laboratory Measurements of Acid-Fracture Conductivity.** It is common to measure acid-fracture conductivity in the laboratory to evaluate the applicability of acid fracturing for a particular formation. The experimental study can examine the reactivity of the acid and rock systems, wormhole propagation patterns and leakoff effects, and conductivity behavior under closure stress. The experimental evaluation of conductivity in general cannot represent the effect of heterogeneity on acid-fracturing treatment results because of the limited size of the samples. However, it can provide important understanding of etching patterns and conductivity decline as a function of closure stress, which are both critical to the success of an acid-fracturing treatment.

1. Equipment apparatus and procedure

   A typical acid-fracture conductivity experiment starts with preparing the samples that resemble a fracture in a conductivity cell, loading the cell in a load frame that can apply overburden stress to the cell (as closure stress), pumping acid through the cell, and then finally measuring conductivity at different closure stresses. Because the statistical variation in properties of the samples strongly affects the etching process and conductivity behavior, the size of the sample has been increased from 1 × 2-in plugs (Nierode and Kruk 1973) to the size of the American Petroleum Institute (API) standard fracture-conductivity test of 2 in. in width and 7 in. in length (Pournik et al. 2007). **Fig. 8.8** shows a sample rock and a modified API conductivity cell. The sample has a 3-in. depth to better simulate acid leakoff.

   With two samples assembled together to form a fracture, acid is injected through the fracture at a constant flow rate chosen to match field velocities in a fracture. Leakoff is controlled by fixing the pressure drop across the rock sample. When the planned injection time is reached, acid is flushed from the system and the cell is disassembled. The etched surface can be examined for etching patterns before the sample is loaded into a load frame for conductivity measurement. During the conductivity test, closure stress is applied by a load frame in increasing increments, water or gas flows through the fracture, and the pressure drop across the cell is measured to calculate the conductivity at each closure stress. Experiments are terminated at the highest designed closure-stress level or when the pressure drop measured reaches a value that indicates flow is through the rock matrix. It is important that at each stress level, the pressure drop that is used to calculate conductivity becomes stable for a reliable calculation. **Fig. 8.9** illustrates the apparatus for measurements.

   A typical result of a conductivity test is a plot of fracture conductivity vs. closure stress, similar to a proppant-fracture conductivity test. Important parameters for acid-fracturing design are acid type and concentration, contact time, injection rate, rock type, and formation temperature. The impact of these parameters on acid-fracturing results can be evaluated with

(a) (b)

**Fig. 8.8—(a) Experimental sample (Al-Momin et al. 2014) and (b) conductivity cell (Pournik et al. 2007).**

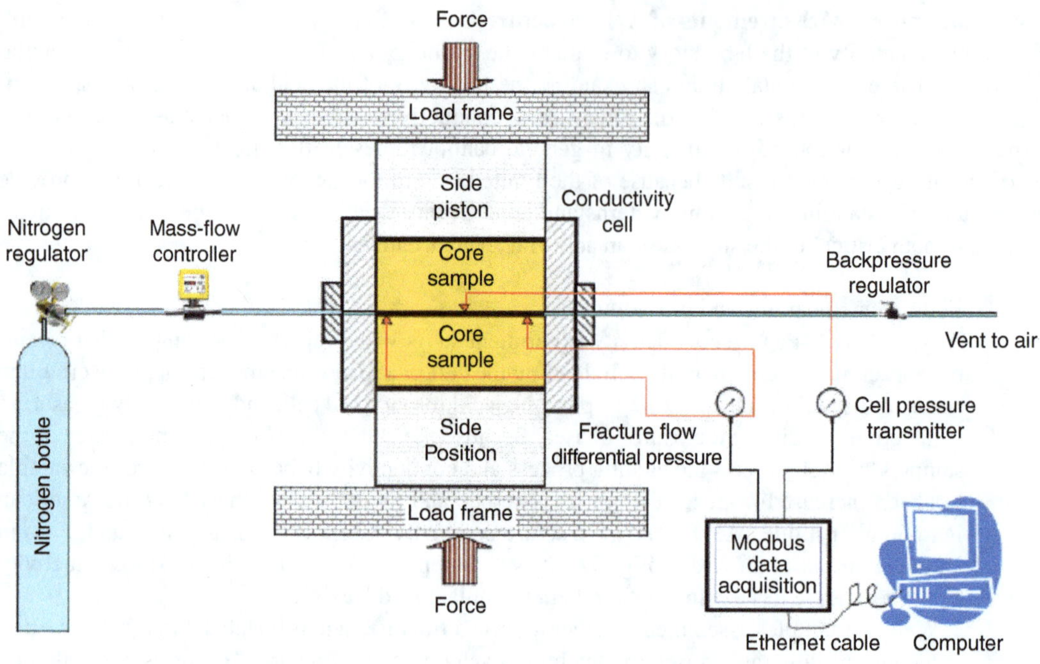


**Fig. 8.9—Schematic of conductivity-measurement experiment.**

experiments before being applied in the field. Because of the uniqueness of formation conditions (temperature, pressure, and mineralogy of the rock), the observation for one formation may not apply to another.

**Fig. 8.10** shows a plot of the effect of contact time and temperature on fracture conductivity for Indiana limestone samples treated with an emulsified-acid system (Pournik et al. 2007).

2. Surface characterization
   Surface characterization identifies the etching pattern of an acid-fracturing experiment and provides information on the etched-fracture width. The sample surface is scanned before and after acid injection, and the difference in the fracture-width distribution after acid etching compared with the initial width profile yields the amount of rock removed from the fracture surface. The topography can be created by a profilometer that consists of a laser sensor and a servo table [**Fig. 8.11** (Nieto et al. 2008)]. The sample is set on the servo table (Fig. 8.11a), which

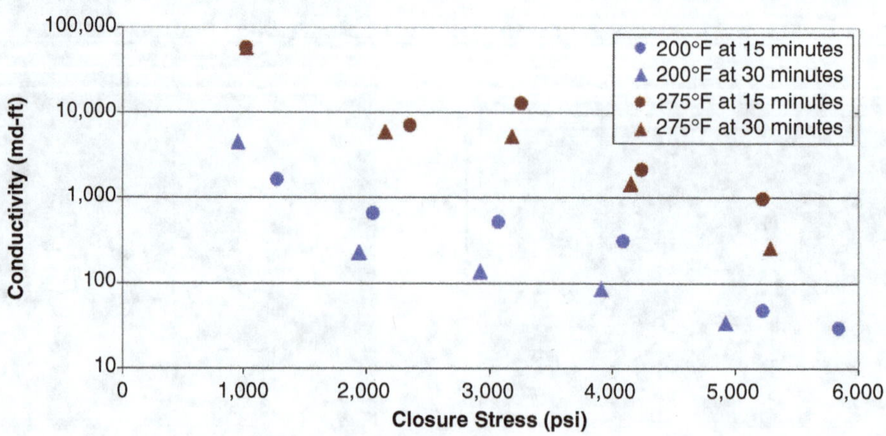


**Fig. 8.10—Conductivity measured from experiments for Indiana limestone with emulsified acid at different temperatures.**

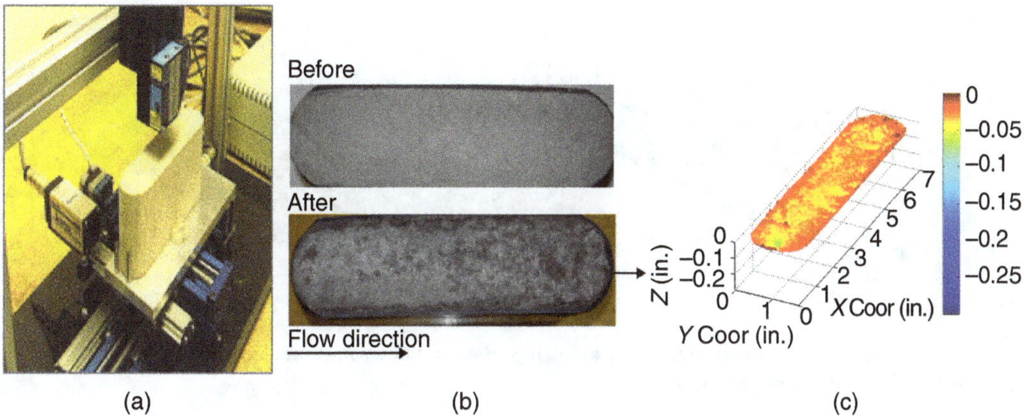


**Fig. 8.11—Laser profilometer for characterizing rock surfaces.**

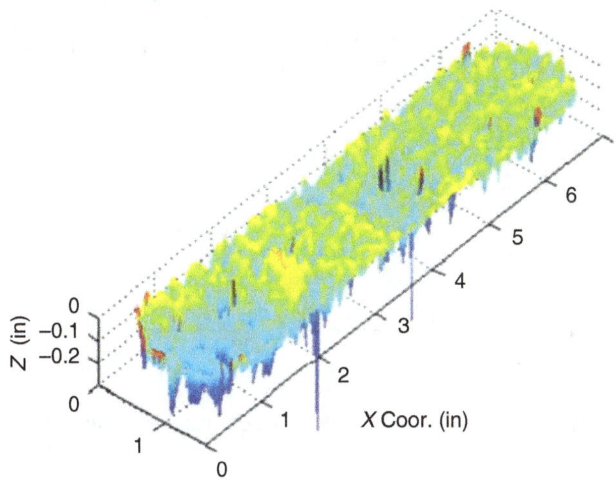


**Fig. 8.12—Acid-etched fracture surface.**

moves back and forth during the scanning. The laser sensor records the distance between the sensor and the sample. From an etched surface (Fig. 8.11b), a 3D image is generated by the profilometer (Fig. 8.11c). A typical acid-etched surface profile is shown in **Fig. 8.12** (Pournik et al. 2007).

3. Surface condition

   Most experimental studies of acid-fracture conductivity were conventionally conducted using two smooth surfaces of rock samples to represent a fracture. Studies (Neumann et al. 2012) have pointed out that, in reality, fracture surfaces are not smooth and the roughness of fracture surfaces before acid injection may have a critical effect on acid etching and fracture conductivity. The roughness of fracture surfaces before acid injection may be smoothed by acid etching, resulting in less conductivity that would be created by acid "roughening" an initially smooth surface. From the experimental study of using rough-surface cores instead of using smooth-surface cores, their results showed less-irregular fracture surfaces after acidizing if the rock samples had rough surfaces originally (**Fig. 8.13**). On the other hand, rough surfaces create more-nonuniform acid flow and, therefore, the conductivity created from etching may improve. The experimental study conducted by Almomen (2013) showed that compared with the smooth-surface samples, the rough-surface samples of San Andres dolomite do exhibit higher conductivity at zero stress (**Fig. 8.14**). However, at higher closure stress, the conductivity eventually declined to the same level, regardless of whether the fractures initially had smooth surfaces or rough surfaces.

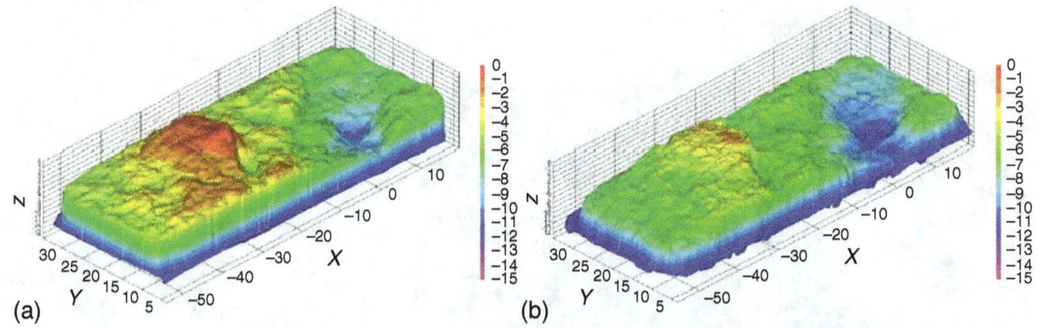


**Fig. 8.13—Fracture surface before and after acidizing of an initially rough-surface fracture: (a) Before acid injection and (b) after acid injection.**

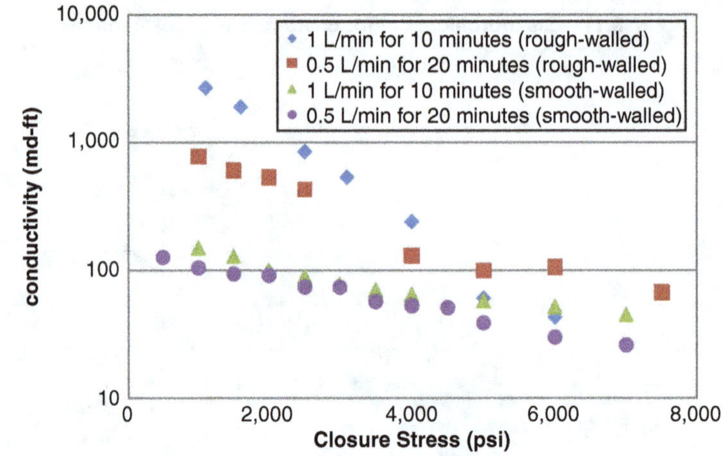


**Fig. 8.14—Conductivities of initially smooth- and rough-walled fracture (Al-Momin et al. 2014).**

**8.4.2 Laboratory-Scale Conductivity Correlations.** ***Nierode-Kruk Correlation.*** From the ideal fracture width, the conductivity of an acid fracture has been most commonly obtained from the correlation of Nierode and Kruk (1973). This correlation is based on extensive laboratory measurements of acid-fracture conductivity and correlates conductivity with the ideal width, the closure stress, $\sigma_c$, and the rock embedment strength, $S_{rock}$. The Nierode-Kruk correlation is

$$k_f w = C_1 e^{-C_2 \sigma_c}, \qquad (8.20)$$

where

$$C_1 = 1.47 \times 10^7 w_i^{2.47}, \qquad (8.21)$$

$$C_2 = (13.9 - 1.3 \ln S_{rock}) \times 10^{-3} \qquad (8.22)$$

for $S_{rock} < 20{,}000$ psi, and

$$C_2 = (3.8 - 0.28 \ln S_{rock}) \times 10^{-3} \qquad (8.23)$$

for $S_{rock} > 20{,}000$ psi.

For the constants given in these equations, $k_f w$ is in md-ft, $w_i$ is in in., and $\sigma_c$ and $S_{rock}$ are in psi. The rock embedment strength is the force required to push a metal sphere a certain distance into the surface of a rock sample. Laboratory-measured values of embedment strength and acid-fracture conductivity for several common carbonate formations from Nierode and Kruk (1973) are given in **Table 8.5.**

Comparisons with field-measured acid-fracture conductivities have shown that the Nierode-Kruk correlation sometimes overpredicts conductivity and sometimes underpredicts conductivity, and the errors can be large (Settari et al. 2001; Bale et al. 2010). This is not surprising because it was based on laboratory measurements using very small core plugs, which cannot be expected to scale up to a field-scale fracture conductivity. Furthermore, the role of heterogeneity in formation properties, which is necessary for effective acid-fracture conductivity to be created, is not considered in the Nierode-Kruk correlation.

**8.4.3 Role of Leakoff and Heterogeneity.** A new correlation for acid-fracture conductivity that explicitly depends on statistical variations in formation properties was presented in Deng et al. (2011). For a formation in which variations in permeability, and thus the distribution of acid leakoff, cause differential etching, the correlation is as follows.

First, the conductivity at zero closure stress for permeability-distribution-dominant cases (Mou et al. 2011) is

$$\left(k_f w\right)_0 = 4.48\times10^9 \bar{w}^3\left[1+\left(a_1 erf\left(a_2(\lambda_{D,x}-a_3)\right)-a_4 erf\left(a_5(\lambda_{D,z}-a_6)\right)\right)\sqrt{(e^{\sigma_D}-1)}\right],$$
$$a_1=1.82, a_2=3.25, a_3=0.12, a_4=1.31, a_5=6.17, a_6=0.03 \quad \text{........ (8.24)}$$

| Reservoir | Maximum Conductivity | $S_{rock}$ | Conductivity (md-in.) vs. Closure Stress (psi) 0 | 1,000 | 3,000 | 5,000 | 7,000 |
|---|---|---|---|---|---|---|---|
| San Andres dolomite | $2.7\times10^6$ | 76,600 | $1.1\times10^4$ | $5.3\times10^3$ | $1.2\times10^3$ | $2.7\times10^2$ | $6.0\times10^0$ |
| San Andres dolomite | $5.1\times10^8$ | 63,800 | $1.2\times10^6$ | $7.5\times10^5$ | $3.0\times10^5$ | $1.2\times10^5$ | $4.7\times10^4$ |
| San Andres dolomite | $1.9\times10^7$ | 62,700 | $2.1\times10^5$ | $9.4\times10^4$ | $1.9\times10^4$ | $3.7\times10^3$ | $7.2\times10^2$ |
| Canyon limestone | $1.3\times10^8$ | 88,100 | $1.3\times10^6$ | $7.6\times10^5$ | $3.1\times10^5$ | $4.8\times10^4$ | $6.8\times10^3$ |
| Canyon limestone | $4.6\times10^7$ | 30,700 | $8.0\times10^5$ | $3.9\times10^5$ | $9.4\times10^4$ | $2.3\times10^4$ | $5.4\times10^3$ |
| Canyon limestone | $2.7\times10^8$ | 46,400 | $1.6\times10^6$ | $6.8\times10^5$ | $1.3\times10^5$ | $2.3\times10^4$ | $4.4\times10^3$ |
| Cisco limestone | $1.2\times10^5$ | 67,100 | $2.5\times10^3$ | $1.3\times10^3$ | $3.4\times10^2$ | $8.8\times10^1$ | $2.3\times10^1$ |
| Cisco limestone | $3.0\times10^5$ | 14,800 | $7.0\times10^3$ | $3.4\times10^3$ | $8.0\times10^2$ | $1.9\times10^2$ | $4.4\times10^1$ |
| Cisco limestone | $2.0\times10^6$ | 25,300 | $1.4\times10^5$ | $6.2\times10^4$ | $1.3\times10^4$ | $2.7\times10^3$ | $5.7\times10^2$ |
| Capps limestone | $3.2\times10^5$ | 13,000 | $9.7\times10^3$ | $4.2\times10^3$ | $7.6\times10^2$ | $1.4\times10^2$ | $2.5\times10^1$ |
| Capps limestone | $2.9\times10^5$ | 30,100 | $1.8\times10^4$ | $6.8\times10^3$ | $9.4\times10^2$ | $1.3\times10^2$ | $1.8\times10^1$ |
| Indiana limestone | $4.5\times10^6$ | 22,700 | $4.6\times10^5$ | $1.5\times10^5$ | $1.5\times10^4$ | $1.5\times10^3$ | $1.5\times10^2$ |
| Indiana limestone | $2.8\times10^7$ | 21,500 | $7.9\times10^5$ | $3.0\times10^5$ | $4.3\times10^4$ | $6.3\times10^3$ | $9.0\times10^2$ |
| Indiana limestone | $3.1\times10^8$ | 14,300 | $7.4\times10^6$ | $2.0\times10^6$ | $1.4\times10^5$ | $1.0\times10^4$ | $7.0\times10^2$ |
| Austin chalk | $3.9\times10^6$ | 11,100 | $5.6\times10^4$ | $1.6\times10^3$ | $1.3\times10^0$ | - | - |
| Austin chalk | $2.4\times10^6$ | 5,600 | $3.9\times10^4$ | $1.2\times10^3$ | $1.2\times10^0$ | - | - |
| Austin chalk | $4.8\times10^5$ | 13,200 | $1.0\times10^4$ | $1.7\times10^3$ | $4.9\times10^1$ | $1.4\times10^0$ | - |
| Clearfork dolomite | $3.6\times10^4$ | 35,000 | $3.4\times10^3$ | $1.7\times10^3$ | $4.1\times10^2$ | $1.0\times10^2$ | $2.4\times10^1$ |
| Clearfork dolomite | $3.3\times10^4$ | 11,800 | $9.3\times10^3$ | $1.6\times10^3$ | $4.5\times10^1$ | $1.3\times10^0$ | - |
| Greyburg dolomite | $8.3\times10^6$ | 14,400 | $2.5\times10^5$ | $4.0\times10^4$ | $1.0\times10^3$ | $2.5\times10^1$ | - |
| Greyburg dolomite | $3.9\times10^6$ | 12,200 | $2.1\times10^5$ | $7.9\times10^4$ | $1.0\times10^4$ | $1.5\times10^3$ | $2.0\times10^2$ |
| Greyburg dolomite | $3.2\times10^6$ | 16,600 | $8.0\times10^4$ | $1.5\times10^4$ | $4.8\times10^2$ | $1.6\times10^1$ | - |
| San Andres dolomite | $1.0\times10^6$ | 46,500 | $8.3\times10^4$ | $4.0\times10^4$ | $9.5\times10^3$ | $2.2\times10^3$ | $5.2\times10^2$ |
| San Andres dolomite | $2.4\times10^6$ | 76,500 | $1.9\times10^4$ | $6.8\times10^3$ | $8.5\times10^2$ | $1.0\times10^2$ | $1.3\times10^1$ |
| San Andres dolomite | $3.4\times10^6$ | 17,300 | $9.4\times10^3$ | $2.8\times10^3$ | $2.5\times10^2$ | $2.3\times10^1$ | - |

Table 8.5—Effective fracture conductivity for well-known reservoirs as a function of rock embedment strength and closure stress (Nierode and Kruk 1973).

where $a_1 = 1.82$, $a_2 = 3.25$, $a_3 = 0.12$, $a_4 = 1.31$, $a_5 = 6.71$, $a_6 = 0.03$, and $\bar{w}$ is average fracture width in inches. In this correlation, conductivity depends on statistical parameters describing the variation of the formation permeability. These are $\lambda_{D,x}$, the dimensionless correlation length in the horizontal direction (along the fracture); $\lambda_{D,z}$, the dimensionless correlation length in the vertical direction; and $\sigma_D$, the dimensionless standard deviation of permeability. The average width is the fracture width at zero closure stress. In practice, ideal fracture width $w_i$, which is defined as dissolved-rock volume divided by fracture-surface area, is easier to obtain than average fracture width. For the high leakoff coefficient (>0.004 ft/(min)$^{0.5}$),

$$\bar{w} = 0.56\text{erf}(0.8\sigma_D)w_i^{0.83} \quad (8.25)$$

For the medium leakoff coefficient [≈0.001 ft/(min)$^{0.5}$] with uniform mineralogy distributions,

$$\bar{w} = 0.2\text{erf}(0.78\sigma_D)w_i^{0.81} \quad (8.26)$$

Then, the correlation for overall fracture conductivity is

$$k_f w = \alpha \exp(-\beta\sigma_c), \quad (8.27)$$

where

$$\alpha = (k_f w)_0 \left[0.22(\lambda_{D,x}\sigma_D)^{2.8} + 0.01\left((1-\lambda_{D,z})\sigma_D\right)^{0.4}\right]^{0.52} \quad (8.28)$$

and

$$\beta = \left[14.9 - 3.78\ln(\sigma_D) - 6.81\ln(E)\right]\times 10^{-4} \quad (8.29)$$

In Eqs. 8.25 through 8.29, $\sigma_c$ is closure stress in psi, $\sigma_D$ is normalized standard deviation, and $E$ is Young's modulus in million psi (Mpsi). In addition, Young's modulus is required to be greater than 1 Mpsi. In general, soft rock with Young's modulus less than 2 Mpsi is not a good candidate for acid fracturing.

Often, the vertical correlation length of permeability distribution is low because the sedimentary carbonates are laminated. When the dimensionless vertical correlation length is low enough (e.g., $\lambda_{D,z}$<0.02) Eqs. 8.26 and 8.27 can be simplified to

$$\alpha = 0.12(k_f w)_0 (\lambda_{D,x}\sigma_D)^{0.1} \quad (8.30)$$

$$\beta = \left[15.6 - 4.5\ln(\sigma_D) - 7.8\ln(E)\right]\times 10^{-4} \quad (8.31)$$

For a discussion of how to obtain the statistical parameters needed in this correlation and to see an example of its application, refer to Oeth et al. (2011).

Application of this model to the numerical simulation results presented in Section 8.2.4 yield the conductivity profile shown in **Fig. 8.15.**

**8.4.4 Overall Acid-Fracture Conductivity.** Once the conductivity variation along the fracture is obtained, an average conductivity for the entire fracture can be calculated in order to estimate the productivity of the acid fractured well. If the variation in the conductivity is not too great, Bennett (1982) has shown that a simple average of the conductivity predicts the well productivity after early production.

This average is

$$\overline{k_f w} = \frac{1}{x_f}\int_0^{x_f} k_f w\,dx \quad (8.32)$$

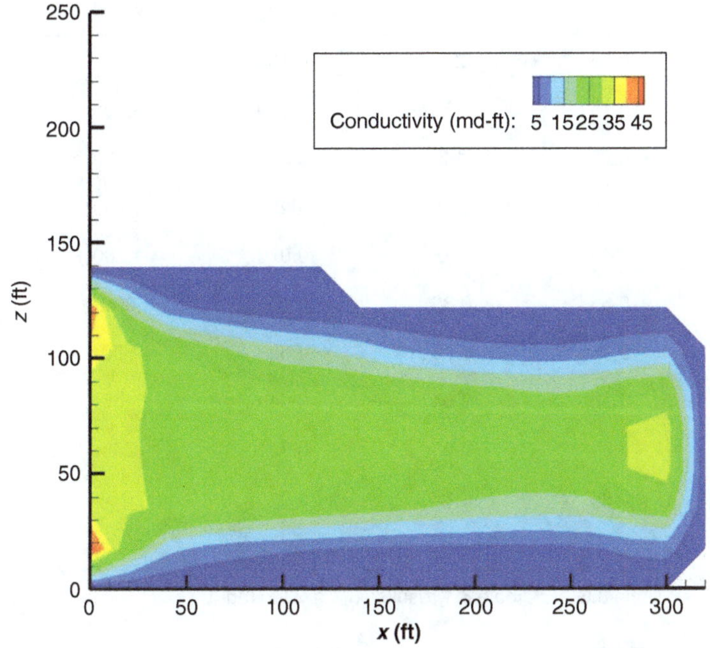


**Fig. 8.15—Mou-Deng conductivity after closure for the gelled-acid case (Oeth et al. 2014).**

This average should be adequate when the fluid-loss Péclet number is greater than 3. For lower values of Péclet number, this average will overestimate the well productivity because it is too strongly influenced by the high conductivity near the well. It is possible in some acid fractures that the conductivity is a maximum at some distance from the wellbore (caused, for example, by increasing diffusion or reaction rates as the acid warms up) (Elbel 1989). For this type of distribution, a harmonic mean better approximates the behavior of the fractured well (Ben-Naceur and Economides 1989):

$$\overline{k_f w} = \frac{x_f}{\int_0^{x_f} dx / k_f w} \quad (8.33)$$

---

**Example 8.2. Average Acid-Fracture Conductivity.** The fracturing treatment described in Example 8.1 is conducted using 15 wt% HCl in a 15%-porosity limestone formation. The formation has an embedment strength of 60,000 psi, and the closure stress is 4,000 psi. Acid is injected for a total of 15 minutes. Calculate the average fracture conductivity after fracturing using the Neirode-Kruk correlation.

***Answer.*** First, the average ideal width is calculated with Eq. 8.19. For a total of 600 bbl of injection [(40 bbl/min)(16 minutes)] into a 100-ft-high, 286-ft-long fracture, the average ideal width is

$$w_i = \frac{(0.082)(600\text{ bbl})(5.615\text{ ft}^3/\text{bbl})}{(2)(1-0.15)(100\text{ ft})(286\text{ ft})}\left(\frac{12\text{ in.}}{1\text{ ft}}\right) = 0.068\text{ in.} \quad (8.34)$$

The fluid-loss Péclet number was found to be approximately 2 in Example 8.1. Interpolating between the curves for $N_{Pe} = 1$ and $N_{Pe} = 3$ in **Fig. 8.7,** values for $\overline{w}_i / w_i$ are read. These are shown in **Table 8.6.**

| $x/x_f$ | $w_i/\overline{w}_i$ | $w_i$(in.) | $k_f w$ (md-ft) |
|---|---|---|---|
| 0.05 | 2.6 | 0.183 | 12,500 |
| 0.15 | 1.55 | 0.109 | 3,500 |
| 0.25 | 1.3 | 0.092 | 2,300 |
| 0.35 | 1.15 | 0.81 | 1,700 |
| .045 | 1.0 | 0.070 | 1,200 |
| 0.55 | 0.85 | 0.060 | 800 |
| 0.65 | 0.75 | 0.053 | 600 |
| 0.75 | 0.65 | 0.046 | 400 |
| 0.85 | 0.45 | 0.032 | 200 |
| 0.95 | 0.15 | 0.011 | 10 |

Table 8.6—Example 8.1 results; $k_f\overline{w}$ = 2300 md-ft.

Using the Neirode-Kruk correlation, $C_2$ is given by Eq. 8.23,

$$C_2 = [3.8 - 0.28\ \ln(60{,}000)]\times 10^{-3} = 7.194\times 10^{-4}\,, \quad (8.35)$$

and combining Eqs. 8.20 and 8.21,

$$k_f w = (1.47\times 10^7)w_i^{2.47} e^{-(7.194\times 10^{-4})(4{,}000)} = (8.27\times 10^5)w_i^{2.47}. \quad (8.36)$$

The fracture conductivities for 10 equal increments along the fracture are shown in Table 8.6. Using the simple average, the average conductivity is 2,100 md-ft. This value is probably high because of the high conductivity near the well. In contrast, using the harmonic average gives a value of 95 md-ft for the average conductivity.

## 8.5 Acid-Fractured-Well Performance

The productivity of an acid fractured well can be predicted in the same manner as for a well with a propped fracture if an average fracture conductivity adequately describes the well flow behavior. One complicating factor is that the conductivity of an acid fracture depends strongly on the closure stress, and the closure stress increases as the bottomhole flowing pressure decreases during the life of the well. Because $p_{wf}$ for a given flow rate depends on fracture conductivity and the fracture conductivity depends on the closure stress, and hence $p_{wf}$, an iterative procedure is required to determine the productivity of an acid fractured well. Ben-Naceur and Economides (1989) presented a series of performance type curves for acid-fractured wells producing at a constant $p_{wf}$ of 500 psi by use of the Nierode-Kruk conductivity correlation. An example of these curves is reproduced in **Fig. 8.16.** In this graph, the dimensionless cumulative production $Q_D$ and the dimensionless time $t_{Dxf}$ are defined as

$$Q_D = \frac{3.73\times 10^{-2} N_p B_o}{\phi h c_t x_f^2 (p_i - p_{wf})}, \text{ for oil}, \quad (8.37)$$

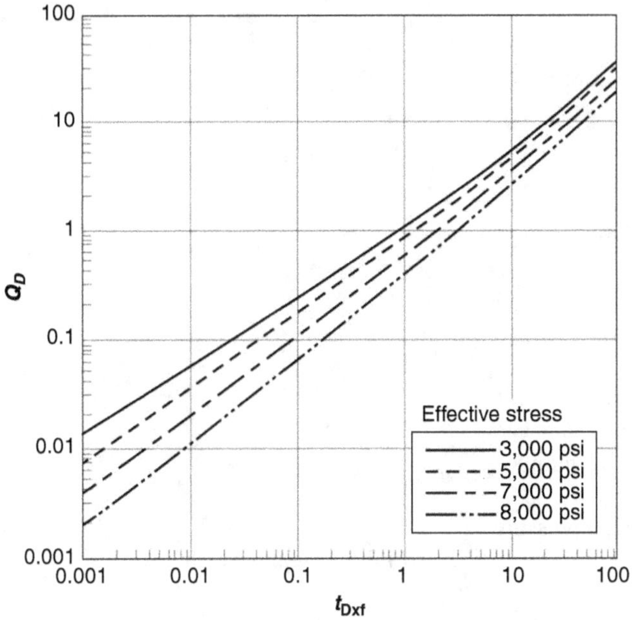


**Fig. 8.16—Performance curves for a 30,000-psi embedment strength formation as a function of closure stress (Ben-Naceur and Economides 1988).**

$$\text{and } Q_D = \frac{0.376 G_p ZT}{\phi h c_t x_f^2 (p_i^2 - p_{wf}^2)}, \text{ for gas}, \qquad (8.38)$$

$$t_{Dxf} = \frac{0.000264 kt}{\phi \mu c_t x_f^2}, \qquad (8.39)$$

where $N_p$ and $G_p$ are the cumulative productions of oil (STB) and gas (Mscf), respectively, and the constants in the equations are for units of ft for lengths, psi for pressure, psi$^{-1}$ for compressibility, °R for temperature, cp for viscosity, md for permeability, and hours for time.

## Nomenclature

$a_i$ = constants
$B_o$ = oil formation volume factor
$C$ = acid concentration
$\bar{C}$ = average concentration
$C_c$ = compressibility leakoff coefficient
$C_i$ = injected acid concentration
$C_L$ = leakoff coefficient
$c_t$ = total compressibility
$C_v$ = viscous leakoff coefficient without wormholes
$C_{v,wh}$ = viscous leakoff coefficient with wormholes
$C_w$ = wall building leakoff coefficient
$C_{wh}$ = leakoff coefficient with wormholes
$C_1$ = constant
$C_2$ = constant
$D_A$ = acid diffusion coefficient
$D_{\text{eff}}$ = effective diffusion coefficient
$E$ = Young's modulus
$E_f$ = reaction rate constant

$\bar{g}_{i,n}$ = constants
$G_n$ = constants, series
$G_p$ = cumulative production of gas
$h$ = reservoir thickness
$h_f$ = fracture height
$h_{i,n}$ = constants
$k$ = permeability
$k_f w$ = fracture conductivity
$(k_f w)_o$ = fracture conductivity at zero closure stress
$\overline{k_f w}$ = Average fracture conductivity
$n$ = order of the reaction
$N_{Pe}$ = Peclet number
$N_p$ = cumulative production of oil
$N_{Re}$ = Reynolds number
$N_{Re}^*$ = fluid loss Reynolds number
$p_i$ = pressure
$p_{wf}$ = bottomhole flowing pressure
$PV_{bt}$ = pore volumes of acid injected at wormhole breakthrough
$q_i$ = injection rate
$q_L$ = fluid loss rate
$Q_D$ = dimensionless cumulative production
$S_{rock}$ = rock embedment strength
$t$ = time
$t_A$ = time of acid injection
$T_{Dxf}$ = dimensionless time
$T$ = Temperature
$u_A^o$ = acid flux at the wellbore
$u_x$ = flux along a fracture
$u_y$ = transverse flux due to fluid loss
$\bar{u}_N$ = average fluid loss flux
$\bar{u}_y$ = average leakoff flux
$V$ = total acid volume
$\bar{w}$ = average fracture width
$w$ = fracture width
$w_i$ = ideal width
$\bar{w}_i$ = average ideal width
$x$ = position along the fracture
$x_f$ = fracture half-length
$X$ = volumetric acid dissolving power
$y$ = position in direction perpendicular to fracture length
$Z$ = gas compressibility factor
$\alpha$ = conductivity coefficient
$\beta$ = gravimetric acid dissolving power
$\beta$ = conductivity correlation parameter
$\lambda_n$ = eigenvalues, series
$\lambda_{D,x}$ = dimensionless horizontal correlation length
$\lambda_{D,z}$ = dimensionless vertical correlation length
$\mu$ = viscosity
$\rho$ = fluid density
$\rho_f$ = density of fracture fluid
$\sigma_c$ = closure stress
$\sigma_D$ = dimensionless standard deviation of permeability
$\phi$ = porosity, fraction

## References

Al-Momin, A., Zhu, D., and Hill, A. D. 2014. The Effects of Initial Condition of Fracture Surfaces, Acid Spending and Acid Type on Conductivity of Acid Fracture. Presented at the Offshore Technology Conference-Asia, Kuala Lumpur, Malaysia, 25–28 March. OTC-24895-MS. http://dx.doi.org/10.4043/24895-MS.

Almomen, A. M. 2013. The Effects of Initial Condition of Fracture Surfaces, Acid Spending, and Type on Conductivity of Acid Fracture. MS thesis, Texas A&M University, College Station, Texas, USA (August 2013).

Alkhaldi, M. H., Sarma, H. K., and Nasr-el-Din, H. A. 2010. Diffusivity of Citric Acid During Its Reation With Calcite. *J Can Pet Technol* **49** (8): 45–52.

Bale, A., Smith, M. B., and Klein, H. H. 2010. Stimulation of Carbonates Combining Acid Fracturing with Proppant (CAPF): A Revolutionary Approach for Enhancement of Final Fracture Conductivity and Effective Fracture Half-Length. Presented at the SPE Annual Technical Conference and Exhibition, Florence, Italy, 19–22 September. SPE-134307-MS. http://dx.doi.org/10.2118/134307-MS.

Bartko, K. M., Nasr-El-Din, H. A., Rahim, Z. et al. 2003. Acid Fracturing of a Gas Carbonate Reservoir: The Impact of Acid Type and Lithology on Fracture Half-Length and Width. Presented at the SPE Annual Technical Conference and Exhibition, Denver, 5–8 October. SPE-84130-MS. http://dx.doi.org/10.2118/84130-MS.

Ben-Naceur, K. and Economides, M. J. 1988. The Effectiveness of Acid Fractures and Their Production Behavior. Presented at the SPE Eastern Regional Meeting, Charleston, West Virginia, USA, 1–4 November. SPE-18536-MS. http://dx.doi.org/10.2118/18536-MS.

Ben-Naceur, K. and Economides, M. J. 1989. Acid Fracture Propagation and Production. In *Reservoir Stimulation*, second edition. ed. M. J. Economides and K. Nolte, G., Vol. Chap. 18. Englewood Cliffs, New Jersey, USA: Prentice Hall.

Bennett, C. 1982. *Analysis of Fractured Wells*. PhD thesis, University of Tulsa, Tulsa.

Conway, M. W., Asadi, M., Penny, G. S., Change, F. 1999. A Comparative Study of Straight/Gelled/Emulsified Hydrochloric Acid Diffusivity Coefficient Using Diaphragm Cell and Rotating Disk. Presented at the Annual Technical Conference and Exhibition, Houston, 3–6 October. SPE-56532-MS.

Deng, J., Mou, J., Hill, A. D. et al. 2011. A New Correlation of Acid Fracture Conductivity Subject to Closure Stress. Presented at the SPE Hydraulic Fracturing Technology Conference, The Woodlands, Texas, USA, 24–26 January. SPE-140402-MS. http://dx.doi.org/10.2118/140402-MS.

Elbel, J. L. 1989. Field Evaluation of Acid Fracturing Treatments Using Geometry Simulation, Buildup, and Production Data. Presented at the SPE Annual Technical Conference and Exhibition, San Antonio, Texas, USA, 8–11 October. SPE-19773-MS.

Furui, K., Burton, R. C., Burkhead, D. W. et al. 2012. A Comprehensive Model of High-Rate Matrix-Acid Stimulation for Long Horizontal Wells in Carbonate Reservoirs: Part II—Wellbore/Reservoir Coupled-Flow Modeling and Field Application. *SPE J* **17** (01): 280–291. SPE-155497-PA. http://dx.doi.org/10.2118/155497-PA.

Gomaa, A. M., Mahmoud, M. A., and Nasr-El-Din, H. A. 2011. Effect of Shear Rate on the Propagation of Polymer-Based In-Situ-Gelled Acids inside Carbonate Cores. *SPE Prod & Oper* **26** (1): 41–54. SPE-142927-PA. http://dx.doi.org/10.2118/142927-PA.

Hill, A. D., Zhu, D., and Wang, Y. 1995. The Effect of Wormholing on the Fluid-Loss Coefficient in Acid Fracturing. *SPE Prod & Fac* **10** (4): 257–264. SPE-27403-PA. http://dx.doi.org/10.2118/27403-PA.

Kung, M. S. 1998. Flow and Reaction of Weak Acids in Carbonate Porous Media. MS thesis, University of Michigan, Ann Arbor, Michigan.

Li, Y., Sullivan, R. B., de Rozieres, J. et al. 1993. An Overview of Current Acid Fracturing Technology with Recent Implications for Emulsified Acids. Presented at the SPE Annual Technical Conference and Exhibition, Houston, 3–6 October. SPE-26581-MS. http://dx.doi.org/10.2118/26581-MS.

Lo, K. K. and Dean, R. H. 1989. Modeling of Acid Fracturing. *SPE Prod Eng* **4** (2): 194–200. SPE-17110-PA. http://dx.doi.org/10.2118/17110-PA.

Mou, J., Zhu, D., and Hill, A. D. 2011. New Correlations of Acid-Fracture Conductivity at Low Closure Stress Based on the Spatial Distributions of Formation Properties. *SPE Prod & Oper* **26** (2): 195–202. SPE-131591-PA. http://dx.doi.org/10.2118/131591-PA.

Mou, J., Zhang, S., Zhang, Y. 2012 Acid Leakoff Mechanism in Acid Fracturing of Naturally Fractured Carbonate Reservoirs. *Transp Pourous Med* **91:** 573–584.

Muller, G. T. and Stokes, R. H. 1957. The Mobility of the Undissociated Citric Acid Molecule in Aqueous Solutions. *J. Trans. Faraday Soc.* **53**: 642–645. http://dx.doi.org/10.1039/tf9575300642.

Neumann, L. F., Sousa, J. L. A. O., Brandao, E. M. et al. 2012. Acid Fracturing: New Insights on Acid Etching Patterns from Experimental Investigation. Presented at the SPE Hydraulic Fracturing Technology Conferenc, The Woodlands, Texas, USA, 6–8 February. SPE-152179-MS. http://dx.doi.org/10.2118/152179-MS.

Nierode, D. E. and Kruk, K. F. 1973. An Evaluation of Acid Fluid Loss Additives, Retarded Acids, and Acidized Fracture Conductivity. Presented at the Fall Meeting of the Society of Petroleum Engineers of AIME, Las Vegas, Nevada, USA, 30 September–3 October. SPE-4549-MS. http://dx.doi.org/10.2118/4549-MS.

Nierode, D. E. and Williams, B. B. 1971. Characteristics of Acid Reaction in Limestone Formations. *SPE J* **11** (04): 406–418. SPE-3101-PA. http://dx.doi.org/10.2118/3101-PA.

Nieto, C. M., Pournik, M., and Hill, A. D. 2008. The Texture of Acidized Fracture Surfaces: Implications for Acid Fracture Conductivity. *SPE Prod & Oper* **23** (3): 343–352. SPE-102167-PA. http://dx.doi.org/10.2118/102167-PA.

Oeth, C., Hill, A. D., Zhu, D. et al. 2011. Characterization of Small Scale Heterogeneity to Predict Acid Fracture Performance. Presented at the SPE Hydraulic Fracturing Technology Conference, The Woodlands, Texas, USA, 24–26 January. SPE-140336-MS. http://dx.doi.org/10.2118/140336-MS.

Oeth, C. V., Hill, A. D., and Zhu, D. 2014. Acid Fracture Treatment Design With Three-Dimensional Simulation. Presented at the SPE Hydraulic Fracturing Technology Conference, The Woodlands, Texas, USA, 4–6 February. SPE-168602-MS. http://dx.doi.org/10.2118/168602-MS.

Oeth, C. V. 2013. *Three-Dimensional Modeling of Acid Transport and Etching in a Fracture*. PhD dissertation, Texas A&M University, College Station, Texas, USA (December 2013).

Pournik, M., Zou, C., Malagon, C. et al. 2007. Small-Scale Fracture Conductivity Created by Modern Acid-Fracture Fluids. Presented at the SPE Hydraulic Fracturing Technology Conference, College Station, Texas, USA, 29–31 January. SPE-106272-MS. http://dx.doi.org/10.2118/106272-MS.

Rabie, A. I. and Nasr-El-Din, H. A. 2011. Measuring the Reaction Rate of Lactic Acid with Calcite Using the Rotating Disk Apparatus. Presented at the SPE Middle East Oil and Gas Show and Conference, Manama, Bahrain, 25–28 September. SPE-140167-MS. http://dx.doi.org/10.2118/140167-MS.

Roberts, L. D. and Guin, J. A. 1974. The Effect of Surface Kinetics in Fracture Acidizing. *SPE J.* **144** (4): 385–395. SPE-4349-PA. http://dx.doi.org/10.2118/4349-PA.

Sayed, M., Nasr-El-Din, H. A., and Nasrabadi, H. 2013. Reaction of Emulsified Acids with Dolomite. *J Can Pet Technol* **52** (3): 164–175. SPE-151815-PA. http://dx.doi.org/10.2118/151815-PA.

Schechter, R. S. 1990. *Oil Well Stimulation*, first edition. Englewood Cliffs, New Jersey, USA: Prentice Hall.

Settari, A. 1993. Modeling of Acid-Fracturing Treatments. *SPE Prod & Fac* **8** (1): 30–38. SPE-21870-PA. http://dx.doi.org/10.2118/21870-PA.

Settari, A., Sullivan, R. B., and Hansen, C. 2001. A New Two-Dimensional Model for Acid-Fracturing Design. **16** (04): 200–209. SPE-73002-PA. http://dx.doi.org/10.2118/73002-PA.

Terrill, R. M. 1965. Head Transfer in Laminar Flow Between Parallel Porous Plates. *Int. J. Heat Mass Transfer* **8:** 1491–1497.

Vitagliano, V. and Lyons, P. A. 1956. Diffusion in Aqueous Acetic Acid Solutions. *J. Amer. Chem. Soc.* **78** (18): 4,538–4,542.

Williams, B. B. and Nierode, D. E. 1972. Design of Acid Fracturing Treatments. *J Pet Technol* **24** (07): 849–859. SPE-3720-PA. http://dx.doi.org/10.2118/3720-PA.

Chapter 9

# Additives for Acidizing Fluids: Their Functions, Interactions, and Limitations

**Hisham A. Nasr-El-Din**, Texas A&M University

## 9.1 Introduction

Acids are commonly used to stimulate oil and gas wells drilled in both carbonate and sandstone reservoirs. The objectives of using these acids depend on the lithology of the formation. For carbonate reservoirs, acids are used to bypass the damaged zone, whereas in sandstone reservoirs, they are used to remove the damaging material.

Acids used to stimulate carbonate reservoirs are, in general, different from those used in sandstones. Regular hydrochloric acid (HCl) is the main acid used to stimulate carbonates, primarily because its salts ($CaCl_2$ and $MgCl_2$) are soluble in water. Mud acids [HCl and hydrofluoric acid (HF)] are used to acidize sandstone reservoirs because of the ability of HF to dissolve aluminosilicates (zeolites, clays, and feldspars) and silica (sand).

HCl is a strong acid ($pK_a$ of –7 to –8) (Frenier 2001). It reacts vigorously with calcite, especially at high temperatures. The fast reaction of HCl with carbonates is not desirable because it causes surface washout, in which the acid is consumed at the surface of the rock and does not penetrate deep into the formation. Another drawback of HCl is its reaction with well tubulars. It can cause severe corrosion to well tubulars, submersible pumps, screens, and smart completions. Corrosion is also a major concern if corrosion-resistant alloys (CRAs) are used in wells where corrosion because of carbon dioxide ($CO_2$) is a concern. In such wells, HCl dissolves the protecting layer of chrome oxide, which causes severe corrosion to these alloys. Jenkins and Cullion (2009) mentioned that HCl can cause hydrogen embrittlement and chloride stress cracking to CRAs. As a result, there are limitations on the use of HCl in wells completed with CRA tubulars.

HCl can also interact with crude oils that contain asphaltenes and form sludges, which can cause severe damage. This is especially true when concentrated HCl solutions are used and when the acid is contaminated with ferric iron (Ali et al. 1994).

Live HCl dissolves anhydrites from the carbonate formations. However, once the acid is spent, calcium sulfate will precipitate in a form that depends on temperature. Obviously, this precipitation is a concern in formations that contain anhydrites (Delorey et al. 1996; Kalfayan 2000). It is also a concern when acidizing a carbonate formation with acids prepared using seawater or aquifer water that contains high sulfate-ion concentration (He et al. 2011).

Hydrochloric acid dissolves rust and other corrosion products that are present in the mixing tanks, pumps, and well tubulars. It will also dissolve iron sulfide ($Fe_xS_y$, where $x/y$ is close to unity) and iron carbonate ($FeCO_3$) in wet wells that contain acidic gases [$CO_2$ and hydrogen sulfide ($H_2S$)]. Live HCl is a good solvent for ferric iron, and it does not present any problems as long the pH is close to zero. However, once HCl reacts with carbonates and the pH increases to 1 to 2, then ferric iron will precipitate as ferric hydroxide in sweet wells or as iron sulfide in sour wells. The presence of ferric iron in the acid is not desirable because it aggravates sludge formation with crude oils that contain asphaltenes. Ferric iron also interacts with viscoelastic surfactants (VESs) and can cause phase separation and even precipitation of VES at high ferric iron and acid concentrations (Al-Nakhli et al. 2008; Shu et al. 2015).

HCl can also cause additional damage in sandstone reservoirs that contain zeolite (Hartman et al. 2006) and illite (Mahmoud et al. 2015). Both are acid-sensitive and they disperse and break down when contacting live HCl solutions.

Simple organic acids (mainly acetic and formic) can replace HCl to overcome some of the problems created by HCl, especially in high-temperature wells. However, these acids are much weaker than HCl, are more expensive, and can only be used up to certain acid concentrations (9 wt% for formic and 13 wt% for acetic). Their reaction with carbonates is reversible and, as a result, there will be live acid present when the well is put back on production. The presence of live acid in well flowback can cause corrosion and facility upsets.

Still at even higher temperatures, chelating agents (mainly aminopolycarboxylic acids) can be used to stimulate carbonate and sandstone reservoirs (Frenier et al. 2000; LePage et al. 2011; Reyes et al. 2013). They are basically used to chelate or complex with di- and trivalent cations and prevent their precipitation. Divalent cations of interest to the oil industry include Ca, Mg, Fe(II), Ba, Sr, and Ra. Trivalent cations include B, Al, and Fe(III).

As briefly explained so far, various acids are used to stimulate carbonate and sandstone reservoirs. Acids in general are active chemical species, and they do react with almost everything downhole including some crude oils, well tubulars, formation rock, inorganic scale if present, and filter cake generated by drilling fluids.

As mentioned before, acids are reactive species, and even if we deal with weaker organic acids, they become relatively reactive at high temperatures. There is a need to include additives when using acids for the following main reasons:

1. To protect downhole completions: well tubulars, submersible pumps, and smart completions. Acid will tend to react with these internals and cause severe corrosion.
2. To prevent sludge formation when stimulating acid-sensitive crude oils.
3. To prepare the rock to react with acids. It is recommended to have a water-wet rock for the acid to react effectively with it.
4. To retard acid reaction with the rock, especially at high temperatures.
5. To minimize precipitation of acid reaction products during and after acid treatments.
6. To restore the original wettability of the formation if one or more additives change the wettability of the rock during acid treatments.
7. To improve sweep efficiency or acid distribution across the target zone during matrix acidizing, or to reduce the leakoff rate during acid fracturing treatments.
8. To minimize trapping of spent acids in tight formations.

In the following sections, we will discuss the main additives that are used during acid treatments, their functions, and side effects. However, there is a need to realize that additives will significantly change some of the physical properties of regular HCl such as the surface tension and viscosity of the acid. Additives can affect the acid reaction rate with the rock, as will be explained later.

The objectives of this chapter are to (1) discuss the role of important additives that are commonly used during acid treatments of carbonate and sandstone reservoirs; (2) address the impact of these additives on acid reaction with the rock; and (3) examine potential formation damage that may result from using these additives. Acid additives are also active chemicals that can interact with each other and result in formation damage or the loss of the performance of other additives (Ali and Hinkel 2000).

## 9.2 Corrosion Inhibitors

**9.2.1 General.** Corrosion inhibitors are an integral part of acid formulations. Current corrosion inhibitors used in the oil field to protect low-carbon steel contain two main chemicals: quaternary amine and acetylenic alcohols or octynol (Frenier et al. 1988; Frenier 1989; Rae and Di Lullo 2003). The first carries positive charges and tends to adsorb on the negatively charged points on the inside surface of the tubulars, whereas the second contains a triple bond and does polymerize forming a protective layer on the tubular surface. This layer isolates the injected acids from the tubular, and thus protects the tubulars from acid attacks. It should be noted that this layer has a specified lifetime beyond which it does not protect the tubulars. This layer, however, should last for a period long enough to pump the acid into the formation. It should be mentioned that no acid should be pumped downhole without using the right concentration of proper corrosion inhibitors.

Typical chemicals (Frenier et al. 1989; Frenier and Growcock 1989; Frenier and Ziauddin 2008; Rostami and Nasr-El-Din 2009) used as corrosion inhibitors vary. **Fig. 9.1** shows some of the commonly

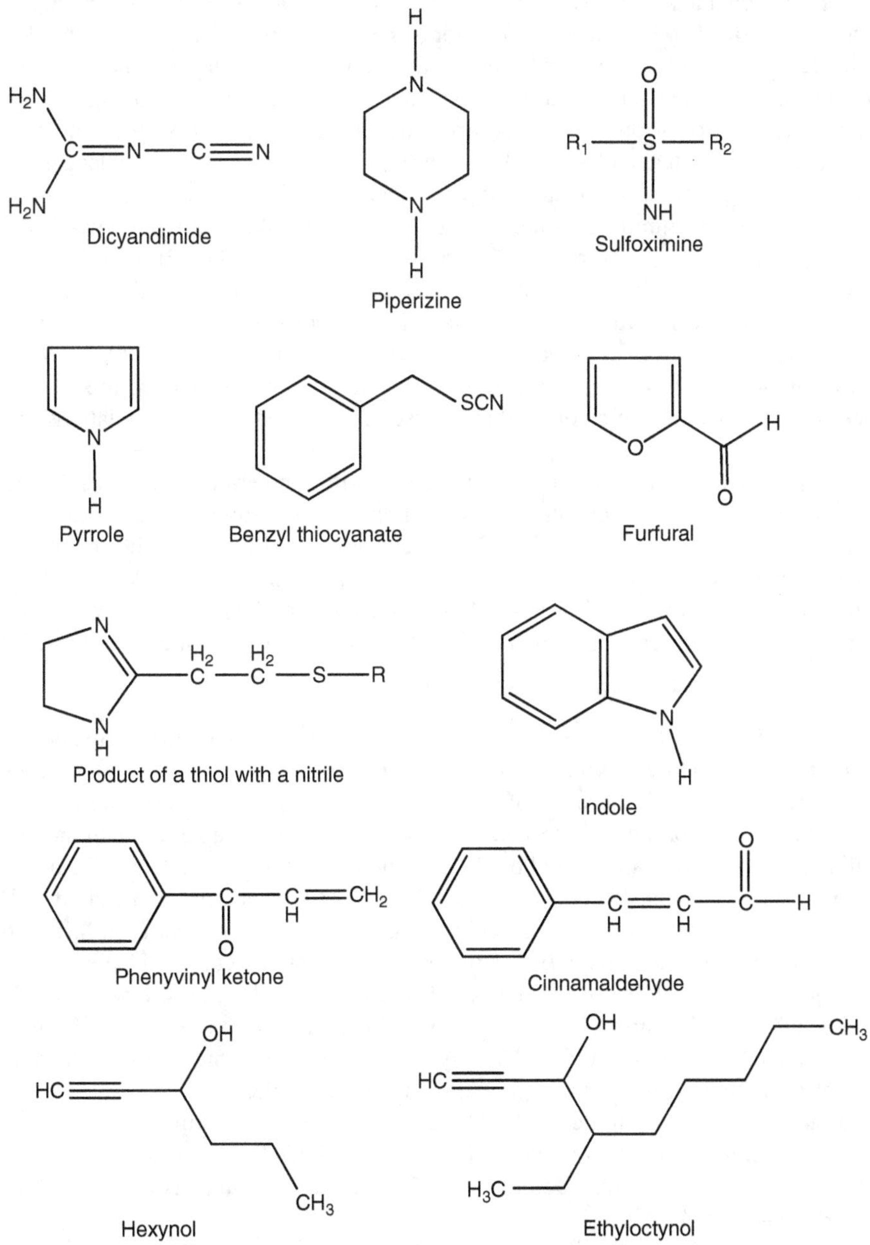


**Fig. 9.1—Corrosion inhibitors for acid treatments.**

used inhibitors including quaternary amines, propargyl alcohols, unsaturated aldehydes (cinnamaldehyde), and monoamines (Malwitz 2010). More examples of various chemicals used as corrosion inhibitors are given by Frenier (2001) and Kelland (2009). It is important to note that there is a continuous development of new corrosion inhibitors. For example, Al Taq et al. (2009) developed a new series of mono/diamine-based corrosion inhibitors for HCl solutions. Sitz et al. (2012) developed corrosion inhibitors with improved environmental profiles.

The concentration of a corrosion inhibitor depends on the acid type and concentration, metallurgy of well internals and tubulars, temperature, concentrations of acidic gases, other acid additives, and the treatment duration. Laboratory tests in autoclaves are needed to perform corrosion testing for a given acid system and metallurgy. It is recommended to conduct these tests at bottomhole temperature and a pressure of at least 1,000–1,500 psi (Smith et al. 1978). The metal coupons should have a composition similar to that of the well tubulars or coiled tubing. The acid should contain all planned additives, and the gas cap should contain acidic gases as per gases produced from the well to be acidized.

It is important to note that the corrosion inhibitor is the only additive that does not dissolve in acids. If it does dissolve in the acids, it will not be able to form this protective layer, which isolates the tubulars from acid attacks. Dispersants are used to prepare corrosion inhibitors to keep them dispersed in the acid. However, if a corrosion inhibitor is mixed with the acid and left without mixing for a long period of time, then it will float and accumulate at the surface of the acid. This is a dangerous situation that can cause serious problems in the field. Typically, acid is withdrawn from the bottom of the mixing tank. If the corrosion inhibitor is floating on the surface, then the acid will be pumped downhole without an ample amount of the corrosion inhibitor to protect the tubulars. If the acid is left to stand in the mixing tank for more than 2–3 hours, it is recommended to roll the acid before pumping into the formation. If the acid is left for a week, for example, then some of the alcohols used to prepare the corrosion inhibitor will evaporate, especially in hot areas. That also can impact the performance of the corrosion inhibitor. Additional amounts of the corrosion inhibitor should be mixed with the acid to compensate for evaporation of alcohols that are usually included in the formulations of corrosion inhibitors.

It is well-known that the rate of corrosion increases with temperature. As a result, the concentration or loading of the corrosion inhibitor increases as the bottomhole temperature increases. However, at temperatures greater than 200–220°F, corrosion inhibitors will not be able to protect well tubulars from corrosion caused by of HCl. An intensifier is needed in addition to the corrosion inhibitor to protect well tubulars. There are four main types of inhibitor aids: potassium iodide, formic acid, antimony chloride, and copper iodide. A brief summary about each of these intensifiers is to follow:

1. Potassium iodide—It is expensive and not effective at temperatures greater than 325°F (Cassidy et al. 2011). Potassium iodide is a concern in sandstone reservoirs, where HF-based acids are used. This is because of the potential precipitation of potassium fluorosilicates ($K_2SiF_6$). It is important to note that other iodide salts are available, including ammonium iodide (Malwitz 2010) and should be used in acid treatments that include HF acid.
2. Formic acid—It is considered a precursor for CO, a known corrosion inhibitor. Details of decomposition reactions of formic acid were given by Cassidy et al. (2007). It is used with 15 wt% HCl over a temperature range of 250–325°F (Cassidy et al. 2011a).
3. Antimony chloride—It is used as an intensifier with 15 wt% HCl only. It works by electroplating the tubulars with antimony. It causes bimetallic corrosion. It reacts with $H_2S$ in sour wells and precipitates antimony sulfide (Delorey et al. 2002), which may cause formation damage. It contains antimony, which has a poor environmental profile.
4. Cuprous Iodide—It has limited solubility in live acids at room temperature. It works by electroplating. Similar to antimony chloride, it causes bimetallic corrosion. It also precipitates copper sulfide in sour wells (Delorey et al. 2002), and it contains copper, which has a poor environmental profile.
5. Combinations of these intensifiers can also be used to mitigate corrosion. Brezinski and Desai (1997) developed an intensifier that is based on an iodide salt and formic acid.

It is important to note that there is a continuous development of new intensifiers. For example, Cassidy et al. (2011a) developed a new intensifier for sour wells using a sulfur and nitrogen cyclic compound. Seth et al. (2011) developed a new intensifier system to protect CRAs from acid attacks.

**9.2.2 Acceptable Corrosion Rates.** Corrosion tests should be performed under downhole conditions. A pressure of least 1,000–1,500 psi should be used. Coupons should be obtained from the supplier and are better from the same batch of well tubulars. Head space should contain acidic gases at the same concentrations as per the composition of produced gases. The duration of the test should at least match the time needed to execute the treatment in the field. The acceptable corrosion rates for low-carbon-steel tubulars are 0.05 and 0.02 $lbm/ft^2$ for coiled tubing for the duration of the treatments (Rae and Di Lullo 2003). The acceptable corrosion rate can vary from one company to another, and may change with temperature (Metcalf 1998). The absence of pitting is also required. However, some companies having a pitting index, which is dependent on the number and depth of pits per unit area (Seth et al. 2011). Discoloration of the coupon is not desirable because it may indicate the onset of corrosion (Nasr-El-Din et al. 2002). More requirements and tests are needed if coiled tubing is used in sour wells (Smuga-Otto et al. 2001; MetCalf et al. 2003).

**9.2.3 Impact on Rock Wettability.** Corrosion inhibitors are designed to adsorb on the surface of the tubulars. They carry positive charges, and as a result, they can also adsorb on the surface of the rock. This is an important issue, especially in sandstone reservoirs. In such cases, the corrosion inhibitor adsorbs on the surface of the clays present in sandstone formations and changes the wettability of the rock to oil-wet, which can adversely impact the oil production from the treated well. Several authors (Hall 1975; Crowe and Minor 1982; King and Lee 1988) have reported on this issue and have recommended using mutual solvent (ethyleneglycol monobutylether) in the preflush and post-flush to remove the adsorbed corrosion inhibitor and change the wettability of the rock into water-wet conditions.

**9.2.4 Impact of Other Additives.** Corrosion inhibitors are usually dispersed in acids. Some additives may enhance their solubility in acids, which will reduce the effectiveness of the inhibitor. Additives such as mutual solvent, short-chain alcohols, antisludge agents (anionic), and anionic surfactants can reduce the effectiveness of the corrosion inhibitor. In such cases, the loading of the corrosion inhibitor should be increased to compensate for the loss of the inhibitor. Corrosion tests should be conducted with all additives that will be used in the field. The acid should be prepared using water similar to field water.

**9.2.5 Organic Acids.** HCl is corrosive, especially at high temperatures. Organic acids are much weaker than HCl and, as a result, they are less corrosive to well tubulars. One way to address high corrosion rates of concentrated HCl solutions is to mix relatively lower concentrations of HCl with organic acids (Chang et al. 2008). Another way is to use organic acids (Harris 1961; Chatelain et al. 1976; van Domelen and Jennings 1995; Huang et al. 2000; Nasr-El-Din et al. 2001; Buijse et al. 2004; Li et al. 2008; Al-Douri et al. 2013). Organic acids commonly used in stimulation treatments include formic and acetic acids. Other organic acids include citric acid (Al-Khaldi et al. 2005, 2007, 2010) and encapsulated citric acid (Blauch et al. 2003; Burgos et al. 2004). The ester of lactic acids (polylactic acid or PLA) were also used in the field (Nasr-El-Din et al. 2009c), as well as combinations of these acids.

There is a need to use suitable corrosion inhibitors when organic acids are used. It is important to note that corrosion inhibitors used for HCl may not be effective for organic acids. Corrosion inhibitors for HCl contain quaternary amine and unsaturated oxygen compounds, whereas inhibitors for organic acids usually contain a nitrogen-based compound and a sulfur-based compound (Frenier 2001).

Another way to address the corrosiveness of HCl is to use HCl in an emulsified form. In this case, HCl is emulsified in diesel (Al-Anazi et al. 1998) or xylene (Abdel Fatah and Nasr-El-Din 2010). In emulsified acids, diesel or xylene will act as a physical barrier that separates the acid from the well tubulars. Laboratory tests and field results confirmed that emulsifying the acid reduced the rate of the reaction of the HCl with the rock and minimized the acid attacks on the tubulars.

In a recent effort to mitigation corrosion of acids at high temperatures, aminopolycarboxylic acids (APCs) (**Fig. 9.2**) were tested (Fredd and Fogler 1998a, b; Mahmoud et al. 2015; Reyes et al. 2013)

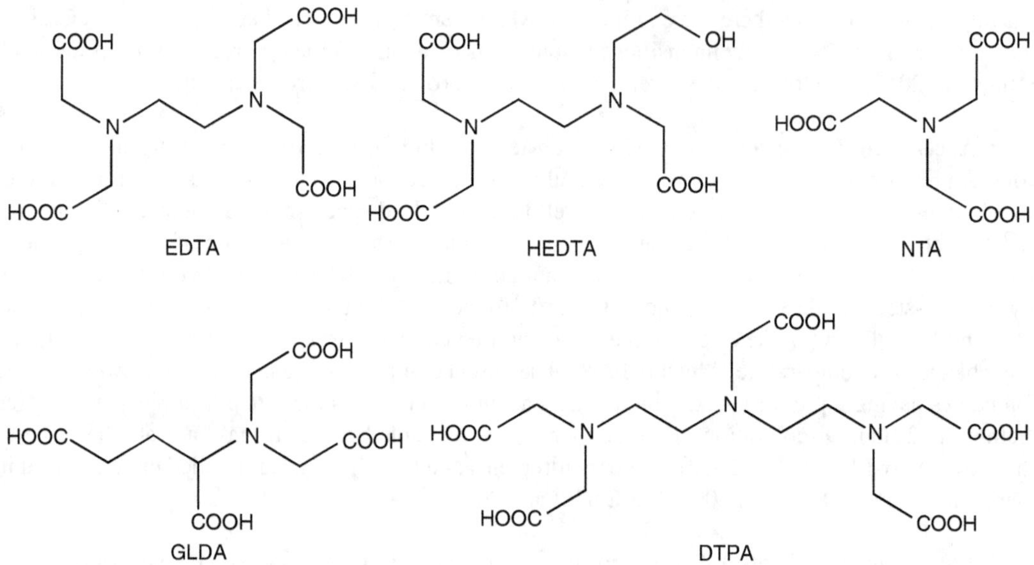


**Fig. 9.2—APCs used in the oil industry.**

and used in the field as standalone stimulation fluids. These included EDTA, HEDTA, and GLDA. The last two chelates were used in the field as standalone simulation fluids: HEDTA (Frenier et al. 2004) and GLDA (Nasr-El-Din et al. 2013, 2015).

**9.2.6 CRAs.** CRAs are used in oil and gas wells, where there is fear of severe corrosion to the tubulars (Mingjie and Boles 2004; Hinshaw et al. 2011). Some examples include gas wells that produce gases with a significant $CO_2$ concentration, enhanced-oil-recovery (EOR) methods that rely on $CO_2$, and wells used for $CO_2$ sequestration. Low-carbon steel will suffer from severe corrosion in the presence of $CO_2$ in wet wells. In such cases, it is advisable to use tubulars made from CRA, which have good resistance to corrosion in such corrosive environments.

Many production engineers think that CRA tubulars will be safe during acid treatments. This is not necessarily the case. CRA tubulars used downhole are based on Cr or Ni. The cost of CRA is much higher than that of low-carbon steel, especially those made of Ni. CRA tubulars that are based on Cr rely on the fact that they contain Cr, which will oxidize in air and form a layer of $Cr_2O_3$. This layer is impermeable, uniform, and adheres strongly to the surface of the tubulars, and as a result, it will protect well tubulars from $CO_2$ corrosion. This layer will not dissolve in carbonic acid, which is present in $CO_2$ environments that contain water. This is why CRA is recommended for sweet corrosion (no or very low $H_2S$). However, this is not the case during well stimulation using HCl. Chrome oxide dissolves in HCl, and as a result, the tubulars will be exposed to severe corrosion. The problem becomes more complicated because downhole conditions are devoid of oxygen, and there is no air to restore this protective layer without possible chemical treatments.

Protection of CRA during acidizing is a difficult task, especially at high temperatures and high acid concentrations. Chrome content plays a key factor as well. The higher the chrome content, the lower the HCl concentration that can be used (Boles et al. 2009). At temperatures greater than 200°F, there will be a need for high loading of corrosion inhibitor and intensifier. A preflush before to acid treatment will help by cooling the downhole temperature. Replacement of HCl, partially or completely, with organic acids will reduce the corrosion of the CRA tubulars. Even organic acids are corrosive at high temperatures. Chelating agents are, in general, less corrosive than simple organic acids. De Wolf et al. (2012) found that GLDA has a low-corrosion rate with CRA tubulars. Their results were confirmed in the field, as was reported by Nasr-El-Din et al. (2013).

**9.2.7 Iron-Control Agents.** Iron is a major concern when conducting acid treatments, especially in water injectors. Several papers addressed the sources of iron present in the pumped acid (Dill and

Fredette 1983; Gougler et al. 1985; Crowe 1986; Hall and Dill 1988). The main source of iron is the mixing tanks and tubulars. Some sandstone formations contain hematite ($Fe_2O_3$), siderite ($FeCO_3$), ankerite [$CaMgFe(CO_3)_2$], and/or chlorite (a clay mineral that is rich in iron). Formation waters can contain up to 100 ppm iron. Iron has two oxidation states: ferrous and ferric. Both types of ions dissolve in live HCl. However, both will precipitate once the acid is spent. The form of the precipitate and the pH at precipitation depend on the presence of $H_2S$, among other factors. In sweet environments, ferrous ions precipitate at pH values greater than 6; however, in the presence of $H_2S$, they precipitate at a pH of 1.9 (Smith et al. 1969; Crowe 1985; Brezinski 1999). In sweet wells, ferric ions precipitate at a pH of 1–2 (Taylor et al. 1999), whereas they precipitate at a pH of 1.9 in sour wells. On the basis of these trends, an iron-control strategy depends on whether a specific well is sour or not.

For sweet wells, there are several techniques. The first one is to add a reducing agent with the live acid (e.g., sodium erythorbate). The reducing agent will convert ferric iron into ferrous iron, which does not precipitate in the formations because the pH of the spent acid following acidization of the carbonate formation is 4.5 (Chang et al. 2008). This strategy will not work in sour environments because $H_2S$ will combine with ferrous ions and precipitate iron sulfide. Therefore, strategies that are based on reducing agents will not work in sour environments. A second strategy is to add a chelating agent with the acid. There are several chelating agents that can be used. There are two main types of chelating agents: -hydroxycarboxylic acids (lactic and citric acids) and APCs (Fig. 9.2). Selection of the iron-control agent depends on the temperature, cost, and environmental profile. For seawater injectors, the bottomhole temperature is low (100–120°F). A mixture of acetic and citric acids can prevent iron precipitation. However, at higher temperatures, there is a need to use APCs or their salts. Some of the commonly used APCs include EDTA, HEDTA, NTA, and GLDA. EDTA has a limited solubility in HCl.

Frenier et al. (2001 and 2003) examined chelating agents with a hydroxyl group to determine their acid solubility, and ability to complex iron and calcium under oilfield conditions. Frenier et al. (2000, 2001, 2003, 2004) advocate the use of chelates, particularly HEDTA for the dissolution of calcite.

EDTA has limited solubility in 15 wt% HCl, and the solubility decreases in concentrated acids (20 and 28 wt% HCl). HEDTA has good solubility in 15 wt% HCl, but a lower solubility in concentrated HCl solutions. NTA has a poor environmental profile and its Fe complex is a carcinogen. NTA is commonly used at temperatures of up to 300°F, but it is not recommended for use because of its poor environmental profile. GLDA is soluble in live HCl over a wide range of acid concentrations (LePage et al. 2011). It minimizes formation damage resulting from iron precipitation (Assem et al. 2013a, b). It has a good environmental profile and is approved for use in the North Sea (Braun et al. 2012). In sandstone formations, the pH of produced fluids depends on the carbonate content of the target zone. This should be taken into account when one decides whether to use an iron-control agent or not. However, in the presence of $H_2S$, there will be a need to use a sulfide scavenger with the acid to prevent iron sulfide precipitation. Another important point is that chelating agents used in a sandstone reservoir should be salts of ammonium (no sodium or potassium). Sodium and potassium ions are not desirable when HF-based fluids are used because of potential precipitation of $Na_2SiF_6$, and $K_2SiF_6$.

For sour wells, iron control is a real challenge. Hydrogen sulfide reacts with ferrous ions and precipitates iron sulfide. It also reacts with ferric iron and precipitates elemental sulfur and iron sulfide (Walker et al. 1991).

$$2Fe^{3+} + S^{2-} = 2Fe^{2+} + S^{o} \quad (9.1)$$

As the pH rises above 1.9, any residual iron can react with sulfide ions to reprecipitate FeS (Crowe 1985).

$$Fe^{2+} + S^{2-} = FeS \quad (9.2)$$

Elemental sulfur does not dissolve in water, oil, or acid, and requires special solvents (e.g., carbon disulfide) to remove it. The question now is would iron-control agents prevent iron precipitation in sour wells? This question was addressed by Brezinski (1999). The bond between iron and sulfide ions is much stronger than that between iron and APCs. As a result, iron will precipitate as iron sulfide, even in the presence of iron control agents (APCs).

It is clear from this discussion that mitigation of iron-precipitation problems in sour wells will not be solved by using a chelating agent. There is a need to remove $H_2S$ from the system first, before iron-control agents can work. $H_2S$ can be removed from the system by using a suitable $H_2S$ scavenger, as will be discussed in the next subsection.

**9.2.8 $H_2S$ Scavengers.** $H_2S$ causes problems during drilling, completion, production, and acidizing. It is a toxic and corrosive gas, and has to be removed from the system. $H_2S$ scavengers are typically used to remove it. There are different types of $H_2S$ scavengers, and the selection of the proper scavenger will depend on the pH of the aqueous phase. In drilling operations using water-based drilling fluids, $H_2S$ is removed from the system by raising the pH to nearly 10. Alkalis (e.g., KOH) and oxides (e.g., CaO and ZnO) are used to remove $H_2S$ from the system. These chemicals cannot be used during acid treatments because they will react with the acid. For acid treatments, potential $H_2S$ scavengers include aldehydes (Frenier 1992), ketones, and triazines. Xiao et al. (2009) examined several $H_2S$ scavengers including amine aldehyde mixtures, and a mixture of aldehydes. They did not disclose the chemical properties of the scavengers they tested. Al-Duailej et al. (2010) examined the performance of triazines and found them not suitable for use in HCl solutions.

The reactions of $H_2S$ and aldehydes or ketones occur by means of the carbonyl group, CO. The simplest aldehyde is formaldehyde. It reacts with $H_2S$ as follows:

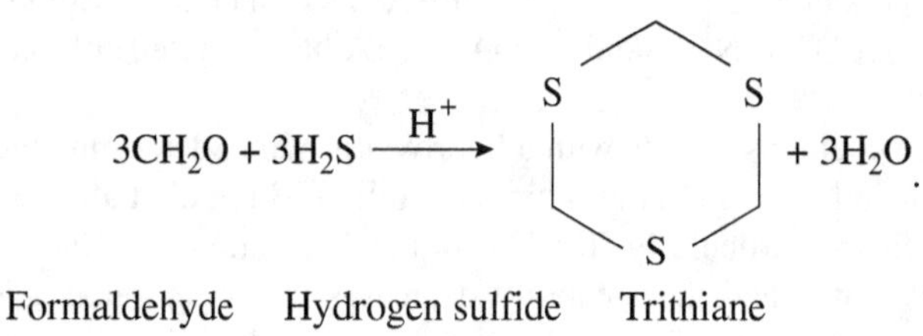

. . . . . . . . . . . . . . . . . . . . . . . . . . . . . . .(9.3)

The reaction products include various sulfur-based species, but the main one is trithiane. It is insoluble in acids, crude oil, or water, and forms a white precipitate at temperatures greater than 60°C. It can be dissolved using carbon disulfide.

Formaldehyde, however, has limited use in the oil industry because it is toxic and a carcinogen. One way to overcome this problem is to use hexamethylenetetramine (HMTA), which reacts with HCl to produce formaldehyde downhole (Frenier 1982).

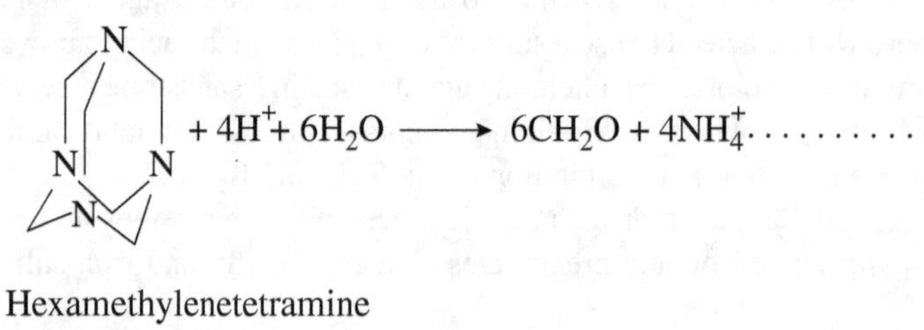

. . . . . . . . . . . . . . . . . . . . . . . . . . . . . .(9.4)

Generation of formaldehyde downhole is not as efficient as using formaldehyde (Al-Humaidan and Nasr-El-Din 1999).

Nasr-El-Din et al. (2000) examined the effect of additives on the performance of formaldehyde. Several additives had adverse effects on the performance of formaldehyde. They also recommended not using the aldehyde types of scavengers at high concentrations, as the scavengers formed a polymeric layer on the surface, which may adversely impact the reaction of HCl with iron sulfide scale.

Aldehydes should not be used at high concentrations. They should be added just before pumping the acid. Some aldehydes can crosslink polyacrylamide-based polymers that are used for acid diversion (Cassidy et al. 2011). This should be considered when designing acid formulas for field treatments.

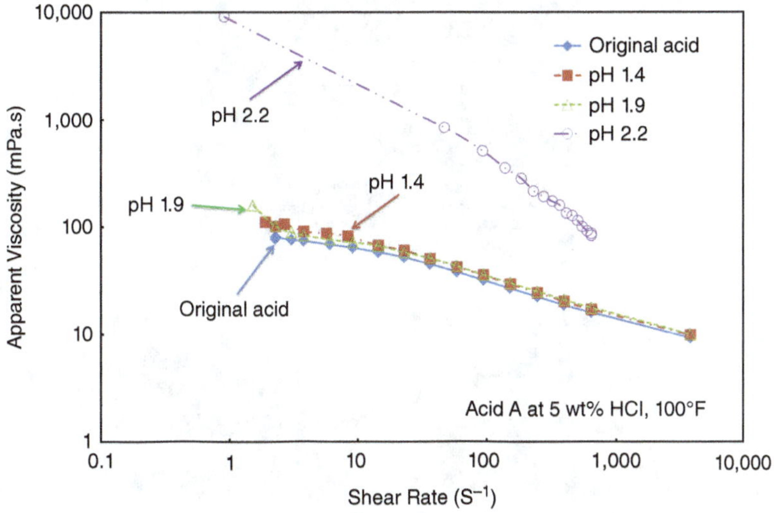


**Fig. 9.3—Effect of pH on the viscosity of in-situ-gelled acids (Taylor and Nasr-El-Din 2003).**

**9.2.9 Polymers.** HCl is a strong acid that vigorously reacts with calcite. When this acid is used, it follows the path of least resistance. If there are permeability variations over the target zone, or the zone contains calcite and dolomite, the acid will flow and react with calcite. There is a need to divert the acid to have better sweep efficiency during matrix-acidizing treatments. Unlike diversion using mechanical means, diversion using chemicals occurs in the formation. The main technique to achieve that in carbonate reservoirs includes using in-situ-gelled acids based on polymers (Taylor and Nasr-El-Din 2003).

The in-situ-gelled acids contain a polymer that has carboxylic groups. The polymer has no charge at pH = 0. When HCl reacts with the rock, and the pH rises to almost 2, the carboxylic acid groups deprotonate, and the carboxylate groups become negatively charged. A crosslinker of Fe(III) or Zr(VI) is used to form the gel, and as a result the viscosity of the acid increases **(Fig. 9.3).** Once a strong gel is formed, the acid is diverted into low-permeability or highly damaged zones, and the process repeats itself until the whole target zone is covered. There is a need to break this gel. In the in-situ polymer system, a reducer is used to convert Fe(III) to Fe(II); the latter does not form a gel. Another breaker is fluoride ions, which extract Zr from the gel, thus breaking it down.

Laboratory testing and field data highlighted several drawbacks of both systems. For the polymer-based system, Nasr-El-Din et al. (2002c) and Gomaa et al. (2011) showed that the polymer filtered out at the surface of the rock **(Fig. 9.4),** causing loss of injectivity. The crosslinker, Fe(III), precipitates on the surface of wormholes, even in the presence of iron-control agents **(Fig. 9.5)** (Lynn and Nasr-El-Din 2001).

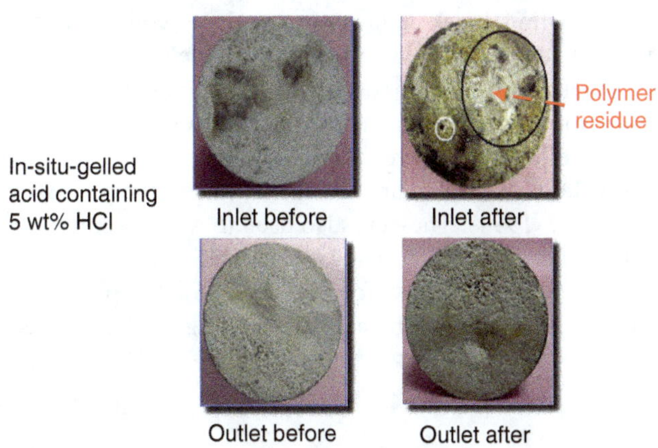


**Fig. 9.4—Accumulation of polymer molecules on the injection face of a core (Nasr-El-Din et al. 2002).**

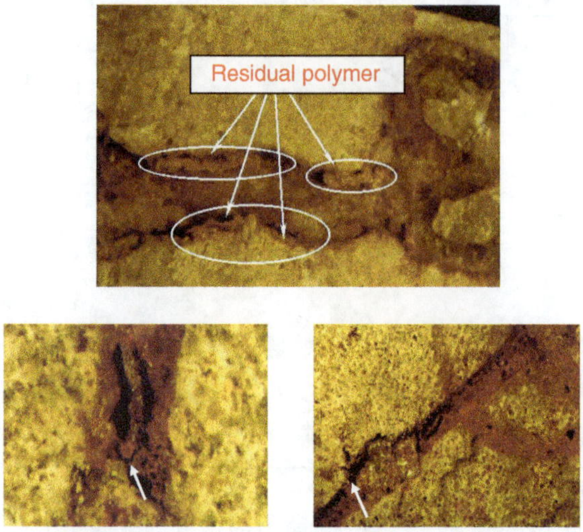


Precipitation in wormhole

Polymer trapped in the wormhole

**Fig. 9.5—Precipitation of iron on the wall of the wormhole created by the acid (Lynn and Nasr-El-Din 2001).**

In sour wells, there are more problems. The crosslinker precipitates in the presence of $H_2S$. The recommended acid concentration for this system is 3–5 wt% HCl (Gomaa and Nasr-El-Din 2010), and the system contains several additives that make its preparation in the field somewhat difficult and time-consuming. Finally, ferric ions are used as a crosslinker in some systems. Ferric iron promotes asphaltene precipitation; therefore, acid compatibility with the native crude oil is recommended before acid treatments.

## 9.3 Surfactants

**9.3.1 General.** Surfactants are chemicals containing an oil-soluble group (lipophilic) and a water-soluble group (hydrophilic). These chemicals have the ability to adsorb at various surfaces and alter their properties. Surfactants are classified according to their ionic nature into four categories: anionic, cationic, nonionic, and amphoteric. Various surfactants from these groups are used in well-acidizing treatments. Some specific applications of these surfactants in well stimulation are discussed here.

Surfactants are used in almost all acidizing treatments to perform one or more of the following functions: water-wet the formation, break emulsions or sludges, reduce surface or interfacial tension, and suspend fine particles. Other functions of surfactants include forming foams, acid diversion using VESs, and preparation of emulsified acids for deep acid penetration in carbonate formations.

The type and concentration of the surfactant used during acid stimulation depends on the characteristics of the treated well: temperature, lithology, salinity, and the function of the surfactant. For oil-producing wells, surfactants are added to prevent the formation of acid/oil sludges (Houchin et al. 1992; Ali et al. 1994). They are also used to water-wet the formation after the acid treatment (McLeod 1984). For gas wells in low-permeability reservoirs, the spent acid can be trapped around the wellbore by the capillary forces, **(Fig. 9.6).** The pressure drop ($\Delta p$) required to mobilize a drop of trapped spent acid can be calculated using the Laplace equation for water-wet formations:

$$\Delta p = 2\sigma\left(\frac{1}{r_1}-\frac{1}{r_2}\right), \qquad (9.5)$$

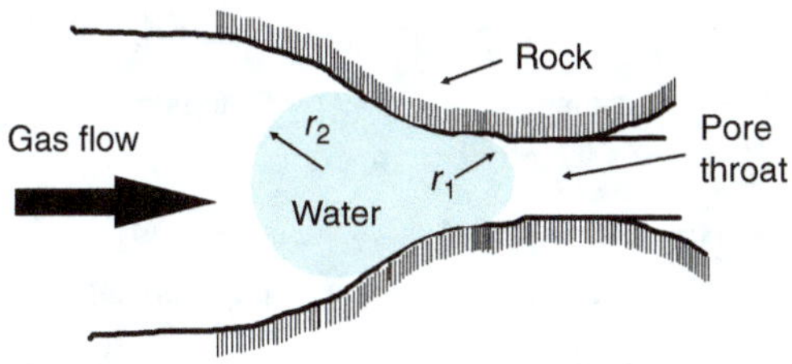


**Fig. 9.6—A drop of water trapped in the formation by the capillary forces.**

where σ is the surface tension and $r_1$ and $r_2$ are the radii of the water droplet. According to Eq. 9.5, the surface tension should be reduced so that the spent-acid droplet can be mobilized and produced from the formation. Lower surface-tension values of the spent acid can be achieved by adding special surfactants to the injected acids. Nonionic surfactants and mutual solvents are commonly added to the acid to lower its surface tension (Paktinat 1991; Gdanski 1995; Dabbousi et al. 1999; Nasr-El-Din et al. 2004; Saneifar et al. 2011).

For water-disposal wells, surfactants are used to lower interfacial tension between the residual oil and treating fluids and, as a result, the trapped oil will be mobilized and the injectivity of the well will be increased (Nasr-El-Din and Al-Taq 1998).

Besides the aforementioned functions, surfactants are used to enhance acid penetration into the formation and improve sweep efficiency during well stimulation. It is well-known that acidizing tight carbonate formations using HCl solutions at low rates will result in surface washout only (Hoefner et al. 1987). One way to enhance acid propagation is to use emulsified acids (Al-Anazi et al. 1998). In this case, a suitable surfactant is used to emulsify the acid (typically 15–28 wt% HCl) and a light-hydrocarbon phase (crude oil, diesel, or xylene). Injection of the acid in this form (acid-in-diesel emulsion) will minimize acid/rock contact, which will result in the formation of deep wormholes (i.e., channels with high permeability). Emulsifying the acid will also increase its viscosity (Al-Mutairi et al. 2008; Sayed el al. 2013) and improve acid distribution in heterogeneous reservoirs (Buijse and van Domelen 2000).

Stimulation of horizontal and multilateral wells presents a very challenging problem. With a very long target zone(s), acid diversion plays a key role in determining the efficiency of the acidizing job. Without a proper means for acid diversion, most of the injected acid will only flow into high-permeability zones. This uneven flow will result in improving the permeability of the already-permeable zone, leaving the tight zones nearly unaffected. One way to improve sweep efficiency during well stimulation is to use foams (Chambers 1994). In this case, a surfactant is used to form viscous foam. The foam will flow into the high-permeability zones, causing the acid to flow into the tight zones.

As mentioned previously, surfactants are included in acid formulations to perform specific tasks. In acid stimulation treatments, surfactants encounter various chemical species. First, the surfactant is mixed with the acid and its additives. Some of these additives are cationic (e.g., corrosion inhibitors and clay stabilizers). Others are anionic or nonionic species. Second, the acid reacts with the formation and releases several cations. HCl reacts with carbonate minerals and, as a result, the spent acid contains calcium, magnesium, and iron. HF-based fluids react with clay minerals and release silicon and aluminum in addition to those elements dissolved by HCl. The presence of these chemicals together with surfactants can cause phase separation of the surfactants. As a result, surfactants will not perform their task as anticipated.

**9.3.2 Anionic Surfactants.** These surfactants carry negative charges. **Fig. 9.7** shows some of these surfactants that are used in the field. One of the main functions of acid stimulation is to remove formation damage, and hence, to enhance the productivity of the oil-producing zones. One of the major problems encountered during stimulation of oil-producing wells is the precipitation of asphaltene particles, which induce sludge formation. These viscous sludges can plug the formation and cause further damage to the well (Moore et al. 1965). They also cause operational problems in the surface facilities following acid-stimulation treatments (Bansal 1993).

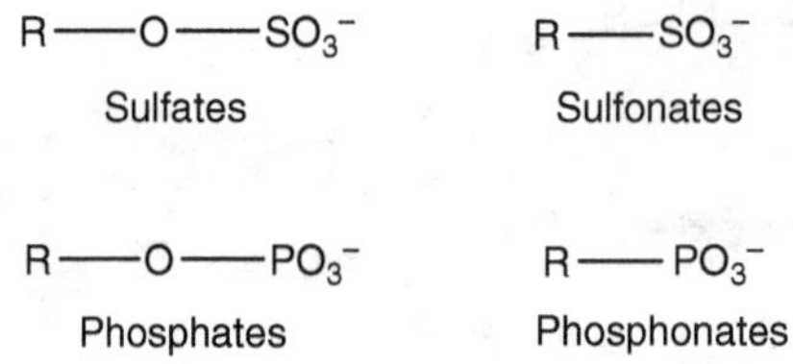


**Fig. 9.7—Examples of anionic surfactants.**

It is important to understand the causes of acid-sludge formation. One mechanism that is frequently cited is the precipitation of asphaltene particles and heavy hydrocarbons when asphaltic crude oils contact strong acids. Asphaltenes are colloidal particles dispersed in crude oils that are composed of condensed aromatic ring structures containing a significant number of hetroatoms such as oxygen, sulfur, and nitrogen (Houchin and Hudson 1986). They exist in a micelle form with natural resins (maltenes) adsorbed on the surface of the asphaltic particles **(Fig. 9.8).** The adsorbed resins stabilize asphaltene micelles (Lichaa and Herrera 1975).

Because asphaltene particles are negatively charged (Moore et al. 1965), the positively charged acid protons ($H^+$) can have a significant effect on the surface charge of asphaltene particles. Neutralization of these particles with $H^+$ allows larger aggregates of asphaltenes to form (Jacobs and Thorne 1986). These aggregates form very rigid emulsions during acidization and also deposit at the pore throats, reducing well productivity **(Fig. 9.9).** Moreover, the problem becomes more complicated if other particulate solids are present. Sand, silt, and metal oxides can also accumulate at the oil/acid interface and stabilize acid-in-oil emulsions (Moore et al. 1965).

Crude oils have different emulsion stability, viscosity, and density. Houchin et al. (1990) studied the effect of acids on 231 crude oils from 27 formations. Depending on oil characteristics, acids can form sludges or rigid film emulsions. Acid-induced sludge occurs primarily in crudes at ≥ 27 °API and asphaltene content ≤ 3 wt%. Houchin et al. (1990) found that rigid film emulsions form when crude oils at ≤ 22 °API and asphaltene content ≥ 4 wt% come into contact with strong acids. Strassner (1968) found that rigid film emulsions occurred when crude containing high asphaltenes contacts an aqueous phase of pH < 6.

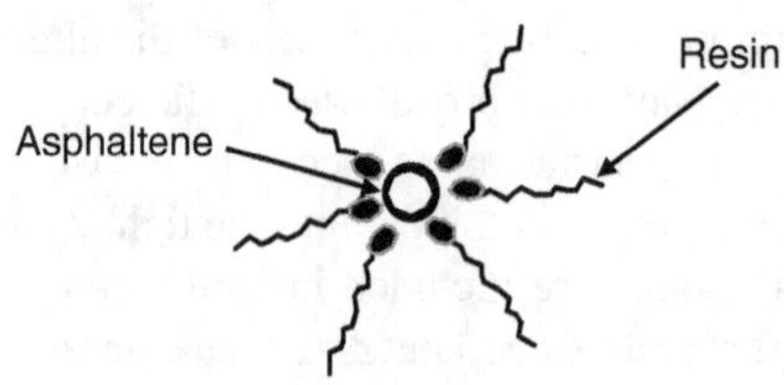


**Fig. 9.8—Asphaltene dispersed in oil.**

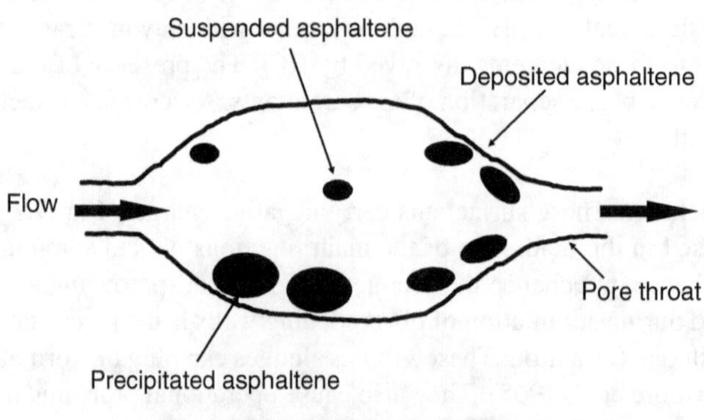


**Fig. 9.9—Asphaltene precipitation.**

There are several factors that can enhance acid sludging. Increasing HCl strength will aggravate sludging of asphaltic crudes, especially at high temperatures. Therefore, asphaltic crudes should not be directly exposed to 28 wt% HCl. Iron contamination, especially ferric ion, also plays an important role in sludge formation (Vinson 1996). The influence of iron on sludge formation can be explained by flocculation of asphaltene particles, which follows charge neutralization. The latter occurs because of the coordination of iron with porphyrin, pyrroles, pyridines, or possibly phenolic hydroxyl groups present in the crude oil. Iron can also catalyze the polymerization processes of heavy hydrocarbons at high temperatures. Iron in acid can originate by dissolution of corrosion products and some iron-rich minerals (chlorite, ankerite, siderite, and hematite).

Full-strength regular mud acid (12 wt% HCl + 3 wt% HF) appears to cause more-severe sludging than HCl at the same acid concentration. Other factors that control sludge formation include reservoir temperature, viscosity of the native oil, and acid additives (e.g., corrosion inhibitors) (Jacobs 1989).

On the basis of the preceding discussion, every effort should be made to avoid the formation of acid/oil sludges. Antisludge agents, surfactants, and mutual solvents are usually added to the injected acid to minimize the impact of acid/oil sludge (Dunlap and Houchin 1990). These agents disperse the asphaltene particles and prevent their precipitation.

The American Petroleum Institute (API) recommends conducting an acid-sludge test (*API RP 42* 1977) before performing acidizing jobs in the field (Rietjens 1997). These tests are used to determine the type and concentration of the surfactant (nonionic or anionic) needed to break acid-in-oil emulsions in a reasonable period of time. It is very important to perform the tests using live and spent acids at reservoir temperature. Spent acid should be prepared in the laboratory using formation rock. The live and spent acids should include acid additives as per field formula. The crude-oil sample should be fresh, and free from water and oilfield chemicals (e.g., scale inhibitors, demulsifiers).

Besides adding a suitable antisludge agent to the injected acid, the concentration of ferric ion in the injected acid should be reduced. This can be achieved by using properly cleaned mixing tanks and pickling the coiled tubing used to place the acid. Moreover, iron-reducing agents should be added to the acid when sweet wells are acidized (Taylor et al. 1999). Another method to reduce sludge formation is to minimize acid contact with the native crude oil. One way to achieve this goal is to use solvent preflush (xylene) that can act as a barrier to separate sensitive crudes and the acid (Houchin et al. 1990; Bansal 1993). A second way is to use acid emulsified in solvent (Figueroa-Ortiz et al. 1996; Abdel Fatah and Nasr-El-Din 2010). The solvent in this emulsion will minimize acid/oil contact.

**9.3.3 Cationic Surfactants.** These surfactants carry positive charges. **Fig. 9.10** shows some of these surfactants that are used in the field. HCl is commonly used to stimulate carbonate (calcite and dolomite) reservoirs. The reaction rate of HCl with calcite is very fast. In conventional stimulation when 15 wt% HCl is used at low flow rates, the acid reacts with carbonate rocks and causes a face dissolution or surface washout. This means that the acid is spent on the formation surface and does not penetrate the damaged zone to enhance the formation permeability. Face dissolution leads to consumption of large volumes of acid. One way to overcome this problem is to use acid-in-diesel emulsions. This emulsion consists of diesel or xylene (continuous phase), HCl acid (dispersed phase), and an emulsifier (a cationic surfactant). This type of surfactant is the preferred one. This is because it has positive charges and the carbonate rocks carry positive charges. This will reduce surfactant loss because of

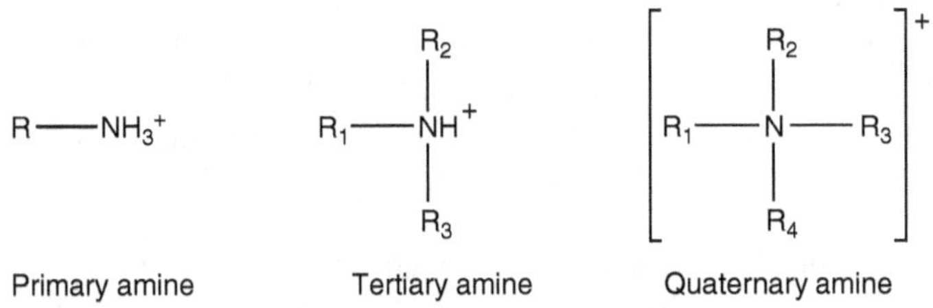


**Fig. 9.10—Examples of cationic surfactants.**

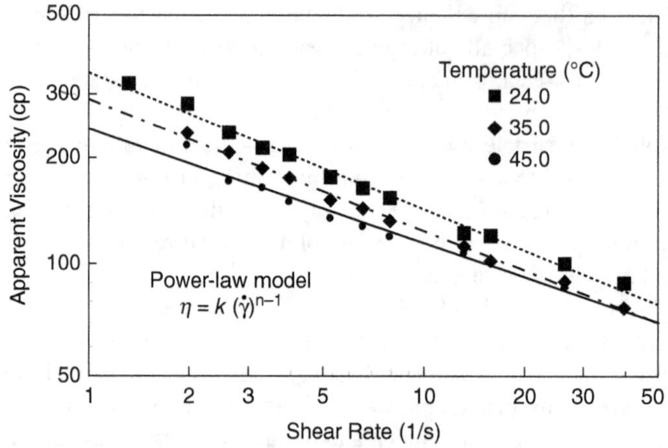


**Fig. 9.11—Apparent viscosity of emulsified acid.**

adsorption on the rock surface and ensure deep acid penetration. Diesel acts as a diffusion barrier between the acid and the rock, slowing the reaction rate of the acid with the carbonate rock. This gives the acid the ability to penetrate deeper into the formation by creating wormholes and enhancing well productivity (Hoefner and Fogler 1985; Sayed et al. 2013).

Sayed et al. (2013) measured the apparent viscosity of the acid-in-diesel emulsion as a function of shear rate. **Fig. 9.11** shows that the apparent viscosity decreased as the shear rate was increased. This result indicates that the acid-in-diesel emulsion is a non-Newtonian fluid (shear-thinning behavior). Crowe and Miller (1974) and Krawietz and Rael (1996) reported a similar behavior. Al-Mutairi et al. (2008) discussed the effect of acid-droplet size on the apparent viscosity of the emulsified acid.

Surfactants are used to form emulsified acid, where acid is the dispersed (or internal) phase. Pumping the acid in this form reduces acid contact with the rock. This enables the acid to form deep wormholes. It also reduces acid contact with the native crude. This will reduce sludge precipitation and associated problems.

The apparent viscosity of this acid is relatively high, which will enhance sweep efficiency during acidizing (Buijse and van Domelen 2000). The acid can be pumped using coiled tubing or can be bullheaded into the target zones. The fact that the acid does not come in contact with the well tubulars minimizes corrosion, which will reduce iron concentration in the acid. Other advantages of this acid are that only small amounts of corrosion inhibitors and iron-control agents are usually needed.

There are specific properties required for using emulsified acids. The emulsified acid should be stable for a long period of time at room temperature, in case of operational problems, and should be stable until it reaches the target zone. Obviously, a longer period of time will be needed when acidizing horizontal wells. Once inside the formation, the acid should be soaked for a long period of time until it completely reacts with the formation. It should be mentioned that the soaking time should be longer when stimulating injectors or water-disposal wells because of their lower temperature.

Because the acid is injected in an emulsified form, extreme care should be taken in the design of the preflush and post-flushes. The presence of demulsifiers or mutual solvent may result in the breaking of the emulsified acid.

Emulsified acid was successfully applied in oil-producing wells, water injectors, water-disposal wells and in acid fracturing of deep gas wells. Its use is recommended when the permeability of the target zone is less than 50 md. Emulsified acid should be used with care when stimulating weak carbonate formations (e.g., chalk) in which there is a potential for producing fine particles. These particles, acid insoluble, can cause loss of well performance.

**9.3.4 Nonionic Surfactants.** These surfactants carry no charges. **Fig. 9.12** shows some of these surfactants that are used in the field. Nonionic surfactants have been added to various acid formulations (Al-Ghamdi and Nasr-El-Din 1997). These surfactants are chosen because they maintain low

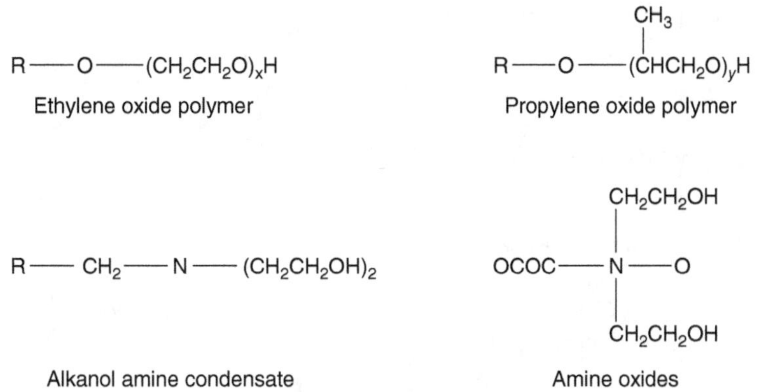


**Fig. 9.12—Examples of nonionic surfactants.**

interfacial tension between the acid and oil. They are inexpensive and can be mixed with other types of surfactants to enhance their properties (Denoyel et al. 1993). However, in common with other types of surfactants, they should be carefully tested before field applications. Nonionic surfactants can separate out of solution under certain conditions, as will be explained next.

Nonionic surfactants owe their solubility in water to the hydration of the polyethylene oxide chains. The solubility of these surfactants in water increases as the length of the hydrophilic part of the surfactant increases. As the temperature of a nonionic surfactant solution is increased, hydrogen bonds break (Hinze and Pramauro 1993) and the solubility of the surfactant in water diminishes. At a certain temperature, known as the cloudpoint, the surfactant molecules separate out of solution, causing it to become cloudy. Ultimately, the surfactant solution separates into two immiscible phases: a surfactant-rich phase and a surfactant-lean phase. In oilfield operations, the cloudpoint of nonionic surfactants is an important parameter that determines the efficiency of such operations. For example, in stimulation, or during drilling operations, separation of surfactant from solution in the injected fluid can plug the formation, and hence the productivity or injectivity of the well will diminish (Jachnik and Green 1995).

Nasr-El-Din and Al-Ghamdi (1997) measured the cloudpoint of ethoxylated octylphenyl alcohols known as the Triton-X series. These surfactants have the general chemical formula of $R'–C_6H_4–(OC_2H_4)_x–OH$, where $R'$ is a branched octyl group and $x$ is the number of ethylene oxide groups. Four surfactants having various numbers of ethylene oxide groups were examined.

The cloudpoint of various surfactant solutions was determined visually by noting the temperature at which the continuously heated solution suddenly became cloudy. In most cases, the surfactant solution under investigation was heated starting from room temperature. The repeatability of the cloudpoint measurements using this method was ± 0.5°C.

To examine the effect of acids on the cloudpoint, various acids were added to several nonionic surfactants at acid concentrations from 0 to15 wt%. The cloudpoint of TX-45 (five ethylene oxide groups) solutions is less than 0°C, and the addition of HCl up to 15 wt% did not raise its cloudpoint above room temperature. On the other hand, the cloudpoint of TX-405 (number of ethylene oxide groups is 40) solutions was higher than 100°C at all acid concentrations examined. Only TX-114 and TX-100 (number of ethylene oxide groups is 7.5 and 10, respectively) have cloudpoints that can be measured with the method employed by Nasr-El-Din and Al-Ghamdi (1997).

**Fig. 9.13** shows the effect of HCl and acetic, citric, and formic acids on the cloudpoint of TX-100. The cloudpoint of neutral TX-100 solutions at a surfactant concentration of 2 wt% is 64.5°C. At HCl concentrations greater than 1 wt%, the cloudpoint monotonically increased with HCl concentration, and exceeded 100°C at an acid concentration of nearly 10 wt%. A similar trend was observed by Travalloni-Louvisse and Gonzalez (1988).

The effect of HCl acid on the cloudpoints of TX-100 can be explained as follows: hydrogen ion causes salting-in of nonionic surfactants (Schott et al. 1984), whereas the chloride ion induces a salting-out

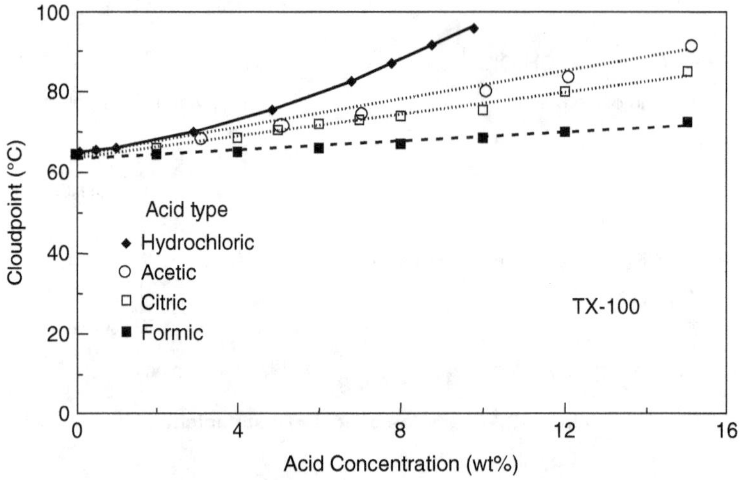


**Fig. 9.13—Effect of acids on the cloudpoint.**

effect (Travalloni-Louvisse and Gonzalez 1988). It appears from the results shown in Fig. 9.13 that the salting-in effect dominates, and becomes important at higher acid concentrations. Obviously, the concentration of hydrogen ions in the case of HCl is much higher than that obtained with the weaker organic acids. Consequently, the cloudpoint is higher with HCl.

Nonionic surfactants have several advantages over other classes of surfactants (Al-Ghamdi and Nasr-El-Din 1997). This explains their extensive use in the area of acid stimulation. However, these surfactants can separate out of solution at temperatures greater than their cloudpoint. Because phase separation can cause formation damage (Jachnik and Green 1995), these surfactants should be used in the field at temperatures below their cloudpoint. Several additives may depress the cloudpoint of nonionic surfactants. Therefore, the cloudpoint of these surfactants should be measured in the presence of all expected additives as per the field formula. Nonionic surfactants should not be used in high-pH fluids (e.g., fracturing using high-pH borate gel and various chemical EOR methods that include an alkali).

**9.3.5 Amphoteric Surfactants.** These surfactants carry charges that depend on the pH of the solution. **Fig. 9.14** shows some of these surfactants that are used in the field. HCl is a strong acid, which vigorously reacts with calcite. When acid is used, it follows the path of least resistance. If there are permeability variations over the target zone, or the zone contains calcite and dolomite, the acid may flow and react with calcite. There is a need to divert the acid to have better sweep efficiency during matrix-acidizing treatments. Unlike the diversion using mechanical means, diversion using chemicals occurs in the formation. The main technique to achieve that in carbonate reservoirs is to use in-situ-gelled acids by using either a polymer or a viscoelastic surfactant (VES). Although the chemistry and gelation mechanisms are completely different, there are similarities.

Amphoteric VESs used for acid diversion contain carboxylic (Fu and Chang 2006; Nasr-El-Din et al. 2008a), amine oxide (Nasr-El-Din et al. 2006a, 2006b; Li et al. 2010), or sulfonated groups Cationic VES were successfully used in the field (Nasr-El-Din et al. 2009b). They have a positive

$RNH—(CH_2)_xSO_3H$

Amine sulfonates

$RNH—(CH_2)_yOPO_3H$

Amine phosphates

**Fig. 9.14—Examples of amphoteric surfactants.**

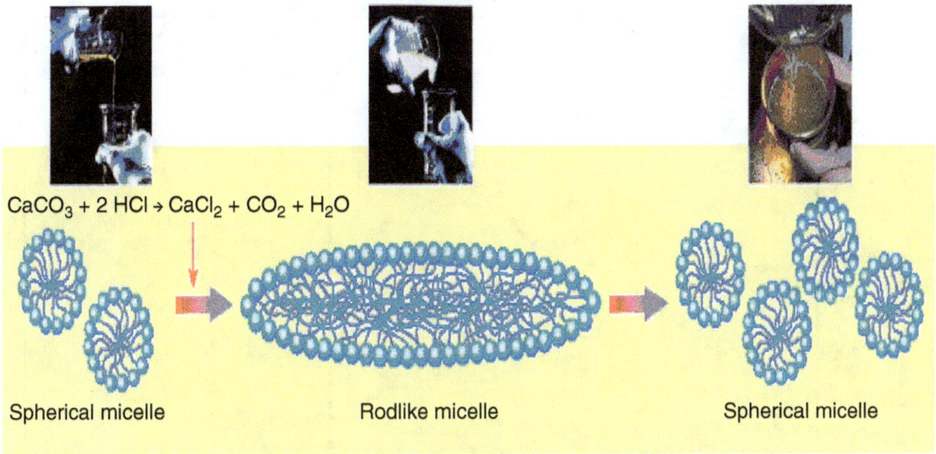


**Fig. 9.15—Gelling of VES.**

charge at a pH of zero. When HCl reacts with the rock and the pH rises to almost 2, the carboxylic groups deprotonate and become negatively charged. Calcium or magnesium can crosslink the long micelles and form a gel **(Fig. 9.15).** Once a strong gel is formed, the acid is diverted into low-permeability or highly damaged zones, and the process repeats itself until the whole target zone is covered. Obviously, there is a need to break this gel once the treatment is finished. VES gel can be broken by using mutual solvent, native crude oil, or condensate in gas wells. Internal breakers are also available for some VESs (Huang and Crews 2009; 2010).

The VES-based system was developed to address problems encountered when the polymer-based system was used in the field. To overcome loss of injectivity, low-molecular-weight VES was used, thus eliminating the loss-of-injectivity problems. There is no need to use Fe(III) or Zr as a crosslinker. The system is simple and easy to mix in the field. The recommended acid concentration for this system is 20 wt% HCl. Even with these advantages, VES-based systems have some drawbacks. For example, if the acid is contaminated with Fe(III), then VES will react with iron and, as a result, an Fe-VES complex can precipitate (Al-Nakhli et al. 2008; Shu et al. 2015). VES can also hydrolyze in acidic conditions (Fu and Chang 2006; Yu et al. 2012, He et al. 2013) and loses its effect. The use of high concentrations of VES can cause formation damage. VES is a surfactant, and there is always a need to run compatibility tests between the crude oil and the acid that will be used in the field. Finally, VES in general is expensive; one way to reduce the concentration of VES is to use nanoparticles (Gurluk et al. 2013).

## 9.4 Clay Stabilizers

Sandstone formations contain varying amounts of clay minerals and fine particles, which are attached to sand grains by weak van der Waals attraction forces. Naturally occurring clay minerals are generally present in an agglomerated or flocculated state. It is well-known that injection of low-salinity or high-pH fluids will disturb the equilibrium state between clay particles and the matrix sand grains. As a result, these fine particles will disperse, block pore throats, and cause severe loss of permeability **(Fig. 9.16).**

Many investigators have examined factors affecting permeability decline because of fines migration and clay swelling, including salinity changes (Gray and Rex 1966; Mungan 1968; Khilar and Fogler 1983; Nasr-El-Din et al. 1991), pH (Kia et al. 1987; El-Monier and Nasr-El-Din 2011b), flow rate (Gruesbeck and Collins 1982) and temperature (Musharova et al. 2012). These studies showed that clay particles are released when clays come into contact with low-salinity fluids (i.e., salinity less than that of the formation water). The dispersed particles remain suspended in fresh water, and are carried with the flowing fluid until they are captured at pore throats, thereby decreasing the permeability. The effect of heavy brines used in completion operations on permeability changes was also investigated (Azari and Leimkuhler 1988; Patton and Phelan 1985; Evans and Ali 1997). The effects of acid on clay minerals were investigated by Simon and Anderson (1990).

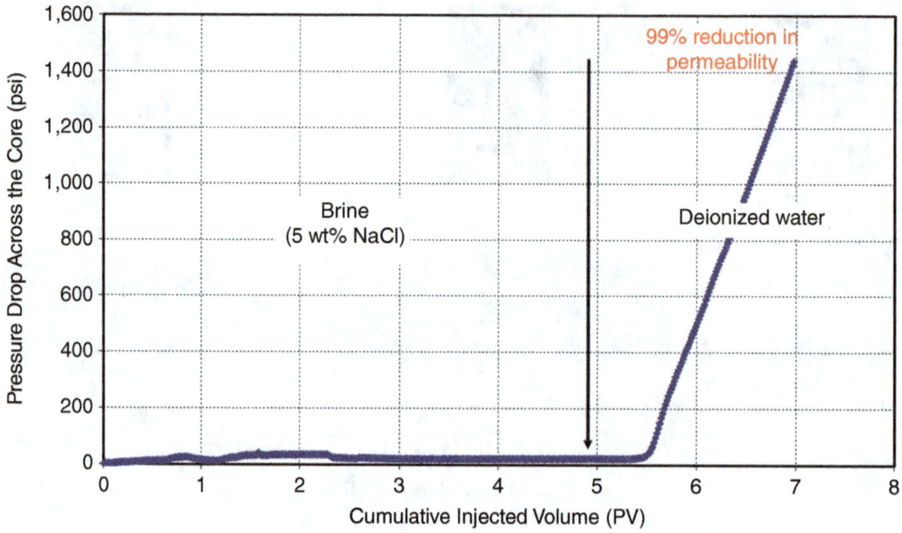


**Fig. 9.16—Fines migration because of freshwater injection.**

Clay minerals are negatively charged at pH values greater than 5 (Zhou et al. 1995). The negative charges on the clay minerals make them sensitive to fluids and provide the mechanism by which clay stabilizers operate. Several chemical treatments are available to mitigate formation damage because of fines migration and clay swelling. A thorough review of these treatments was given by Crowe (1991) and Zhou et al. (1995). According to their work, clay stabilizers can be divided into the following classes: simple inorganic salts, cationic inorganic polymers, cationic organic polymers, anionic organic polymers, and nonionic organic polymers. Zhou et al. (1995) discussed the advantages and disadvantages of each class of these clay stabilizers.

Clay stabilizers are commonly used in acidizing and fracturing treatments. They should be thermally stable under reservoir pressure and temperature. They must be chemically stable in acids used for stimulation, and in the high-pH environment encountered in fracturing treatments (in case of using high-pH borate gels).

Cationic organic polymers have functional groups of positive polarity. They adsorb onto the clay surfaces at multiple sites (McLaughlin et al. 1976). The adsorbed polymer with multiple attachments to the rock surface will resist exchange with other cations in solution. This will increase the lifetime of the clay-stabilization treatment. Cationic organic polymers have been shown to be excellent clay stabilizers (Behenna 1994; Borchardt 1989; Hill 1982; Himes et al. 1991; Zaiton and Berton 1992, 1996; Behenna 1994; Masliyah and Bhattacharjee 2006).

For clay-stabilizer applications, a suitable cationic charge, a low molecular weight, and the use of nonwetting chemicals should be maintained (Himes et al. 1991). Advances in the treatment of clay-bearing formations led to the development of numerous clay stabilizers. Temporary clay stabilizers include potassium chloride, sodium chloride, and ammonium chloride. The primary concerns of these salts are their temporary effect, the difficulties in handling them, the mixing of large quantities of salts, the disposal of unused material, and the produced salt solutions (Patel 2009). Veley (1969) and Zhou et al. (1995) investigated the effects of highly charged cations on formation-water sensitivity. Their laboratory work concluded that hydrolyzable metal ions, such as zirconium oxychloride, combat water sensitivity and they should be at least partially hydrolyzed to stabilize clays.

Ross (1967; 1968) and Reed (1972) described the use of hydroxylaluminum as a treatment for fines migration. They concluded that a hydroxylaluminum solution is a relatively inexpensive and commercially available chemical that effectively inhibits clay damage in normally water-sensitive formations. It adsorbs very tightly on clay surfaces and stabilizes them by decreasing the dispersive forces and eliminating structural expansion. According to Hesterberg and Reed (1991), hydroxylaluminum treatment and polyquaternary amine had equal volumetric treatment efficiencies on Berea sandstone cores. Their results highlighted the importance of clay type and content on determining the amount of clay stabilizer needed to mitigate fines migration. Coppel et al. (1973) presented a field study with a hydroxylaluminum treatment. They concluded that some wells experienced a rapid production decline after acid stimulations. Treatment of these wells by reacidizing and then applying a hydroxylaluminum solution reduced the rate of decline markedly. Himes et al. (1991) showed that cationic polymers are ineffective in microdarcy to low-millidarcy sandstones (low-permeability formations). Similar results were reported by Nasr-El-Din et al. (1999). To address this problem, Weaver et al. (2011) developed ultralow-molecular-weight cationic polymers.

Nguyen et al. (2007) described the use of an ultrathin tackifying agent (UTTA) for stabilizing fines at their sources in high-rate production or injecting wells. Berry et al. (2008) investigated the performance of an ionic liquid as a clay stabilizer in comparison with the industry-standard clay stabilizers. Different tests were conducted, and their results showed that ionic liquids were effective as a KCl substitute and a clay stabilizer in preventing clay swelling. Patel (2009) developed a quaternary amine compound in shale inhibition with an improved environmental profile. He also discussed several environmental issues for previous inhibitors such as oligomeric cationic amines, polymeric quaternary amines, and monocationic amines. Huang et al. (2008) and Belcher et al. (2010) introduced the use of nanotechnology and showed that the nanoparticles had the ability to retain fine particles that consist of clays, nonclays, and charged and noncharged particles.

El-Monier et al. (2013) introduced a new clay stabilizer that is based on an inorganic aluminum-zirconium-based compound. The performance of their stabilizer was compared with tetramethyl ammonium chloride (TMAC) and choline chloride by El-Monier and Nasr-El-Din (2011a). This work showed that the Al-Zr stabilizer is more effective than these two products. The use of this stabilizer with high-pH fracture fluids was also examined by El-Monier and Nasr-El-Din (2011b), as was the mechanism of its fines stabilization (El-Monier and Nasr-El-Din 2011c).

**9.4.1 Mutual Solvents.** Mutual solvents are multifunctional, nonionic agents soluble in oil, water, and acids. They contain strong ether and alcohol groups, which provide a wide range of solvent properties. Ethylene glycol monobutyl ether (EGMBE) is one of the mutual solvents that is commonly used in the oil industry. It is used to achieve several tasks, including reducing surface tension (discussed later) and enhancing water-wetting characteristics of formation rocks after acidizing.

Mutual solvents are commonly included in acid formulations (Hall 1975; King and Lee 1988). They are used as water-wetting agents, demulsifiers, and surface/interfacial-tension reducers.

**9.4.2 Alcohols.** Short-chain alcohols are added to the acid for several reasons: to remove water blockages, enhance recovery of spent acid, and retard acid reactivity (Ali and Hinkel 2000). Alcohols most used in the oil and gas industry include methanol and isopropanol. Water blocks are severe in formations with permeability less than 120 md. The use of alcohol will reduce the capillary forces, thus allowing the trapped fluid to flow into the wellbore. To be effective, alcohol should be added at concentrations greater than 20 wt%.

The use of alcohol with acids has several disadvantages. The flash point of short-chain alcohols is low. They are expensive and incompatible with some crude oils. In addition, alcohols react with HCl as follows:

$$CH_3OH + HCl ===== CH_3Cl + H_2O \qquad (9.6)$$

Organic chlorides (e.g., $CH_3Cl$) are not desirable because they can poison catalysts used in the refinery. This reaction becomes significant at temperatures greater than 175°F (Ali and Hinkel 2000; Kalfayan 2000). Therefore, it is not advisable to use alcohol with the acid-in-oil wells. Methanol can be used in gas wells up to 20% concentration.

## 9.5 Impact of Additives on the Properties of Acids and Their Reactions With the Rock

Additives are used to achieve some specific objectives during acid treatments. For example, corrosion inhibitors are used to protect surface equipment and downhole completion. However, these inhibitors can affect some of the physical properties of the acid. It is important to realize that additives will change the chemical and physical properties of the acid used. HCl is widely used in carbonate reservoirs and in some sandstone reservoirs. The following two sections will discuss how additives affect the properties of HCl and the way it reacts with the rock.

## 9.6 Impact of Additives on Key Physical Properties of HCl

**9.6.1 General.** Additives can impact two important physical properties of HCl: surface tension and viscosity. There are two main types of additives. The first type includes polymers and VESs. The addition of a polymer or a VES will increase the viscosity of the acid and the acid will behave as a shear-thinning non-Newtonian fluid. The second type includes additives with surface activities. These additives include short-chain alcohols, wetting agents, foaming agents, and emulsifiers. These additives will reduce the surface tension of the acid. **Fig. 9.17** is a schematic that shows that surfactant is the most effective way to reduce surface tension at low-additive concentrations. Low surface tension is needed because it helps the flowback of spent acid to depart from the formation. This is especially true when we deal with tight formations.

Nasr-El-Din (2005) measured surface tension of HCl as a function of temperature, acid concentration, and pressure. They found that the surface tension of HCl decreases at high temperatures **(Fig. 9.18).** Unlike HCl, organic acids reduce surface tension gradually **(Fig. 9.19).** Corrosion inhibitors, surfactants, and mutual solvents reduced surface tension significantly **(Figs. 9.20 through 9.22).** At temperatures greater than 200°F, there is a need to use intensifiers that include formic acid because it reduces surface tension of HCl solutions **(Fig. 9.23).**

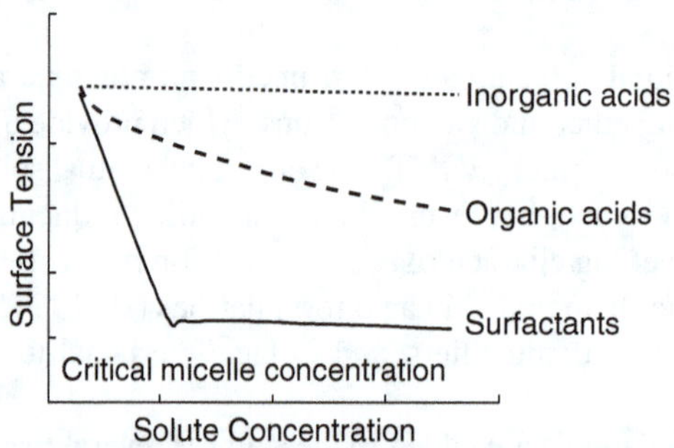


**Fig. 9.17—Surface tension of various aqueous solutions.**

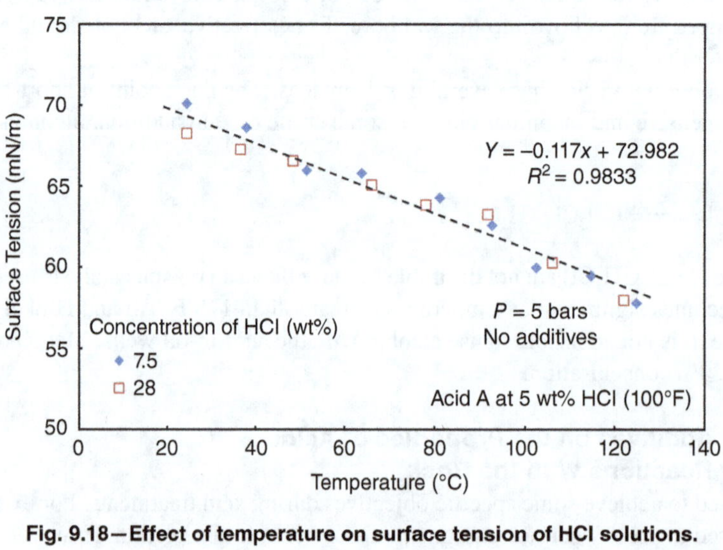


**Fig. 9.18—Effect of temperature on surface tension of HCl solutions.**

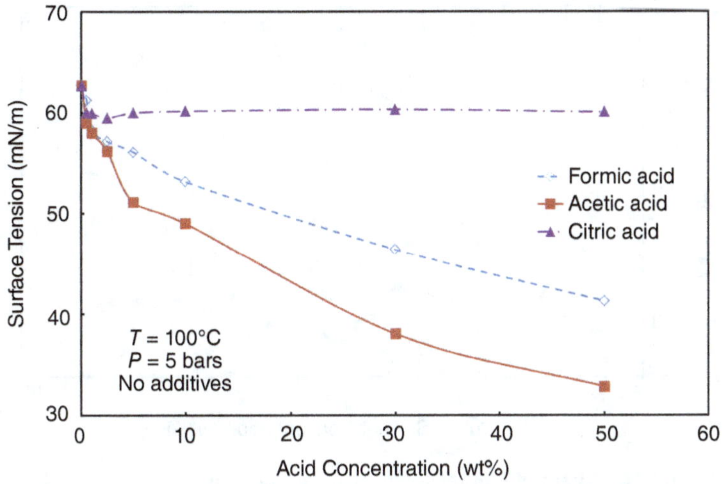


**Fig. 9.19—Surface tension of organic acids.**

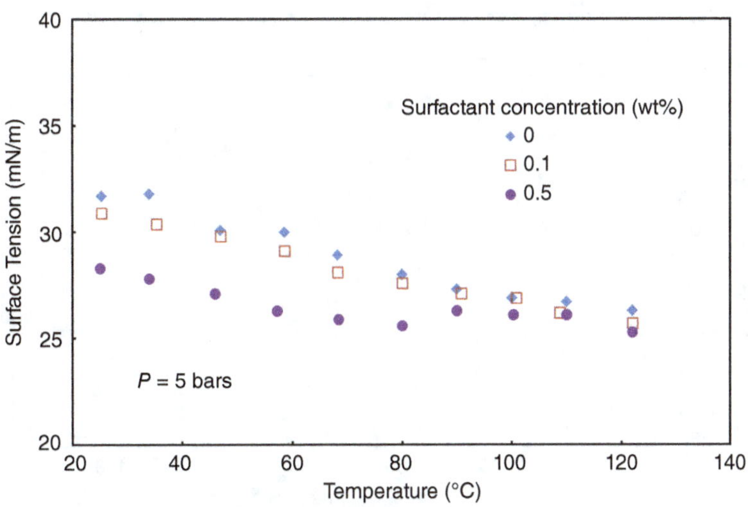


**Fig. 9.20—Effect of nonionic surfactant on surface tension of HCl.**

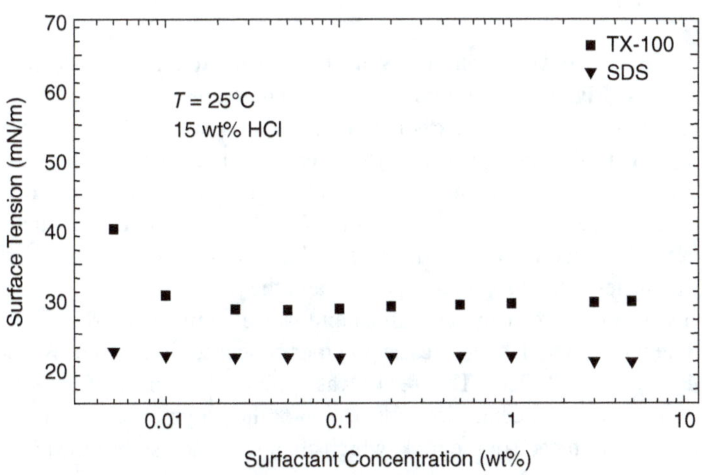


**Fig. 9.21—Effect of anionic surfactant on surface tension of HCl solution.**

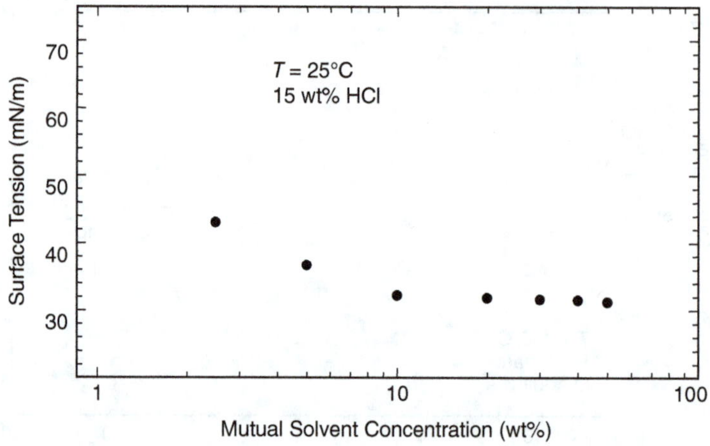


**Fig. 9.22—Effect of mutual solvent on surface tension of 15 wt% HCl.**

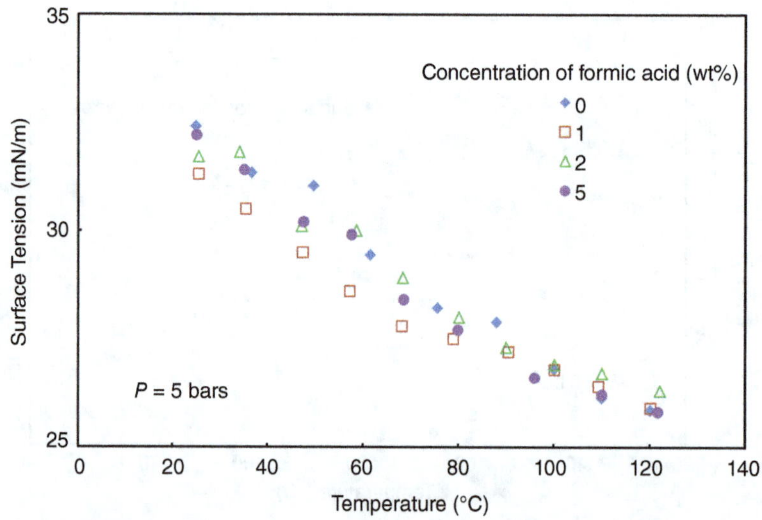


**Fig. 9.23—Effect of intensifier on surface tension of 15 wt% HCl.**

**9.6.2 Impact of Additives on Acid Reaction With the Rock.** Acid additives can also affect the way HCl reacts with the rock **(Fig. 9.24)** (Taylor et al. 2004). There are two mechanisms in which an additive may impact the acid reaction with carbonate rocks. The first occurs when an additive increases the viscosity of the acid. Typical additives used to increase acid viscosity are polymers, **Fig. 9.25,** and VESs, **Fig. 9.26.** In both cases, increasing the viscosity of the acid reduces the diffusion coefficient of $H^+$ (Nasr-El-Din et al. 2008b, 2009a; Rabie et al. 2012). As a result, the reaction rate decreases **(Figs. 9.27 and 9.28).** The second occurs when the additive adsorbs on the rock surface. In this case, the additive forms a barrier, which separates the acid and the rock.

The effect of polymers on acid reaction with calcite was examined by Nasr-El-Din et al. (2008b). The acid reaction rate with the rock decreases as polymer concentration increases from 0.5 to 1.5 vol% (Fig. 9.25) (Nasr-El-Din et al. 2008b). This result was explained in terms of the viscosity of the acid, which increased with polymer concentration, adsorption of the polymer on the surface of the rock, and the flow pattern close to the surface of the rock, which changed as a result of the presence of polymers.

VESs increase the viscosity of HCl (Fig. 9.26). They also reduced the acid reaction rate with carbonates (Fig. 9.27) (Nasr-El-Din et al. 2009). This trend was found to be a function of acid viscosity and

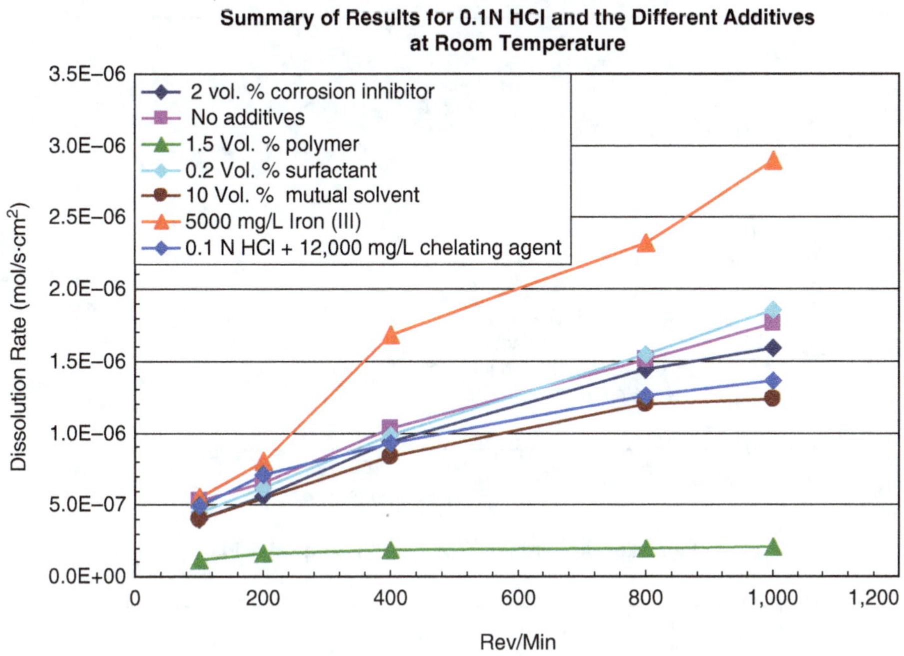


**Fig. 9.24—Effect of additives on acid reaction rate.**

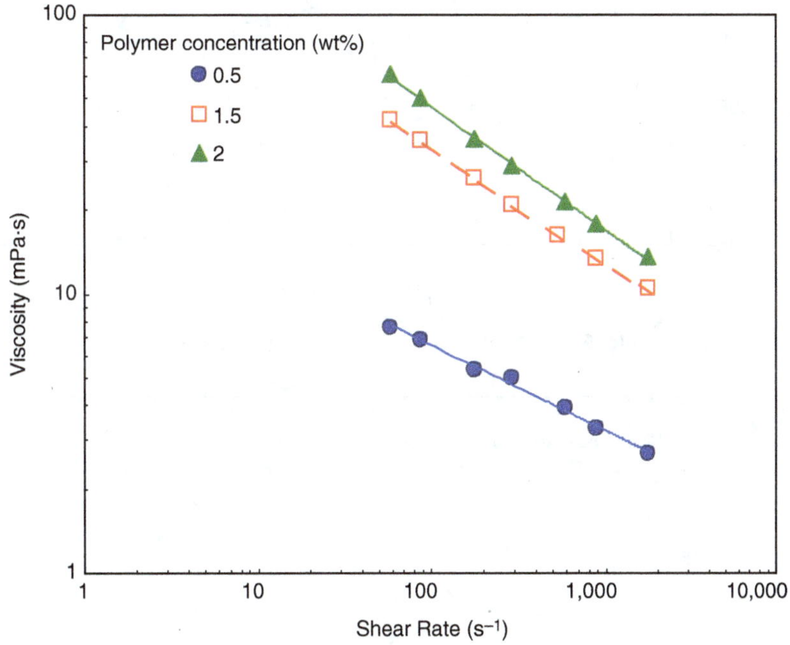


**Fig. 9.25—Apparent viscosity of gelled acids.**

temperature. Temperature affects this reaction in two ways: It increases the reaction rate and changes the viscosity of the VES solution.

Several additives can form a layer on the surface of the rock. It is well-known that a corrosion inhibitor adsorbs and forms a layer on the surface of rocks. Surfactants can also adsorb on the surface. This layer will isolate the rock and reduce the acid reaction rate.

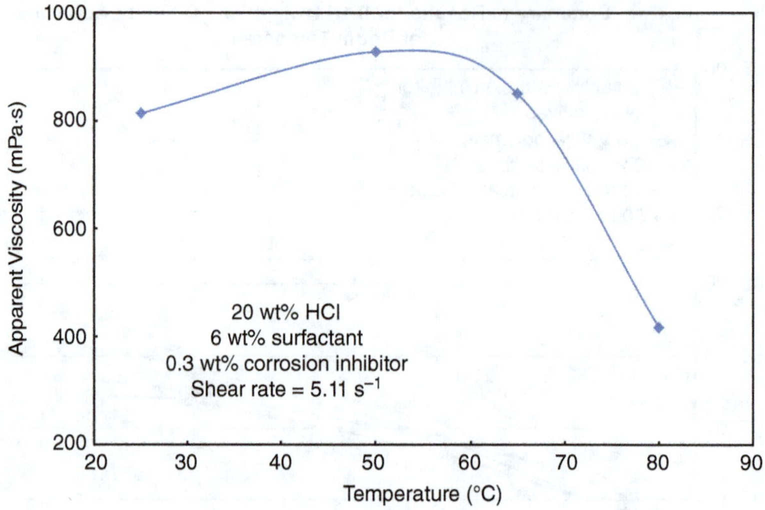


**Fig. 9.26—Apparent viscosity of VES-based acids.**

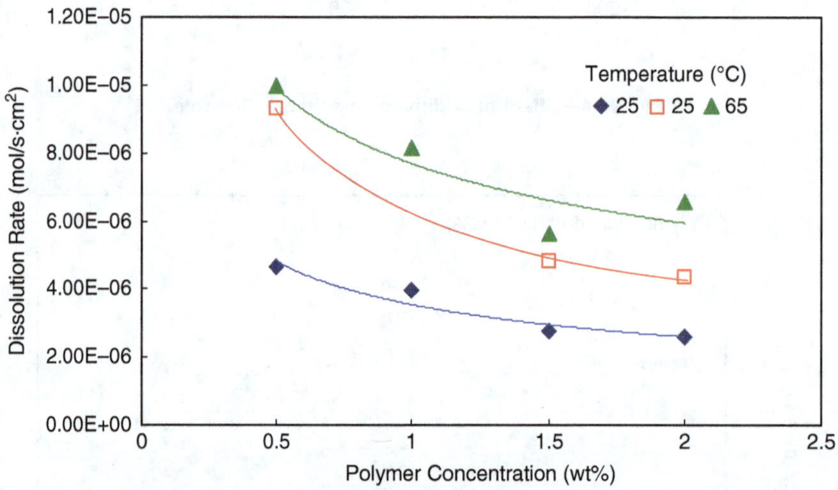


**Fig. 9.27—Calcite dissolution using gelled acids.**

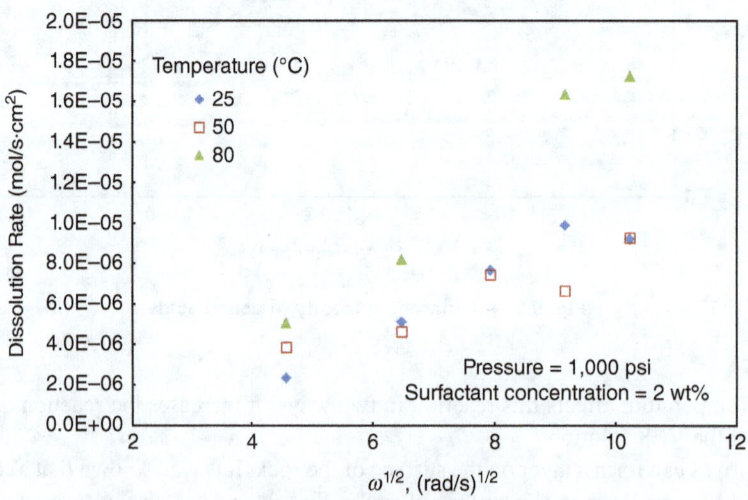


**Fig. 9.28—Effect of VES on acid reaction rate with calcite.**

Obviously, several additives are used in acid matrix treatments. These additives will increase viscosity and may form a layer on the surface of the rock. Either way, additives that increase viscosity or adsorb on the surface will tend to reduce the acid reaction with the rock. The effect is significant and should be considered in the modeling of acid flow in porous media. Using rate equations based on regular acid without additives will result in erroneous results. It should be mentioned that the impact of this layer can be minimized or eliminated by using mutual solvents or similar chemicals in the preflushes, the post-flushes, or both.

**9.6.3 Potential Formation Damage Because of Acid Additives.** Improper use of additives can cause operational problems and formation damage in some cases. Additives should be used only as needed. Compatibility of additives with each other and with downhole fluids should be determined before field application. Some additive interactions can cause a reduced performance of another additive. For example, the use of antisludge agents (mostly, anionic species) can reduce the efficiency of corrosion inhibitors (carry positive charges). Another example occurs when mutual acid is added to the acid. Mutual solvent (e.g., EGMBE) and short-chain alcohols (methanol, ethanol, and isopropyl) increase the solubility of corrosion inhibitors in acid, and thus reduce their effectiveness. Mutual solvents reduce the ability of VES to increase the viscosity of the acid. As a matter of fact, mutual solvents are used to break the gel formed by VESs.

Additive interactions present serious concerns because of the presence of $H_2S$. $H_2S$ reacts with corrosion-inhibitor intensifiers that contain heavy metal. Some of these intensifiers include copper iodide and antimony chloride. If there is a mix of $H_2S$ and these intensifiers, then copper sulfide and antimony sulfide will precipitate. This can cause operational problems and formation damage if these precipitate inside the formation. For in-situ-gelled acid that is based on ferric chloride, $H_2S$ will precipitate iron sulfide, even if an iron-control agent is included in the acid formulation. It was thought that iron-control agents will prevent iron precipitation; however, this is not true in the presence of $H_2S$. To overcome this problem, it was suggested to use $H_2S$ scavengers (aldehyde-based). However, these scavengers react with the polymer used (copolymer of polyacrylamide), and as a result, the performance of the polymer and scavenger diminishes and the product can cause formation damage.

## 9.7 Compatibility Testing

An important step needed before acidizing in the field is to run compatibility tests. Acid formulations contain several surface-active species, which can interact with the native crude oil and create extremely viscous emulsions. One of these tests is the *API Recommended Practices for Laboratory Testing of Surface Active Agents for Well Stimulation* (*RP 42* 1977). For sandstone formations, Ali et al. (1994) modified this method to make it more representative to downhole conditions. The objective of this test is to ensure that the acid formulation is compatible with downhole fluids. Coreflood tests are needed to evaluate compatibility of acid formulation with the formation rock. These tests should be performed using reservoir cores, reservoir fluids, and temperature. Finally, corrosion tests should conducted to evaluate corrosion rate of the acid formulation with downhole completions. These tests should be conducted using coupons from all alloys present downhole, and the acid formation should contain all planned additives and be prepared using field mixing water. Additionally, these tests should be performed at pressures greater than 1,000 psi.

## 9.8 Concluding Remarks

Acids are a reactive species. Additives are needed to control acid reactions with the well tubulars, well completion, reservoir rocks, and crude oil. However, these additives change the physical and chemical properties of these acids. The following are examples of these changes:

1. Acid-soluble polymers and VESs increase the viscosity of regular HCl solutions. However, they change the rheological properties of the acid into non-Newtonian shear-thinning behavior. More importantly, they reduce the rate of the acid reaction with the rock.
2. Corrosion inhibitors are used to protect the well completion. However, they can also coat the rock and scale and change the wettability of the rock. This will tend to reduce the reaction rate of the acid with scale or rock.

3. Corrosion inhibitors contain short-chain alcohols and dispersants, which can reduce the surface tension of the injected acids.
4. Short-chain organic acids (e.g., formic acid) are used as intensifiers. However, they can also reduce surface tension of HCl solutions. This can help in the pumping and recovery of spent acids from the formations.

Acid additives can interact with each other and with reservoir fluids, and can have adverse effects on the performance of other additives. The following are examples of these effects:

1. Antisludge agents, anionic surfactants, mutual solvents, and short-chain alcohols reduce the performance of corrosion inhibitors, and when they are used, the loading of corrosion inhibitor should be increased.
2. Aldehydes scavengers interact with polyacrylamide-based copolymers. This can have a serious impact on field treatments.
3. High concentrations of corrosion inhibitors and aldehyde-based $H_2S$ scavengers can form a polymeric layer on the scale or formation. This will reduce the acid reaction with scale or formation.

## References

Abdel Fatah, W. and Nasr-El-Din, H. A. 2010. Acid Emulsified in Xylene: A Cost Effective Treatment to Remove Asphaltene Deposition and Enhance Well Productivity. *SPE Prod & Oper* **25** (2): 151–154.

Al-Anazi, H. A., Nasr-El-Din, H. A., and Mohamed, S. K., 1998. Stimulation of Tight Carbonate Reservoirs Using Acid-in-Diesel Emulsions—Field Application. Paper SPE 39418 presented at the SPE Int. Symposium on Formation Damage Control, Lafayette, Louisiana, 18–19 February.

Al-Duailej, Y. K., Al-Mutairi, H., and Al-Humaidan, A. Y. 2010. Evaluation of Triazine-Based Scavengers for Stimulation Treatments. Paper 136915 presented at the Annual Technical Symposium and Exhibition, Al-Khobar, Saudi Arabia, 4–7 April.

Al-Ghamdi, A. and Nasr-El-Din, H. A. 1997. Effect of Oilfield Chemicals on the Cloud Point of Nonionic Surfactants *J. Colloids and Surfaces A: Physicochemical and Engineering Aspects* **125:** 5–18.

Al-Humaidan, A. Y. and Nasr-El-Din, H. A. 1999. Optimization of Hydrogen Sulfide Scavengers Used During Well Stimulation. Paper SPE 50765 presented at the SPE International Symposium on Oilfield Chemistry, Houston, 16–19 February.

Ali, S. A. and Hinkel, J. 2000. Additives in Acidizing Fluids. In *Reservoir Stimulation*, third edition. ed. M. J. Economides and K. G. Nolte. Chap. 15. New York: John Wiley & Sons, Inc.

Ali, S. A., Durham, D. K., and Elphingstone, E. A. 1994. Test Identifies Acidizing-Fluid/Crude Compatibility Problems. *Oil Gas J* **92** (13): 47–51.

Alkhaldi, M. H., Nasr-El-Din, H. A., and Hemanta, S. 2010. Kinetics of the Reaction of Citric Acid with Calcite. *SPE J.* **15** (3): 704–713. SPE-118724-MS.

Al-Khaldi, M. H., Nasr-El-Din, H. A., Blauch, M. E., and Funkhouser, G. P. 2005. New Findings on Damage Potential, Geochemical Reaction Mechanisms, and Production Enhancement Applications for Citric Acid. *SPE J.* **10** (3): 267–275.

Al-Khaldi, M. H., Nasr-El-Din, H. A., Mehta, S., and Al-Aamri, A. D. 2007. Reaction of Citric Acid with Calcite. *Chem. Eng. Sci.* **62** (21): 5,880–5,896.

Al-Mutairi, S. H., Hill, A. D., and Nasr-El-Din, H. A. 2008. Effect of Droplet Size and Acid Volume Fraction on the Rheological Properties and Stability of Emulsified Acid. *SPE Prod & Oper* **23** (4): 484–497.

Al-Mutairi, S. H., Nasr-El-Din, H. A, Hill, A. D., and Al-Aamri, A. D. 2009. Effect of Droplet Size on the Reaction Kinetics of Emulsified Acid with Calcite. *SPE J.* **14** (4): 606–616.

Al-Nakhli, A., Nasr-El-Din, H. A., and Al-Baiyat, A. A. 2008. Interactions of Iron and Viscoelastic Surfactants: A New Formation-Damage Mechanism. Paper SPE 112465 presented at the SPE

International Symposium and Exhibition on Formation Damage Control, Lafayette, Louisiana, 13–15 February.

Al-Taq, A. A., Ali, S. A., and Nasr-El-Din, H. A. 2009. Inhibition Performance of a New Series of Mono-/Diamine-Based Corrosion Inhibitors for HCl Solutions. *SPE J.* **14** (04): 627–633. SPE-114087-PA. http://dx.doi.org/10.2118/114087-PA.

*API RP42, API Recommended Practices for Laboratory Testing of Surface Active Agents for Well Stimulation*, second edition. 1977. Washington, DC: API.

Assem, A. I., Nasr-El-Din, H. A., and de Wolf, C. A. 2013a. Formation Damage Due to Iron Precipitation in Carbonate Rocks. Paper SPE 165203 presented at the SPE European Formation Damage Conference and Exhibition held in Noordwijk, The Netherlands, 5–7 June.

Assem, A. I., Nasr-El-Din, H. A., and de Wolf, C. A. 2013b. A New Finding in the Interaction Between Chelating Agents and Carbonate Rocks During Matrix Acidizing Treatments. Paper SPE 164130 presented at the SPE International Symposium on Oilfield Chemistry, The Woodlands, Texas, USA, 8–10 April.

Azari, M. and Leimkuhler, J. 1988. Permeability Changes Due to Invasion of Sodium and Potassium-Based Completion Brines in Berea and Casper Sandstones. Paper SPE 17149 presented at the 1988 SPE Formation Damage Control Symposium, Bakersfield, California, 8–9 February.

Bansal, K. M. 1993. Effect of Nonproduced Fluids on Produced Water Treatment Equipment Efficiency. SPE paper 25199 presented at the Int. Symposium on Oilfield Chemistry, New Orleans, 2–5 March.

Behenna, R. 1994. Tests Show Organic Clay Stabilizers Prevent Permeability Loss. *Oil & Gas J.* (September 12): 86–88.

Belcher, C. K., Seth, K., Hollier, R. et al. 2010. Maximizing Production Life With the Use of Nanotechnology to Prevent Fines Migration. Presented at the International Oil and Gas Conference and Exhibition, Beijing, 8–10 June. SPE-132152-MS. http://dx.doi.org/10.2118/132152-MS.

Berry, S. L., Boles, J. L., Brannon, H. D. et al. 2008. Performance Evaluation of Ionic Liquids as a Clay Stabilizer and Shale Inhibitor. Paper SPE112540 presented at the SPE International Symposium and Exhibition on Formation Damage Control, Lafayette, Louisiana, 13–15 February. http://dx.doi.org/10.2118/112540-MS.

Blauch, M. E., Cheng, A., Rispler, K., and Khallad, A. 2003. Novel Carbonate Well Production Enhancement Application for Encapsulated Acid Technology: First-Use Case History. Paper SPE 84131 presented at the SPE Annual Technical Conference and Exhibition, Denver, 5–8 October.

Boles, J. L., Ke, M., and Parker, C. 2009. Corrosion Inhibition of New 15 Chromium Tubulars in Acid Stimulation Fluids at High Temperatures. Presented at the SPE Annual Technical Conference and Exhibition, New Orleans, 4–7 October. SPE-124887-MS. http://dx.doi.org/10.2118/124887-MS.

Borchardt, J. K. 1989. Cationic Organic Polymer Formation Damage Control Chemicals: A Review of Basic Chemistry and Field Results. In *Oil-Field Chemistry, Enhanced Recovery and Production Stimulation*, eds. J. K. Borchardt and T. F. Yen, ACS Symposium Series, **396:** 204–221.

Braun, W., De Wolf, C., and Nasr-El-Din, H. A. 2012. Improved Health, Safety and Environmental Profile of a New Field Proven Stimulation Fluid. Presented at the SPE Russian Oil and Gas Exploration and Production Technical Conference and Exhibition, Moscow, 16–18 October. SPE-157467-MS. http://dx.doi.org/10.2118/157467-MS.

Brezinski, M. M. and Desai, B. 1997. Method and Composition for Acidizing Subterranean Formations Utilizing Corrosion Inhibitor Intensifiers. US Patent No. 5,697,443.

Brezinski, M. M. 1999. Chelating Agents in Sour Well Acidizing: Methodology or Mythology. Paper SPE 54721 presented at the SPE European Formation Damage Conference, The Hague, 31 May–1 June.

Buijse, M., de Boer, P., Breukel, B., and Burgos, G. 2004. Organic Acids in Carbonate Acidizing. *SPE Prod & Fac* **19** (3): 128–134.

Buijse, M. A. and van Domelen, M. S. 1998. Novel Application of Emulsified Acids to Matrix Stimulation of Heterogeneous Formations. *SPE Prod & Fac* **15** (3): 208–213.

Buijse, M. A. and van Domelen, M. S. 2000. Novel Application of Emulsified Acids to Matrix Stimulation of Heterogeneous Formations. *SPE Prod & Fac* **15** (3): 208–213.

Burgos, G., Birch, G., and Buijse, M. 2004. Acid Fracturing with Encapsulated Citric Acid. Paper SPE 86484 presented at the SPE International Symposium and Exhibition on Formation Damage Control, Lafayette, Louisiana, 18–20 February.

Cassidy, J. 2006. Design and Investigation of a North Sea Acid Corrosion Inhibition System. Presented at the Corrosion, San Diego, California, USA, 12–16 March. NACE-06482.

Cassidy, J. M., Kiser, C. E., and Lane, J. L. 2011a. Corrosion Inhibitor Intensifier Compositions and Associated Methods. US Patent No. 7,960,316 B2.

Cassidy, J. M., McNeil, R. I., and Kiser, C. 2007. Understanding Formic Acid Decomposition as a Corrosion Inhibitor Intensifier in Strong Acid Environments. Presented at the International Symposium on Oilfield Chemistry, Houston, 28 February–2 March. SPE-106185-MS. http://dx.doi.org/10.2118/106185-MS.

Cassidy, J. M., Kiser, C. E., and Smith, C. 2011b. Corrosion-Inhibitor Design for Compatibility with Acid-Gelling polymers. Paper SPE 141435 presented at the SPE International Symposium on Oilfield Chemistry, The Woodlands, Texas, 11–13 April.

Chambers, D. J. 1994. In *Foams: Fundamentals and Applications in the Petroleum Industry*, Schramm, L. L., ed., American Chemical Society, Washington, DC, p. 355.

Chang, F., Nasr-El-Din, H. A., Lindvig, T., and Qiu, W. 2008. Matrix Acidizing of Carbonate Reservoir Using Organic Acids and Mixture of HCl and Organic Acids. Paper SPE 116601 presented at the SPE Annual Technical and Exhibition, Denver, 21–24 December.

Chatelain, J. C., Silberberg, I. H., and Schechter, R. S. 1976. Thermodynamic Limitations in Organic Acid/Carbonate Systems. *SPE J.* **16** (4): 189–195.

Coppel, C. P., Jennings, H. Y. Jr., and Reed, M. G. 1973. Field Results From Wells Treated With Hydroxy-Aluminum. *J Pet Technol* **25** (9): 1,108–1,112. SPE-3998-PA. http://dx.doi.org/10.2118/3998-PA.

Crowe, C. W. 1985. Evaluation of Agents for Preventing Precipitation of Ferric Hydroxide from Spent Treating Acid. *J Pet Technol* **37** (4): 691–695.

Crowe, C. W. 1986. Prevention of Undesirable Precipitates from Acid Treating Fluids. Paper SPE 14090 presented at the International Meeting on Petroleum Engineering, Beijing, China, 17–20 March.

Crowe, C. W. 1991. Laboratory Study Provides Guidelines for Selecting Clay Stabilizers. Paper SPE 21556 presented at the SPE Oilfield Chemistry, Anaheim, California, 20–22 February.

Crowe, C. W. and Miller, B. D. 1974. New, Low-Viscosity Acid-in-Oil Emulsions Provide High Degree of Retardation at High Temperature. Presented at the SPE Rocky Mountain Regional Meeting, Billings, Montana, USA, 15–16 May. SPE-4937-MS. http://dx.doi.org/10.2118/4937-MS.

Crowe, C. W. and Minor, S. S. 1982. Acid Corrosion Inhibitor Adsorption and Its Effect on Matrix Stimulation Results. Presented at the SPE Formation Damage Control Symposium, Lafayette, Louisiana, USA, 24–25 March. SPE-10650-MS. http://dx.doi.org/10.2118/10650-MS.

Dabbousi, B. O., Nasr-El-Din, H. A., Al-Muhaish, A. S. 1999. Influence of Oilfield Chemicals on the Surface Tension of Stimulating Fluids. Paper SPE 50732 presented at the Oilfield Chemistry Conference, Houston, 16–19 February.

De Wolf, C., Nasr-El-Din, H. A., Bouwman, A. et al. 2012. A New, Low Corrosive Fluid to Stimulate Deep Wells Completed with Cr-Based Alloys. Presented at the SPE International Conference and Exhibition on Oilfield Corrosion, Aberdeen, 28–29 May. SPE-152716-MS. http://dx.doi.org/10.2118/152716-MS.

Delorey, J. R., Allen, S., and McMaster, L. 1996. Precipitation of Calcium Sulphate During Carbonate Acidizing: Minimizing the Risk. Presented at the Annual Technical Meeting, Calgary, 10–12 June. PETSOC-96-84. http://dx.doi.org/10.2118/96-84.

Delorey, J. R., Vician, D. N., and Metcalf, A. S. 2002. Acid Stimulation of Sour Wells. Presented at the SPE Gas Technology Symposium, Calrgary, 30 April–2 May. SPE-75697-MS. http://dx.doi.org/10.2118/75697-MS.

Denoyel, R., Giordano, F., and Rouquerol, J. 1993. *J. Colloids and Surfaces A: Physicochemical and Engineering Aspects* **76:** 141.

Dill, W. R. and Fredette, G. 1983. Iron Control in the Appalachian Basin. Paper SPE 12319 presented at the Eastern Regional Meeting, Pittsburgh, Pennsylvania, 9–11 November.

Dunlap, D. D. and Houchin, L. R. 1990. Evaluation of Acid System Design and Formation Damage Using Polarized Microscopy. Presented at SPE Formation Damage Control Symposium, Lafayette, Louisiana, 22–23 February. SPE-19425-MS.

El-Monier, E. A. and Nasr-El-Din, H. A. 2011a. A Study of Several Environmentally Friendly Clay Stabilizers. Paper SPE 142755 presented at the SPE Project and Facilities Challenges Conference at METS, Doha, Qatar, 13–16 February. http://dx.doi.org/10.2118/142755-MS.

El-Monier, E. A. and Nasr-El-Din, H. A. 2011b. A New Al-Based Stabilizer for High pH Applications. Paper SPE 143260 presented at the Brazil Offshore, Macaé, Brazil, 14–17 June.

El-Monier, E. A. and Nasr-El-Din, H. A. 2011c. Mitigation of Fines Migration Using a New Clay Stabilizer: A Mechanistic Study. Paper SPE 144180 presented at the SPE European Formation Damage Conference, Noordwijk, The Netherlands, 7–10 June. http://dx.doi.org/10.2118/144180-MS.

El-Monier, I. A., Nasr-El-Din, H. A., Haper, T., Rossen, R. 2013. A New Environmentally Friendly Clay Stabilizer. *SPE Prod & Oper* **28** (2): 145–153.

Evans, B. and Ali, S. 1997. Selecting Brines and Clay Stabilizers to Prevent Formation Damage. *World Oil* (May): 65–68.

Figueroa-Ortiz, V., Cazares-Robles, F., and Fragachan, F. E. 1996. Controlling Organic Deposits and Sludge in a Severe Hostile Environment. Presented at the SPE Formation Damage Control Symposium, Lafayette, Louisiana, USA, 14–15 February. SPE-31124-MS. http://dx.doi.org/10.2118/31124-MS.

Fredd, C. N. and Fogler, H. S. 1998a. Alternative Stimulation Fluids and Their Impact on Carbonate Acidizing. *SPE J.* **13** (1): 21–41.

Fredd, C. N. and Fogler, H. S. 1998b. The Influence of Chelating Agents on the Kinetics of Calcite Dissolution. *J. Colloid and Interface Science* **204** (1): 187–197.

Frenier, W. W. 1989. Acidizing Fluids Used to Stimulate High Temperature Wells Can Be Inhibited Using Organic Chemicals. Presented at the SPE International Symposium on Oilfield Chemistry, Houston, 8–10 February. SPE-18468-MS. http://dx.doi.org/10.2118/18468-MS.

Frenier, W. W. 2001. *Technology for Chemical Cleaning of Industrial Equipment*, first edition. Houston: NACE International.

Frenier, W. W. 1982. Method and Composition for Removing Sulfide-Containing Scale From Metal Surfaces. US Patent 4,310,435 issued on 12 January.

Frenier, W. W., Brady, M., Al-Harthy, S., Arangath, R., Chan, K. S., Flamant, N., and Samuel, M. 2004. Hot Oil and Gas Wells Can Be Stimulated without Acids. Paper SPE 86522 presented at the SPE International Symposium and Exhibition on Damage Control, Lafayette, Louisiana, 18–20 February.

Frenier, W. W., Fredd, C. N., and Chang, F. 2001. Hydroxyaminocarboxylic Acids Produce Superior Formulations for Matrix Stimulation of Carbonates. Paper SPE 68924 presented at the SPE European Formation Damage Conference, The Hague, 21–22 May.

Frenier, W. W., Rainey, M., Crump, D. W., and Jones, L. 2003. A Biodegradable Chelating Agent Is Developed for Stimulation of Oil and Gas Formations. Paper SPE 80597 presented at the SPE/EPA/DOE Exploration and Production Environmental Conference, San Antonio, Texas, 10–12 March.

Frenier, W. W., Wilson, D., Crump, D. W., and Jones, L. 2000. Use of Highly Acid-Soluble Chelating Agents in Well Stimulation Services. Paper SPE 63242 presented at the SPE Annual Technical Conference and Exhibition, Dallas, 1–4 October.

Fu, D. and Chang, F. 2006. Compositions and Methods for Treating a Subterranean Formation. US Patent No. 7,028,775 B2, issued on 18 April.

Gdanski, R. D. 1995. Molecular Modeling Gives Insight Into Nonionic Surfactants. Paper SPE 28971 presented at the Int. Symp. Oilfield Chemistry, San Antonio, Texas, 14–17 February.

Gomaa, A. M. and Nasr-El-Din, H. A. 2010. New Insights into the Viscosity of Polymer-Based In-Situ Gelled Acids. *SPE Prod & Oper* **25** (3): 367–375.

Gougler, P. D. Jr., Hendrick, J. E., and Coulter, A. W. 1985. Field Investigation Identifies Source and Magnitude of Iron Problems. Paper SPE 13812 presented at the SPE Production Operations Symposium, Oklahoma City, Oklahoma, 10–12 March.

Gray, D. H. and Rex, R. W. 1966. Formation Damage in Sandstones Caused by Dispersion and Migration. *Clays and Clay Minerals* **15:** 355–366.

Gruesbeck, C. and Collins, R. E. 1982. Entrainment and Deposition of Fine Particles in Porous Media. *SPE J.* **22** (6): 847–856.

Gurluk, M. R., Nasr-El-Din, H. A., and Crews, J. B. 2013. Enhancing the Performance of Viscoelastic Surfactant Fluids Using Nanoparticles. Paper SPE 164900 presented at the 75th EAGE Conference & Exhibition in SPE EUROPEC, London, 10–13 June.

Hall, B. E. 1975. The Role of Mutual Solvents on Adsorption in Sandstone Acidizing. *J Pet Technol* **27** (12): 1,439–1,442.

Hall, B. E. and Dill, W. R. 1988. Iron Control Additives for Limestone and Sandstone Acidizing of Sweet and Sour Wells. Paper SPE 17157 presented at the SPE Formation Damage Control Symposium, Bakersfield, California, 8–9 February.

Harris, F. N. 1961. Application of Acetic Acid to Well Completion, Stimulation and Reconditioning. *J Pet Technol* **13** (7): 637–639.

Hartman, R. L., Lecerf, B., Frenier, W. W. et al. 2006. Acid-Sensitive Aluminosilicates: Dissolution Kinetics and Fluid Selection for Matrix-Stimulation Treatments. *SPE Prod & Oper* **21** (2): 194–204. SPE-82267-PA. http://dx.doi.org/10.2118/82267-PA.

He, J., Mohamed, I. M., and Nasr-El-Din, H. A. 2011. Mixing Hydrochloric Acid and Seawater for Matrix Acidizing: Is It a Good Practice? Presented at the SPE European Formation Damage Conference, Noordwijk, The Netherlands, 7–10 June. SPE-143855-MS. http://dx.doi.org/10.2118/143855-MS.

He, Z., Wang, G., Nasr-El-Din, H. A., and Holt, S. 2013. Hydrolysis Effect on the Properties of a New Class of Viscoelastic Surfactant-Based Acid and Damage Caused by the Hydrolysis Products. Paper SPE 165161 presented at the SPE European Formation Damage and Exhibition, Noordwijk, The Netherlands, 5–7 June.

Hesterberg, D. and Reed, M. G. 1991. Volumetric Treatment Efficiencies of Some Commercial Clay Stabilizers. *SPE Prod Eng* **6** (1): 57–62. SPE-18466-PA. http://dx.doi.org/10.2118/18466-PA.

Hill, D. G. 1982. Clay Stabilization – Criteria for Best Performance. Paper SPE 10656 presented at the SPE Formation Damage Control Symposium, Lafayette, LA, 24-25 March.

Himes, R. E., Vinson, E. F., and Simon, D. E. 1991. Clay Stabilization in Low-Permeability Formations. *SPE Prod Eng* **6** (3): 252–258. SPE-18881-PA. http://dx.doi.org/10.2118/18881-PA.

Hinze, W. L. and Pramauro, E. 1993. *Critical Reviews Analytical Chemistry* **24** (2): 133.

Hoefner, M. L. and Fogler, H. S. 1985. *Chem. Eng. Prog.* **81** (4): 40.

Hoefner, M. L., Fogler, H. S., Stenius, P. et al. 1987. Role of Acid Diffusion in Matrix Acidizing of Carbonates. *J Pet Technol* **39** (2): 203–208. SPE-13564-PA. http://dx.doi.org/10.2118/13564-PA.

Houchin, L. R. and Hudson, L. M. 1986. The Prediction, Evaluation, and Treatment of Formation Damage Caused by Organic Deposition. Paper SPE 14818 presented at the Int. Symposium on Formation Damage Conference, Lafayette, Louisiana, 26–27 February.

Houchin, L. R., Dunlap, D. D., Arnold, B. D., and Domke, K. M. 1990. The Occurrence and Control of Acid-Induced Asphaltene Sludge. Paper SPE 19410 presented at the Formation Damage Control Symposium, Lafayette, Louisiana, 22–23 February.

Houchin, L. R., Foxenburg, W. E., Usie, M. J., and Zhao, J. 1992. A New Technique for the Evaluation of Acid Additive Packages. Paper SPE 23817 presented at the International Symposium on Formation Damage Control, Lafayette, Louisiana, 26–27 February.

Huang, T. and Crews, J. B. 2009. Fluid-Loss Control Improves Performance of Viscoelastic Surfactant Fluids *SPE Prod & Oper* **24** (1): 60–65.

Huang, T., Crews, J. B., and Agrawal, G. 2010. Nanoparticle Pseudo Crosslinked Micellar Fluids: Optimal Solution for Fluid-Loss Control with Internal Breaking. Paper SPE 128067 presented at the SPE International Symposium and Exhibition on Formation Damage Control, Lafayette, Louisiana, 10–12 February

Huang, T., Crews, J. B., Willingham, J. R. 2008. Using Nano-Particle Technology to Control Fine Migration. Paper SPE 115384 presented at the Annual technical Conference and Exhibition, Denver, 21–24 September.

Huang, T., Ostensen, L., and Hill, A. D. 2000. Carbonate Matrix Acidizing with Acetic Acid. Paper SPE 58715 presented at the SPE International Symposium on Formation Damage Control, Lafayette, Louisiana, 23–24 February.

Jachnik, R. P. and Green, P. 1995. Horizontal Well Drill-In Fluid Utilising Alcohol Ethoxylate. Presented at the SPE International Symposium on Oilfield Chemistry, San Antonio, Texas, USA, 14–17 February. SPE-28963-MS. http://dx.doi.org/10.2118/28963-MS.

Jacobs, I. C. 1989. Chemical Systems for the Control of Asphaltene Sludge During Oilwell Acidizing Treatments. Paper SPE 18475 presented at the International Symposium on Oilfield Chemistry, Houston, 8–10 February.

Jacobs, I. C. and Thorne, M. A. 1986. Asphaltene Precipitation During Acid Stimulation Treatments. Paper SPE 14823 presented at the Symposium on Formation Damage Control, Lafayette, Louisiana, 26–27 February.

Jenkins, A. and Cullion, P. 2009. High Temperature Environmentally Friendly Corrosion Inhibitor for Organic Acids. Paper NACE 9232 presented at NACE Conference and Expo.

Kalfayan, L. 2000. *Production Enhancement With Acid Stimulation*, first edition. Tulsa: PennWell Corporation.

Kelland, M. A. 2009. Production Chemicals for the Oil and Gas Industry. CRC press, pp. 153–160.

Khilar, K. C. and Fogler, H. S. 1983. Water Sensitivity of Sandstones. *Society of Petroleum Engineers Journal* **23** (1): 55–64. SPE-10103-PA. http://dx.doi.org/10.2118/10103-PA.

Khilar, K. C. and Fogler, H. S. 1984. The Existence of a Critical Salt Concentration for Particle Release. *J. Colloid Interface Science* **101:** 214–224.

Kia, S. F., Fogler, H. S., and Reed, M. G. 1987. Effect of pH on Colloidally Induced Fines Migration. *J. Colloid Interface Science* **118:** 158.

King, G. E. and Lee, R. M. 1988. Adsorption and Chlorination of Mutual Solvents Used in Acidizing. *SPE prod eng* **3** (2): 205–209.

Krawietz, T. E. and Rael, E. L. 1996. Horizontal Well Acidizing of a Carbonate Formation: A Case History of Lisburne Treatments, Prudhoe Bay, Alaska. *SPE Prod & Fac* **11** (4): 238–243.

LePage, J. N., de Wolf, C. A., Bemelaar, J. H., and Nasr-El-Din, H. A. 2011. An Environmentally Friendly Stimulation Fluid for High Temperature Applications. *SPE J.* **16** (1): 104–110. SPE-121709-MS.

Li, L., Nasr-El-Din, H. A., and Cawiezel, K. 2010. Rheological Properties of a New Class of Viscoelastic Surfactant. *SPE Prod & Oper* **25** (3): 355–366.

Li, L., Nasr-El-Din, H. A., Chang, F., and Lindvig, T. 2008. Reaction of Simple Organic Acids and Chelating Agents with Calcite. Paper SPE IPTC 12886 presented at the International Petroleum Technology Conference, Kuala Lumpur, Malaysia, 3–5 December.

Lichaa, P. M. and Herrera, L. 1975. Electrical and Other Effects Related to the Formation and Prevention of Asphaltene Deposition Problem in Venezuelan Crudes. Presented at the Int. Symposium on Oilfield Chemistry, Dallas, Texas, 13 -14 January. SPE 5304-MS.

Lynn, J. D. and Nasr-El-Din, H. A. 2001. A Core-Based Comparison of the Reaction Characteristics of Emulsified and In-Situ Gelled Acids in Low-Permeability, High Temperature, Gas Bearing Carbonates. Paper SPE 65386 presented at the SPE Oilfield Chemistry, Houston, TX, 13–16 February.

Mahmoud, M. A., Nasr-El-Din, H. A., and de Wolf, C. A. 2015. High-Temperature Laboratory Testing of Illitic Sandstone Outcrop Cores With HCl-Alternative Fluids. *SPE Prod & Oper* **30** (1): 43–51. SPE-147395-PA. http://dx.doi.org/10.2118/147395-PA.

Mahmoud, M. A., Nasr-El-Din, H. A., de Wolf, C. A., and LePage, J. N. 2011. Optimum Injection Rate of a New Chelate that can be used to Stimulate Deep Carbonate Reservoirs. Paper SPE 133497, *SPE J.* **16** (4): 968–980.

Malwitz, M. A. 2010. Corrosion Inhibitor Composition Comprising a Built-In Intensifier. Patent US 7,842,127, B2 issued on November 30.

Masliyah, J. H. and Bhattacharjee, S. 2006. *Electrokinetic and Colloid Transport Phenomena*. Hoboken, New Jersey: John Wiley & Sons.

McLaughlin, H. C., Elphingstone, E. A., and Hall. B. 1976. Aqueous Polymers for Treating Clays in Oil and Gas. Paper SPE 6008 presented at the Annual Fall Technical Conference and Exhibition, New Orleans, 3–6, October.

McLeod, H. O. J. 1984. Matrix Acidizing. *J Pet Technol* **36** (12): 2,055–2,069. SPE-13752-PA. http://dx.doi.org/10.2118/13752-PA.

Metcalf, A. S. 1998. Environment and Coiled Tubing: Both Can Be Protected. Paper SPE 51095 presented at the SPE Eastern Regional Meeting held in Pittsburgh, Pennsylvania, 9–11 November.

Metcalf, A. S., Delorey, J. R., and Allen, S. A. 2003. Coiled Tubing Acid Corrosion Protection: Laboratory vs. Field Performance. Paper SPE 80483 presented at the SPE Asia Pacific Oil and Gas Conference and Exhibition, Jakarta, 15–17 April.

Mingjie, K. and Boles, J. 2004. Corrosion Behavior of Various 13 Chromium Tubulars in Acid Stimulation Fluids. Paper SPE 87561 presented at the SPE International Oilfield Corrosion.

Moore, E. W., Crowe, C. W., and Hendrickson, A. R. 1965. Formation, Effect and Prevention of Asphaltene Sludges During Stimulation Treatments. *J Pet Technol* **17** (9): 1,023–1,028. SPE-1163-PA. http://dx.doi.org/10.2118/1163-PA.

Mungan, N. 1968. Permeability Reduction Due To Salinity Changes. *J. Cdn. Pet. Tech.* **7:** 113.

Musharova, D., Mohamed, I. M., and Nasr-El-Din, H. A. 2012. Detrimental Effect of Temperature on Fines Migration in Sandstone Formations. Presented at the SPE International Symposium and Exhibition on Formation Damage Control, Lafayette, Louisiana, USA, 15–17 February. SPE-150953-MS. http://dx.doi.org/10.2118/150953-MS.

Nasr-El-Din H. A., Al-Mohammad, A. M., Al-Aamri, A. D., and Al-Fahad, M. A., Chang, F. 2009a. Quantitative Analysis of Reaction Rate Retardation in Surfactant-Based Acids. *SPE Prod & Oper* **24** (1): 107–116.

Nasr-El-Din, H. A. 2005. Formation Damage Induced by Chemical Treatments: Case Histories. *J. Energy Resources Technology* **127:** 214–224.

Nasr-El-Din, H. A. and Al-Ghamdi, A. 1997. Accepted for publication, *SPE Prod & Fac*.

Nasr-El-Din, H. A. and Al-Taq, A. A. 1998. Water Quality Requirements and Restoring the Injectivity of Waste Water Disposal Wells. Paper SPE 39487 presented at the Int. Symposium on Formation Damage Control, Lafayette, Louisiana, 18–19 February.

Nasr-El-Din, H. A., Al-Ghamdi, A. H., Al-Qahtani, A. A., and Samuel, M. M. 2008a. Impact of Acid Additives on the Rheological Properties of Viscoelastic Surfactants. *SPE J.* **13** (1): 35–47.

Nasr-El-Din, H. A., Al-Habib, N. S., Al-Mumen, A. A., Jemmali M., and Samuel, M. 2006. A New Effective Stimulation Treatment for Long Horizontal Wells Drilled in Carbonate Reservoirs. *SPE Prod & Oper* **21** (3): 330–338.

Nasr-El-Din, H. A., Al-Humaidan, A., Fadel, B., Frenier, W., and Hill, D. 2002a. Investigation of Sulfide Scavengers in Well Acidizing Fluids. *SPE Prod & Fac* **17** (4): 229–235.

Nasr-El-Din, H. A., Al-Humaidan, A., Fadel, B. A., and Saleh, R. 2000. Effect of Acid Additives on the Efficiency of Dissolving Iron Sulfide Scale. NACE #439 presented at the NACExpo, Orlando, Florida, 26–31 March.

Nasr-El-Din, H. A., Al-Khuraidah, A., Krizler, T., and Cassidy, J. 2002b. Recent Development in High-temperature Acidizing With Super Cr-13 Completions: Laboratory Testing. Paper SPE 78557 presented at the SPE/ADIPEC Conference, held in Abu Dhabi, UAE, 13–16 October.

Nasr-El-Din, H. A., Al-Mohammed, A., Al-Aamri, A., and Al-Fuwaires, O. 2008b. Reaction of Gelled Acids With Calcite. *SPE Prod & Oper* **23** (3): 353–361.

Nasr-El-Din, H. A., Al-Mulhem, A. A., and Lynn, J. D. 1999. Evaluation of Clay Stabilizers for a Sandstone Field in Central Saudi Arabia. *SPE Prod & Oper* **14** (4): 253–259. SPE-57393-PA. http://dx.doi.org/10.2118/57393-PA.

Nasr-El-Din, H. A., Al-Nakhli, A., Leopoldo, S., Walton, T., and Van Domelen, M. 2009b. Application of Cationic Surfactant-Based Fluids for Acid Diversion.*SPE Prod & Oper* **24** (1): 124–134.

Nasr-El-Din, H. A., Al-Othman, A., Taylor, K. C. et al. 2004. Surface Tension of Acid Stimulating Fluids at High Temperatures, *J. Pet. Science Eng.* **43** (1–2): 57–73.

Nasr-El-Din, H. A., Chesson, J. B., Cawiezel, K. E. et al. 2006a. Lessons Learned and Guidelines for Matrix Acidizing and Diversion Techniques in Carbonate Formations. Paper SPE 102468 presented at the Annual Technical Conference & Exhibition, San Antonio, Texas, 24–27 September.

Nasr-El-Din, H. A., Chesson, J. B., Cawiezel, K. E. et al. 2006b. Field Success in Carbonate Acid Diversion, Utilizing Laboratory Data Generated by Parallel Flow Testing. Paper SPE

102828 presented at the Annual Technical Conference & Exhibition, San Antonio, Texas, 24–27 September.

Nasr-El-Din, H. A., Dana, H., Tomos, V., Stanitzek, T., de Wolf, C. A., 2015. Field Treatment to Stimulate an Oil Well in a Sandstone Formation Using a Novel, Low Corrosive, Environmentally Friendly Fluid. Paper SPE 168163 presented at the Formation Damage Conference and Exhibition held in Lafayette, Louisiana, 26–28 February. Accepted for publication in *J Can Pet Technol.*

Nasr-El-Din, H. A., de Wolf, C. A., Stanitzek, T., Alex, A., Gerdes, S., and Lummer, N. R. 2013. Field Treatment to Stimulate a Deep, Sour, Tight Gas Well Using a New, Low Corrosive and Environmentally Friendly Fluid. Paper SPE 163332, *SPE Prod & Oper* **28** (3): 277–285.

Nasr-El-Din, H. A., Lynn, J. D., and Taylor K. C. 2001. Lab Testing and Field Application of a Large Scale Acetic Acid Based Treatment in a Newly Developed Carbonate Reservoir. Paper SPE 65036 presented at the SPE Oilfield Chemistry, Houston, 13–16 February.

Nasr-El-Din, H. A., Maini, B. B., and Stanislav, P. 1991. Fines Migration in Unconsolidated Sand Formations. *AOSTRA J. Research* **7:** 1–15.

Nasr-El-Din, H. A., Taylor, K. C., and Al-Hajji, H. H. 2002c. Propagation of Cross-linkers Used in In-Situ Gelled Acids in Carbonate Reservoirs. Paper SPE 75257 presented at the SPE/DOE Thirteenth Symposium on Improved Oil Recovery, Tulsa, 13–17 April.

Nasr-El-Din, H. A., Zahrani, A. A., Garzon, F. O., Giraldo, C. A. F., Al-Hekami, I. M., and Al-Marri, H. M. 2009c. Acid Fracturing of Gas Wells Using Solid Acid: Lessons Learned From First Field Application. *SPE Prod & Oper* **24** (2) 320–335.

Nguyen, P. D., Weaver, J. D., Rickman, R. D. et al. 2007. Controlling Formation Fines at Their Sources To Maintain Well Productivity. *SPE Prod & Oper* **22** (2): 202–215. SPE-97659-PA. http://dx.doi.org/10.2118/97659-PA.

Paktinat, J. 1991. Reduced Adsorption and Emulsion Tendencies in Sandstone Formation Through the Use of Ethoxylates. Paper SPE 21011 presented at the Int. Symp. Oilfield Chemistry, Anaheim, California, 20–22 February.

Patel, A. D. 2009. Design and Development of Quaternary Amine Compounds: Shale Inhibition with Improved Environmental Profile. Paper SPE 121737 presented at the International Symposium on Oilfield Chemistry, The Woodlands, Texas, 20–22 April. http://dx.doi.org/10.2118/121737-MS.

Patton, J. T. and Phelan, P. F. 1985. Well Damage Hazards Associated With Conventional Completion Fluids. Paper SPE 13800 presented at the Production Operations Symposium, Oklahoma City, Oklahoma, 10–12 March.

Rae, P. and Di Lullo, G. 2003. Matrix Acid Stimulation—a Review of the State-of-the-Art. Presented at the SPE European Formation Damage Conference, The Hague, 13–14 May. SPE-82260-MS. http://dx.doi.org/10.2118/82260-MS.

Reed, M. G. 1972. Stabilization of Formation Clays With Hydroxy-Aluminum Solutions. *J Pet Technol* **24** (7): 860–864. SPE-3694-PA. http://dx.doi.org/10.2118/3694-PA.

Reyes, E. A., Smith, A. L., and Beuterbaugh, A. 2013. Properties and Application of an Alternative Aminopolycarboxylic Acid for Acidizing Sandstones and Carbonates. Paper 165142 presented at the SPE European Formation damage Conference and Exhibition, Noordwiijk, The Netherlands, 5–7 June.

Rietjens, M. 1997. Sense and Non-Sense About Acid-Induced Sludge. Paper SPE 38163 presented at the European Formation Damage Control Conference, The Hague, 2–3 June.

Ross, G. J. 1967. Kinetics of acid dissolution of an orthochlorite mineral. *Can. J. Chem.* **45** (24): 3,031–3,034. http://dx.doi.org/10.1139/v67-491.

Ross, G. J. 1968. Structural Decomposition of an Orthochlorite During Its Acid Dissolution. *The Canadian Mineralogist* **9** (4): 522–530.

Rostami, A. and Nasr-El-Din, H. A. 2009. Review and Evaluation of Corrosion Inhibitors Used in Well Stimulation. Paper SPE 121726 presented at the Oilfield Chemistry Conference, The Woodlands, Texas, 20–22 April.

Sayed, M. A., Nasr-El-Din, H. A., Nasrabadi, H., 2013. Reaction Rate of Emulsified Acids and Dolomite. Paper SPE 151815, *J Can Pet Technol* **52** (3): 164–175.

Seth, K., Evans, B., and Gabrysch, A. 2011. Development and Testing of a Novel Corrosion Inhibitor Technology for Acid Corrosion. SPE paper 142675 presented at the SPE Middle East Oil and Gas Show and Conference, Manama, Bahrain, 25–28 September.

Shu, Y., Wang, G., Nasr-El-Din, H. A. et al. 2015. Interactions of Fe(III) and Viscoelastic-Surfactant-Based Acids. *SPE Prod & Oper*. SPE-165149-PA. http://dx.doi.org/10.2118/165149-PA. (in press; posted 10 August 2015).

Simon, D. E. and Anderson, M. S. 1990. Stability of Clay Minerals in Acid. Paper SPE 19422 presented at the SPE Formation Damage Control Symposium, Lafayette, Louisiana, USA, 22–23 February. SPE-19422-PA. http://dx.doi.org/10.2118/19422-MS.

Sitz, C., Frenier, W., Vallejo, C. 2012 Acid Corrosion Inhibitors with Improved Environmental Profiles. Paper SPE 155966 presented at the SPE International Conference on Oilfield Corrosion, Aberdeen, 28–29 May.

Smith, C., Crowe, C., and Nolan, T. J., III. 1969. Secondary Deposition of Iron Compounds Following Acidizing Treatment. *J Pet Technol* **21** (9): 1,121–1,129.

Smith, C. F., Dollarhide, F. E., and Byth, N. S. 1978. Acid Corrosion Inhibitors—Are We Getting What We Need? *J Pet Techol* **30** (5): 737–746.

Smuga-Otto, I., Nowinka, J., Wilde, G. et al. 2001. Coiled Tubing Performance Evaluation for Sour Service under Variable Loading Conditions. Paper NACE 01080 presented at the NACE Annual Corrosion Meeting, Houston, 12–16 March.

Strassner, J. E. 1968. Effect of pH on Interfacial Films and Stability of Crude Oil-Water Emulsions. *J. Pet. Technol.* **20** (3): 303–312.

Taylor, K. C. and Nasr-El-Din, H. A. 2003. Laboratory Evaluation of In-Situ Gelled Acids for Carbonate Reservoirs. *SPE J.* **8** (4): 426–434.

Taylor, K. C., Al-Ghamdi, A., and Nasr-El-Din, H. A. 2004. Effect of Additives on the Acid Dissolution Rates of Calcium and Magnesium Carbonates. *SPE Prod & Fac* **19** (3): 122–127.

Taylor, K. C., Nasr-El-Din, H. A., and Al-Alawi, M. 1999. Systematic Study of Iron Control Chemicals Used During Well Stimulation. *SPE J.* **4** (1): 19–24.

Travalloni-Louvisse, A. M., Gonzalez, G. 1988. Adsorption of Nonionic Surfactants on Quartz in the Presence of Ethanol, HCl or $CaCl_2$: Its Effect on Wettability. In *Surfactant-Based Mobility Control, Progress in Miscible-Flood Enhanced Oil Recovery*, ed. D. H. Smith, Chap. 11, 220. Washington, DC: ACS Symposium Series 373: American Chemical Society.

Van Domelen, M. S. and Jennings, A. R. 1995. Alternate Acid Blends for HPHT Applications. Paper SPE 30419 presented at the Offshore European Conference, Aberdeen, 5–8 September.

Veley, C. D. 1969. How Hydrolyzable Metal Ions React With Clays to Control Formation Water Sensitivity. *J. Pet. Technol.* **21** (9): 1,111–1,118. SPE-2188-PA. http://dx.doi.org/10.2118/2188-PA.

Vinson, E. F. 1996. A Novel Reducing Agent for Combatting Iron-Induced Crude Oil Sludging: Development and Case Histories. Paper SPE 31127 presented at the Formation Damage Control Symposium, Lafayette, Louisiana, 14–15 February.

Walker, M. L., Dill, W. R., Besler, M. R., and McFatridge, D. G. 1991. Iron Control in West Texas Sour-Gas Wells Provides Sustained Production Increases. *J Pet Technol* **43** (5): 603–607.

Weaver, J. D., Nguyen, P. D., Loghry, R. A. 2011. Stabilizing Fracture Faces in Water-Sensitive Shale Formation. Paper SPE 149218 presented at the Eastern Regional Meeting, Columbus, Ohio, 17–19 August.

Xiao, Z., Morgenthaler, L. N., Cowan, M. K., and Adams, J. G. 2009. Evaluation of $H_2S$ Scavengers for Matrix Acidizing Treatment. Paper SPE 121698 presented at the International Symposium on Oilfield Chemistry, Woodlands, Texas, 20–22 April.

Yu, M., Mu, Y., Wang, G., and Nasr-El-Din, H. A. 2012. Impact of High Temperatures on the Rheological Properties of Carboxybetaine Visco-Elastic Surfactant Based Acid: Experimental and Molecular Dynamics Simulation Studies. *SPE J.* **17** (4): 1,119–1,130.

Zaiton, A. and Berton, N. 1992. Stabilization of Montmorillonite Clay in Porous Media by High-Molecular-Weight Polymers. *SPE Prod Eng* **7** (2): 160–166.

Zaiton, A. and Berton, N. 1996. Stabilization of Montmorillonite Clay in Porous Media by Polyacrylamides. Paper SPE 31109 presented at the SPE Formation Damage Control Symposium, Lafayette, Louisiana, 14–15 February.

Zhou, Z. J., Gunter, W. O., and Jonasson, R. G. 1995. Controlling Formation Damage Using Clay Stabilizers: A Review. Presented at the Annual Technical Meeting, Calgary, 7–9 June. PETSOC-95-71. http://dx.doi.org/10.2118/95-71.

## General References

Alkhaldi, M. H., Nasr-El-Din, H. A., and Hemanta, S. 2010. Kinetics of the Reaction of Citric Acid with Calcite. *SPE J.* **15** (3): 704–713. SPE-118724-MS.

Al-Khaldi, M. H., Nasr-El-Din, H. A., Blauch, M. E., and Funkhouser, G. P. 2005. New Findings on Damage Potential, Geochemical Reaction Mechanisms, and Production Enhancement Applications for Citric Acid. *SPE J.* **10** (3): 267–275.

Al-Khaldi, M. H., Nasr-El-Din, H. A., Mehta, S., and Al-Aamri, A. D. 2007. Reaction of Citric Acid with Calcite. *Chem. Eng. Sci.* **62** (21): 5,880–5,896.

Blauch, M. E., Cheng, A., Rispler, K., and Khallad, A. 2003. Novel Carbonate Well Production Enhancement Application for Encapsulated Acid Technology: First-Use Case History. Paper SPE 84131 presented at the SPE Annual Technical Conference and Exhibition, Denver, 5–8 October.

Crowe, C. W. 1991. Laboratory Study Provides Guidelines for Selecting Clay Stabilizers. Paper SPE 21556 presented at the SPE Oilfield Chemistry, Anaheim, California, 20–22 February.

Dabbousi, B. O., Nasr-El-Din, H. A., Al-Muhaish, A. S. 1999. Influence of Oilfield Chemicals on the Surface Tension of Stimulating Fluids. Paper SPE 50732 presented at the Oilfield Chemistry Conference, Houston, 16–19 February.

Duan, Z. and Sun, R. 2003. An Improved Model Calculating $CO_2$ Solubility in Pure Water and Aqueous NaCl Solutions From 273 to 533 K and From 0 to 2000 Bar. *Chem. Geol.* **193** (3–4): 257–271.

Frenier, W. W. 1989. Acidizing Fluids Used to Stimulate High Temperature Wells Can Be Inhibited Using Organic Chemicals. Presented at the SPE International Symposium on Oilfield Chemistry, Houston, 8–10 February. SPE-18468-MS. http://dx.doi.org/10.2118/18468-MS.

Kane, R. D. and Wilhelm, S. M. 1986. Compatibility of Stainless and Nickel Base Alloys in Acidizing Environments. Paper No. 89481, CORROSION/89, Houston, NACE International.

Li, L., Nasr-El-Din, H. A., and Cawiezel, K. 2010. Rheological Properties of a New Class of Viscoelastic Surfactant. *SPE Prod & Oper* **25** (3): 355–366.

Lynn, J. D. and Nasr-El-Din, H. A. 1998. Evaluation of Formation Damage Due To Frac Stimulation of a Saudi Arabian Clastic Reservoir. *J. Pet. Sci. Eng.* **21**: 179–201.

Masliyah, J. H. and Bhattacharjee, S. 2006. *Electrokinetic and Colloid Transport Phenomena.* Hoboken, New Jersey: John Wiley & Sons.

Metcalf, A. S., Delorey, J. R., and Allen, S. A. 2003. Coiled Tubing Acid Corrosion Protection: Laboratory vs. Field Performance. Paper SPE 80483 presented at the SPE Asia Pacific Oil and Gas Conference and Exhibition, Jakarta, 15–17 April.

Mingjie, Ke, M. and Boles, J. 2004. Corrosion Behavior of Various 13 Chromium Tubulars in Acid Stimulation Fluids. Paper SPE 87561 presented at the SPE International Oilfield Corrosion Conference, Aberdeen, 28 May.

Musharova, D., Mohamed, I. M., and Nasr-El-Din, H. A. 2012. Detrimental Effect of Temperature on Fines Migration in Sandstone Formations. Presented at the SPE International Symposium and Exhibition on Formation Damage Control, Lafayette, Louisiana, USA, 15–17 February. SPE-150953-MS. http://dx.doi.org/10.2118/150953-MS.

Nasr-El-Din, H. A., Al-Ghamdi, A. H., Al-Qahtani, A. A., and Samuel, M. M. 2008. Impact of Acid Additives on the Rheological Properties of Viscoelastic Surfactants. *SPE J.* **13** (1): 35–47.

Chapter 10

# Acid Corrosion and Its Control

**Donald G. Hill,*** Nalco Champion, and **Juanita Cassidy,**** Halliburton

## 10.1 Introduction: The Importance of Corrosion Evaluation

The development and use of acid-corrosion inhibitors form an important means of controlling metal corrosion during acidizing operations. For oilwell acid stimulation and chemical-scale removal, the metallurgy selected can easily corrode on exposure to strong acids, such as hydrochloric acid (HCl). Metallurgy for a producing well is selected on the basis of the produced fluids and the environment to which it will be exposed during daily operations. Acid stimulation applications are less frequent, and inhibitors must provide inhibition for a variety of stimulation fluids in all metallurgy that might be encountered. The appropriate corrosion test should be designed to simulate well-treating conditions. The selection of the most-effective inhibitor should be based on its performance, but environmental impact should also be considered.

**10.1.1 Acid-Corrosion Process.** The acidizing of petroleum reservoirs involves pumping inhibited acid through pumps and tubular goods during the treatment. When the pipe surface comes in contact with HCl, the steel pipe readily corrodes, dissolving the metal and releasing ferrous ions and hydrogen gas, as shown in Eq. 10.1. Oxidation of the metal surface readily occurs at the anode (Eq. 10.2), and hydrogen ions are reduced at the cathode, forming hydrogen gas (Eq. 10.3). This process is illustrated in **Fig. 10.1a.**

Reaction Sequence for the Corrosion of Iron in HCl Acid:

Basic Reaction: $Fe^{\circ} + 2HCl \rightarrow Fe^{++} + H_2 + 2Cl^-$ ...................... (10.1)

Anode Reaction (Oxidation): $Fe^{\circ} \rightarrow Fe^{++} + 2e^-$ ............................ (10.2)

Cathode Reaction (Reduction): $2H^+ + 2e^- \rightarrow 2H^{\circ} \rightarrow H_2$ ...................... (10.3)

As also illustrated in Fig. 10.1b, the corrosion process is inhibited by forming a tenacious inhibitor film over the metal surface at the anodic and cathodic sites. The corrosion process is a fast reaction, which becomes even faster at higher temperatures and with higher-strength acids. The addition of a corrosion inhibitor to the acid, resulting in the formation of an inhibitor film, is the most cost-effective process to effectively control acid corrosion. Many corrosion inhibitors use an anodic as well as a cathodic component in the inhibitor to form a protective film that is adsorbed on the metal surface. If this film is not uniformly adsorbed over the surface, pitting will result, forming pinhole pits that can eventually result in tubing failure. Also, adding inadequate inhibitor concentrations can result in excessive, but uniform, corrosion. Examples of failures caused by severe acid corrosion and pitting are illustrated in **Fig. 10.2.** These failures are the result of insufficient corrosion-inhibitor use and are not caused by the inability of the corrosion inhibitor for the particular acidizing environment.

---

* Now retired.

** Now with Cassidy Corrosion Control.

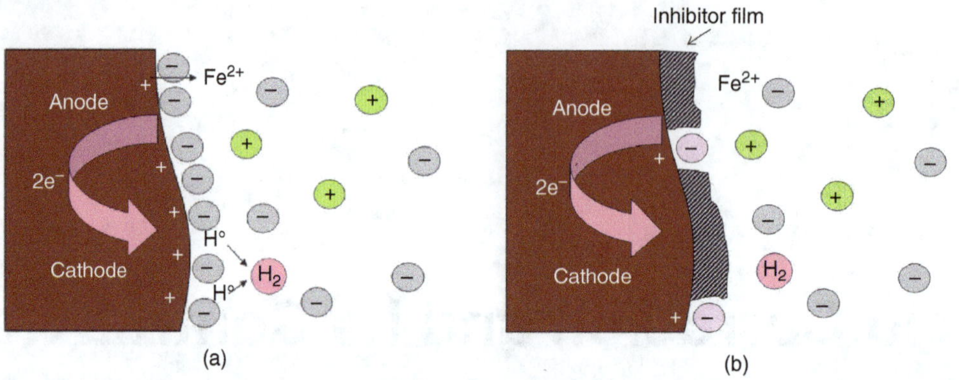


**Fig. 10.1—(a) Iron metal corrosion (uninhibited) and (b) iron metal corrosion (inhibited).**

**Fig. 10.2—Examples of excessive acid corrosion of coiled tubing.**

**10.1.2 Parameters That Affect Acid Inhibition.** Before performing acidizing treatments, inhibitor loadings are determined by corrosion testing to establish required concentrations, which are dependent on acid type and strength, anticipated time for acid treatments, and differing metallurgy. The performance of varied inhibitor concentrations is evaluated until acceptable inhibition is achieved, as defined by the testing criteria. Thus, the use of the selected inhibitor at the specified concentration should help prevent future occurrences of failures caused by the use of inadequate inhibitor. The actual corrosion observed downhole for an acid treatment, in many cases, will not be known unless field test coupons are used in the field process. This is not always possible, and laboratory testing becomes necessary. All factors that affect the corrosion process must be considered to simulate field operations.

The corrosion measured in the laboratory is influenced by the test operational parameters and the design of the test apparatus. Duplicating the corrosion process observed in field applications is ideal; however, this is often very difficult to achieve, and the laboratory procedure most closely representative of the field application should be used. The next objective is to ensure consistency in the corrosion-test procedures used to generate the data. Using the selected procedures should aid in the development of inhibitor-loading recommendations and schedules over a temperature range, which allows for comparisons with other inhibitor products. This is also important for comparing corrosion data from one laboratory with those from another using the same test conditions.

Using metal coupons obtained from oilfield tubulars is highly recommended for evaluating acid corrosion in oilfield applications. Three main classifications of metallurgy are encountered in oilfield operations—low-alloy steels, stainless steels, and high-nickel alloys. Low-alloy steels are steels that are iron-based and are alloyed with 0.1 to 1% carbon and other minor alloying elements. These steels are not made to meet a specific chemical composition; rather, they are made according to specific mechanical properties, such as yield strength and tensile strength (API 2011; Degarmo et al. 2003; Oberg 1996). Variable grain size of the steel will affect the corrosive behavior in the acid. Because of these factors, the same type of oilfield tubular, such as N-80 steel, from different manufacturers can corrode differently in strong acid. The mechanical properties and corrosion resistance of the steel depend significantly on the manufacturing process and the trace elements added for improved strength and workability.

High-chrome stainless steels and high-nickel alloys are also used extensively but for specific applications in which hydrogen sulfide (H2 S) and carbon dioxide (CO2 ) are encountered. A full range of oil-country-tubular-goods (OCTG) grades of pipe for oilfield applications are shown in **Fig. 10.3** (Vallourec 2015). The compositions of select chrome steels with the chrome content ranging from 13 to 25% chrome are compared with the composition of N-80 steel and CT-HS-80 (coiled tubing). The inhibition performance of these metallurgies will be discussed in more detail in this chapter.

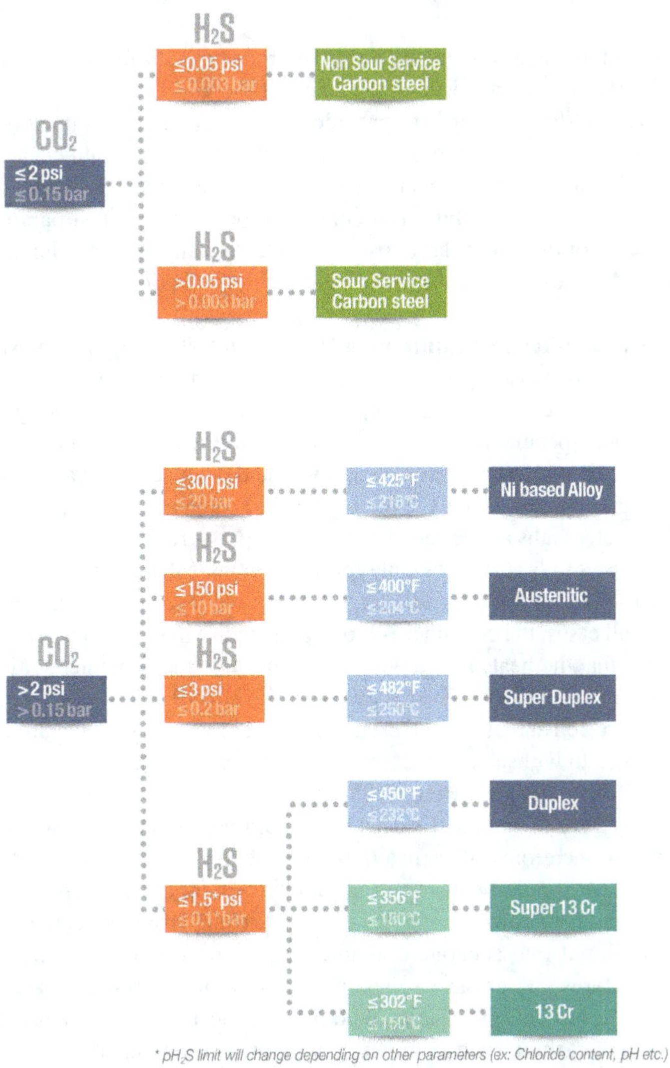


**Fig. 10.3—Application of oilfield steels for oilfield application. A full range of OCTG grades for all applications (after Vallourec 2015).**

**10.1.3 Corrosion-Testing Protocol.** In corrosion testing, the most important factor is designing and implementing test procedures that best simulate the preferred environment and operations. Procedures described in this chapter have been used extensively by service companies for the past 40 years to predict the effect of the fluid on the selected metallurgy and determine the best corrosion inhibitor for the acidizing applications. The procedures for completing corrosion tests in various industries have been extensively reviewed most recently by Baboian (1995) and previously by Ailor (1971). These works cover most, if not all, of the important aspects of corrosion-test parameters, reviewing numerous test environments and chemical industries. Also, a list of standards is provided by the American Society for Testing Materials (ASTM), National Association of Corrosion Engineers (NACE), and International Standards Organization (ISO). ASTM developed the majority of current standardized corrosion-test procedures and evaluations:

- *Standard Practice for Preparing, Cleaning, and Evaluating Corrosion Test Specimens* [*ASTM G1-03* 2011]
- *Standard Guide for Laboratory Immersion Corrosion Testing of Metals* (*ASTM G31-12a* 2012)
- *Standard Guide for Corrosion Tests in High Temperature or High Pressure Environment, or Both* [(*ASTM G111-97* 2013]

Other organizations might have equivalent or similar procedures; unfortunately, these procedures do not apply to the inhibition of acid corrosion during acid stimulation.

The primary objective of this chapter is to provide a summary of the parameters that can affect the evaluation of the corrosion process and to provide procedures that best simulate acid inhibition during the acidizing process. All of the procedures investigated provide a good basic method for understanding the corrosion process and the inhibition of corrosion; however, the test parameters selected will depend on the application of the particular corrosive fluid. No standard procedures can be written that would predict the actual corrosiveness of the fluid in a field application.

**10.1.4 Required Corrosion-Testing Equipment.** Corrosion-testing equipment should replicate the particular oilfield conditions as closely as possible, while taking into consideration the controlling factors that influence the outcome of the test. Temperature, pressure, time, metallurgy, and acid-volume/surface-area ratio of the test specimen all influence corrosion-test results. In immersion tests, a metal coupon is placed in acid with the preferred corrosion inhibitor and other acidizing additives. To perform tests at bottomhole temperatures (BHTs) below the boiling point of water and up to 200°F (91°C), glass bottles are often placed in hot-water baths and tested at atmospheric pressure. The container must be nonreactive with the test fluid. The use of glass is acceptable for all HCl fluids that do not contain hydrofluoric acid (HF). HF, commonly used in sandstone acidizing, will react with glass, so a suitable substitute, such as Teflon®, is required. In all cases, the container is loosely capped so that pressure generated during the test can be vented, but such that the heated fluid will recondense into the container so that the fluid volume is not lost over the duration of the test. For temperatures >200°F (91°C) or tests that require additional pressure, pressurized corrosion autoclaves are used. Autoclaves must be manufactured from acid-resistant materials that are designed to withstand the temperatures and pressures of corrosion tests as specified by American Society of Mechanical Engineers (ASME) safety standards (ASME 2010a, b). High-nickel alloys, such as Hastelloy B or C, are most frequently used. Large-sized autoclaves of Hastelloy can be cost prohibitive to obtain; therefore, other materials, such as low-carbon steel (ASME 4340 steel), can also be used. The autoclave should be designed to rapidly heat up at an acceptable rate, with minimal temperature overshoot, and cool down should be fast and efficient. External heating is preferable.

Acids solutions are placed into separate containers (glass or Teflon®) in the autoclave and tested at the preferred temperatures and pressures. The glass and Teflon® containers are used to isolate the acid or test fluid from the autoclave so that no acid reaction occurs with the autoclave. The test autoclaves are filled with an inert fluid, such as mineral oil, to provide a temperature- and pressure-transfer medium for the autoclave. This minimizes acid contact with the autoclave, which helps ensure autoclave integrity and that no contamination of the test fluid with corrosion products occurs.

## 10.2 Simulation of Well Conditions—High-Pressure/High-Temperature (HP/HT) Evaluation

**10.2.1 Corrosion-Test Autoclaves.** Autoclaves of varied sizes can be used. The autoclaves used for acid-corrosion evaluations can be divided into two main types—small NACE autoclaves **(Fig. 10.4)** and large Chandler autoclaves **(Fig. 10.5).** The industry is not restricted to the use of these autoclaves for the evaluation of corrosion. Similar types of autoclaves can be used, provided they afford rapid heat-up and cooldown and operate under sufficient pressures to maintain the fluid composition at varied temperatures.

**Fig. 10.4—NACE corrosion test setup. Courtesy of NALCO Champion.**

**Fig. 10.5—Chandler Autoclave, Model 56 (Chandler Engineering 2012).**

***NACE Corrosion Test Cell.*** Small NACE cell autoclaves are used for testing a single test fluid, with one coupon per autoclave test (API 1978). The 4-oz container will hold 100 mL of test fluid per test. NACE cell autoclaves hold only one container with one coupon for each test. Multiple cells are required to complete multiple corrosion tests at the same time. The NACE cell was designed for temperatures up to 350°F and pressures up to 1,500 psi, although actual limitations are based on the alloy of construction, wall thicknesses, and fittings. The cell is externally heated, and a heat-transfer fluid (such as mineral oil) is required to fill the void space between the cell wall and containers to provide a rapid temperature increase. When corrosive gases are a factor in corrosion, such as in a sour-gas situation in which hydrogen sulfide ($H_2S$) and $CO_2$ are present, the heat-transfer fluid is omitted. As a result, heat-up and cooldown times are greater than when mineral oil is used. Because of these differences in heating, errors can be introduced when comparing a sweet test to a sour test. This NACE test cell was initially designed by a NACE task group to evaluate the possibility of creating a standard corrosion test that could be used by multiple service and oil companies to obtain comparable results. Unfortunately, test results varied greatly when the same test fluid was evaluated by multiple service and oil companies. Attempts to develop a standard test were abandoned because of the wide variation; however, this cell is still used and provides good results when a consistent procedure is employed.

***Chandler Corrosion Apparatus.*** The Chandler corrosion apparatus is a large-capacity autoclave manufactured by Chandler Engineering, Tulsa (Model 56) (Chandler Engineering 2012). These autoclaves are designed to hold up to 20 tests with a 100-mL volume for each test, each containing one coupon specimen per 4-oz container. The Chandler autoclave can also be equipped to handle an 8-oz container to accommodate larger volumes of acid and larger coupon(s). Agitation of the glass containers can also be achieved with the Chandler autoclave. The glass containers are placed on a rack, and the rack is then placed on a sealed shaft that is mechanically agitated by reciprocation of the rack containing the test fluids. The rack can be agitated at 60 to 80° with 35 to 100 cycles/min. Chandler test autoclaves are heated by external heating of the autoclave-body wall. To increase heat transfer, the remaining space inside of the autoclave is filled with mineral oil. The use of nitrogen, $H_2S$, and $CO_2$ as pressurizing gases is not recommended for this application because the large volume of gas required for pressurizing the autoclave poses a significant hazard potential.

Heating and cooling characteristics of the two types of autoclaves can significantly affect results when comparing tests, not only between the NACE and Chandler autoclaves, but also between the same types of autoclaves from different laboratories.

**10.2.2 General Corrosion-Test Procedures.** The most common means of corrosion testing use the weight-loss and electrochemical methods. For weight-loss testing procedures, a metal specimen (coupon), of known dimensions and surface area, is cleaned of surface deposits, such as mill scale and oils, is preweighed, and then is placed into the test acid with the preferred acid inhibitors and other additives. This sample is placed into a corrosion-test autoclave and heated to the preferred temperature and held at the preferred pressure for the chosen test time. After cooling, the test coupon is removed, it is rinsed in organic solvents to remove the organic inhibitor film, adherent corrosion deposits are removed, and then the coupon is cleaned with soap and water. After a final rinse in hot water and in acetone to facilitate drying, the coupon is then reweighed to determine the resultant weight loss per unit area, or corrosion rate.

***Acceptable Corrosion-Rate-Test Criteria.*** In each standardized testing procedure, there is no acceptable corrosion rate specified. The limits currently used are set predominantly by the industry's unwritten standards for processes involved. For example, with acidizing applications for oilfield stimulation, the usually acceptable corrosion rate is 0.05 lbm/ft²/test period, with no unacceptable pitting. The corrosion rate is calculated by weight loss per unit area and per test period, as shown in the following equations. The mean and standard deviation can also be calculated, if preferred.

$$\text{Corrosion Rate (lbm/ft}^2\text{/test period)} = \frac{(W_o - W_f)}{A} \times CF\text{/test period}, \quad (10.4)$$

where

$W_0$ = initial weight (g),
$W_f$ = final weight (g),
$A$ = coupon-surface area (cm$^2$), and
$CF$ = conversion factor = 2.05 lbm/ft$^2$/g/cm$^2$.

Alternatively, corrosion rates based on mils/yr (mpy) can also be calculated, as follows:

$$\text{Corrosion Rate (mpy)} = \frac{W_o - W_f}{\rho \times A \times T} \times CF, \quad (10.5)$$

where

$W_0$ = initial weight (g),
$W_f$ = final weight (g),
$\rho$ = metal density (g/cm$^3$),
$A$ = coupon-surface area (cm$^2$),
$T$ = exposure time (hours), and
$CF$ = conversion factor = $3.45 \times 10^6$ (hr/cm/yr/mils).

The preferred corrosion-rate unit in acidizing operations is lbm/ft$^2$/test period for varied acid/steel exposure times. Because of short exposure times, such as 6 to 24 hours, these rate results are much more meaningful than mpy, depending on the application and time of exposure. A corrosion rate of 0.05 lbm/ft$^2$/6 hr would equate to a 0.610-g weight loss for N-80 steel with a surface area of 25 cm$^2$ and result in a corrosion rate of 1,799 mpy, 4.89 mils/D (mpd), or 1.22 mils/6-hr test period. These stimulation treatments range from 1 to 2 hours and up to weeks, depending on the application of the acidic fluid. A corrosion loss limit of 0.05 lbm/ft$^2$ is typically used for pipe-wall losses; not as much loss can be tolerated for smaller-diameter metallic materials, such as mesh screens or spring elements in tools.

Other industries will have different maximum corrosion rates to match their requirements. For example, in the chemical cleaning industry, the metal loss with reactive fluids is typically 0.01 lbm/ft$^2$/D (0.24 mpd)*. Brines are significantly less corrosive than acids, but they can be in contact with the wetted metal parts for months; therefore, their acceptable limit is significantly lower. For oilwell production and pipeline industries, the recommended corrosion rate is 1.0 mpy; however, as low as <0.1 mpy might be required**. Under these circumstances, exposure times are significantly longer than those observed for the stimulation fluids discussed previously. It is obvious that a standard weight loss cannot be recommended.

***Acceptable Pitting.*** Pitting is an equally important consideration when determining the acceptability of materials for corrosion inhibition because a localized type of corrosion can penetrate a pipe wall more rapidly than general corrosion. Weight-loss measurements are worthless without an evaluation of the size and depth of pitting. Control of pitting can be achieved through the addition of corrosion inhibitors and intensifiers. Several methods are used to report pitting results, including the maximum pit depth and number of pits per unit area. The maximum pit depth is the most important parameter, regardless of the number of pits, because it provides an indication of the time to potential failure. Edge effects tend to confuse this issue. Pitting observed on the edge of test specimens is sometimes considered acceptable because these areas are generated when steel pipe is cut to obtain the preferred size of coupon. In actual practice, these edges are not exposed to the stimulation acidic fluid. To evaluate pitting, the recommended procedure is to first expose a specimen of reasonable size (25 cm$^2$) to a test fluid for the purposes of measuring corrosion. After the corrosion measurement, the coupon is examined for the number and size of pits. There are several pitting scales in use currently. The pitting scale used by most service companies is illustrated in **Table 10.1** (Billings and Morris 1961). As an example, **Fig. 10.6** shows the differences in varied degrees of pitting for N-80 and 13Cr steels. Others in the petroleum industry use variations of the numerical scale for pit size and depth; a standard also exists as *ASTM G46-94* (2013). Pitting can be eliminated by the addition of more inhibitor or by changing the inhibitor type.

* Personal communication with W. Frenier. October 2003. The maximum allowable corrosion for the chemical cleaning industry.
** Personal communication with K. Bartrip. October 2003. Acceptable corrosion limits for production and pipeline operations.

| Description | Pitting Index |
|---|---|
| None | 0 |
| Minor edge corrosion | 1 |
| Pitting on edge only | 2 |
| Pinpoint pits on surface <25 | 3 |
| Pinpoint pits on surface >25 | 4 |
| Pits 16 to 31 mils diameter and 8 to 16 mils depth. Total 10 or less. | 5 |
| 11 to 25 pits of pitting index 5 | 6 |
| More than 25 pits of pitting index 5 | 7 |

Table 10.1—Pitting scale designation (Smith et al. 1978).

**Fig. 10.6—Examples of pitting observed at conclusion of test (Smith et al. 1978).**

**10.2.3 Corrosion-Testing Parameters.** There are many factors that influence the acid corrosion of oilfield tubulars when performing immersion tests in the laboratory, from specimen preparation to the test equipment required and the evaluation of the results. The major test parameters and procedures that can affect the outcome of a corrosion test are discussed in this subsection.

Initial studies in the 1960s examined the effects of fluid-volume/surface-area ratios on corrosion rates (Billings and Morris 1961) and the development of pressure autoclave test equipment that would evaluate the effect of velocity on corrosion (Billings et al. 1963). Smith et al. (1978) evaluated the effect of test parameters on corrosion inhibition and observed interesting effects. Each factor was evaluated to understand the effect of varied parameters on the corrosion results observed for highly corrosive systems. These papers helped to establish a protocol for laboratory evaluations to determine inhibitor-loading recommendations.

The parameters used for the simulation of downhole corrosive environments included temperature, time of test (protection time), selection of representative metallurgy for testing, metal-coupon preparation, pressure, volume of test fluid, and metal-coupon surface. Each of these factors can affect the overall outcome of a corrosion inhibitor's performance.

***Effect of Time and Temperature.*** The major contributing parameters of the corrosion process that occur when a metal is in contact with test fluid are time and temperature. The obvious effects of time and temperature are that the longer the time the coupon is at temperature, the greater the corrosion; also, the higher the temperature, the greater the corrosion. This is the test "time at temperature." Additional effects of time and temperature are factors during the heat-up and cooldown periods necessary for most of the common testing methods. These periods also contribute to the overall corrosion loss, because the corrosion process for a weight-loss test is an average of the temperature from the start to the finish of the test. The longer the testing time at the preferred temperature, the less effect heat-up time has on the evaluation. When reporting corrosion results, both the time at temperature and the heat-up and the cooldown rates should be fully described.

| Temperature (°F) | Protection Time (hours) | | |
|---|---|---|---|
| | At Temperature | Contact | Increase |
| 150 | 16 | 16.5 | 0.5 |
| 200 | 12 | 13.0 | 1.0 |
| 250 | 8 | 9.5 | 1.5 |
| 300 | 6 | 8.3 | 2.3 |
| 350 | 4 | 7.0 | 3.0 |
| 400 | 2 | 6.0 | 4.0 |

Table 10.2—Comparison of autoclave heatup times for protection times at temperature vs. acid steel contact time (Smith et al. 1978).

For autoclave testing, the heat-up rate and cooldown times are dependent on the autoclave design. The size and thickness of the pressure cell and the power wattage required by the heaters used are key factors that vary among autoclaves. The difference in time for a test completed at temperature vs. total contact time is shown in **Table 10.2** for a programmed heat-up rate. Use of equipment and procedures must be consistent to obtain reproducible data. The objective is to have as rapid as possible heat-up and cooldown times. The maximum average rate, approximately 5°/min, is a result of equipment limitations. Care must be exercised to be consistent with heat-up times, depending on preferred temperatures. It is advised that a programmable temperature controller be used for heat up to help minimize temperature overshoots.

***Pressure Effects.*** Pressure effects at high temperature can also affect the results observed. The corrosion process generates hydrogen, which acts as an inhibitor by minimizing the accessibility of the metal surface to the reactive fluid when under a minimum pressure (≈1,500 psi) (Jasinski et al. 1988). As another example, formic acid is frequently used as an intensifier because it generates carbon monoxide (CO) gas through decomposition in the presence of HCl. See Reaction Sequence 2 (Eqs. 10.6 through 10.10).

Reaction Sequence 2—Decomposition Mechanism of Formic Acid:

$$HCOOH + HCl \longleftrightarrow HCOOH_2^+ + Cl^- \quad (10.6)$$

$$HCOOH_2^+ + \longleftrightarrow HCO^+ + H_2O \quad (10.7)$$

$$Cl^- + HCO^+ \longleftrightarrow CO\,(sol) + HCl \quad (10.8)$$

$$CO\,(sol) \longleftrightarrow CO\,(gas) \quad (10.9)$$

$$HCOOH \longrightarrow CO + H_2O \quad (10.10)$$

The decomposition occurs both at the metal surface and in the HCl fluid itself. If the pressure is not adequate, the loss of CO will ultimately result in significantly higher corrosion rates. Corrosion inhibitors formulated with formic acid or precursors, such as formates or formamides, will thus be pressure sensitive. This is observed primarily at temperatures above 200°F. The increase in corrosion inhibition is not observed when potassium iodide is used **(Fig. 10.7),** as noted with Acid System D compared with Systems A, B, and C.

***Selection of Metallurgy.*** The selection of metal coupons should correspond to the metallurgy of the well to be treated. The application of these oilfield steels is demonstrated in Fig. 10.3. Miller and Iob (1986) studied the effect of metallurgical factors, such as normalization and quenching, on corrosion inhibition; a significant acid-corrosion effect was observed. These factors include the grain size of the steel, addition of alloyed metals and carbon, as well as content of phosphorus, sulfur, and nitrogen. Low concentrations of alloying metals can be added to iron or chromium steel to increase the strength of the steel or resistance to corrosive downhole environments, such as to $H_2S$ and $CO_2$. Depending on

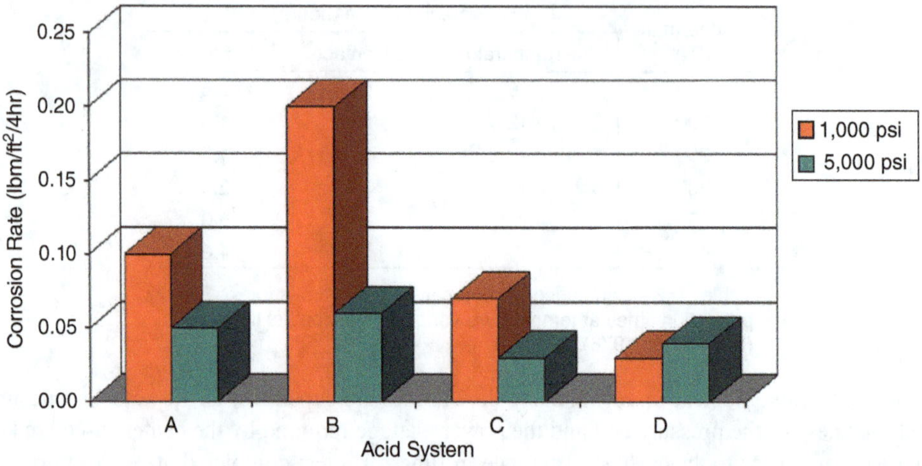


**Fig. 10.7—Effect of pressure on corrosion—test: 280°F, 13% Cr, 4 hours; A–28% HCl + 1.5% CI + 4% formic acid; B–28% HCl / EGMBE + 1.5% CI + 4% formic acid; C–15% HCl / MeOH + 2.0% CI + 6% formic acid; D–5% HCl / MeOH + 2.0% CI + 0.6% KI (Jasinski et al. 1988).**

the concentration of the alloying element, the corrosivity in acid can be dramatically affected. Typically, these steels can be easily protected in an acid environment by selecting the proper corrosion inhibitor and intensifier. When higher concentrations of these alloying elements (such as 1% chrome) are added to N-80 steels, their corrosion dramatically increases in HCl fluids. As a result, higher concentrations of the acid-corrosion inhibitor are required to provide acceptable inhibition.

The same grade of steel, such as N-80, can also vary greatly, depending on the manufacturer of the steel and because the chemical specifications only control the concentrations of a few elements. It is strongly recommended to obtain pipe from the same manufacturer for a project. For a new string of pipe, a review of the mill test report is recommended to select pipe of approximately the same composition. Alternatively, a corrosion test should be performed to verify the inhibitor's performance with alternative sources of steel pipe.

***Metal-Coupon Preparation.*** The initial preparation of the coupon surface for testing and the cleaning of the coupon after use are critical for obtaining useful data. Both the relevance and reproducibility of weight-loss data are highly sensitive to these procedures. Representative samples of the metal coupon for testing should be prepared as outlined in standard metal-coupon-preparation procedures, such as *Standard Practice for Preparing, Cleaning, and Evaluating Corrosion Test Specimens* (*ASTM G1-03* 2011). Cleaning of the specimens before they are weighed and exposed to the test fluid is important for removing any contaminants that could affect test results. Surface-finishing methods vary across a broad range for specific applications. Blasting with glass beads, sand, or other aggregates can provide an acceptable finish for most oilfield applications, where an unpolished surface is acceptable. Bead-blasting applications maintain the pipe in its original condition while removing mill scale. Alternatively, sanding with abrasive belts also provides an excellent clean surface for research evaluation. Reference *NACE Standard RP-0775-09* (1999) and *ASTM G4-01* (2014) for further details.

***Effect of Acid-Volume/Surface-Area Ratio.*** The reactive-fluid-volume/surface-area ratio can affect the corrosion-test outcome. The greater the volume of reactive fluid used, the greater the amount of inhibitor being added per unit area of metal coupon, provided that the same percentage of inhibitor is used for the tests. A greater fluid-volume/surface-area ratio, thus, can result in lower corrosion losses. On the other hand, using too small of a fluid volume not only decreases the per unit area of inhibitor, but also causes a concentration of corrosion products near the metal surface, which can potentially increase corrosion, especially when the fluid is not agitated. It should also be kept in mind that if a low solution-volume/surface-area ratio is used, the corrodent might be completely depleted during the course of the test. **Fig. 10.8** illustrates the effect of decreased corrosion for a greater solution-volume/surface-area ratio for metal specimens of equivalent area evaluated over 6 hours (Smith et al.

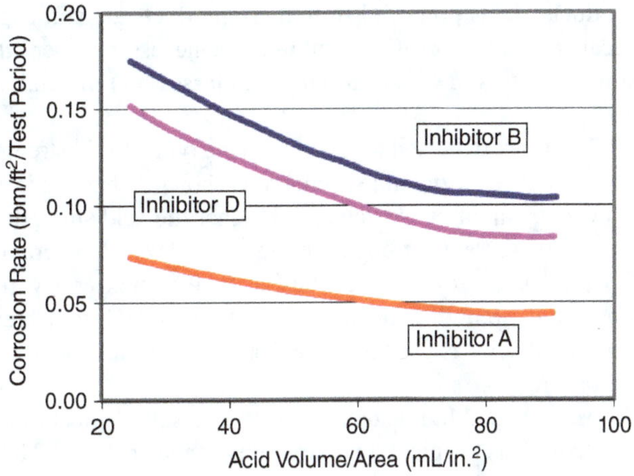


**Fig. 10.8—Effect of acid volume per unit area of coupon on observed corrosion (Smith et al. 1978).**

1978). However, this trend is not always observed. Billings et al. (1963) observed that when different acid-volume/surface-area ratios are used with the same inhibitor percentage and monitored over longer time periods, the results were not always consistent—with more total inhibitor resulting in less corrosion—and appeared to depend on the specific inhibitor being tested. One example is shown in **Fig. 10.9.** It appears from these studies that, at minimum, corrosion results will likely vary depending on the performance of the inhibitor with a given ratio of solution volume to surface area. The fluid volume corresponding to the diameter size of the pipe should be used when deciding the most appropriate testing ratio.

## 10.3 Fluids and Additives

**10.3.1 Typical Fluids Encountered and Their Inhibition.** Many different acids are used in downhole applications, and the choice of where and when to use each is discussed in other chapters. The most commonly used acids are the mineral acids (HCl and HCl/HF acid) and the organic (formic and acetic acids). Chelants are also used for certain applications. When placing these acids downhole, the only known method for circumventing the acid flow through pumps, manifolds, and pipes is to use

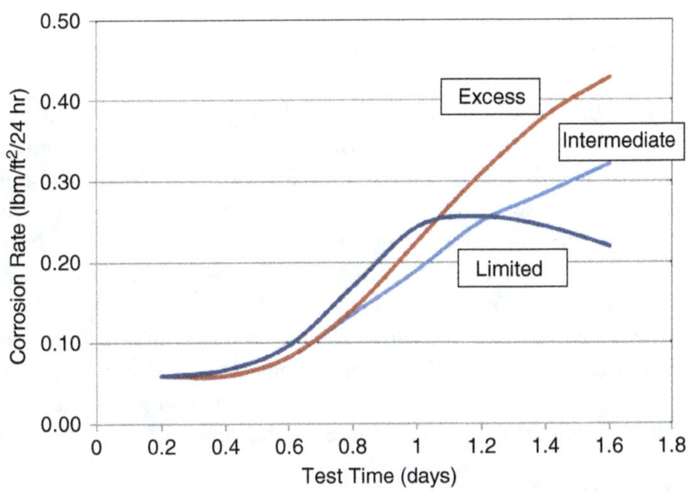


**Fig. 10.9—Effect of time on the inhibition of 15% HCl with varied volumes (Billings and Morris 1961).**

delayed-release acid products; acids produced from these products are still capable of corroding any metals they contact. Because of this, corrosion inhibitors are necessary whenever acids are pumped.

In acidizing, corrosion inhibitors are divided into two classes: (1) mineral-acid inhibitors and (2) organic-acid inhibitors.

***HCl.*** Mineral-acid inhibitors are designed to function in the typical HCl strengths of 5 to 28%, which is sufficient to stimulate carbonate formations, remove carbonate deposits in sandstone formations, and remove scale. However, as might be expected, the higher the acid strength, the more challenging it is to inhibit corrosion. Increased acid strength and increased well temperature limit application at temperatures above 325°F. Mixtures of organic acids, as well as mixtures with HCl (15% HCl), are recommended to increase acid strength equivalent to 20 to 28% HCl to achieve inhibition. When a mixture of organic and mineral acids is used in an acid stimulation treatment, a mineral-acid inhibitor is recommended for inhibition.

***Mud Acids (HCl/HF Mixtures).*** Mud acids are used to dissolve silicate materials, such as clays, deposited in sandstone formations. Typical acid strengths range from 9% HCl/1% HF acid to 12% HCl/3% HF. The HCl completely dissociates in water, whereas the pKa for HF is 3.14 at 25°C. The greater the acidity of the acid, the more difficult the acid is to inhibit. Typically, 12% HCl/3% HF is as difficult to inhibit as 12 to 15% HCl, as determined by the inhibitor loadings required. The presence of HF is not problematic unless special metallurgy is used that is more susceptible to corrosion by HF.

***Acetic Acid and Formic Acid.*** Organic-acid inhibition is much easier to achieve than in mineral acids, such as HCl, and therefore organic acids are applicable in higher temperatures. Acetic acid is applicable at temperatures up to 500°F but requires different chemistries than mineral-acid inhibitors. For organic acids, the most-effective inhibitors are typically those that contain amines and sulfur compounds, such as mercaptans and thiol compounds, in the inhibitor product.

Similar to mineral-acid inhibitors, the greater the acidity of the organic acid, the more difficult it is to inhibit. Formic acid and acetic acid are the most commonly used organic acids. Formic acid has a greater dissociation in water compared to acetic acid, resulting in a greater concentration of dissociated $H^+$. The pKa of formic acid and acetic acid is 3.75 and 4.76 at 25°C, respectively. Therefore, formic acid is more difficult to inhibit than acetic acid but much easier than mineral acids, such as HCl, which are completely dissociated. Even mixtures of the organic acids are easier to inhibit than HCl fluids. These organic acids have been effectively inhibited up to 500°F (static), although at very high temperatures, decomposition of the organic acid itself can occur. Greater inhibition times are observed for acetic acid than for formic acid at these temperatures.

Typical acid concentrations used are 9% formic acid and 10% acetic acid. These concentrations were initially used because these were believed to be the maximum concentrations of acids that would not result in calcium-salt deposition on spending of the acid. More recently, modeling of the solubility of the acid vs. temperature and pressure indicates concentrations up to 15 to 20% can be used, depending on well conditions.

***Weighted Acids.*** In some usages, acids are weighted to control hydrostatic pressure or to segregate fluids according to densities in the wellbore. Brines are used to weight the fluids, with the particular brine used being dependent on the solubility limits of the brine in the fluid and the weight required. Some typical brines used are NaCl, NaBr, KCl, KBr, $CaCl_2$, $CaBr_2$, $ZnBr_2$, and mixtures of these salts. The corrosivity of various weighted acids has been studied (Ke et al. 2005). Although these brines do impart some acidity, inhibitor dispersability is predominantly affected in an HCl weighted fluid. High concentrations of some of these brines can cause phase separation and/or precipitation of the corrosion inhibitor. It is imperative that the dispersability of the corrosion inhibitor in the weighted fluid be checked in addition to the inhibitor performance.

Brines themselves can be corrosive and can require a corrosion inhibitor. For instance, the pH of halide brines normally ranges from 2 to 6, which can be further influenced by dissolved gases that might be present, such as $CO_2$ and $H_2S$. Formate brines, on the other hand, are alkaline, with a pH of 9 to 11. Brine inhibitors should be checked for these applications.

**10.3.2 Special Acids and Their Challenges.** Certain unique acid formulations have been used and continue to be used today. These are fluids that have high concentrations of surfactants and oils and polymers, including alcoholic acid, emulsified acid, surfactant-based viscoelastic (VES) fluids, and polymer-gelled acids.

***Alcoholic Acids.*** Alcoholic acids are used primarily to remove gas blocks and to remove or prevent organic deposits. They consist of 15% HCl, which can contain up to 40% methanol or isopropanol. These fluids are more difficult to inhibit because the formation of inhibitor films is prevented. The addition of a corrosion inhibitor with an inhibitor intensifier is required to obtain inhibition. Frequently high inhibitor loadings (≥3%) are required with high loadings of a metallic intensifier, such as cuprous chloride. The maximum temperature at which they can be used is typically less than 250°F.

***Emulsified Acids.*** Emulsified acids consist of 15 to 28% HCl emulsified with 20 to 40% oil. To achieve this, a surfactant emulsifier is required, which can be oil-internal or oil-external. The primary purpose of these fluids is either to remove tarlike deposits or to retard the acid reaction with the carbonate formation. The limits of inhibition are primarily determined by the effect of temperature on phase separation. Because of this phase separation, corrosion inhibition can be very difficult. For oil-external emulsions, the addition of a corrosion inhibitor is very important. The steel surface of the tubular is still susceptible to acid corrosion because of the equilibration of the acid into the oil phase. Care must be exercised in selecting the appropriate inhibitor. Excessive inhibitor will cause a break of the emulsion, resulting in phase separation and causing excessive corrosion. The oil-soluble inhibitor will migrate to the oil phase, leaving uninhibited acid. This was shown by Hill and Jones (2003), who demonstrated the amount of emulsifier required to keep a stable emulsion, often increased with increased corrosion-inhibitor loading to achieve a balance of corrosion inhibition and emulsion stability.

Hill and Jones (2003) also evaluated the effect of emulsified-acid systems on acid-corrosion inhibition. These emulsion systems were prepared using 70% acid internal phase and 30% hydrocarbon or aromatic solvents. Initially, it was believed that an oil-external phase emulsion system containing 28% HCl should be easily inhibited, provided that the emulsion was stable. However, even stable emulsions exhibited increased corrosion rates (**Fig. 10.10**). This was attributed to the presence of an acidic aqueous interlayer between the oil-external phase and the metal coupons. Temperature and the presence of other acidizing additives also affect the stability of the emulsion. As shown in **Table 10.3,** excessive corrosion occurs, which is attributed to emulsion stability. When the emulsified system is maintained under constant shear and pressure, higher corrosion rates are observed even with increased corrosion-inhibitor loading, up to the point at which the sufficient inhibitor concentration is determined. This further illustrates that the corrosion process of the reactive fluid must be duplicated for accurate evaluation. These studies were completed in a flow loop similar to the one shown in **Fig. 10.11.**

***VES Fluids.*** VES fluids are used primarily to viscosify fluids, such as brines and acids. VES fluids also provide acid diversion, which helps open restricted flow channels during an acidizing treatment. The brine produced aids in the formation of viscosifying micelles. These surfactants are very detrimental to corrosion inhibition because of their long fatty carbon chain, which prevents the development of inhibitor films. Therefore, inhibition of these fluids is restricted typically to temperatures

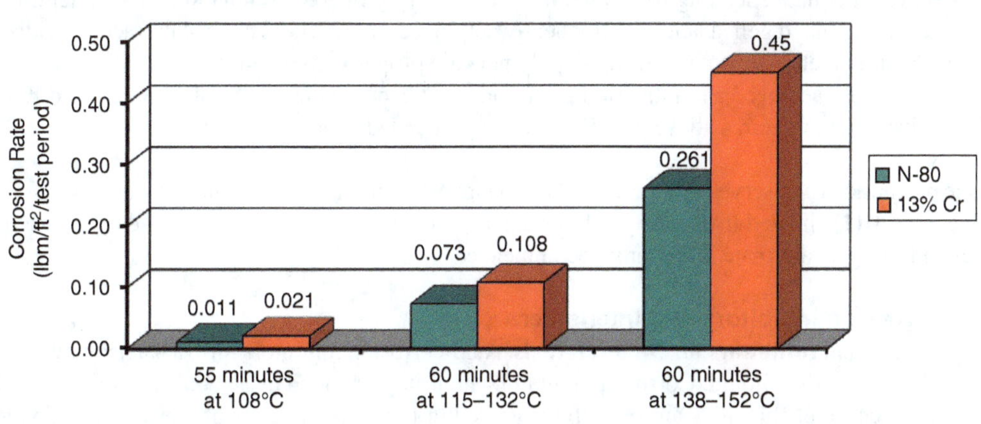


**Fig. 10.10—Corrosion inhibition of emulsified 28% HCl—dynamic test loop: 1,200 psi, 2.0% emulsifier, 0.2% acid-corrosion inhibitor (Hill and Jones 2003).**

| Inhibitor Loading | | | | | |
|---|---|---|---|---|---|
| ACI | AI | Emulsifier | Circulation Time (hours:minutes) | Corrosion Rate lbm/ft²/Test Time (PI) | Emulsion Stability |
| 0.2% | — | 2% | 1:00 | 0.264 (7) | Unstable |
| 0.4% | 1.0% | 3% | 3:55 | 0.170 (1) | Stable |
| 0.4% | 1.0% | 3% | 3:15 | 0.329 (1) | Stable |
| 0.8% | 1.0% | 3% | 3:00 | 0.129 (1) | Stable |
| 0.8% | 3.0% | 3% | 3:00 | 0.036 (1) | Stable |

Table 10.3—Corrosion inhibition of emulsified 28% HCl—dynamic test loop: N80 steel, 149°C, 1,000 psi; ACI = acid-corrosion inhibitor; AI = acid intensifier; PI = pitting index (Hill and Jones 2003).

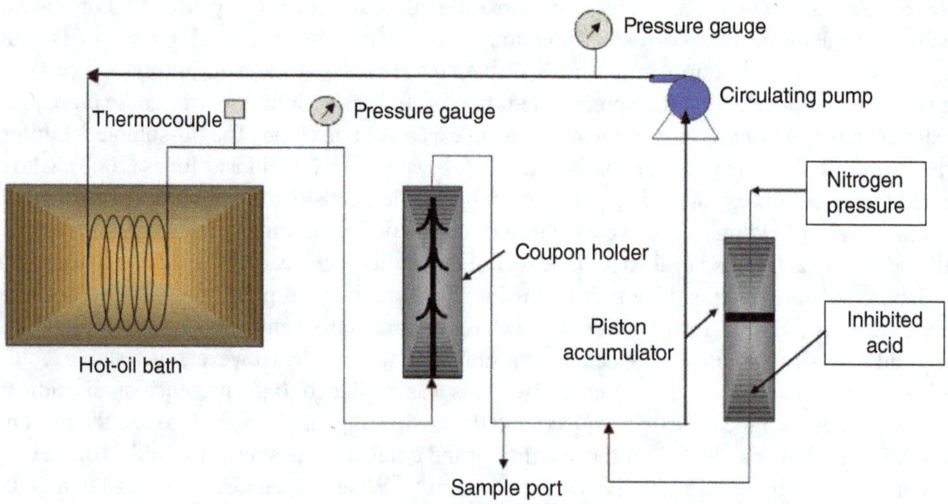


**Fig. 10.11—Flow-loop schematic to test emulsified acids (Hill and Jones 2003).**

≤ 200°F for short protection times, depending on the structure of the VES fluid and corrosion inhibitors used in the acid systems.

***Polymer Fluids (Gelled Acids).*** These gelled acids are easily mixed and can be inhibited with corrosion inhibitors and intensifiers. Typically, these fluids are based on a polyacrylamide polymer or copolymers in 5 to 28% HCl. The ease at which these polymers disperse and hydrate in the acid and the effect of carbonate spending determine their applicability in acid stimulation. Gelled-acid fluids and in-situ-crosslinked gelled acids are frequently employed for fluid-loss control in fracture acidizing and diversion in matrix treatments. If these polymers are properly mixed in the acid, adequate inhibition can be easily achieved and inhibition performance is comparable to that of ungelled acid. Using gelled acids at temperatures above 300°F is not recommended because of temperature limitations of the polymer.

One precaution to observe with these fluids is that they can be easily crosslinked by aldehydes. Cassidy et al. (2011) showed that if the inhibitor contains an aldehyde, crosslinking of the polymer can occur and using an alternate inhibitor is recommended.

## 10.4 Corrosion Inhibitors and Intensifiers

**10.4.1 Corrosion Inhibitors for Mineral Acids.** Acid-corrosion inhibitors form a barrier between the alloy and the corrosive liquid. Corrosion inhibitors are usually applied as an additive in the corrosive fluid, and because of this, they are typically triggered to form the inhibitor barrier by the oxidation/reduction reaction at the metal surface or by the electrostatics set up by the metal-surface charge at the metal/solution interface (Fig. 10.1).

For example, with corrosion inhibitors containing monomer molecules, evidence shows that the hydrogen radical produced by reduction of $H^+$ from the acid at the metal surface could serve as a catalyst for the development of a polymer layer at the surface; alternatively, the corrosion potential of the system could be such that a reduction polymerization could occur (Jasinski and Frenier 1995). Potential monomers in acid-corrosion-inhibitor formulations are α,β-unsaturated carbonyl compounds, such as cinnamaldehyde, or acetylenic alcohols (Growcock and Lopp 1988a, b), although whether polymerization or merely surface adsorption occurs with these molecules has not been definitively established (Bartos et al. 1993; Aramaki and Fujioka 1997; Bockris and Yang 1991). In HCl, large cationic aromatic ringed compounds, such as alkyl pyridine or benzyl quinoline quaternary ammonium chlorides, are effective corrosion inhibitors, presumably for several reasons: (1) the permanent positive charge on the N, which attracts the molecule to the negatively charged chloride layer at the metal; (2) the potential for the aromatic rings to form pi-bonds with the iron or other metals at the surface; (3) the ability of these molecules to cover a large surface area when the rings are oriented parallel to the surface; and (4) their ability to repel an aqueous-based solution resulting from the hydrophobicity of the organic hydrocarbon groups.

**10.4.2 HCl Intensifiers.** Intensifiers, also called inhibitor aids, are additives used to improve the performance of the primary acid-corrosion inhibitor, usually when the inhibitor has reached its limits, such as the maximum length of time or maximum temperature at which it can inhibit. Intensifiers do not inhibit by themselves but require the primary inhibitor to depress the corrosion rate. Intensifiers fall into two main groups: (1) those that inhibit the metal oxidation reaction (i.e., anodic inhibitors) and (2) those that interfere with the $H^+$ reduction (i.e., cathodic inhibitors). Iodides, from salts such as potassium iodide, are believed to adsorb at the metal-solution interface, forming a negatively charged iodide layer, which supplants the chloride layer in HCl solutions. Iodides have been shown to be synergistic with many heteroatomic organic compounds and are thought to form a favorable intermediate layer that allows for adsorption of a protonated or positively charged organic molecule (Bockris and Yang 1991; Hackerman et al. 1966). Similarly, CO, formed from the decomposition of formic acid in HCl, is an example of an intensifier that bonds with the metal surface (Cassidy et al. 2007; Uhlig 1940; Brezinski 1999c). Because CO solubility is greater with increased pressure, formic acid intensification improves with higher pressures. Examples of intensifiers that interact with the reduction of hydrogen ions at the surface include the Periodic Table Group 15 elements arsenic (As), antimony (Sb), and bismuth (Bi). If the alloy metal oxidation electrical potential is in a favorable range, then rather than the $H^+$ being reduced at the surface, the As, Sb, or Bi ion will be reduced, forming a protective metal coating or plating. One of the first types of acid-corrosion inhibitors used was an arsenic compound (Gravell 1928). Arsenic ion reduction is limited to acid strengths of no greater than 15% HCl. For this reason, 15% HCl strengths became very common in the oil industry. Exception for some usages of As, as noted by Cizek (1994), an organic-compound-based primary corrosion inhibitor is required with these Group 15 salts to form a proper, adherent plate on the alloy to be inhibited (Grebe and Sanford 1932; Walker 1985; Jasinski and Frenier 1992; Walker 1997; Cassidy 2006).

**10.4.3 Corrosion Inhibitors for Organic Acids.** Organic-acid inhibitors are formulated differently from HCl mineral inhibitors to be effective in the organic acids. In HCl, the chloride anion forms the negative charge at the electric double layer at the metal surface, but organic-acid counteranions, which would be the formate or acetate anions for formic or acetic acids, respectively, are not as strongly adsorbed to the metal surface, as shown by Eq. 10.11 (Hackerman et al. 1966).

$$I^- > Br^- > Cl^- > F^- \approx SO_4^- > RCO_2^- > ClO_4^- . \quad (10.11)$$

Thus, the electrostatic bonding that is believed to support adsorption of, for example, quaternized nitrogen compounds, is not as favorable. However, the addition of sulfur sources, such as thiourea, thiosulfate, or a mercaptan, as part of the organic-acid inhibitor formulation improves performance, even for inhibitors containing quaternized nitrogen compounds (Brezinski 1999a, b). Studies of the mechanism of mixtures of quaternary ammonium compounds used with sulfur compounds produced the theory that the sulfur compounds form $HS^-$ ions at the iron surface, thus attracting the organic cationic inhibitor ($In^+$) (see Reaction Sequence 3, Eqs. 10.12 through 10.14) (Iofa 1965; Iofa and Batrakov et al.1964).

Reaction Sequence 3—Formation of Iron Sulfide/Inhibitor Films:

$$RSH = HS^- \quad (10.12)$$

$$Fe + HS^- = Fe(HS)^- \quad (10.13)$$

$$In^+ + Fe(HS)^- = Fe(InSH) \quad (10.14)$$

Currently, there is no complete explanation for the specificity of these metals in a nitrogen/sulfur combination.

Acid intensifiers that function in HCl are also not as applicable in organic-acid environments. Commonly used plating intensifiers do not have adequate solubility in organic acids and/or the reduction/oxidation potentials are shifted too far to allow for good reduction of the plating elements.

**10.4.4 Corrosion Inhibitors for Brine.** Corrosion inhibitors for brines are often referred to as filming amines but can also be based on heteroatoms other than nitrogen (Kelland 2009). These corrosion inhibitors can contain quaternized amines, amine polycarboxylates, amides, thiocyanates, polyphosphate esters, and thiophosphates, and they will also frequently contain oxygen scavengers to mitigate the pitting tendency caused by the combination of chlorides and oxygen in these fluids. The effectiveness of corrosion inhibitors in brines is dependent on the brine type, brine density, pH, and temperature. The pH of the system influences the nature of the metal surface itself, which further dictates the type of corrosion-inhibitor molecules that will be effective. The pH of brine saturated with $CO_2$ will be 4 to 6. In this range, the metal surface will be a mixture of metal oxides and hydroxides. In particular, inhibitor surfactant molecules with polar head groups and alkyl chain tails are often used. Imidazoline is a commonly used example. These molecules have been shown to align with the metal oxide surface, with the surfactant head group of the molecule directed to the surface and the hydrocarbon tail pointed toward the aqueous fluid. By its hydrophobicity, this positioning not only provides a barrier to corrosion but also an attractant to hydrocarbons in multiphase flow; it is referred to as a "self-assembled monolayer" (Ramachandran et al. 1996). Further accumulations of corrosion inhibitor can occur if the hydrocarbon tail is long enough, with formation of bilayers and trilayers.

**10.4.5 Environmental Considerations.** Chemistry formulations of acid-corrosion inhibitors can be restricted by environmental considerations for the areas of the world in which they are applied. North Sea standards have been set for the use and discharge of oilfield chemicals offshore under the Oslo Paris Commission (OSPAR) and are focused on the effects on aquatic organisms by comparing component chemical biodegradation, bioaccumulation, and toxicity. The UK also monitors chemical biodegradation, bioaccumulation, and toxicity but uses a different assessment method called the Chemical Hazard and Risk Management (CHARM) model. The European community further operates under Registration, Evaluation, Authorisation, and Restriction of Chemicals (REACH), which limits chemical usage to protect human health and the environment. In the US, chemical use can be limited by the Environmental Protection Agency Priority Pollutants regulations. Several areas worldwide also restrict the concentrations of benzene, ethylbenzene, toluene, and xylene (BETX) allowed in products. All of these standards and regulations are constantly being reviewed and updated by the different governing authorities.

## 10.5 Acidizing Additives

All of the previously discussed fluids require additional additives to function in reservoir stimulation treatments. These additives are discussed in Chapter 9, and no further discussion here is necessary other than to say that these additives are required for their designed purpose; therefore, the inhibition of acid containing these additives must be properly determined.

**10.5.1 Surfactants.** Most acidizing additives contain surfactants. These include corrosion inhibitors and intensifiers to improve dispersibility in acidizing fluids, formation wetting agents to improve injection of the acid and return of the acid fluids, and nonemulsifiers to prevent and break acid/crude emulsions. Acid containing additives must be physically and chemically compatible with the corrosion inhibitor and intensifiers used. If it is incompatible, phase separation or changes in the activity of the inhibitor compounds will occur, potentially resulting in excessive corrosion. Surfactants that are oil soluble or highly water soluble can adversely affect corrosion inhibition.

**10.5.2 Solvents.** Solvents can dramatically affect the inhibition of acid because they provide improved dispersion for surfactants. Solvents also aid in well cleanup on return of the acid and for removing heavy-oil deposits on the rock matrix. Alcohols are the most commonly used solvents and include methanol, isopropanol, and ethylene glycol monobutyl ether (EGMBE). These are added to the acid in concentration ranges of 5 to 20% solvent. Some acid blends use up to 40% methanol. Again, excessive volumes will dramatically affect the inhibitor's dispersibility and solubility.

**10.5.3 Formulating a Stable Acid Blend.** The first issue encountered when formulating a stable acid blend is with the corrosion inhibitor itself. Surfactants, in combination with selected solvents, are used to provide optimum dispersibility of active inhibitor components. If the hydrophilic/lipophilic balance (HLB) of these surfactants is high (>14), dispersions or emulsions are often produced that are too stable, preventing the formation of an effective inhibitor film. The preferred HLB for corrosion inhibitors and other acidizing additives is typically 8 to 10.

Since most acidizing additives contain surfactants and solvents, each of the acidizing additives can individually affect the solubilization of the inhibitor in the acid and the inhibitor film formed. Often, combinations of high- and low-HLB surfactants are also used to improve dispersibility.

**10.5.4 Effect of Additives on Corrosion Inhibition.** The effect of these additives on the inhibition of HCl systems was also demonstrated by Smith et al. (1978) as well as Jasinski et al. (1988). Fig. 10.7 shows the influence of surfactant additives on inhibition in Acid Systems B and C compared to Acid System A. As demonstrated by Woodruff and Baker et al. (1975), the use of glycol solvents can also have a dramatic effect on corrosion; however, these effects are attributed to changes in the dispersibility of the corrosion inhibitor in the acid.

Acids containing these additives can be inhibited, provided adequate inhibitor and intensifier are used as determined through testing.

An example of this is shown in **Fig. 10.12,** where these additives were individually tested and their effect on the corrosion inhibition is demonstrated. Typically, nonionic surfactants and chelants used at nominal concentrations have very little or no effect, as shown with the first bar. The addition of other

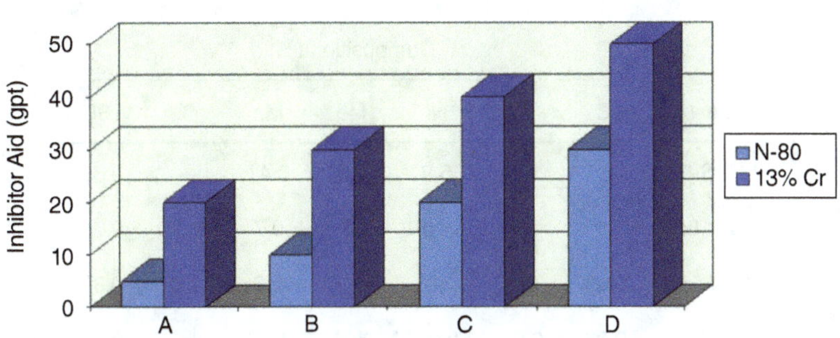


A–No additives or nonionic surfactant or chelant
B–Nonemulsifying agents or foaming agents
C–Mutual solvent
D–Antisludge additives

**Fig. 10.12—Effect of acid-stimulation additives on corrosion inhibition.**

surfactants and nonemulsifiers requires more inhibitor (second bar). Similarly, mutual solvents (e.g., EGMBE) require greater concentrations of inhibitor (third bar). The anionic surfactant dodecylbenzene sulfonic acid (DDBSA), commonly used as an antisludging additive, was the most difficult to inhibit because of its interaction with cationic corrosion inhibitors. This evaluation was completed at 300°F. At higher temperatures, the effect of corrosion and inhibitor loadings would be even greater.

## 10.6 Metallurgy

Three main classifications of metallurgy are encountered in oilfield operations—low-alloy steels, stainless steels, and high-nickel alloys. Typical compositions for select alloys are shown in **Tables 10.4 and 10.5** (Jasinski et al. 1988).

**10.6.1 Low-Alloy Steels.** Oilfield steels are an alloy of iron, but can contain many other alloying elements. They have a carbon content between 0.05 to 0.25% to retain formability and weldability. Other alloying elements include up to 2.0% manganese and small quantities of copper, nickel, vanadium, chromium, molybdenum, titanium, and other elements, such as rare earth elements and zirconium, depending on the physical properties preferred and the application (API 2011; Degarmo et al. 2003). Examples of low-carbon steel include the API grades N-80, L-80, J-55, P-105, P-110, and AISI 4130 and API grades CT80, CT90, and CT100 coiled tubing. With low-alloy steels, the presence of sulfur impurities is known to be a contributing factor in both general and pitting corrosion. Low-alloy steels are encountered in pumps, manifolds, casing, production tubulars, screens, downhole tools, and coiled tubing.

**10.6.2 Stainless Steels.** Stainless steels are iron-based and alloyed with carbon but contain at least 10% chromium. Stainless steels can be further classified according to their crystalline structure as austenitic, ferritic, or martensitic. Common austenitic stainless steels encountered in the oil industry are SAE grades 304 and 316. Austenitic stainless steels are susceptible to stress corrosion cracking in the presence of chloride and oxygen and for this reason are not recommended for long-term use

| | | Composition (%) | | |
|---|---|---|---|---|
| Steel | UNS No. | Cr | Ni | Mo |
| SM90S | G41300 | 1 | – | 0.3 |
| 414SS | S41623 | 13 | – | – |
| 13% Cr | S42000 | 13 | 0.35 | 0.02 |
| SM25Cr | S31803 | 25 | 6 | 3 |
| SM2535 | N0828 | 25 | 35 | 3.5 |

Table 10.4—Nominal composition of select chrome alloys (Jasinski et al. 1988).

| | Composition (%) | | | | | | |
|---|---|---|---|---|---|---|---|
| Steel | C | Cr | Ni | Mo | Mn | Cu | Si |
| 13CrS-95 | 0.01 | 12.1 | 5.9 | 1.9 | 0.41 | – | 0.33 |
| 13Cr-85 | 0.19 | 12.7 | 0.09 | – | 0.7 | 0.01 | 0.26 |
| Inc-825 | 0.03 | 21.5 | 40 | 2.5 | 0.6 | 2.0 | 0.30 |
| L80 (Variable) | 0.28 | 0.06 | 0.01 | 0.01 | 1.35 | – | 0.23 |
| CT-HS-80 | 0.12 | 0.57 | 0.16 | – | 0.78 | 0.27 | 0.33 |

Table 10.5—Composition of select CRAs.

in HCl environments. Ferritic stainless steels contain less nickel and higher chromium content than the austenitic steels, and they show good corrosion resistance with respect to chloride stress cracking and pitting; however, they are less tough and cannot be hardened by heat treatment. For this reason, pure ferritic-structured steels are not used very commonly in oilfield applications. Martensitic steels are less corrosion resistant than ferritic and austenitic steels but are tough and hardenable. Oilfield grade AISI 420, commonly called 13Cr-L80, is iron-based and contains approximately 13% chromium. It was developed to be resistant to $CO_2$ corrosion. However, it has proved to be more challenging to inhibit in HCl environments than low-alloy steels (Miyasaka and Ogawa 1990). Ke and Boles (2004) demonstrated the ability to inhibit corrosion of the different varieties of 13Cr steel, which was found to be dependent on the hardness and alloying elements present. For example, Mack (1992) and Joia et al. (2001) observed that adding molybdenum and nickel to create a modified 13Cr makes it more difficult to inhibit and increases the pitting tendency in an HCl-based fluid. 13Cr martensitic is used in a variety of oilfield applications, including tubulars and sand screens. Austenitic and ferritic crystal types are also combined to make duplex stainless steels. These two-phase steels, such as ASTM 2205 and 2507, show very good resistance in general fluids to $CO_2$ and $H_2S$ corrosion and are resistant to localized corrosion. However, they are a challenge to inhibit in HCl and HCl/HF fluids, demonstrating preferential ferritic phase dissolution. As such, when duplex alloys are used, acid applications can be limited to using organic acids at temperatures >300°F. High-performance cationic inhibitors typically demonstrate the best performance on duplex alloys in HCl, along with the use of iodide and plating-type intensifiers (Walker et al. 1994).

**10.6.3 Nickel-Based Alloys.** Nickel-based alloys are used in oilfield applications when corrosive conditions are more severe, such as when temperatures are high and/or $H_2S$ is present. Commonly encountered nickel alloys are 625, 718, 825, 925, G-3, and C-276. These alloys all contain more than 40% nickel and are alloyed with differing concentrations of chromium, copper, iron, and molybdenum. Generally, high-nickel alloys show good resistance to oilfield fluids containing HCl and can be inhibited using typical inhibitors.

Jasinski et al. (1988) observed the order of inhibiting these steels **(Fig. 10.13).** The order observed is as follows:

Nickel alloys>>carbon steel>martensitic chromes>>duplex.

## 10.7 Evaluation of Corrosion by Electrochemical Techniques

**10.7.1 Electrochemical Techniques.** Electrochemical techniques can be used for monitoring and measuring corrosion, with the advantage that electrochemical measurements can provide corrosion rates throughout the specimen's immersion in the test fluid, whereas weight-loss tests provide corrosion as an average over the total time period.

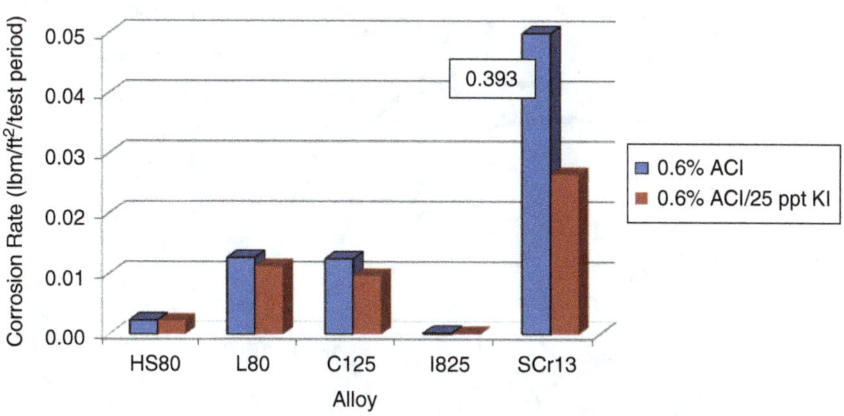


**Fig. 10.13—Comparison of corrosion inhibition of steel alloys. Test: 300°F, 3 hours, 3,000 psi (Jasinski et al. 1988).**

In acidizing, electrochemical measurements take advantage of the fact that corrosion is an electrochemical reaction in which the metal is oxidized (i.e., electrons are removed) in the anodic reaction and the proton of the acid is reduced in the cathodic reaction. An example of iron corrosion is shown in Eq. 10.1.

***Direct-Current Polarization.*** Direct-current polarization methods are typically used to characterize the active, passive, and transpassive potential regions for a given alloy in a particular fluid **(Fig. 10.14).** Linear polarization resistance (LPR) measuring techniques are used to provide in-situ corrosion rates but are not suitable for analysis of localized corrosion. Corrosion also can be monitored using an alternating-current method called electrochemical impedance spectroscopy (EIS). EIS data are modeled to fit an electrical circuit and provide polarization resistance values. Electrochemical methods require three electrodes—the test specimen as the working electrode, a counter electrode, and a reference electrode. Typical reference electrodes are limited in temperature, and for this reason most electrochemical measurements are performed only up to approximately 150°F at ambient pressures. Pseudoreference electrodes can be used to provide relative corrosion values at higher temperatures. Standard practices are available for acquiring these types of electrochemical measurements (ASME 2010a; *ASTM G106-89* 2010; *ASTM G5-14* 2014).

***Electrochemical Noise (ECN).*** ECN is another type of electrochemical technique that measures fluctuations in currents or potential over time at the corrosion potential and provides information concerning both general and pitting corrosion. Statistical methods are used to obtain information from the output. ECN is generally more suited to providing information on pitting than LPR and electrical-resistance (ER) techniques (discussed next). Literature is available regarding the standardization of ECN experiments (Kearns et al. 1996; *ASTM G199-09* 2014; Goellner et al. 1999).

***Electrochemical Techniques With Flow Conditions.*** Flow conditions are believed to be an important parameter in simulating field corrosion. Several specialized laboratory methodologies are available for testing corrosion with regard to flow conditions, depending on the flow characteristics preferred. These methodologies still use weight-loss and/or electrochemical measuring procedures for determining corrosion. Flow characteristics can range from laminar to turbulent flow and are described by the dimensionless Reynolds number, defined as the product of velocity, density, and pipe diameter divided by viscosity. LPR can be used for "moderately turbulent conditions" (NACE 1996) but is not

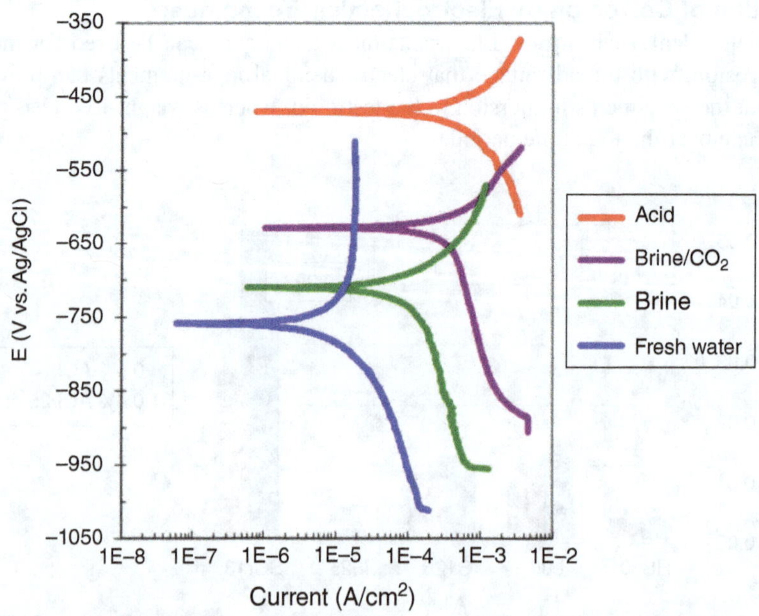


**Fig. 10.14—Polarization scans of the active potential region of cycled CT-90 in different test environments at 95°F (from Van Arnam et al. 2000).**

used for high-shear simulations, where shear is defined as fluid frictional force per unit area on pipe. For high-flow rates, flow loops, jet impingement, rotating-cylinder electrode (RCE), and rotating-cage methods can be used. RCE and rotating-cage methods can be used to evaluate flow at high-shear stress or turbulent conditions. The RCE method allows turbulent-flow conditions to be simulated at low Reynolds numbers, and jet impingement enables high-turbulence and multiphase systems to be simulated. High-pressure/high-temperature (HP/HT) testing is possible using the rotating-cylinder and -cage techniques by housing each system within an autoclave. Flow loops can also be used at HP/HT conditions. Use of jet impingement is not recommended under HP/HT conditions because of inconsistencies in the results caused by acid corrosion of the jet nozzle. Standards for these techniques have been published in the literature (*ASTM G170-06* 2012; NACE 1995).

***Alternative Corrosion-Evaluation Techniques.*** While the majority of corrosion-measuring methodologies use weight loss and electrochemistry, a few do not. ER has been used for decades and relies on the ER changes of metal probes that are in contact with the corrosive medium. As the cross-sectional area of the probe decreases as corrosion proceeds, the resistance value increases. The disadvantage of using these probes in highly corrosive environments is that as the sensitivity of the probe is increased, the life of the probe is reduced. However, this type of measurement has proved useful for less-corrosive environments and in environments that have low conductivity.

Imaging techniques can also be used for the quantification of corrosion. Atomic force microscopy (AFM) and vertical scanning interferometry (VSI) are two techniques that scan the surface of the specimen and compare it to a known reference to determine a corrosion value (Koyuncu et al. 2006). AFM and VSI have both been used to image and measure pit depths; VSI has been used to obtain general corrosion loss rates as well by using programming to calculate the volume of the metal surface removed and then converting that volume to a corrosion rate (Menendez et al. 2011). AFM uses a combination of scanning tunneling microscopy and profilometry, whereas VSI is an optical profiler that uses white-light interferometry. In AFM, the surface is mapped by running a cantilever-mounted tip over the surface while recording the force in relation to the surface features. A limitation of using AFM is its scan size, with a typical area scanned of 100 $\mu m^2$; whereas, VSI scan sizes can be in square millimeters.

Finally, ultrasonic testing can be used to continuously monitor metal wall thickness. In this method, sensors are mounted on the exterior of the pipe wall and acoustic waves are propagated through the wall. The difference in the echoes at the surface and the interior wall are converted to wall loss. Ultrasonic testing has compared favorably to the weight-loss method in side-by-side testing, with some limitations attributed to corrosion deposits on the interior wall and surface roughness (Ström and Strannhage 2006; Rannou et al. 2010).

## 10.8 Special Applications

**10.8.1 Velocity Effects.** Most studies concerning the effects of flow on the corrosion process and its inhibition were performed with aqueous systems, which can contain corrosive dissolved gases, such as $O_2$, $CO_2$, or possibly $H_2S$. Industrial cleaning applications have used flow-loop studies for years, where corrosion is evaluated by the weight loss of pipe samples or segments.

For circulating aqueous systems, flow and circulation (recirculation or one pass), aeration of water, chemical composition, and temperature must be considered. Circulation water is aerated and consequently becomes the primary source of corrosion. The water can be analyzed for metal content.

Only a few studies simulating pipe flow in strongly acidic fluids have been published (Frenier 1990; Unsal et al. 1988). Frenier (1990) performed a study using three different acid-corrosion inhibitors in 15% HCl at different temperatures and using a rotating cylinder. At the lower temperatures, an increase in corrosion rate with flow was observed, which was attributed to mass transport. At higher temperatures, no flow effect was observed up to 2 m/s. Generally, acid treatments through coiled tubing are not pumped at a sufficient rate to achieve turbulent flow because of the excessive pumping pressures that would be required. However, turbulent flow can be an issue for bullheaded acid treatments, including when acid-fracturing treatments are performed.

**10.8.2 Testing of Physical Properties.** Evaluating the metallurgy mechanical properties in relation to corrosion can be just as important as determining the metal loss caused by corrosion.

Mechanical-property testing can include stress corrosion, fatigue testing, cracking, and hydrogen permeation tests. These tests can be considered secondary for the development of the preferred inhibitor product, depending on the application conditions for which the product is designed. Discussion of these applications is beyond the scope of this chapter; however, the development of a suitable testing procedure, as discussed next, should be researched closely before the treatment begins. For recommended procedures, consult ASTM and NACE recommended standards.

**10.8.3 Effect of Dissolved Gases ($H_2S$ and $O_2$) on Corrosion Inhibition.** It is well-known that acids are significantly more corrosive than brine fluids to oilfield equipment. Brines are more easily inhibited against corrosion than acids. As a result, strong mineral acids, such as HCl, provide a significant challenge; however, the corrosion process is detrimentally affected by dissolved gases, such as $O_2$ and $H_2S$, for both acids and brines, as shown in Eqs. 10.15 through 10.20.

Oxidation of Iron Metal:

$$5\,Fe^0 + 4O_2 + H_2O \longrightarrow Fe_2O_3\downarrow + Fe_2O_4\downarrow + 2OH^- \quad (10.15)$$

Reaction of Iron and Iron Scale With $H_2S$

$$16\,Fe^{3+} + 8\,H_2S \longrightarrow S_8\downarrow + 16\,Fe^{2+} + 16H^+ \quad (10.16)$$

$$Fe^{2+} + H_2S \longrightarrow FeS\downarrow + 2H^+ \quad (10.17)$$

$$Fe^0 + H_2S \longrightarrow FeS\downarrow + H_2 \quad (10.18)$$

$$Fe_2O_3 + 3H_2S \longrightarrow 2\,FeS\downarrow + 3\,H_2O + S^0 \quad (10.19)$$

$$FeS + 2HCl \longrightarrow H_2S\uparrow + FeCl_2 \quad (10.20)$$

When providing acid-corrosion-inhibitor recommendations for metallurgy exposed to these gases, corrosion studies should be completed with mixtures of the acid containing these gases to simulate the downhole environment. $CO_2$ exhibits little to no effect on the HCl corrosion of oilfield metallurgy. Testing acid fluid with a gas ($O_2$ or $H_2S$) atmosphere will not produce the same result as the direct addition of these gases to the fluid. The time required to achieve equilibration at temperature is important for simulating the fluid chemistry effect and the corrosion process. The quantity of $H_2S$ in the acid fluid can be controlled by generating the $H_2S$ in situ by adding thioacetamide, which hydrolyzes to $H_2S$, acetic acid, and ammonium chloride, as shown in Eq. 10.21. Pyrite, the reactive form of iron sulfide, can also be used; however, cleaner and more complete reactions are observed with thioacetamide.

Generation of $H_2S$ In-Situ With Thioacetamide:

$$CH_3C(=S)NH_2 + HCl + 2H_2O \longrightarrow H_2S\uparrow + CH_3C(=O)OH + NH_4Cl \quad (10.21)$$

***Effect of Oxygen on HCl Corrosion.*** Dissolved oxygen in the test environment, such as dissolved oxygen in the acid, can be a concern for some treating fluids. Oxygen can be displaced by sparging the fluid with nitrogen to expel the oxygen, or oxygen scavengers can be used. When acidizing fluids are used with nitrogen to promote well cleanup and flowback of the spent acid, significant quantities of air (oxygen) can be present in the nitrogen stream, resulting in significant oxidation and excessive corrosion of the tubulars and downhole equipment. Attempting to duplicate field conditions should be considered if an open flow loop is used in the process. Even the simple act of bubbling oxygen into

| Test Number | Corrosion Inhibitor | Inhibitor Aid | Test Procedure | Corrosion Rate (lbm/ft²/4 hr) | Pitting Index |
|---|---|---|---|---|---|
| 4 | 1.2% | 2.5% FA | 50:50 $O_2$:$N_2$ | 0.057 | 1 |
| 6 | 1.2% | 2.5.% FA | 50:50 $O_2$:$N_2$ 65 sccm | 0.404 | 9 |
| 7 | 1.2% | 2.5% FA + 50 ppt Sodium Erythorbate | 50:50 $O_2$:$N_2$ 65 sccm | 0.245 | 9 |
| 8 | 1.2% | 30 ppt KI | 50:50 $O_2$:$N_2$ 65 sccm | 0.180 | 2 |
| 9 | 1.2% | 20 ppt KI + 10 ppt $CuCl_2$ | 50:50 $O_2$:$N_2$ 65 sccm | 0.117 | 2 |

Table 10.6—Effect of oxygen on corrosion inhibition of 10% HCl systems; test: coiled tubing, 80 steel, 300°F, 18 hours, 3,000 psi, 4-mL/cm²; additives: 5% EGMBE + 5% acetic acid + 0.2% surfactant + 2.0% EDTA; PI = pitting index.

the test fluid would not be representative of the actual fluid or environment because of variability of the oxygen content. As shown in **Table 10.6,** sparging oxygen into 15% HCl under nitrogen pressure dramatically increases acid corrosion, which proves to be difficult to inhibit.

Oxygen is best eliminated at the source because no good inhibitor exists for acidizing applications. Oxygen scavengers can be used, but excessive quantities would be required, which would be cost prohibitive.

***Effect of $H_2S$ on HCl Corrosion.*** The effect of $H_2S$ on corrosion inhibition in HCl was evaluated by Hill and DeMott (1977), who observed that $H_2S$ exhibited a detrimental effect on the inhibition **(Fig. 10.15).** The mechanism for the effect of $H_2S$ on this corrosion process is not fully understood; however, the observed increase in corrosion can be attributed to (1) the reaction of $H_2S$ with the inhibitor components, which produces a multitude of thiols, mercaptans, and sulfides or (2) sulfide adsorbed onto the steel surface. Obviously, reactions do occur under autoclave test conditions, which is apparent because of the observed foul odor typical of thiols and mercaptans, but the odor of $H_2S$ is not present after testing. The potential reaction of $H_2S$ with inhibitor components, such as propargyl alcohol, was studied by Delorey et al. (2002). A mixture of products containing chloride and sulfide could be identified. Either of these proposed mechanism pathways could alter the efficiency of the corrosion inhibitor.

Aldehydes, such as formaldehyde, have been used extensively and are recommended to help control the release of $H_2S$ and to reduce sulfide contamination in produced fluids (Eqs. 10.22 and 10.23). Formaldehyde also has been used in acid fluids to remove iron sulfide scales from producing tubulars.

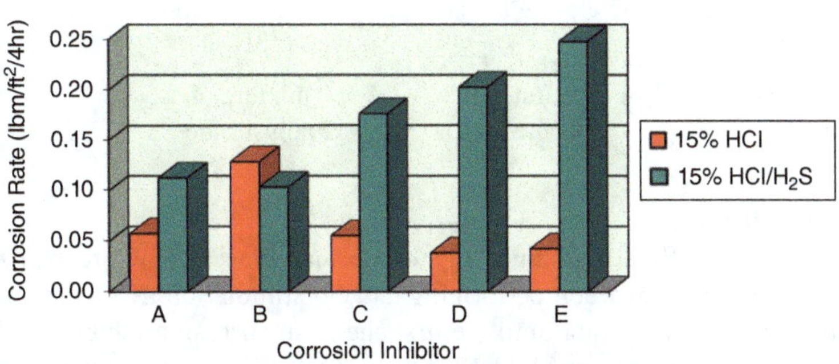


A—Amine-based inhibitor that does not include acetylenic alcohols
B—Amine-based inhibitor formulated with 40% mixture of acetylenic alcohols and formaldehyde
C—An ethoxylated amime and a quaternary amine inhibitor formulated with a 30% mixture of acetylenic alcohols
D—A quaternary amine inhibitor formulated with a 30% mixture of acetylenic alcohols and a copper complex
E—A quaternary amine-based inhibitor formulated with hydrocarbon solvent and 30% acetylenic alcohol
F—An unformulated acetylenic alcohol

**Fig. 10.15—Effect of $H_2S$ on HCl acid corrosion inhibition with select acid corrosion Inhibitors. Test: 300°F for 4 hours with 2% ACI and 0.5% formic acid in 15% HCl (Hill and DeMott 1977).**

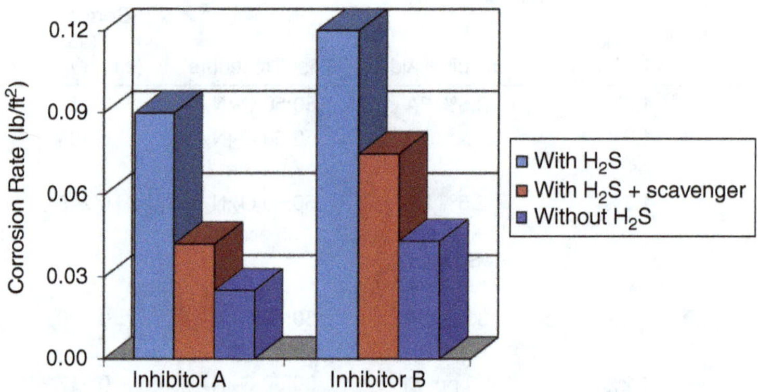


**Fig. 10.16—Benefit of $H_2S$ scavengers on HCl corrosion inhibition (Jasinski et al. 1988).**

Initially, this process was used for industrial cleaning applications. The addition of formalin (37% formaldehyde) eliminated the release of $H_2S$ during the cleaning process (Frenier 1982; Frenier et al. 1980). For acidizing applications, the scope was broadened to minimize the effect of $H_2S$ on corrosion inhibition for use when acidizing wells produce $H_2S$. The $H_2S$ released during an acid treatment dramatically increases the corrosion rate of oilfield tubulars when exposed to HCl fluids. These rates are dramatically improved when formaldehyde, or products that liberate formaldehyde, are added into the acid solution (**Fig. 10.16**). Formaldehyde has also been added to inhibitor formulations to enhance this corrosion inhibition when acidizing sour wells (Coffey et al. 1985). Numerous other aldehydes can be used for this application; however, some do not effectively reduce corrosion (Knox et al. 1972).

Reactions of $H_2S$ With Aldehydes to Scavenge $H_2S$:

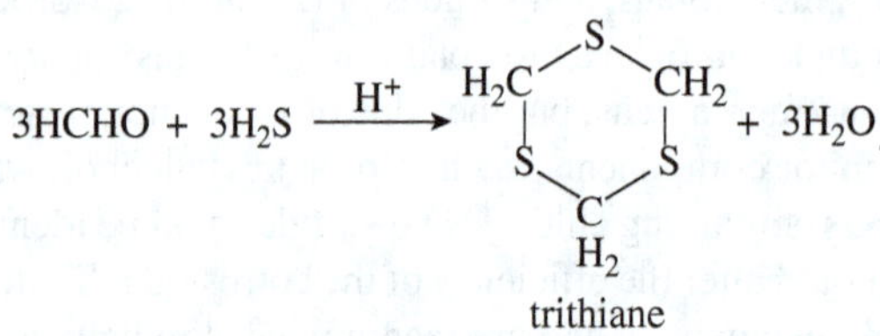


.......... (10.22)

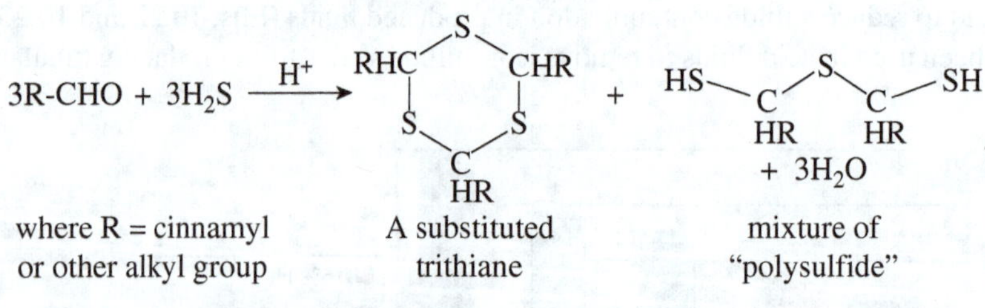


.......... (10.23)

## 10.9 Concluding Remarks

In the very first patent on oilfield acidizing in 1896, Herman Frasch (1896) recognized the problem of corrosion on oilwell tubulars when performing an acid stimulation treatment and offered methods to control it. However, it was not until the first chemical corrosion inhibitor was introduced in 1928 that the acidizing process became widely used (Gravell 1928). Corrosion inhibitors are now an integral component in every acid treatment and, as demonstrated in this chapter, must be designed for myriad, complex conditions. These conditions are constantly changing with the introduction of new oilfield technology, development of new alloys, drilling of higher-temperature wells, and enforcement of more-stringent environmental restrictions. These factors ensure that the challenges associated with acceptable corrosion control will continue into the foreseeable future. Also, corrosion-testing techniques must be developed to ensure that the corrosive fluids can be adequately inhibited.

## References


ASME. 2010a. Part A: Ferrous Material Specifications (Beginning to SA-450). In 2010 ASME Boiler and Pressure Vessel Code: An International Code—Section II: Materials. New York: American Society of Mechanical Engineers.

ASME. 2010b. Part B: Nonferrous Material Specifications. In *2010 ASME Boiler and Pressure Vessel Code: An International Code—Section II: Materials*. New York: American Society of Mechanical Engineers.

Ailor, W. H. ed. 1971. Handbook on Corrosion Testing and Evaluation. New York: Corrosion Monograph Series, John Wiley & Sons, Inc.

*API Unpublished Second Draft, Recommended Practices for Evaluation of Oilfield Acid Corrosion Inhibitors*. 1978. Washington, DC: API.

*API SPEC 5CT, Specification for Casing and Tubing, ninth edition*. 2011. Washington, DC: API.

Aramaki, K. and Fujioka, E. 1997. Spectroscopic Investigations on the Inhibition Mechanism of Propargyl Alcohol for Iron Corrosion in Hydrochloric Acid at Elevated Temperatures. Corrosion **53** (4): 319–326. http://dx.doi.org/10.5006/1.3280474.

*ASTM G111-97, Standard Guide for Corrosion Tests in High Temperature or High Pressure Environment, or Both*. 2013. West Conshohocken, Pennsylvania, USA: ASTM International. http://dx.doi.org/10.1520/G0111.

*ASTM G199-09, Standard Guide for Electrochemical Noise Measurement*. 2014. West Conshohocken, Pennsylvania, USA: ASTM International. http://dx.doi.org/10.1520/G0199-09.

*ASTM G3-89, Standard Practice for Conventions Applicable to Electrochemical Measurements in Corrosion Testing*. 2014. West Conshohocken, Pennsylvania, USA: ASTM International. http://dx.doi.org/10.1520/G0003-89R10.

*ASTM G106-89, Standard Practice for Verification of Algorithm and Equipment for Electrochemical Impedance Measurements*. 2010. West Conshohocken, Pennsylvania, USA: ASTM International. http://dx.doi.org/10.1520/G0106-89R10.

*ASTM G1-03, Standard Practice for Preparing, Cleaning, and Evaluating Corrosion Test Specimens*. 2011. West Conshohocken, Pennsylvania, USA: ASTM International. http://dx.doi.org/10.1520/G0001-03R11.

*ASTM NACE/ASTM G31-12a, Standard Guide for Laboratory Immersion Corrosion Testing of Metals*. 2012. West Conshohocken, Pennsylvania, USA: ASTM International. http://dx.doi.org/10.1520/G0031-12A.

*ASTM G170-06, Standard Guide for Evaluating and Qualifying Oilfield and Refinery Corrosion Inhibitors in the Laboratory*. 2012. West Conshohocken, Pennsylvania, USA: ASTM International. http://dx.doi.org/10.1520/G0170-06R12.

*ASTM G4-01, Standard Guide for Conducting Corrosion Tests in Field Applications*. 2014. West Conshohocken, Pennsylvania, USA: ASTM International. http://dx.doi.org/10.1520/G0004-01.

*ASTM G46-94, Standard Guide for Examination and Evaluation of Pitting Corrosion*. 2013. West Conshohocken, Pennsylvania, USA: ASTM International. http://dx.doi.org/10.1520/G0046.

*ASTM G5-14, Standard Reference Test Method for Making Potentiodynamic Anodic Polarization Measurements*. 2014. West Conshohocken, Pennsylvania, USA: ASTM International. http://dx.doi.org/10.1520/G0005-13.

Baboian, R. ed. 1995. Corrosion Tests and Standards: Application and Interpretation, second, 20. ASTM Manual Series, ASTM International. http://dx.doi.org/10.1520/MNL20-2ND-EB.

Bartos, M., Kapusta, S. D., and Hackerman, N. 1993. A Study of Polymerization of Propargyl Alcohol on Steel. J Electrochem Soc **140** (9): 2604–2605. http://dx.doi.org/10.1149/1.2220870.

Billings, W. E. and Morris, D. 1961. Effect of Acid Volume and Inhibitor Quantity on Corrosion of Steel Oil Field Tubing in Hydrochloric Acid. Corrosion **17** (5): 208t–214t. http://dx.doi.org/10.5006/0010-9312-17.5.86.

Billings, W.E., Knox, J.A., and Morris, D. 1963. Laboratory Apparatus Tests Pressure and Velocity Effects on Corrosion of Oil Field Tubing in Inhibited Hydrochloric Acid. Mater Performance **2:** 59–62.

Bockris, J. O. M. and Yang, B. 1991. The Mechanism of Corrosion Inhibition of Iron in Acid Solution by Acetylenic Alcohols. J Electrochem Soc **138** (8): 2237–2252. http://dx.doi.org/10.1149/1.2085956.

Brezinski, M. M. 1999a. Metal Corrosion Inhibited Organic Acid Compositions. US Patent No. 5,916,484.

Brezinski, M. M. 1999b. Corrosion Inhibited Organic Acid Compositions and Methods. US Patent No. 5,976,416.

Brezinski, M. M. 1999c. New Environmental Options for Corrosion Inhibitor Intensifiers. Presented at the SPE/EPA Exploration and Production Environmental Conference, Austin, Texas, USA, 1–3 March. SPE-52707-MS. http://dx.doi.org/10.2118/52707-MS.

Cassidy, J. 2006. Design and Investigation of a North Sea Acid Corrosion Inhibition System. Presented at the Corrosion, San Diego, California, USA, 12–16 March. NACE-06482.

Cassidy, J. M., McNeil, R. I., and Kiser, C. 2007. Understanding Formic Acid Decomposition as a Corrosion Inhibitor Intensifier in Strong Acid Environments. Presented at the International Symposium on Oilfield Chemistry, Houston, 28 February–2 March. SPE-106185-MS. http://dx.doi.org/10.2118/106185-MS.

Cassidy, J. M., Kiser, C., and Tucker, R. 2011. Corrosion-Inhibitor Design for Compatibility with Acid-Gelling Polymers. Presented at the SPE International Symposium on Oilfield Chemistry, The Woodlands, Texas, USA, 11–13 April. SPE-141435-MS. http://dx.doi.org/10.2118/141435-MS.

Chandler Engineering. 2012. http://www.chandlerengineering.com (accessed October 24, 2012).

Cizek, A. 1994. A Review of Corrosion Inhibitors Used in Acidizing. *Mater. Performance* **33** (1): 56–61.

Coffey, M. D., Kelly, M. Y., and Kennedy, W. C. Jr. 1985. Method and Composition for Corrosion. US Patent No. 4,493,775.

Degarmo, E. P., Black, J. T., and Kohser, R. A. 2003. Materials and Processes in Manufacturing, ninth edition. New York: John Wiley & Soncs, Inc.

Delorey, J. R., Vician, D. N., and Metcalf, A. S. 2002. Acid Stimulation of Sour Wells. Presented at the SPE Gas Technology Symposium, Calrgary, 30 April–2 May. SPE-75697-MS. http://dx.doi.org/10.2118/75697-MS.

Frasch, H. 1896. Increasing the Flow of Oil Wells. US Patent No. 559,669.

Frenier, W. W., Coffey, M. D., Huffines, J. D. et al. 1980. Composition and Method for Removing Sulfide-Containing Scale from Metal Surfaces US Patent No. 4,220,550.

Frenier, W. W. 1982. Method and Composition for Removing Sulfide-Containing Scale from Metal Surfaces. US Patent No. 4,310,435.

Frenier, W.W. 1990. Effect of Flow on Corrosion Inhibition of Steel in HCl. Presented at the 1990 NACE Corrosion, Houston.

Goellner, J., Burkert, A., Heyn, A. et al. 1999. Using Electrochemical Noise to Detect Corrosion: Evaluation of a Round-Robin Experiment. Corrosion **55** (5): 476–492. http://dx.doi.org/10.5006/1.3284010.

Gravell, J. H. 1928. Storage and Transportation of Acid Mixtures in Steel Drums. US Patent No. 1,678,775.

Grebe, J. T. and Sanford, R. T. 1932. Treatment of Deep Wells. US Patent No. 1,877,504.

Growcock, F. B. and Lopp, V. R. 1988a. Film Formation on Steel in Cinnamaldehyde-Inhibited Hydrochloric Acid. Corrosion **44** (4): 248–254. http://dx.doi.org/doi:10.5006/1.3583933.

Growcock, F. B. and Lopp, V. R. 1988b. The Inhibition of Steel Corrosion in Hydrochloric Acid with 3-Phenyl-2-Propyn-1-Ol. Corros Sci **28** (4): 397–410. http://dx.doi.org/10.1016/0010-938X(88)90059-5.

Hackerman, N., Snavely, E. S., and Payne, J. S. Jr. 1966. Effects of Anions on Corrosion Inhibition by Organic Compounds. J Electrochem Soc **113** (7): 677–681. http://dx.doi.org/10.1149/1.2424089.

Hill, D. G. and DeMott, D. N. 1977. Effect of Hydrogen Sulfide on the Inhibition of Oil Field Tubing in Hydrochloric Acid. Presented at the SPE Symposium on Sour Gas and Crude, Tyler, Texas, USA, 14–15 November. SPE-6660-MS. http://dx.doi.org/10.2118/6660-MS.

Hill, D. G. and Jones, A. 2003. An Engineered Approach to Corrosion Control During Matrix Acidizing of HTHP Sour Carbonate Reservoir. Presented at the NACE Corrosion, San Diego, California, USA, 16–20 March. NACE-03121.

Jasinski, R., Frenier, W. W., and Grannan, S. 1988. Inhibiting HCl Corrosion of High Chrome Tubular Steels. Presented at the 1988 NACE Corrosion, Houston.

Jasinski, R. J. and Frenier, W. W. 1992. Process and Composition for Protecting Chrome Steel. US Patent No. 5,120,471.

Jasinski, R. and Frenier, W. 1995. Corrosion Rate Correspondence for Steels in Inhibited HCl. *Proc.*, 8th European Symposium on Corrosion Inhibitors, Ferrara, Italy, 1,193–1,202.

Iofa, Z. A. 1965. Adsorption of Inhibitors of the Acid Corrosion of Iron and the Mechanism of Their Action. *Proc.*, 2nd European Symposium on Corrosion Inhibitors, Ferrara, Italy, 93–106.

Iofa, Z. A., Batrakov, V. V., and Cho Ngok, B. 1964. Influence of Anion Adsorption on the Action of Inhibitors on the Acid Corrosion of Iron and Cobalt. *Electrochim. Acta* **9** (12): 1,645–1,653. http://dx.doi.org/10.1016/0013-4686(64)80091-8.

Joia, C., Moraes, F., de Brito, R. F. et al. 2001. Performance of Corrosion Inhibitors for Acidizing Jobs in Horizontal Wells Completed with CRA Laboratory Tests. Presented at the 2001 NACE Corrosion, Houston, 11–16 March.

Ke, M., Boles, J., and Lant, K. E. 2005. Understanding Corrosivity of Weighted Acids. Presented at the SPE International Symposium on Oilfield Chemistry, The Woodlands, Texas, USA, 2–4 February. SPE-93271-MS. http://dx.doi.org/10.2118/93271-MS.

Ke, M. and Boles, J. 2004. Corrosion Behavior of Various 13 Chromium Tubulars in Acid Stimulation Fluids. Presented at the SPE International Symposium on Oilfield Corrosion, Aberdeen, 28 May. SPE-87561-MS. http://dx.doi.org/10.2118/87561-MS.

Kearns, J. R., Scully, J. R., Roberge, P. R. et al. ed. 1996. Electrochemical Noise Measurement for Corrosion Applications. West Conshohocken, Pennsylvania, USA: ASTM International. http://dx.doi.org/10.1520/STP1277-EB.

Kelland, M. A. 2009. Production Chemicals for the Oil and Gas Industry, first edition. Boca Raton, Florida: CRC Press.

Knox, J. A., Keeney, B. R., and Lasater, R. M. 1972. Method for Inhibiting Sulfide Cracking of Metals with Hydrogen Sulfide-Aldehyde Reaction Product. US Patent No. 3,669,613.

Koyuncu, I., Brant, J., Lüttge, A. et al. 2006. A Comparison of Vertical Scanning Interferometry (VSI) and Atomic Force Microscopy (AFM) for Characterizing Membrane Surface Topography. *J Membr Sci* **278** (1–2): 410–417. http://dx.doi.org/10.1016/j.memsci.2005.11.039.

Mack, R.D. 1992. Corrosion Inhibition of 13Cr, Super 13Cr, and 15Cr Stainless Steels in HCl-HF Acidizing Fluids. Presented at 1992 NACE Corrosion.

Menendez, C. M., Jovancicevic, V., Zhu, Z. et al. 2011. New Method for Assessing Corrosion Under Iron Sulfide Deposits and $Co_2/H_2S$ Conditions. Presented at the 2011 NACE Corrosion, Houston, 13–17 March. NACE-11265.

Miller, T.G. and Iob, A. 1986. Metallurgical Effects on Corrosion and Inhibition in Acidizing Stimulation. Presented at the 1986 NACE Corrosion, Houston.

Miyasaka, A. and Ogawa, H. 1990. Corrosion Behaviors of Corrosion Resistant Alloys in Acid Stimulation Environments. Presented at the 1990 NACE Corrosion, Las Vegas, Nevada, USA.

*NACE 5A195, State-of-the-Art Report on Controlled-Flow Laboratory Corrosion Tests*. 1995. Houston: NACE International.

*NACE 1D196, Laboratory Test Methods for Evaluating Oilfield Corrosion Inhibitors*. 1996. Houston: NACE International.

*NACE RP-0775-99, Preparation, Installation, Analysis, and Interpretation of Corrosion Coupons in Oilfield Operations*. 1999. Houston: NACE International.

Oberg, E. 1996. Machinery's Handbook, twenty fifth edition. New York: Industrial Press, Inc.

Ramachandran, S., Tsai, B.-L., Blanco, M. et al. 1996. Self-Assembled Monolayer Mechanism for Corrosion Inhibition of Iron by Imidazolines. Langmuir **12** (26): 6419–6428. http://dx.doi.org/10.1021/la960646y.

Rannou, G., Thierry, D., and LeBozec, N. 2010. Ultrasonic Monitoring of Steel Corrosion During Accelerated Corrosion Testing and Outdoor Field Exposures. Presented at the Corrosion, San Antonio, Texas, USA, 14–18 March. NACE-10170.

Smith, C. F., Dollarhide, F. E., and Byth, N. J. 1978. Acid Corrosion Inhibitors—Are We Getting What We Need? *J Pet Technol* **30** (5): 737–746. SPE-5644-PA. http://dx.doi.org/10.2118/5644-PA.

Ström, M.G. and Strannhage, G. 2006. Making the Best of Corrosion Testing. Presented at Eurocorr 2006, Maastricht, Netherlands.

Uhlig, H. H. 1940. Carbon Monoxide as an Inhibitor for Stainless Steel. Ind Eng Chem **32** (11): 1490–1494. http://dx.doi.org/10.1021/ie50371a020.

Unsal, A., Selcuk, N., and Doruk, M. 1988. Effect of Cleaning Medium Velocity and Temperature on the Corrosion of Steel. Mater Performance **27** (2): 26–30.

Vallourec. 2015. Material Selection Guide. http://www.vallourec.com/OCTG/EN/Products/material/Pages/default.aspx (accessed 21 September 2015).

Van Arnam, W. D., McCoy, T., Cassidy, J. et al. 2000. The Effect of Corrosion in Coiled Tubing and Its Prevention. Presented at the SPE/ICoTA Coiled Tubing Roundtable, Houston, 5–6 April. SPE-60744-MS. http://dx.doi.org/10.2118/60744-MS.

Walker, M. L. 1985. Method and Composition for Acidizing Subterranean Formations US Patent No. 4,498,997.

Walker, M.L, Cassidy, J.M., Lancaster, K.R. et al. 1994. Acid Inhibition of CRA's: A Review. Presented at NACE Corrosion 1994, Houston.

Walker, M. L. 1997. Corrosion Inhibiting Compositions and Methods. US Patent No. 5,591,381.

Woodroof, R. A. Jr., Baker, J. R., and Jenkins, R. A. Jr. 1975. Corrosion Inhibition of Hydrochloric—Hydrofluoric Acid/Mutual Solvent Mixtures at Elevated Temperatures. Presented at the Fall Meeting of the Society of Petroleum Engineers of AIME, Dallas, 28 September–1 October. SPE-5645-MS. http://dx.doi.org/10.2118/5645-MS.

Chapter 11

# Economics of Matrix Stimulation

**Michael J. Economides*** and **Matteo Marongiu-Porcu,**** University of Houston

## 11.1 Introduction

Production enhancement and, indeed, any petroleum-well activity are subjected to a physical as well as an economic optimization. The physical optimization such as the unified fracture design (UFD) (Economides et al. 2002) balances fracture geometry such as length and width to maximize the well productivity index. This optimization also includes constraints that are imposed not only by physical but also by logistical concerns.

In matrix stimulation, the injection of reactive fluids brings along with it side effects that become limiting considerations to theoretical expectations. These include lithological constituents that may dictate the strength and makeup of the fluid formulation or the injection time duration, limited by corrosion concerns or other factors. Other chapters of this monograph tackle these issues.

Superimposed on all is the economic optimization that starts with a range of physically robust designs and identifies the economically optimum among all options.

## 11.2 General Concepts of Acidizing Economics

The intent of this chapter is to focus on the economics pertinent to the specifics of acidizing treatments. The discussion will address directly the approach to the acidizing treatments design and the methodologies for conducting economic investigations and will present examples of various parametric studies.

The proper design of an acidizing treatment must follow four basic steps:

1. Identify the degree of damage removal that needs to be achieved and the consequent final value of skin.
2. Determine the acidizing treatment requirements (i.e., injection volumes, rates, and acid-mixture composition) to achieve the desired damage removal.
3. Evaluate what oil- and/or gas-production rates and recovery might be expected from the proposed acidizing treatment.
4. Apply a proper economic criterion to appraise the incremental production performance to be achieved by virtue of the acidizing treatment against execution costs.

Steps 1 and 2 have been exhaustively discussed throughout this monograph, while Step 4 is the main topic of this chapter and will be illustrated with several numerical examples. Step 3, in turn, is the production forecast (i.e., the estimation of both the production rate and cumulative recovery of the considered well vs. time).

---

* Deceased.

** Now with Schlumberger.

The methodology presented here for the evaluation of such production forecast for dry gas, undersaturated oil, and two-phase reservoirs with solution gas drive is based on analytical material-balance schemes coupled with proper inflow performance relationships (IPRs). An extensive description of this topic is provided in Economides et al. (2012).

In the case of a dry-gas reservoir, for instance, its cumulative production under the simple fluid-expansion mechanism can be written as

$$G_p = G_i - G = G_i - G_i \frac{B_{gi}}{B_g}, \qquad (11.1)$$

where $G_i$ and $G$ are the initial and current gas in place within a drainage area, respectively, and $B_{gi}$ and $B_g$ are the corresponding formation volume factors.

These formation volume factors can be expressed in terms of pressure, temperature, and gas-compressibility factor (Economides et al. 2012). Assuming isothermal reservoir conditions, easy algebraic manipulations yield

$$G_p = G_i \left( 1 - \frac{\bar{p}/Z}{p_i/Z_i} \right). \qquad (11.2)$$

Eq. 11.2 implies a linear relation between the cumulative production $G_p$ and the variable $\bar{p}/Z$. Thus, for $G_p$=0, then $\bar{p}/Z = p_i/Z_i$, and for the condition $\bar{p}/Z = 0$, then $G_p = G_i$. During the production history of a gas well, any value of the reservoir pressure corresponds to a cumulative volume of gas produced $G_p$.

Coupling Eq. 11.2 with a suitable gas-well IPR expression, a forecast of well performance vs. time can be developed readily, by discretizing the pressure depletion in a number of steps, determining the corresponding amount of gas produced for the $\Delta p$ considered, determining the expected gas-flow rate from the IPR, and then obtaining the time interval of the considered $\Delta p$.

The simplest IPR applicable for a dry-gas reservoir is the one based on Darcy's law for radial flow under a pseudosteady-state-flow regime,

$$q_g = \frac{\left[ m(\bar{p}) - m(p_{wf}) \right] kh}{1424T \left( \ln \frac{0.472 r_e}{r_w} + s \right)}, \qquad (11.3)$$

where $\bar{p}$ and $p_{wf}$ are reservoir and well bottomhole flowing pressures in psi, $m(\bar{p})$ and $m(p_{wf})$ are the corresponding real-gas pseudopressure functions is psi²/cp, $r_e$ and $r_w$ are reservoir drainage and wellbore radii in ft, $q_g$ is the gas rate in Mscf/D, $k$ is the formation permeability in md, $h$ is the pay thickness in ft, $T$ is the reservoir temperature in ºF, and $s$ is the skin factor.

In the case of an undersaturated-oil reservoir, the fluid recovery depends entirely on the fluid expansion as a result of underground withdrawal and associated pressure reduction. This can be evaluated by starting with the differential expression for isothermal compressibility, and with proper integration within the reservoir-pressure-depletion range (Economides et al. 2012), the following relationship for a volumetric recovery ratio is derived:

$$r = \frac{V_p}{V_i} = e^{c_t(p_i - \bar{p})} - 1, \qquad (11.4)$$

where $V_i$ is the original oil in place and $V_p$ is the oil produced at the considered pressure step of depletion. The total compressibility $c_t$ is equal to the weighted sum of oil, water, and gas compressibilities with oil, water, and gas saturations.

Again, Eq. 11.4 provides a relationship between the cumulative oil production $G_p$ and the reservoir pressure during depletion, and coupling it with a suitable oilwell IPR expression, a forecast of well performance vs. time can be developed readily in the same way as illustrated for the dry-gas well.

The simplest IPR applicable for a dry-gas reservoir is the one based on Darcy's law on radial flow under a pseudosteady-state-flow regime,

$$q_o = \frac{(\bar{p} - p_{wf})kh}{141.2 B_o \mu_o \left( \ln \frac{0.472 r_e}{r_w} + s \right)}, \quad (11.5)$$

where $\bar{p}$ and $p_{wf}$ are reservoir and wellbore face pressures in psi, $r_e$ and $r_{wf}$ are reservoir drainage and wellbore radii in ft, $q_o$ is the oil rate in STB/D, $\mu_o$ is the viscosity in cp, $k$ is the formation permeability in md, $h$ is the pay thickness in ft, and $B_o$ is the formation volume factor in RB/STB.

In the case of a two-phase reservoir with solution gas drive, the fluid-recovery calculation scheme is slightly more complicated and it is based on the Tarner (1944) method. This scheme will not be presented in this monograph, but it has been comprehensively described in Economides et al. (2012).

Whenever the value of the reservoir permeability or the drainage area is such that the initial transient flow regime is not negligible and lasts for a relevant period before the boundaries are felt and the pseudosteady-state-flow regime ensues, the use of a numerical reservoir simulator might be recommended. Nonetheless, once the skin effect in the IPR is properly modeled and accounted for, then the forecasted production profiles can be used within the selected economic criteria. The next section describes the most-acknowledged economic criteria in the oil and gas industry, and the following sections provide some numerical examples for actual carbonates, and sandstone oil reservoirs and point out some relevant differences between the two lithologies in terms of economic optimum for acidizing stimulations.

## 11.3 Main Economic Criteria

The crux of matrix stimulation is production enhancement, often meaning accelerating production in a depleting drainage. The question of course is whether this acceleration of the production can be balanced against the costs of the treatment.

In addition to the increase in production, an evaluation of the economics of a stimulation treatment must consider many factors, including treatment costs, additional reserves that may be produced before the well reaches its economic limit, and reservoir risks associated to mechanical problems that could cause the treatment to be unsuccessful.

While there are other criteria for assessing economic attractiveness or lack thereof, in wide use is the net-present-value (NPV) criterion for the optimization and the evaluation of the desirability of the specific treatment. The NPV is simply the algebraic sum of the cash flows discounted to a specific time using a discount rate that expresses the economic importance of the time factor. This rate can be interpreted in terms of "cost of capital" or as a rate of interest on the borrowed capitals. The "actualization time," can be either the initial "time-zero" or any time to which the evaluation may refer.

A fundamental tenet of any economic analysis is the relation between the sum of future revenues (income), adjusted for the time value of money and the capital expenditure (investment). Discounting a revenue stream for the time value of money gives the present value of the income as

$$PV = \sum_{n=1}^{N} \frac{R_n}{(1+i)^n}, \quad (11.6)$$

where $PV$ is the present value of the revenue stream, $R_n$ is the annual revenue (after payment of royalties and taxes and accounting for depreciation) for each of the $N$ years for which the project is considered, and $i$ is the discount rate. The reference time for the evaluation of the discounted annual revenues can be either the initial "time-zero" or any time to which the evaluation may refer.

For a given investment $I$, made to generate the revenue flow, if a discount rate $i$ is selected and a decision is made for the time span $N$ upon which a project is to be assessed, then the NPV is defined as

$$NPV = \sum_{n=1}^{N} \frac{R_n}{(1+i)^n} - I \quad (11.7)$$

For instance, for a constant net annual revenue, expected for a time frame of 5 years, and a discount rate of 10%, the present value of the revenue of a unit value would be (1+0.91+0.83+0.75+0.68) = 4.17 units (rather than 5). Then, if the initial investment were equivalent to, for examle, 2 units, subtracting from the present value of the revenue gives an NPV equal to 2.17 units.

In the specific case of an acidizing treatment, the capital investment $I$ is basically associated to the total costs for the treatment, which can be summarized into the following main categories:

- Stimulation fluids and additives: the volume of main treatment acid, any preflush fluids, additives such as corrosion inhibitors, iron-sequestering agents, clay stabilizers, diverters, and others.
- Service-company mobilization/demobilization costs: service-company presence, operating, and engineering personnel.
- Stimulation pumping equipment: on-location service of pumps, storage vessels, mixing tanks, and others. In contrast to fracturing, matrix stimulation treatments use far less and much-simpler field equipment.
- Stimulation technical support: field operators for treatment monitoring and technicians for mobile-laboratory analyses.

The annual revenue stream $R_n$ in Eq. 11.7 is given by

$$R_n = V_H \cdot \$_H \cdot (1-f_r)\cdot(1-f_o)\cdot(1-f_t), \quad (11.8)$$

where $V_H$ is the cumulative volume of hydrocarbons produced in the reference year (in bbl or in Mscf), $\$_H$ is the unitary revenue for the produced hydrocarbon (in USD/bbl or in USD/Mscf), $f_r$ is the fraction of gross cash flow because of the lease owners and/or to the foreign-nation governments as royalties, $f_o$ is the fraction of gross cash flow to be allocated as operative expenditures, and $f_t$ is the fraction of gross cash flow because of taxes in the relative fiscal regime.

**Fig. 11.1** shows the impact on the NPV calculations of two important parameters, the discount rate $i$ and the project lifetime $N$.

The discount rate expresses the economic value of the time factor. This rate can be interpreted in terms of "cost of capital," or as a rate of interest on the borrowed capitals. The discount rate reflects a number of considerations, one of which is inflation but others include the escalation in the value of competing investments, the least of which is the interest rate paid by banking institutions. An additional major consideration is risk, which can have many manifestations, from the elements to geopolitics and the infrastructure of a location. As evident from Fig. 11.1, a large discount rate substantially reduces the NPV of future revenue.

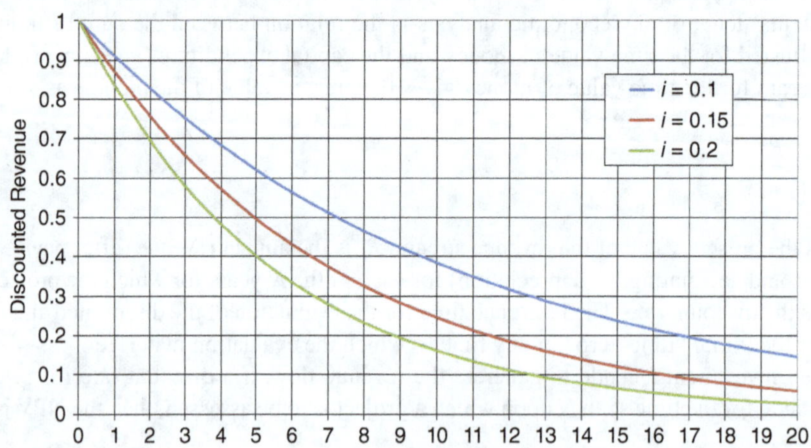


**Fig. 11.1—Impact of time and discount rate on the discounting process of the *NPV* criterion.**

The other important issue associated with any NPV calculations is that the time valuc of future revenue decreases exponentially as time increases. As evident from Fig. 1, the contribution to the NPV from the early years would be considerably larger than the contribution from the late years. This indicates that the actual number of years over which the NPV calculation is performed should be selected rationally: an NPV calculation for 100 years is worthless, even if we would be able to forecast the revenue stream generated by the project in such a far future.

If both $i$ and $N$ are known, imposed, or determined by past experience, then the NPV calculated can be compared with other NPV values and therefore the most attractive option is the one with the largest NPV. Needless to say, a negative NPV is never acceptable.

Eq. 11.7 can be used to determine two other different economic indicators that would add valuable information in assessing the attractiveness or the lack thereof of the investment (in our case, the acidizing treatment). If the NPV value is set equal to zero, then for a given or required discount rate $i$, the calculated $N$ is the *payout time* (PO). If, again, the NPV is set equal to zero and $N$ is specified (e.g., 5- or 10-year economic evaluation) the calculated $i$ is the *internal rate of return* (IRR). The IRR represents a lower limit for the discount rate $i$.

## 11.4 NPV Characterization of Carbonate Acidizing

The acidizing process in carbonate formations is fundamentally different from that in sandstones. In a clastic formation, the surface reaction rates are slow and a relatively uniform acid front moves through the porous medium. In carbonates, surface reaction rates are very high, so mass transfer often limits the overall reaction rate, leading to highly nonuniform dissolution patterns. Often, a few large channels, called wormholes, are created, caused by the nonuniform dissolution of limestone by hydrochloric acid (HCl) in a large block experiment. The structure of these wormhole patterns depends on many factors, including (but not limited to) flow geometry, injection rate, reaction kinetics, and mass-transfer rates.

In carbonate matrix acidizing, knowledge or estimation of the depth of penetration of wormholes allows a prediction of the effect of acidizing on the skin effect. In fact, the objective of a carbonate acidizing treatment is to create wormholes penetrating deep enough into the formation to bypass any damaged region and, often, to create a significantly negative skin factor. Thus, unlike sandstone acidizing where the objective is to overcome formation-damage effects, matrix stimulation of carbonates creates sufficient productivity enhancement to make it an attractive procedure even in the absence of formation damage.

For the sake of our economic analysis, the first thing to do is to have available a model that predicts the volume of acid required to propagate wormholes at a given distance. The simplest model that can be used for this purpose was first presented by Economides et al. (1994) and was called the *volumetric model.* When only a few wormholes are formed, a small fraction of the rock is dissolved; more-branched wormhole structures dissolve larger fractions of the matrix. Defining $\eta$ as the fraction of the rock dissolved in the region penetrated by acid, for radial flow it can be shown that the radius of wormhole propagation is

$$r_{wh} = \sqrt{r_w^2 + \frac{V}{PV_{bt}\pi\phi h}}, \quad (11.9)$$

where $PV_{bt}$ is the number of pore volumes of acid injected at the time of wormhole breakthrough at the end of a linear coreflood experiment, and it is the only parameter needed concerning wormhole propagation for this model.

This model does not consider the dependence of pore volumes to breakthrough on acid flux. A reasonable average of the $PV_{bt}$ over the range of fluxes that will occur in the wormhole region must be used for this model to correctly predict wormhole-propagation distances.

Because the wormholes created in carbonates are such large channels, it is generally assumed that the pressure drop through the wormhole region is negligible, so that the effect of the wormholes on the well skin effect is the same as enlarging the wellbore. With this assumption, the skin evolution in a carbonate matrix acidizing treatment can be predicted with the models of wormhole propagation. In a well with a damaged region having a permeability $k$, extending to a radius $r_s$, the skin effect during acidizing as a function of the radius of wormhole penetration is

$$s = \frac{k}{k_s} \ln \frac{r_s}{r_{wh}} - \ln \frac{r_s}{r_w} \quad (11.10)$$

Eq. 11.10 applies until the radius of wormhole penetration exceeds the radius of damage. If the well is originally undamaged or the wormhole radius is greater than the original damage radius, the skin effect during acidizing is obtained from Hawkins' formula assuming that $k_s$ is infinite, which gives

$$s = -\ln \frac{r_{wh}}{r_w} \quad (11.11)$$

Using Eqs. 11.10 and 11.11, if the injection rate is held constant throughout the treatment, the evolution of the skin effect predicted by the volumetric model during acid injection is (with a damaged zone)

$$s = -\frac{k}{2k_s} \ln \left[ \left( \frac{r_w}{r_s} \right)^2 + \frac{V}{PV_{bt} \pi r_s^2 \phi h} \right] - \ln \frac{r_s}{r_w} \quad (11.12)$$

and, with no damage or the wormholes penetrating beyond the damaged region,

$$s = -\frac{1}{2} \ln \left( 1 + \frac{V}{PV_{bt} \pi r_w^2 \phi h} \right) \quad (11.13)$$

Suppose we now consider the case of an undersaturated-oil-bearing carbonate formation (1-md), with a damaged zone with a permeability of 0.1 md extending 3 ft into the formation. Other main parameters are described in **Table 11.1.**

The first necessary step is to calculate the skin-factor profile as a function of injected volume per unit of formation thickness ($V/h$) by means of Eqs. 11.12 and 11.13.

| | |
|---|---|
| Drainage area, $A$ (acres) | 320 |
| Reservoir permeability, $k$ (md) | 1 |
| Reservoir thickness, $h$ (ft) | 100 |
| Permeability of damaged zone, $k_s$ (md) | 0.1 |
| Initial skin factor | 20.85 |
| Porosity, $\phi$ (fraction) | 0.15 |
| Water saturation, $S_w$ (fraction) | 0.30 |
| Reservoir pressure, $p_R$ (psi) | 6,000 |
| Bottomhole flowing pressure, $p_{wf}$ (psi) | 2,000 |
| Reservoir temperature, $T_R$ (°F) | 220 |
| Oil gravity (°API) | 30 |
| Well radius, $r_w$ (ft) | 0.328 |
| Pore volume to breakthrough, $PV_{bt}$ | 1.5 |
| Damage depth (ft) | 3 |

Table 11.1—1-md undersaturated oilwell example.

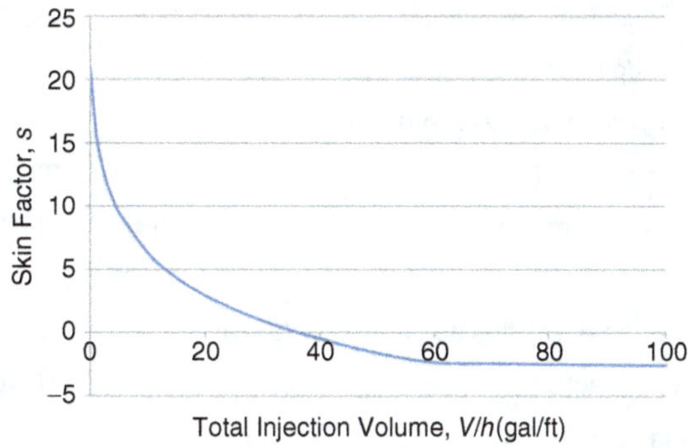


**Fig. 11.2—Skin-factor profile as a function of injected volume per unit of formation thickness predicted by the volumetric model for the 1-md carbonate reservoir.**

This first calculation is presented in **Fig. 11.2.**

Once the skin-factor profile during the acidizing treatment is available, it can be conveniently plugged into the selected IPR (i.e., Eq. 11.5), and use of the previously described material-balance scheme allows the generation of the set of production-forecast profiles illustrated in **Fig. 11.3.**

This set of production-forecast profiles simulates the performance of the 1-md oil well with damage skin partially removed ($V/h < 35$ gal/ft) with wormholes penetrating deep enough into the formation to bypass the damaged region (i.e., 3 ft) and create a significantly negative skin factor ($V/h < 35$ gal/ft).

Applying now the selected NPV criteria to this set of production-forecast profiles and selecting realistic economic parameters such as the ones presented in **Table 11.2,** a 5-year NPV characterization is finally presented in **Fig. 11.4.**

First, it is interesting to note in Fig. 11.4 that the total treatment costs are virtually insignificant (i.e., slightly negative initial NPV value) compared to the overall economic value of the stimulation project, and as such the PO time is extremely short (i.e., intersection with abscissa in less than 1 month for all cases illustrated).

Second, the base $V/h$ curve, corresponding to the unstimulated well ($s = 20.85$) provides a base value of NPV of some USD 3 million; the $V/h$ curve corresponding to zero skin (i.e., 35 gal/ft) happens to provide a 5-year NPV of some USD 9 million (thus, with a net profit from the stimulation treatment of USD 6 million). Prolonging the acidizing treatment and letting the wormholes penetrate deep enough into the formation bypasses the damaged region. The resulting achieved negative skin factor impacts

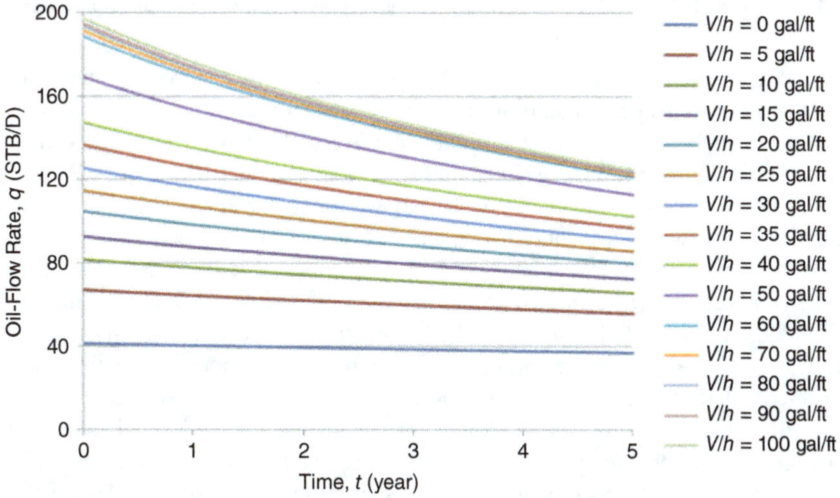


**Fig. 11.3—Production-rate forecast for different volumes of injected acid in the 1-md carbonate reservoir.**

| | |
|---|---|
| Cost of acid (USD/1,000 gal) | 900 |
| Cost of additives (USD/1,000 gal) | 180 |
| Mobilization/demobilization (USD) | 30,000 |
| Pumping charge (USD) | 30,000 |
| Discount rate (fraction) | 0.10 |
| Discount year | 0 |
| Oil price (no escalation applied) (USD/STB) | 93.0 |
| Royalties (fraction) | 0.10 |
| Working interest (%) | 100 |
| Net revenue interest (%) | 100 |
| Operating costs (fraction) | 0.20 |
| Taxes (fraction) | 0.30 |

Table 11.2—Economic parameters for the 1-md undersaturated oilwell example.

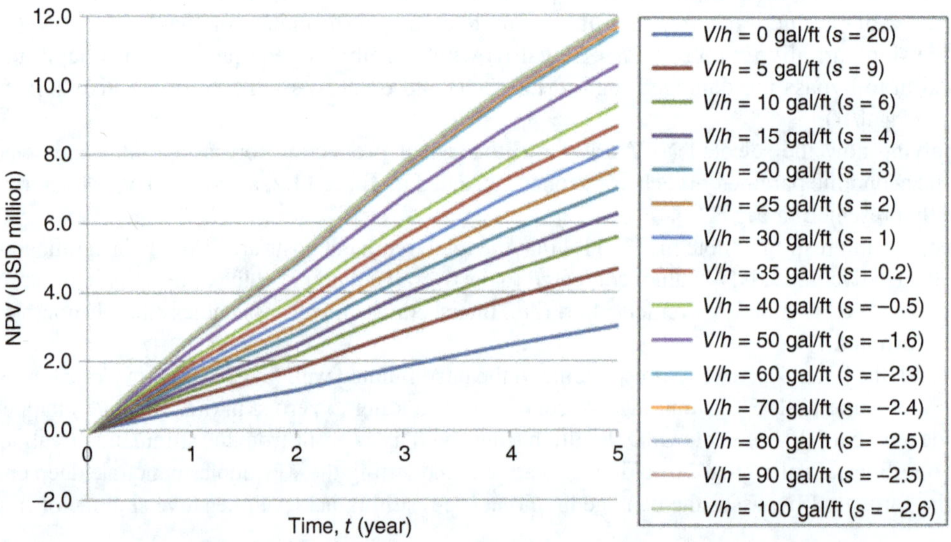


**Fig. 11.4—5-year NPV for different volumes of injected acid in the 1-md carbonate reservoir.**

the 5-year NPV; for instance, a skin factor of –2.6 provides a 5-year NPV of some USD 12 million (thus, with a net profit from the stimulation treatment of USD 9 million).

Third, and most important, replotting explicitly the 5-year NPV values vs. the different injected volumes of acid (see **Fig. 11.5**), it is possible to notice a clear plateau trend of the NPV, corresponding to the plateau of skin effects in Fig. 11.2 for the largest injected volumes. This trend implies a very significant difference between the economics of matrix acidizing and the economics of hydraulic fracturing. In fact, Marongiu-Porcu et al. (2008) and Marongiu-Porcu et al. (2009) have studied in detail the economics of hydraulic fracturing, by superimposition of a physical design optimization scheme, the UFD (Economides et al. 2002) with an NPV optimization procedure, and they have determined that, generally, economics dictates an optimal value for the size of fracturing treatments (i.e., mass of proppant injected), as shown in **Fig. 11.6** (from Marongiu-Porcu et al. 2008).

Because of this trend, generally, matrix stimulation treatments are not just subjected to the obvious physical requirement to minimize or reduce to zero the post-treatment damage skin value. Other considerations include logistical and operational constrains, such as in-situ handling of large volumes of acid and the major issues associated with corrosion within the tubular string during large injection times.

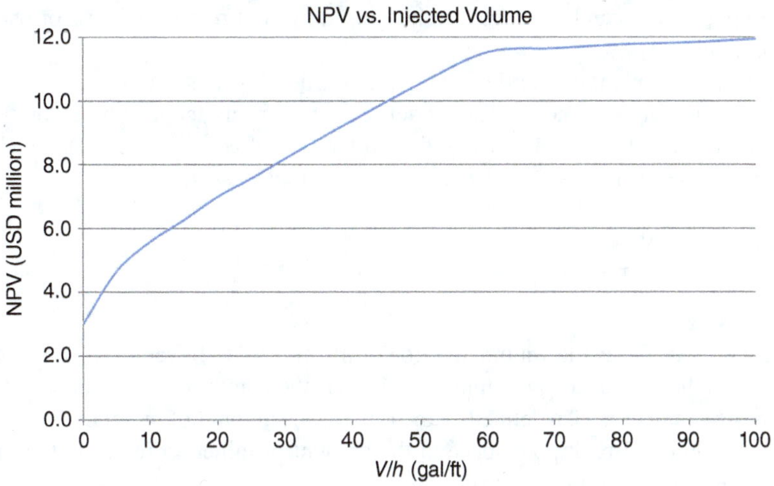


**Fig. 11.5—5-year NPV vs. volume of injected acid in the 1-md carbonate reservoir.**

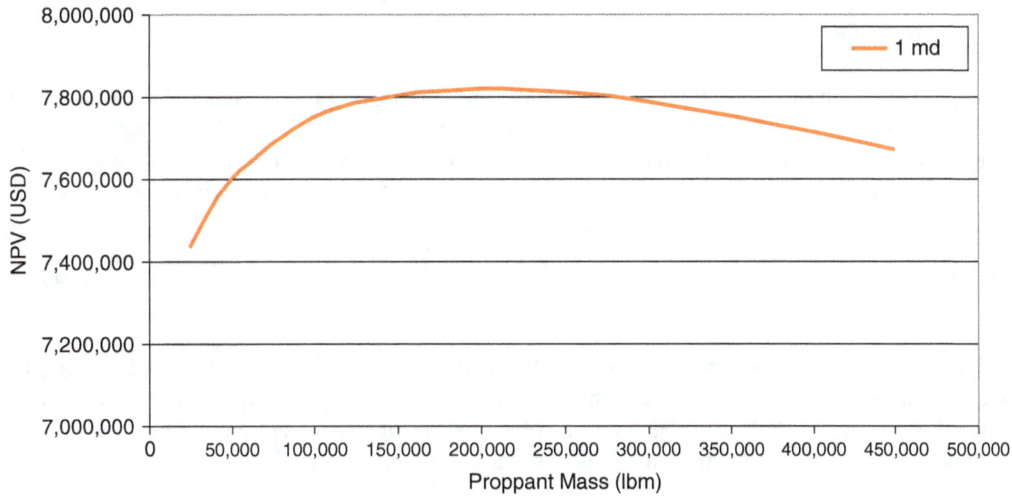


**Fig. 11.6—NPV vs. mass of proppant injected example for a 1-md 2-phase reservoir (from Marongiu-Porcu et al. 2008).**

## 11.5 NPV Characterization of Sandstone Acidizing

Economides et al. (1994, 2012) presented a procedure for the physical optimization of a sandstone acidizing treatment by using well-known relationships but combining them uniquely. The first relationship is the Hawkins (1956) formula, relating the skin effect with the penetration and severity of damage:

$$s = \left( \frac{k}{k_s} - 1 \right) \ln \frac{r_s}{r_w}, \qquad (11.14)$$

from which, given the reservoir permeability and assuming a penetration of damage, the damage permeability $k_s$ can be calculated.

The Labrid (1975) formula can then be used to relate the initial permeability and porosity (with the subscript $i$) and the permeability and porosity after acidizing:

$$\frac{k_i}{k} = M \left( \frac{\phi_i}{\phi} \right)^n \qquad (11.15)$$

The damaged porosity $\phi_l$ can be assumed to be caused by acid-removed debris or particles such as drilling mud or native clays.

Then, an acid stimulation model for describing the reaction kinetics and the acid convection is necessary during the acidizing treatments. The model selected for this scope is based on the Damköhler number (the ratio of the rate of acid consumption to the rate of acid convection), which for the fast-reacting mineral (such as betonies mud particles or native feldspar) is

$$N_{\text{Da},F} = \frac{(1-\phi_0)V_F^0 E_{f,F} S_F^* L}{u}, \qquad (11.16)$$

where $\phi_0$ is the porosity, $V_F^0$ is the volume fraction of the mineral to be removed, $E_{f,F}$ is the reaction-rate constant, $S_F^*$ is the surface area per unit volume, $L$ is the length of a core, and $u$ is the fluid flux.

The acid capacity number is also defined as the ratio of the amount of mineral dissolved by the acid occupying a unit volume of rock pore space to the amount of mineral present in the unit volume of rock, which for the fast-reacting mineral is

$$N_{\text{Ac},F} = \frac{\phi \beta_F C_{\text{HF}}^0 \rho_{\text{acid}}}{(1-\phi)V_F^0 \rho_F} \qquad (11.17)$$

where $\beta_F$ is the volumetric dissolving power and $\rho_{\text{acid}}$ and $\rho_F$ are the densities of the acid solution and the rock, respectively.

The acid/rock reaction model is rather complex and requires numerical solution. Schechter (1992) presented an approximate solution that provides the location of the acid front, given by

$$\theta = \frac{\exp(N_{\text{Da},S}\varepsilon_f)-1}{N_{\text{Ac},F} N_{\text{Da},S}} + \varepsilon_f, \qquad (11.18)$$

which relates dimensionless time (or, equivalently, acid volume) to the dimensionless position of the front, $\varepsilon_f$, defined as the position of the front divided by the core length for linear flow in a laboratory coreflood experiment. The subscripts $S$ and $F$ refer to slow- and fast-reacting minerals, respectively. The dimensionless acid concentration behind the front is given by

$$\psi = \exp(-N_{\text{Da},S}\varepsilon) \qquad (11.19)$$

For radial-flow applications, the Damköhler number can then be related to the linear core-determined value by

$$(N_{\text{Da},S})_{\text{radial}} = (N_{\text{Da},S})_{\text{linear}} \left(\frac{\pi r_w^2 h}{q_i}\right)_{\text{well}} \left(\frac{u}{L}\right)_{\text{core}} \qquad (11.20)$$

Using Eq. 11.18 and the dimensionless position of the acid front $\varepsilon_f$, the location of the acid front can be calculated. Then the required volume of acid can be determined from volumetric considerations,

$$\frac{q_i t}{h} = \theta \pi r_w^2 \varphi_o \qquad (11.21)$$

A similar expression can be shown for the ellipsoidal geometry, present in the stimulation through perforations (see Economides et al. 2012).

Penetration of a certain distance implies removal of all fast-reacting minerals and recovery of the corresponding portion of original porosity that then is related through the Labrid formula into a new permeability value and, henceforth, a new skin effect.

| | |
|---|---|
| Volume fraction of fast-reacting minerals | 0.1 |
| Volume fraction of slow-reacting minerals | 0.85 |
| HF concentration, lbmHF/lbm solution | 0.03 |
| HCl concentration, lbmHCl/lbm solution | 0.12 |
| Acid-solution density ($lbm/ft^3$) | 67.08 |
| Acid-injection rate (bbl/min/ft) | 0.1 |
| Damage depth (ft) | 1; 3 |
| Perforation density (shots/ft) | 4 |
| Perforation Length (ft) | 0.5 |

Table 11.3—Additional sandstone-acidizing parameters.

The original and the resulting values of the skin effect lead to pre- and post-treatment production forecasts, and, thus, an NPV characterization can be promptly performed.

Consider now the case of an undersaturated-oil-bearing sandstone 1-md formation, with a damaged zone with a permeability of 0.1 md extending 3 ft into the formation. Other main reservoir parameters are kept equal to the carbonate example studied in the preceding section and described in Table 11.1, while a set of additional treatment information specific for this sandstone example is provided in **Table 11.3.**

Again, the first necessary step is to calculate the skin-factor profile as a function of injected volume per unit of formation thickness (*V/h*), by means of Eqs. 11.14 through 11.21. This first calculation is presented in **Fig. 11.7.** Interestingly, we notice that the combination of initial skin factor of 20.85 and the 3-ft damage depth is quite severe, and even with an asymptotically infinite injected volume it would not be possible to remove all damage, not actually being possible to achieve skin factors less than 8. **Fig. 11.8** shows an equivalent skin-factor profile as a function of injected volume per unit of formation thickness, generated for 1-ft damage depth. It appears to be possible to remove almost all damage by means of the acidizing treatment.

Once the skin-factor profile during the acidizing treatment is available, it can be conveniently plugged into the selected IPR (i.e., Eq. 11.5), and use of the previously described material-balance scheme allows the generation of the two sets of production-forecast profiles, illustrated in **Figs. 11.9 and 11.10.**

Again, applying now the selected NPV criteria to this set of production-forecast profiles and using again the economic parameters presented previously in Table 11.2, a 5-year-period characterization is finally presented in **Figs. 11.11 and 11.12.**

It is interesting to note that, as for the carbonate-reservoir example, the total treatment costs are virtually insignificant (i.e., slightly negative initial NPV value) compared to the overall

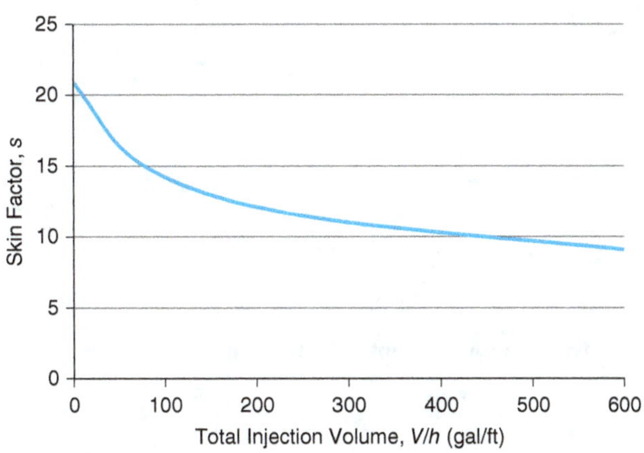


**Fig. 11.7—Skin-factor profile as a function of injected volume per unit of formation thickness predicted for the 1-md sandstone reservoir with 3-ft damage depth.**

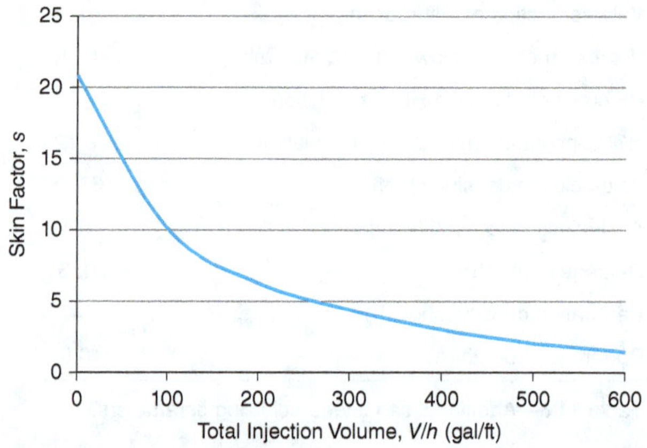


**Fig. 11.8—Skin-factor profile as a function of injected volume per unit of formation thickness predicted for the 1-md sandstone reservoir with 1-ft damage depth.**

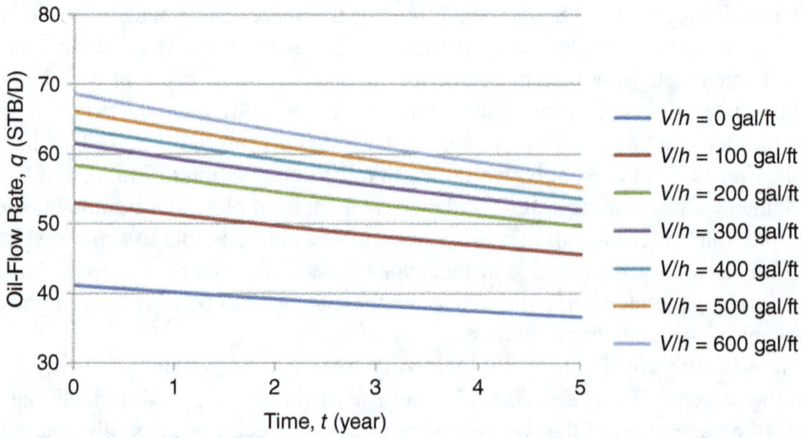


**Fig. 11.9—Production-rate forecast for different volumes of injected acid in the 1-md sandstone reservoir with 3-ft damage depth.**

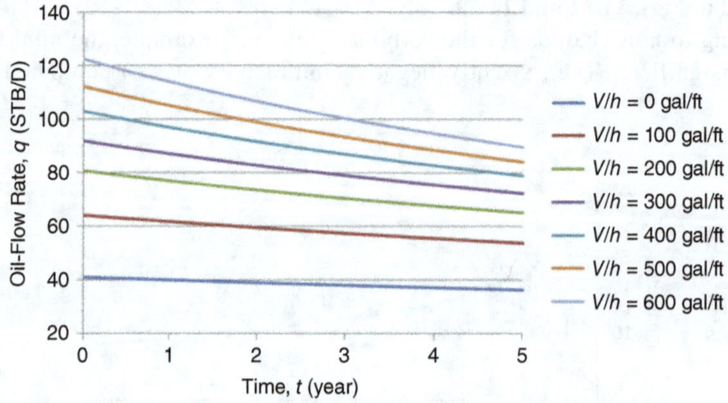


**Fig. 11.10—Production-rate forecast for different volumes of injected acid in the 1-md sandstone reservoir with 1-ft damage depth.**

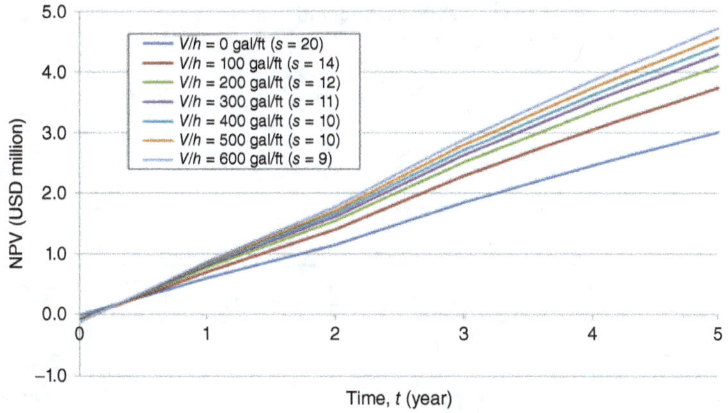


**Fig. 11.11—5-year NPV for different volumes of injected acid in the 1-md sandstone reservoir with 3-ft damage depth.**

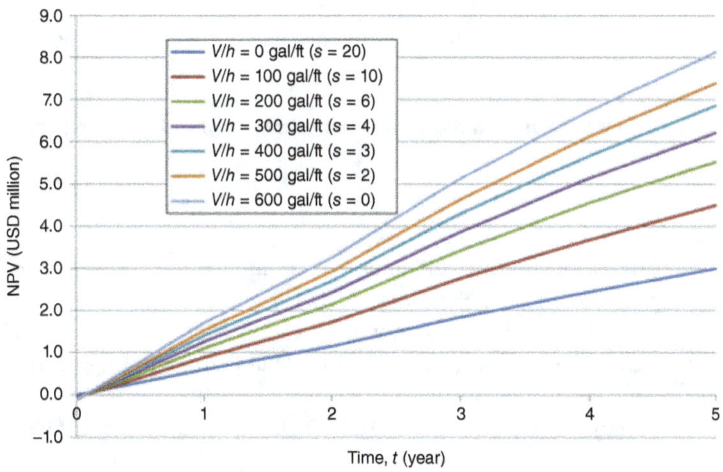


**Fig. 11.12—5-year NPV for different volumes of injected acid in the 1-md sandstone reservoir with 1-ft damage depth.**

economic value of the stimulation project, and as such, the PO time is extremely short (i.e., intersection with abscissa in less than 1 month for all cases illustrated). Second, the magnitude of the 5-year NPV from Fig. 11.11 appears to be about half of the 5-year NPV from Fig. 11.12 (with only 1-ft damage depth), and not surprisingly, the 600-gal/ft curve, corresponding to zero skin, provides the same value of NPV for the zero-skin case in the carbonate reservoir (i.e., some USD 9 million).

Third, replotting explicitly the 5-year NPV values vs. the different injected volumes of acid (see **Fig. 11.13**), it is possible to notice that both the 3-ft- and 1-ft-damage-depth curves would achieve a plateau similar to the curve shown for the carbonate reservoir in Fig. 11.5, as a consequence of the relatively low impact of treatment costs vs. benefits from the incremental NPV achieved. Again, these treatments would be subjected to logistic and operational constraints, such as in-situ handling of large volumes of acid and the major issues associated with corrosion within the tubular string during large injection times.

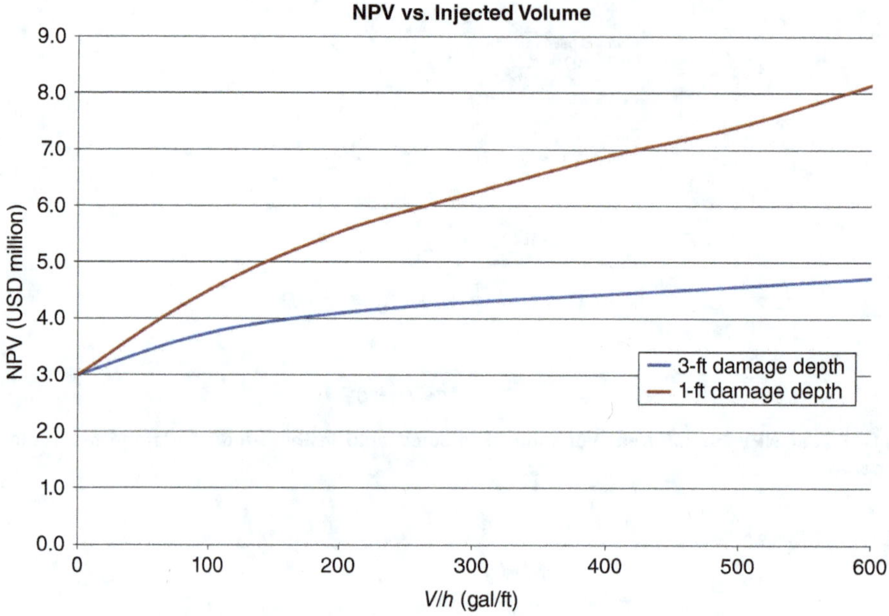


**Fig. 11.13—5-year NPV vs. volume of injected acid in the 1-md sandstone reservoir.**

# References


Economides, M. J., Hill, A. D., Ehlig-Economides, C. et al. 2012. *Petroleum Production Systems*, second edition. Upper Saddle River, New Jersey, USA: Prentice Hall.

Economides, M. J., Hill, A. D., and Ehlig-Economides, C. 1994. *Petroleum Production Systems*, first edition. Englewood Cliffs, New Jersey, USA: Prentice Hall Petroleum Engineering Series, Prentice Hall.

Economides, M. J., Oligney, R. E., and Valko, P. P. 2002. *Unified Fracture Design*. Alvin, Texas, USA: Orsa Press.

Hawkins, M. F. Jr. 1956. A Note on the Skin Effect. *J Pet Technol* **8** (12): 65–66. SPE-732-G. http://dx.doi.org/10.2118/732-G.

Labrid, J. C. 1975. Thermodynamic and Kinetic Aspects of Argillaceous Sandstone Acidizing. *Society of Petroleum Engineers Journal* **15** (2): 117–128. SPE-5156-PA. http://dx.doi.org/10.2118/5156-PA.

Marongiu-Porcu, M., Economides, M. J., and Holditch, S. A. 2008. Economic and Physical Optimization of Hydraulic Fracturing. Presented at the SPE International Symposium and Exhibition on Formation Damage Control, Lafayette, Louisiana, USA, 2008/1/1/. SPE-111793-MS. http://dx.doi.org/10.2118/111793-MS.

Marongiu-Porcu, M., Wang, X., and Economides, M. J. 2009. Delineation of Application and Physical and Economic Optimization of Fractured Gas Wells. Presented at the SPE Production and Operations Symposium, Oklahoma City, Oklahoma, USA, 4–8 April. SPE-120114-MS. http://dx.doi.org/10.2118/120114-MS.

Schechter, R. S. 1992. *Oil Well Stimulation*, first edition. Englewood Cliffs, New Jersey, USA: Prentice Hall.

Tarner, J. 1944. How Different Size Gas Caps and Pressure Maintenance Programs Affect Amount of Recoverable Oil. *Oil Weekly* **144** (2): 32–34.

Chapter 12

# Acidizing Safety, Quality, and Environmental Considerations

**Phil Rae,** InTuition Energy Associates, and
**Leonard Kalfayan,** Hess Corporation

## 12.1 Introduction

Acidizing has established itself as an indispensable technique for enhancing production from hydrocarbon reservoirs worldwide. Yet, to be frank, acidizing is almost certainly used less than it could be for well remediation and production enhancement. Instead, its use has increasingly been directed to small-volume prehydraulic-fracturing perforation-breakdown treatments, especially in tight rock and shale fracturing. The general reluctance to apply acidizing in well remediation and production enhancement can be partly attributed to discrepancies between results and expectations, but it is certainly also because of concerns over safety of personnel and integrity of well components during the handling and high-pressure pumping of acids and dealing with potentially corrosive returns from the well after treatment.

Over the years, research scientists and engineers have developed and improved many chemicals and additives to enhance the efficiency of these acid systems, as well as reduce the health, safety, and environmental (HSE) impact they have at the operational level. However, despite huge improvements in the last decade or two, many issues and concerns, both real and imagined, still exist today. Handling, mixing, storage, pumping, and disposal of acid and acid additives remain critical concerns in the industry.

In response to these concerns, new operational methods using specialized equipment, improved mixing techniques and new acid systems have also been developed; although, perhaps surprisingly, these techniques have not been adopted on a wide scale.

This chapter discusses current acidizing practices, from both chemical and operational perspectives, reviews their benefits and shortcomings, and introduces some of the improved methods for enhancing HSE practices and optimizing efficiency in acidizing operations. Finally, actual implementation of these formulations, methods, and techniques in various scenarios is discussed, describing the HSE and logistical advantages of the processes and materials involved.

## 12.2 Current Practices

As has been discussed elsewhere in this monograph, the practice of matrix acidizing to improve well productivity in both sandstone and carbonate reservoirs has been in use for more than a century. As we have also learned, a wide variety of acid systems ranging from inorganic acid, organic acid, and hydrochloric acid (HCl)/hydrofluoric acid (HF) mixtures (mud acid) has been used. Inorganic acids were first used in the 1890s, mud acid mixtures debuted in the 1930s, and organic acids were introduced in the early 1960s. Over the years, despite significant investment in research and development of acidizing additives and fluid systems, many of these same acids are still in widespread use.

This is not because there has been no progress. In fact, these efforts have resulted in significant improvements in the systems used for acid stimulation, but chemically, there are limited options available on the types of acid that can be used practically, and at commercial scale, to dissolve reservoir rocks and mineral scales.

The least expensive, most-efficient mineral acid for most applications is HCl, an aqueous solution of gaseous hydrogen chloride, which presents as a liquid that fumes in moist air at concentrations greater than 12.5% by weight. The solution is strongly acidic and corrosive to most metals and to human tissue, even when dilute. It is available commercially at concentrations ranging from 30 to 35% w/w but is normally used in wells at concentrations ranging from 3 to 28% w/w, 15% w/w HCl being most common. Alternatives to HCl include formic acid and acetic acid, the two simplest organic acids, but these are much less effective in dissolving minerals such as calcium carbonate, the main component of limestone and also a frequent cause of scaling in well tubulars. These organic acids are also significantly more expensive than HCl and, on the contrary, despite their organic nature, they are not much safer if at all. While dilute acetic acid may be familiar to us as a component of vinegar (approximately 5% concentration), in oilfield operations the normal concentration used is 10% acetic acid and, in any case, this is made by handling and diluting extremely corrosive (to human tissue) glacial acetic acid (100% concentration). Formic acid is even less friendly. The concentrated acid (typically 85–95% formic acid) is a powerful vesicant (blistering agent), in contact with human skin, and the fumes are extremely irritating to the eyes and lungs. Concentrated formic acid is used to mix treatment solutions containing typically 5–10% formic acid (alone or in combination with HCl or acetic acid).

Sandstone acidizing involves the use of HF—or more correctly, a dilute solution of hydrogen fluoride—almost invariably combined with an excess of HCl. This mixture, called mud acid, is used at various ratios and concentrations of the two main constituents, HCl and HF. Combinations such as 12:3, 6:1.5, 9:1, 10:5 are among the more common in the energy industry, the latter being used primarily in geothermal-well work. Even at low concentrations, HF is an extremely dangerous material. It diffuses rapidly across intact skin causing no initial discomfort yet can cause deep, penetrating burns with significant tissue destruction. At higher concentrations, it can quickly be lethal. HF toxicity is related not to its acidity per se but rather its ability to enter the body silently and interact with tissue with disastrous consequences. For this reason, it is unusual to see concentrated HF (available commercially as a 70% solution of HF) in use in the oil field. Instead, HF is usually prepared on location in dilute form by using soluble salts (ammonium bifluoride and ammonium fluoride) and dissolving them in excess HCl. This liberates the HF from the salt and provides a convenient and somewhat safer method to prepare mud-acid (HCl/HF) formulations for field use. However, this technique is certainly not risk-free and all materials involved must be handled with appropriate personnel safeguards [full personal protective equipment (PPE), including suit, face mask, gloves, rubber boots]. Spill-containment contingencies (bunding, neutralization, disposal), safety equipment (deluge showers, eyewash stations), decontamination procedures for personnel and equipment, and emergency medical kits (e.g., calcium gluconate gel) are all essential considerations for any acidizing treatment where HF, in any form, is involved.

## 12.3 Acidizing Additives

For obvious reasons, acids, such as the common ones mentioned in the preceding section, are usually corrosive materials and have specific dangers associated with their handling and use. However, there is no conceivable scenario, in typical oilfield operations, where one (or any combination) of these acids would be pumped alone and unmodified into any well, under any circumstances. The reason is simple—acids are corrosive and can attack well tubulars, corroding and dissolving the steel and potentially threatening well integrity. So, any acid formulation must contain, at the very least, some kind of corrosion inhibitor to minimize damage to well architecture. Corrosion inhibitor, however, is only one example of a bewildering array of acidizing additives that are used to modify the properties and characteristics of acid formulations for oilfield use. Other typical additives that might be used include surfactants and demulsifiers (to modify surface tensions, breakout emulsions), mutual solvents (to reduce emulsion problems and improve wettability), iron-control agents (to avoid precipitates and sludge formation), and scale inhibitors (to reduce precipitates and post-treatment scaling). So, when

| Additive Type | Example Chemistries |
|---|---|
| Corrosion inhibitor | Quaternary ammonium compounds |
| Corrosion-inhibitor intensifier | Potassium iodide, antimony salts, formic acid |
| Iron-control agent | Citric acid, erythorbic acid, carboxylic acid salts |
| Water-wetting agent | Surfactant (multiple chemistries) |
| Mutual solvent | Ethylene glycol monobutyl ether |
| Alcohols | Methanol, isopropanol |
| Demulsifier | Surfactant (multiple chemistries) |
| Antisludging agent | Surfactant (multiple chemistries) |
| Clay stabilizer | Quaternary ammonium salt |
| Fines-stabilizing agent | Organosilane, polyelectrolytes |
| Foaming agent | Surfactant (multiple chemistries) |
| Scale inhibitor | Phosphate esters and salts, phosphonates |
| Friction reducer | Polyacrylamide |

Table 12.1—Acidizing additives.

considering the hazards of acidizing, it is appropriate to consider not just the raw acids themselves, but also the individual, constituent additives in the acid formulation as well as the final composite formulation containing all materials. Each additive has its own specific hazards and must be treated accordingly. Typical acidizing additives are presented in **Table 12.1.**

Thanks to the efforts of many scientists, entrepreneurs, and environmentally conscious individuals both in and out of the oil field, many of these acidizing additives have been reformulated in recent years to account for more-stringent regulatory regimes. In some cases, the initiative has been a competitive or commercial one, and in others it has simply been the clever application of incremental improvements in our chemical knowledge to produce a better, safer product. One such approach is to improve the biodegradability of an additive so that even in the case of a spill, the material is quickly rendered harmless by natural processes and causes no long-term environmental damage. Others include the replacement of toxic constituents of an additive—for example, organic solvents such as aromatic naphtha, often used as a carrier for nonpolar components—with low-toxicity substitutes such as terpenes (natural hydrocarbons derived from plants) or, even better, the complete reformulation of the product into an aqueous additive with no solvents present at all. Obviously, such outcomes are not always achievable, but they are indications of the ongoing efforts to improve environmental stewardship in our industry.

## 12.4 Acidizing Equipment

The equipment used in the acidizing process has changed little over the years. A typical acidizing package consists of pumps, batch mixers, several acid-storage tanks, transfer pumps for additive transfer, and in some cases, nitrogen equipment. In addition, a large number of acid- and additive-storage drums and sacks may need to be stored on location. Offshore, large supply vessels or barges are usually required to mobilize equipment to and from location, in those cases where the equipment and materials are sited on the rig or platform. However, in some areas, customized dedicated marine stimulation vessels are in use. These self-contained "factory ships" are equipped with storage space, chemicals, and mixing and pumping equipment, allowing them to perform acid treatments quickly and efficiently without the need to load/unload materials to the rig. All that is required is a special high-pressure, flexible steel-composite hose to connect the vessel to the rig/platform and, ultimately, the well through which the acid is pumped.

Batch mixing is the process most widely used in preparing treatment fluids. A batch mixer is generally equipped with a means of adding dry and liquid chemicals, some type of agitation or circulation system, and a manifold system to deliver the prepared fluid to storage tanks. The type of formation, coupled with the interval size being treated, determines the chemical composition of the acid system as well as volumes required. This, in turn, dictates the selection of equipment (especially tank sizes) to execute the job.

The number of pieces of equipment varies, depending on the volumes to be pumped. Space limitations on offshore rigs, barges, or platforms can prove daunting for large treatments, providing more impetus for the use of stimulation vessels. Many acidizing jobs involve several stages composed of different acid systems. These stages may include preflushes, main acid stages, after-flushes, and displacement fluids. In large treatments, the picture may be further complicated by the need for diverter stages to improve zonal coverage. Depending on well configuration, coiled tubing may also be required to allow more-accurate fluid placement, among other things.

## 12.5 Current Practice—Disadvantages

The current processes and systems, although used successfully throughout the industry, have much room for improvement. In any acidizing operation, the planning, movement, and storage of materials and equipment from point of origin to point of usage are major considerations and can represent a significant fraction of the total operation cost. In current acidizing spreads, a substantial quantity of pumping equipment and storage tanks is required. The mobilization of all this equipment is time consuming and requires multiple trucks (or multiple journeys) for land operations or larger supply vessels for transport to and from offshore locations. The greater the equipment requirement, the greater the footprint needed to place the equipment.

Additional space is also required for chemical storage. Small land locations may be unable to accommodate the required equipment and, offshore, many unmanned platforms, satellite jackets, and workboats are already very small and unable to meet these work-space requirements.

More equipment requires more space and, in offshore scenarios, a greater number of crane lifts. This is not only time consuming, but limited by crane availability and lifting capacity. Many acidizing treatments have been deemed impractical because of such space and/or crane limitations. This can result in wells that chronically underperform and do not attain their full productive potential, despite the fact that their problem can be easily remedied with proven technology.

Other inefficiencies abound, often thanks to typical conservative oilfield practices. For example, it is routine, in many instances, for chemical additives in excess of the amounts required to be mobilized to location to account for dead volumes in tankage. This practice reduces supply-chain efficiency, increases mobilization cost, takes extra space at the wellsite, and involves the unnecessary transportation of hazardous materials.

HSE is the most widely used term for a management philosophy that encompasses three key requirements for all oilfield operations—the health and safety of all human life and the protection of the environment. The concept is not unique to the petroleum industry, of course, and relates just as much to everyday life as it does to any industrial setting. However, it is vigorously implemented and policed in today's industry and uses various metrics to assess performance and track trends, and to compare contractors, operators, and other industries. Seen in that light, current acidizing operations may, at first glance, look like an HSE manager's worst nightmare because they potentially involve the risk of exposure to many toxic and corrosive chemicals.

Preparation of the treatment fluids requires the physical handling of large quantities of powdered and liquid additives. Powdered additives are packaged in sacks and must be cut manually, then dumped into the batch-mixing tub. Liquid additives are delivered to the mixing equipment by means of small transfer pumps. Although the use of PPE is mandatory, exposure to acid fumes and chemical dust still occurs. Exposure to these chemicals can pose real and serious health risks, both acute and chronic.

Acid systems are usually mixed before pumping, stored in tanks on deck, and then pumped downhole. Preparing and holding, though temporarily, many thousands of gallons of fuming acid on surface creates a potentially hazardous situation. The possibility of spillage or exposure of personnel because of inhalation or skin contact is not insignificant. Environmental damage to water courses and groundwater are probably less of a risk today than they once were because of the mandatory use of bunding and other containment measures to account for even catastrophic failure of any tank, pump, or line.

Because of greater equipment requirements, the time and complexity of rig up is increased. The need for additional treating lines, hoses, and other connections creates a cluttered deck space, increasing the probability of accidents. Many storage tanks have dead volumes that cannot be used and these unused fluids have to be neutralized and disposed of subsequent to every job. Clearly, then, conventional acidizing operations leave much to be desired from the perspective of HSE compliance. Large volumes of hazardous chemicals, complex and bulky equipment rig ups, and numerous disposal issues can temper the enthusiasm for any acid treatment.

In current acidizing operations, large-scale, batch-mixing technology is widely used. In conventional batch mixing, volumes of water, concentrated acid, liquid additives, and solid materials are loaded into the mixer to be mixed and homogenized. The resulting mixture is then pumped to a bulk-storage tank to permit preparation of the next batch in the mixer. This process is very labor and time intensive. Time for preparation of the acid system may span from several hours to more than a day, depending on the volumes required. Inaccuracies, because of human error, can result in a final mixed-acid-system quality that does not meet the required specifications. Furthermore, any unforeseen well problems (e.g., poor injectivity) that might occur after preparation of the acid could potentially involve the costly backloading and eventual disposal of thousands of gallons of unused acid.

## 12.6 Continuous Mixing and Process Control

An alternative method to batch mix preparation is "continuous mix" (also called "mixing on the fly"), which involves the preparation of treating fluids, from concentrates plus water, on an as-needed basis, without the intermediate steps of large-scale batch mixing and storage in surface tanks. The control of this process may be manual (i.e., human operator) or automatic, whereby the input flow of chemicals is automatically controlled and regulated by various output sensor measurements. Continuous blending of acid, chemical additives, and water into a finished fit-for-purpose acid formulation, whether fully automated or not, makes good sense in matrix acidizing operations. Regardless of operational and profitability targets, the health and safety of personnel and protection of the environment are paramount. When materials being used are toxic and corrosive, potentially dangerous operating conditions exist. Continuous-mix/process-controlled acidizing is inherently safer and addresses many of these concerns (Sinanan and Rae 2005), as will be discussed in more detail.

For many years, chemical and process industries have successfully employed continuous mix/process control as a tool for streamlining, monitoring, and maintaining the consistency and operation of mixing processes. Indeed, such an approach is already used in various oilfield operations. Sophisticated blenders equipped with flowmeters and other instrumentation are used to accomplish the continuous mixing of fluids (plus solid components such as proppants) for hydraulic-fracturing operations, particularly those conducted from dedicated special-purpose marine stimulation vessels. At the other end of the spectrum, many cementing operations are continuously mixed using much simpler equipment—metering tanks equipped with sight glasses, displacement tanks, and liquid-additive pumps. Such equipment can be configured to operate in manual, semiautomatic, or automatic mode.

Despite the success of this "continuous mix" approach for fracturing and cementing, it is seldom applied to the process of acidizing (except, perhaps, from marine stimulation vessels that are already equipped with such facilities for fracturing, as noted previously). Yet, such an approach has many obvious advantages both operationally and safetywise. Furthermore, in the same manner as many other oilfield activities, it can be adapted to employ varying degrees of complexity and sophistication in response to different environments and equipment availability.

## 12.7 Advantages of Continuous Mixing/Process Control

The continuous-mixing process eliminates the need for batch mixers, storage tanks, and numerous connections and fittings. The mobilization to and from location is simplified tremendously. Less equipment to transport means fewer trucks or smaller supply boats, both of which reduce costs.

Offshore, no additional cranes are required, and lifting times are reduced. Smaller platforms and jackets can be accessed and smaller workboats can be used because of reduced work-space requirements. Since less equipment space is needed, more area on deck can be used for storing chemicals, thus reducing the number of replenishment trips during big acid campaigns. Work-space requirements are typically reduced by 30–40%, so more wells can be acidized in locations that were formerly inaccessible.

Continuous-mix/process-controlled acidizing uses only liquid additives. The additives and concentrated acid are stored in tote tanks and carboys that are manifolded together, and the liquids are fed directly either to metering tanks or to an automated blending system. There is no cutting of sacks and no transferring of liquid chemicals from drums. This eliminates dust generated when unloading sacked chemicals and spills while transferring liquid additives from drums. Inhalation, physical contact, and exposure to these chemicals are significantly reduced.

Large, pre-prepared volumes of acids are no longer stored in tanks on surface. The possibility of leaking tanks and significant spills is dramatically reduced or eliminated. Fewer fittings, hoses, and connections are needed for this improved acidizing system. The work area is less cluttered, allowing for a safer, more-productive environment.

Dead volumes are eliminated in the continuous-mix process. Exact volumes are mixed, as required, and simply pumped directly downhole. This eliminates the need for neutralizing and disposing of unused or excess acid, hence minimizing environmental fallout and disposal costs.

Because the acid is continuously mixed and pumped, there is tremendous savings in time and labor when compared to premixing in acid tanks. Should well conditions and requirements change, the acid treatment can be immediately redesigned before, up to and even during job execution. The accuracy of this approach ensures all quality-assurance and -control standards are met. The use of this continuous-mixing process is also ideal for jobs performed in real time, allowing several variables to be altered during execution.

## 12.8 Environmental Aspects of Continuous-Mix Acidizing

Meeting and exceeding environmental standards and regulations has become increasingly important for the petroleum industry. The industry has a high profile from an environmental perspective and has a history of being targeted, relentlessly, by activists and nongovernmental organizations. As a result, environmental stewardship is a priority for major oil and gas companies. Operators and legislators are continuously pressing for reduced emissions and better waste disposal, as well as lower consumption of raw materials. Using improved chemistry and continuous-mix/process-control methodology is certainly one way of addressing these issues because these techniques improve acid efficiency while helping to reduce toxic emissions and effluents. However, perhaps the most important thing of all in an energy-hungry world is the overall benefit of a properly managed and technically competent acidizing campaign. Using acidizing to stimulate appropriate candidate wells, thereby increasing production from existing assets at relatively low cost, is, perhaps, the greatest environmental benefit of all.

## 12.9 Field Implementation of Continuous Mix for Routine Matrix Acidizing

As mentioned previously, the continuous mixing and/or process control of matrix acid treatments can be accomplished accurately with varying degrees of sophistication and automation. In its simplest form, the process can borrow from the techniques already used in cementing and adapt these to take account of the additional safety requirements when dealing with acids. Thus, metering tanks for concentrates, covered displacement tanks, and a venting system for head-space air are all that is required for an entry-level continuous-mix acid system. At the other end of the spectrum, one can design a fully automated, closed-loop blending system for field application of process-controlled acidizing (PCA). The differences between the two approaches in terms of output quality are probably less important than the inherent advantages implicit in adopting a strategy of continuous-mix acidizing, regardless of its exact method of implementation. Obviously, the majority of the benefits are still present even when adopting the simple manual approach, but the cost of implementation is dramatically reduced. Therefore, the propagation of the concept through the industry is likely to be faster and more widespread. As benefits accrue and costs per acid treatment decline because of improved efficiency, safer operations, and ever-greater application, investment in more-sophisticated fully automated PCA equipment may become practical.

## 12.10 Obstacles to Continuous Mixing of Acid

Given the numerous advantages for process-controlled/continuous-mix acidizing, it may seem surprising that it has not been widely implemented in the field. There are several potential reasons for this, including a lack of vision by the industry and a tendency to eschew technology in favor of simpler, more-direct methods. However, there are some practical considerations in adopting continuous mix, and one of the most obvious is the volume of components required to conduct matrix acidizing on the fly. In the case of simple HCl, it requires 406 gal of 34% HCl to yield 1,000 gal of 15% HCl. In other words, the volumetric premium, in terms of space saved, may not appear to be so significant. With conventional mud-acid (HCl/HF) systems, the volumes of liquid additives required have historically been too great to justify such an approach. The volume of liquid ammonium fluoride and concentrated (34%) HCl to prepare 12:3 mud-acid amounts to 63% of the final acid volume. It should be noted here that newer, low-acidity HF systems have been developed that achieve the same rock-dissolving power with much lower amounts of additives, typically only 25% or less of the final acid volume. Regardless of any other claimed benefits, such systems are obviously much better suited for continuous-mix applications. See **Table 12.2.**

However, this rather misses the point. The continuous-mix/acidizing PCA approach offers many benefits beyond simple volumetric comparisons and should be adopted more widely to take advantage of these.

| | Per 1,000 gal Solution | | |
|---|---|---|---|
| Acid System | Water | Acid Blend | Ammonium Fluoride (36%) |
| 12:3 mud acid | 385 | 462 | 153 |
| Low-acidity 3% HF formulation | 767 | <80* | 153 |

Table 12.2—Water and acid volumes of mud acid vs. low-acidity HF formulation.

## 12.11 Conclusions

Acidizing is an important technical discipline that provides the means to recover or even enhance production (or injection), often in older or damaged wells. Its application, however, is not without risk because it involves the use of acids mixed and pumped at an industrial scale, and more often than not, injected into wells at high pressure. Safety of personnel is paramount in acidizing so the provision and use of full PPE, safety showers, eyewash stations, and decontamination procedures (mainly for HF) are mandatory. Safety training on the hazards involved as well as on the effective use of safety equipment is also essential. Apart from the acids themselves, many acidizing additives are also hazardous or potentially damaging to the environment, although the latter problem continues to be addressed by successive generations of improved additives with much lower environmental impact.

Acidizing has historically been a batch-mixing process where concentrated chemicals are mixed with water in large tanks to arrive at the final formulation that is pumped in the well. This process, while relatively simple and effective, exposes individuals to risks that might be avoided by the use of improved methods such as continuous-mix acidizing or, ultimately, automated acidizing using process control.

Indeed, continuous mixing offers a number of further advantages over conventional large-scale batch mixing. Implementing continuous-mix (or process-control) techniques in acidizing, even at a rudimentary level using manual control, can help to reduce costs and improve

1. Health, safety, and the environment
2. Operational flexibility and agility
3. Operational reliability
4. Real-time decision making
5. Regulatory compliance

6. Fluid-output quality
7. Fluid-output yields (no dead volumes)
8. Productivity

One of the hallmarks of continuous mix/process control is the smoothness of its operation. Continuous mixing tends to be more technically efficient than conventional large-scale batch mixing because it can be automated and is, thus, more predictable and easier to control. It is more cost- and time-efficient, enhances the quality of fluid systems, and significantly improves HSE conditions. In conclusion, this approach ensures reliability, accuracy, and flexibility while improving safety and environmental compliance in the field.

## References

Ely, J. W. 1994. *Stimulation Engineering Handbook*, first edition. Tulsa: Pennwell Books.

Hebert, P. B., Khatib, Z. I., Norman, W. D. et al. 1996. Novel Filtration Process Eliminates System Upset Following Acid Stimulation Treatment. Presented at the SPE Annual Technical Conference and Exhibition, Denver, 6–9 October. SPE-36601-MS. http://dx.doi.org/10.2118/36601-MS.

Kalfayan, L.J. 2008. *Production Enhancement with Acid Stimulation*, second edition. Tulsa: PennWell Corporation.

King, G. E. and Holman. G. B. 1982. *Acidizing Quality Control at the Wellsite*. Tulsa: Amoco Production Research Co.

Sinanan, H. B. and Rae, P. J. 2005. New Operational Technique Improves HSE in Offshore Acidizing Operations. Presented at the SPE Asia Pacific Health, Safety and Environment Conference and Exhibition, Kuala Lumpur, 19–21 September. SPE-96606-MS. http://dx.doi.org/10.2118/96606-MS.

Watkins, D. R. and Roberts, G. E. 1983. On-Site Acidizing Fluid Analysis Shows HCl and HF Contents Often Varied Substantially from Specified Amounts. *J Pet Technol* **35** (5): 865–871. SPE-10770-PA. http://dx.doi.org/10.2118/10770-PA.

### Appendix 12A—Safety Checklist

Pressure-pumping companies providing acidizing services are expected to maintain all requisite safety standards and equipment. Treating and safety equipment must be tested and inspected regularly, subject to a certification program, and its integrity should never be taken for granted in advance of a well-treatment operation. All joints and connections must be pressure tested and evaluated ultrasonically for integrity. Because concentrations of hydrogen sulfide ($H_2S$) gas may exceed 10 ppm on the site of an acid stimulation treatment, all crews should be provided with training (e.g., exposure risk and effects, $H_2S$ safety equipment, $H_2S$ detection, recognition of and proper response to $H_2S$ warnings on the work site, rescue techniques and first-aid procedures, proper use and maintenance of personal protective equipment and demonstrated proficiency in using PPE) and must have passed through any applicable certification.

The following is a list of items to address in acid stimulation operations, based on the *Stimulation Engineering Handbook* (Ely 1994) with modification and additional commentary. The guidelines, presented by Ely in 1985, still apply today.

First is the prejob safety meeting, which is most important in setting the tone and expectations for the treatment. In the safety meeting, all safety considerations should be discussed in detail. The main purpose of the meeting is to ensure that each individual present at the job site knows his or her role and responsibilities and what to do and where to be at all times during the job. For example, one lead service company representative and one oil company representative (to whom the service company leader will respond) should be appointed. These two individuals should make the final decisions at the job site. All others should provide suggestions or point out any errors of which they are aware, or even be given "stop work" authority should any potentially or actual unsafe practice be observed.

Communication during the job is crucial. All individuals present should be familiar with the service company mode of communication. Service company representatives should report their means of communication to the oil company representative during the treatment. They should plan appropriate action to take if that line of communication fails.

Establish before and continue during the job:

1. *Take all safety rules very seriously.*
   Do not consider any safety rules or expectations to be trivial. Ensuring safety requires a certain mindset—one that consciously places all safety rules and considerations first.
2. *Designate a gathering area in the event of an incident or disaster.*
   This should be established and well-understood before commencing treatment. If offshore, it may be on another level of the platform or on a treating vessel.
3. *Establish a clearly understood maximum treating pressure.*
   This is based on wellbore and tubular conditions, surface equipment limitations, or, in the case of matrix acidizing, formation-fracture pressure limitations. Stepping up injection rate solely to shorten the job time should be prohibited.
4. *Each person should be assigned a specific responsibility.*
   If anyone is on-site to observe only, that should be established as his or her *only* role. If someone is asked to take on additional responsibility or to help in some capacity once the job has started, even if a minor endeavor, it must be done with the knowledge and explicit understanding of all on-site. It must also be understood that no one should take on any responsibilities or activities that are not explicitly established beforehand or during a job.
5. *Assign leadership (one person from the oil company and one from the service company).*
   These two should work together to mutually agree on procedures and responsibilities. Any suggestions should be made to both parties. Both should be kept in the loop on all communication regarding job execution.
6. *The oil company representative should designate the persons responsible for each task.*
   The service company job supervisor should establish the responsibilities of his or her own crew. These should be made clear to the oil company representative for his understanding. The responsibilities of other parties present on-site should be designated or approved of by the oil company representative with communication of such to the service company lead person.
7. *Make sure an effective and reliable communication network is set up. Have alternative visual commands.*
8. *Have fire-extinguishing equipment readily available and placed at ground or immediate-access level.*
9. *Have at least two individuals given the responsibility of transporting injured persons to the nearest clinic/hospital.*
10. *For onshore operations, have at least one designated vehicle set aside for transporting injured personnel.*
11. *Perform a complete head count of all personnel on location.*
    The lead persons should know where everyone is at all times. One should inform the leads of one's whereabouts if leaving the site or one's post, even if temporarily.
12. *Set up a gathering area in case an accident occurs.*
    This may be a different area than that established in Step 2. A disaster, such as an explosion or a major equipment failure or leak, may require evacuation to a safe area. In the event of an accident, the gathering area may, therefore, be different and more immediately accessible.

To ensure the safety of all on-site as well as in surrounding areas, all individuals must be fully aware of all on-site safety requirements and policies and follow them. This starts with being aware of one's surroundings and what is going on at all times on-site while there. Ensuring the safety of the environment and everyone on-site, and preventing injury or loss of life, comes down to the attitude and commitment of individual employees and contractors themselves, not company written practices.

## Appendix 12B—Quality-Control Checklist

Incorporating quality-control measures during all aspects of an acid job can make the difference between success and failure. Quality-control monitoring during the actual pumping of the treatment only is not sufficient. In addition to the pumping stages, quality-control steps must also be planned and executed during rig up of equipment, before pumping, and after pumping.

In 1982, George King and George Holman, at that time of the Amoco Production Company, produced a booklet titled *Acidizing Quality Control at the Wellsite* (King and Holman 1982). Included in that booklet is a section on the steps that should be taken to ensure quality control of an acidizing treatment. Because the majority of acidizing treatments are still batch operations, these guidelines still apply today.

Following is a modification to the quality-control steps originally presented in that booklet, with additional commentary from *Production Enhancement With Acid Stimulation* (Kalfayan 2000).

### Quality Control During Rig Up of Equipment

A. *Inspect all tanks that will be used to hold acid or water. The tanks must be clean. Small amounts of solids can easily ruin an acid job.*
Acid mix tanks can contain debris, for example, from a cementing operation if cement tanks are used to mix acid. Tanks must be regularly cleaned, and the service company should demonstrate a regular cleaning program.

B. *Make sure the service company has equipment to circulate the acid tank before pumping*. This is important for the protection of the tubulars and to prevent emulsion problems. Acid corrosion inhibitors and other additives can separate to the top of the tank in a couple of hours or less.

C. *The line to the flowback pit or tank, if applicable, should be laid and ready to connect to the wellhead so the acid can be backflowed immediately after the end of the overflush stage.*
Sometimes this is left undone, especially when there is a race against the clock to complete a treatment. Advanced planning is required to avoid such oversights that will otherwise result in delays or rushed work that is inherently unsafe. Flowing back an acidized well as soon as possible, with minimal delay, is an underrated concern in sandstone acidizing in particular. Unnecessary shut-in time can lead to reprecipitation of acid reaction products near enough to the wellbore to cause radial permeability damage. This is especially true in cases in which acid was not or could not be displaced far away from the wellbore.

Injection wells, particularly in low-pressure reservoirs, are exceptions to this general guideline. The difficulty in recovering injected acids and their reaction products from such wells makes it much more practical to simply recommence injection as soon as practical after the treatment and continue to overflush. This strategy ensures that spent acid is carried far from the wellbore where any precipitates have insignificant effects on effective injectivity.

Extended shut-in is usually not a major concern in carbonate acidizing, at least not from the standpoint of unwanted reprecipitation reactions. However, it is always best to prepare a well to be returned to production as soon as possible if there is no reason to shut the well in or to allow for a soak period. Soak periods are sometimes necessary in scale-removal treatments or in slowly reacting carbonate formations such as low-temperature dolomites.

### Quality Control Before Pumping

A. *Check service company ticket to be sure all additives for the job are on location.*
It is not unreasonable, and should not be taken as a personal affront to anyone, to check the service company ticket. It is important to make sure that additives included in the treatment design are also included in the delivery of chemical components to be used and mixed on-site. Not only do additives once in a while fail to show up, sometimes an incorrect additive, or at least an unacceptable substitute, may be delivered instead of what had been requested or specified.

B. *Circulate the acid-storage tank(s) just before the acid is injected into the well.*
When batch mixed, it is known that certain additives, particularly corrosion inhibitor, can settle to the top of the acid tank, leaving the body of acid inadequately inhibited. It is important to mix acid on delivery and continuously, if possible, up to injection. If that is not possible, then acid should be mixed immediately before injection.

C. *Check the concentration of acids with a test or titration kit.*
In the late 1970s and early 1980s, Watkins and Roberts of the Union Oil Company of California conducted a survey of on-site acid concentrations measured in tanks (Watkins and Roberts 1983). The results were alarming. Wide variations existed in acid concentrations, as well as acid-concentration gradients within individual tanks. Attention to actual acid concentrations relative to specifications is still largely absent. Fortunately, this issue can be eliminated with continuous-mix operations.

D. *Make sure everyone on-site, especially the service company personnel, knows the maximum "safe" surface pressure, and stay below that pressure throughout the pumping operation.*
It is important to stay below the maximum surface injection pressure for equipment-safety-limit and well-tubular-integrity considerations. It is also generally important in sandstone acidizing to stay below fracturing pressure.

E. *Check the pressure/time recorder, or any other on-site treatment monitoring or evaluation mechanism, for proper operation.*
Evaluation of an acidizing treatment, especially removal of skin damage or effectiveness of diverter stages, is not possible if pressure devices and monitoring equipment are not working properly. The working order of such equipment should be ensured on-site.

F. *Acid clean (pickle) the tubing.*
Use 5% HCl or a nonacid pickling solution. If tubing is not pickled, debris, iron scale, pipe dope, and other solids can be inadvertently injected into the formation at the outset of an acid treatment. This can irreparably damage the formation, ruining an otherwise well-designed treatment. Avoid using overly strong HCl solution for pickling, unless it is necessary because of space (tank) limitations to use the same HCl solution as in the main acid treatment.

**Quality Control During Pumping**

A. *Control injection rate. Maintain surface annulus pressure at or below 500 psi during treatment.*

B. *Watch the pressure response when acid reaches the formation.*
The surface pressure should slowly decrease if the rate is held constant, indicating skin removal. If the surface pressure rises sharply or rises continuously for several barrels of acid, the acid may not be removing the damage and may actually be damaging the formation. Treatment should be stopped and the well flowed back. Samples of backflowed acid should be collected for analysis.

C. *Note the pressure response when the diverting agent reaches the formation.*
The surface pressure should rise slightly. If there is no diverter response, more diverter, or a different diverter, may be needed in the future. If diverter response is not noted, the injection rate may be increased, if the pressure limit permits, in order to enhance zone coverage. This should only be considered if it does not otherwise compromise design.

D. *Do not exceed the breakdown (fracturing) pressure of the formation in a sandstone acidizing treatment if not intended in the design, or unless absolutely necessary initially to break down perforations, or to bypass severe damage to initiate injectivity.*

E. *In the final flush, make sure acid is displaced from the wellbore.*
The full overflush volume should be pumped as designed. Spent acid, especially in a sandstone, must be sufficiently displaced from the wellbore to minimize the damaging effects of precipitation of acid reaction products. In carbonate stimulation, it also makes sense to displace acid effectively, to avoid possible corrosion problems during flowback.

Also a sensible practice, especially when pumping an acid treatment through coiled tubing, is to include corrosion inhibitor (minimal concentration, for example, 0.1–0.2%) in a flush stage to treat the backside of the coiled-tubing string, preventing corrosion during spent-acid flowback.

**Quality Control After Pumping / During Flowback**

A. *Do not design to shut the well in after acid injection. Flow the well back to the tank or pit as soon as the flowline is connected.*

As mentioned earlier, it is always desirable to flow spent acid back after treatment in carbonates and sandstones, unless there is a specific benefit to shutting the treatment in. However, even then, flowback execution is often treated casually. The spent-acid flowback period is not routine production. If care is not taken, severe damage can be caused during this initial period. Three of the most common problems during flowback are

- Fines migration
- Reprecipitation of acid reaction products
- Back production and processing of additives

Fines are generated during sandstone acidizing, or released during carbonate acidizing, and are carried more efficiently by the denser spent-acid mixtures. To avoid potentially damaging effects, the flow rate should be set low at first. The flow rate then should be gradually stepped up to planned production conditions after spent acid is produced back.

Reprecipitation of reaction products is inevitable. In sandstone acidizing, the only way to minimize the problem once the acid treatment has been pumped is to return the well to production as soon as possible. The process of flowing spent acid back must be initiated immediately.

Most additives included in acid mixtures are produced back to varying degrees. They can cause upsets to production equipment and facilities. Depending on the circumstances, spent acid must be collected and properly disposed of, meeting all regulatory and environmental requirements. Fluid-filtration methods may also be incorporated (Hebert et al. 1996).

B. *Collect at least three 1-qt samples of backflowed acid for analysis, if possible. Sample the acid backflow at the beginning, in the middle, and near the end of the flow. If on swab, get a sample from every other swab run.*

If possible, get a laboratory analysis for

1. Amount, size, and type of solids
2. Strength of returned acid
3. Total iron content
4. Presence of emulsions and/or sludges
5. Formation of any precipitates (besides iron)

This information can indicate the nature of acid spending, whether overtreatment or undertreatment occurred, and whether iron and potential precipitation of reaction products are a concern. It also can indicate if acid mixtures injected might have been incompatible with formation fluids. It is especially important to patiently evaluate return samples before any possibly incorrect conclusions are drawn about the treatment pumped.

If possible, a complete analysis of acid returns can be made by the service company using spectroscopic methods. Concentrations of the following ions should be determined:

- Aluminum (Al)
- Magnesium ($Mg^{2+}$)
- Chloride ($Cl^-$)
- Barium ($Ba^{2+}$)
- Potassium ($K^+$)
- Fluoride ($F^-$)
- Calcium ($Ca^{2+}$)
- Sodium ($Na^+$)
- Iron ($Fe^{2+}$, $Fe^{3+}$)
- Silicon (Si)

Such analysis can indicate on what the acid stages were spent and whether the spent-acid stages were produced back in the expected and desired manner. This information can be important in designing future treatments in the same field. For example, if spent HF returns in a sandstone acidizing treatment showed high calcium-ion content, this might suggest the need for a larger acid (HCl) preflush. Or, it may suggest the need to reduce the HF stage. High dissolved silicon-/aluminum-ion ratios may indicate reprecipitation of aluminum fluoride salts, suggesting the need to modify the designed HCl/HF ratio. Or the modifications could include lowering HF concentration, or using a special HF system designed to minimize reprecipitation of such reaction products.

C. *Ensure that the treatment report and the pressure charts (electronic or otherwise) are available for post-treatment evaluation and permanent retention in the well file.*

Allow the service company to have copies for future reference, in order to facilitate improvement in subsequent-treatment design and execution. Both the operator and the service company must make an effort to exchange and discuss all relevant treatment information to give each other the best chance for future success.

# AUTHOR INDEX

# SUBJECT INDEX

www.ingramcontent.com/pod-product-compliance
Lightning Source LLC
LaVergne TN
LVHW080028210926
848481LV00011B/3131